Estimating Building Costs

Estimating Building Costs

for the Residential and Light Commercial Construction Professional

Third Edition

WAYNE J. DEL PICO

WILEY

Library of Congress Cataloging-in-Publication Data Applied for:

Paperback ISBN: 9781394150113

Cover Design: Wiley
Cover images: © ribeirorocha/Getty Images; © photovs/Getty Images

SKY10054968_090823

This book is dedicated to my wife, Krisanne, and my daughters Maria Laina and Kristina, my son-in-law Joe, and my beautiful, perfect, granddaughter Gloria Rose . . . my family, my life!

Contents

About the Author

Wayne J. Del Pico is president of W. J. Del Pico, Inc., where he provides construction management and litigation support services for construction-related matters dealing with cost and schedule. He has more than 43 years of experience in construction project management and estimating and has been involved in projects throughout most of the United States. His professional experience includes private commercial construction, public construction, retail construction, and residential land development and construction.

Mr. Del Pico holds a degree in civil engineering from Northeastern University in Boston, where he taught construction-related curricula in cost estimating, project management, and project scheduling from 1992 until 2006. He is also a member of the adjunct faculty at Wentworth Institute of Technology in Boston, where he currently teaches programs in construction cost analysis, estimating, project control, and construction scheduling.

Mr. Del Pico holds the designation of RSMeans Data Certified Professional. He is a regular seminar presenter for the RSMeans Company, where he lectures on estimating topics.

Mr. Del Pico is the author of *Plan Reading and Material Takeoff* (1994) and *Estimating Building Costs* (2004, 2012, and 2023), *Project Control – Integrating Cost and Schedule in Construction* (2013 and 2023), and *Electrical Estimating Methods* (2015) and is a coauthor of *The Practice of Cost Segregation Analysis* (2005).

He has served as the 2010 president of the Builders Association of Greater Boston. He is also a practicing Neutral for the American Arbitration Association since 2009, where he hears construction-related arbitration cases.

Preface

Estimating has always and will continue to have a prominent place in the construction marketplace. It is the first, and some would even say the most important part of the process of building anything. Simply stated, without a proper estimate, there is very little hope in salvaging the financial well-being of a project. The best field personnel and project management team cannot resurrect a winner from the ashes of a poorly estimated project.

The estimate must represent the value of the project, not only for the purposes of the contract but also for project cost control once the project has started. The estimate, with slight changes, will serve as the metric for cost performance measurement. We compare the actual costs to what was anticipated in the estimate. When actual costs are less than estimated costs, the performance is, in short – favorable. When actual costs exceed estimated costs, well . . . there is a problem.

Over the life of the project the estimate will serve as a continual reference point. It will be measured against the schedule and serve as the basis for any purchasing. It will also dictate how long the project should last relative to direct and indirect overhead.

It is paramount to the success and even the survival of our industry to develop competent estimators and accurate estimates. Throughout my entire career I have made an effort to help educate and pass on what I have learned to further the goal of competency and professionalism in this crucial aspect of the industry. While computers are indispensable in estimating, they cannot, under any circumstances, take the place of estimating skill. They are a tool and not a panacea. Education that creates a solid foundation and years of practice make good estimators. Industry professionals who have a fundamental knowledge of estimating will be better equipped to advance their careers.

Since the first edition of *Estimating Building Costs* was released in 2004, this text has enjoyed an unanticipated popularity, and with the updated second edition, that popularity has grown. In each successive edition, I have endeavored to keep up with changes to the industry and to add those sections that represent new topics or existing topics that required more detail based on reader suggestions.

To that end, I truly hope that you will find *Estimating Building Costs, 3rd Edition* an educational resource or a reference tool that will help create that foundation in estimating.

—WAYNE J. DEL PICO – SPRING 2023

Introduction

Estimating has always been one of the weakest links in the construction process. Most contractors can recall one or two craftspeople whose talents have been enviable but who have ultimately failed miserably in business. Many of these failures were the result of poor estimating practices. As common a problem as estimating appears to be, contractors are reluctant to spend the time to learn the fundamental skills that are needed to produce accurate estimates.

This book was created as a reference for professional contractors – in both the new and the remodeling markets—who want to master organized, efficient industry standards for estimating residential and light commercial projects. It is designed and presented for use by the professional contractor or homebuilder who has some experience in the industry and is familiar with the materials and tasks typical of building projects. The chapters provide step-by-step guidelines – ideal for learning industry-accepted standards for estimating.

For novices, *Estimating Building Costs*, third edition, explains the fundamentals of the estimating process in a clear and concise format, which will become an essential part of their work. It has been updated since the second edition to include prominent changes to the process. Experienced estimators will find the book useful for reviewing their own methods and enhancing their expertise. Regardless of whether the "estimating staff" is a dedicated person among a company's many employees or the same individual who performs the work, the principles are the same. The text is written in what is referred to as general knowledge estimating practice, from the point of view of the general contractor's estimator, yet it is equally helpful as a foundation for subcontractors' estimators. Regardless of the market in which a contractor works, the estimating practice and procedures are the same. The third edition has been reorganized to reflect the new CSI MasterFormat® 2020 structure. Readers should find this beneficial in locating specific topics.

Chapters 1 and 2 explain what estimators can expect to find on the drawings and in the specifications – the information and details necessary to create an accurate estimate. These chapters have been revised to reflect changes in the contract document language and practice since the original publication. Chapters 1 and 2 still reinforce that plans and specifications are complementary tools that must be reviewed and understood before attempting to prepare a project estimate. Chapter 2 explores the new CSI breakdowns of work and the impact that a contract delivery method can have on the estimate. The discussion has been expanded to reflect the changes to the AIA A201. Chapter 3 presents the time-tested mathematical principles needed to accurately take off material quantities, by calculating area and volume, and the extension of the units into the final estimating units. While nobody expects one to memorize formulas as long as the internet exists, the basic relationships are important to understand.

Chapter 4 reviews the accepted rules for quantity survey, or takeoff, and the mechanics of the takeoff process. This chapter also explores the pricing part of the estimating process. It has been retitled to reflect its contents more accurately. There have been various enhancements and clarifications to help the reader. Chapter 5 explains in detail how materials costs are derived, how to determine allowances for waste, and how to assign markups typical for materials and labor. These are essential to understanding how a unit price estimate is built. The chapter also covers labor productivity, crew and individual tasks, and budgeting for subcontractor work. It is a cornerstone of the book and has remained virtually the same except for minor updating and an enhancement of the discussion of direct and indirect overhead.

The next group of chapters, 6 through 25, have been rewritten and reorganized according to the 2020 edition of the Construction Specifications Institute (CSI) MasterFormat® – the most widely recognized and used system for organizing construction information in specifications and estimates. Learning estimating techniques for each building system in this format helps estimators understand and apply them according to the industry's most professional standards. Sections have been expanded and new topics added to reflect industry changes since the second edition publication date. Each chapter describes specific estimating tasks for that particular trade— including special cost considerations and standard quantity takeoff guidelines for each building system and material. The examples and easy-to-follow steps can be referred to time and again. These chapters include all of the work one might encounter on a typical residential or light commercial project—from demolition, concrete, and masonry to windows and doors, siding, roofing, mechanical and electrical systems, finish work, and site construction.

Chapter 26 introduces two topics crucial to finalizing a reliable estimate yet rarely addressed by books or courses on estimating: profit and contingencies. This chapter considers both the tangible and the intangible aspects of residential and light commercial projects that must be evaluated before assigning a suitable profit or adding a contingency amount to cover unknowns. Some of the factors to evaluate are risk, scheduling impacts, and contractual obligations.

Chapter 27 explains computerized estimating, including benefits and cautions. It includes tips for developing and customizing simple spreadsheet estimating applications, rather than relying on "canned" software alone. New to the third edition is an in-depth discussion of developing a historical database.

Chapter 28 is a revised discussion of conceptual estimating and its most common methodologies. It covers what can be expected for accuracy and how to determine when conceptual estimating is appropriate. It also makes suggestions for using saved data to develop parametric costs.

Two of the divisions of the Site and Infrastructure subgroup and all of the divisions of the Process Equipment subgroup have specifically been omitted, as they are beyond the scope of this text.

Overall, the third edition has updated and supplemented the graphics as a result of feedback from readers and students since the original text was published. Clarifications and enhancements have been added as needed. The author has added tables to suggest possible labor-hours for specific tasks when appropriate. These are not foregone conclusions as to the time it takes to execute a task but merely guidance to help the estimator. *The labor-hours for each and every task must be evaluated on their own merits on a project-by-project basis.* Each chapter now contains a series of questions or problems the reader can use to measure the learning objectives and outcomes of the chapter. The solutions are at the rear of the text. The third edition has been updated to be more user friendly for the student, professor, or instructor learning or teaching estimating in a university environment.

Accurate construction estimating is an essential skill for every successful professional contractor. *Estimating Building Costs,* third edition, builds on the foundation of estimating principles in the previous editions and adds topics addressing changes to the industry as time progresses.

—Wayne J. Del Pico
Spring 2023

1 The Basis of the Estimate

An estimate requires that there be a basis for that estimate – in other words, a document or set of documents that the estimator can use to derive the necessary tasks to determine the scope and value of the work. This document or set of documents will serve as a foundation for the estimate. Clearly, small projects may not require a full set of plans or specifications as an entire structure would. For small projects, one can develop a Scope of Work or Statement of Work (SOW) that identifies what the owner wants. Realistically, the documents that serve as the basis of the estimate run the full gamut from a single-page SOW to a fully developed, well-coordinated set of plans and specifications that identify in detail the scope of the project.

Estimating construction projects requires fluency in the language and symbols used in construction plans. This chapter provides an overview of a project's bid drawings. It does not offer detailed instruction in plan reading but reviews the organization of the plans and the information necessary for estimating. This chapter will also provide a glimpse into the contents necessary for SOW to be considered sufficient.

ROLE OF THE DRAWINGS IN ESTIMATING

The three terms most often used to refer to the graphic portion of the documents for a building project are:

- Plans
- Drawings
- Blueprints

While the last is a fairly antiquated term, plans are still frequently referred to as blueprints. For the purpose of this text, these terms are synonymous and can be used interchangeably. They are the graphic representation or

Estimating Building Costs for the Residential and Light Commercial Construction Professional,
Third Edition. Wayne J. Del Pico.
© 2023 John Wiley & Sons, Inc. Published 2023 by John Wiley & Sons, Inc.

illustration of the project and comprise the lines, symbols, and abbreviations printed on paper that represent the owner's wishes, as interpreted by the architect. The majority of plans today are black line drawings plotted on white sheets of paper of varying sizes. Plans are the quantitative representation of the project. Plans provide the information necessary to calculate lengths, areas, volumes, and quantities of individual items (counts) that serve as the basis of the take-off and ultimately the estimate. These plans combined with the technical specifications (discussed further in Chapter 2) will comprise the bid documents. If the project goes forward, the bid documents will become the contract documents and form the basis of the contract for construction.

DESIGN DEVELOPMENT

Most drawings develop over several generations of review and modification as a result of owner input, coordination with other design disciplines, building code compliance, and general fine-tuning. This process is referred to as *design development,* and it occurs before the release of the final version of drawings, called the *working drawings*. Working drawings are the completed design – a code-compliant representation of the project, ready for bidding and, ultimately, construction. They are the focus of this chapter and are the prerequisite for preparing a detailed unit price estimate. (*Note:* "Preliminary" drawings created early in the design development process may be used as a basis for budget estimates, but budget estimating requires specific skills of seasoned professional estimators who have years of experience developing unit price estimates. Preliminary drawings require that the estimator fill in the blanks as to what is not shown on the plans but is required for a complete estimate.)

The completed drawings become a "set" that incorporates all adjustments, changes, and refinements made by the architect or engineer as the final step in design development. Working drawings should comply with all applicable building codes, including any local ordinances having jurisdiction over the project. Drawings should include all the information one will need to prepare a detailed estimate and eventually build the project. The set of working drawings consists of various disciplines of design that use the architectural or *core drawings* as their basis. The core drawings illustrate the owner's desire or requirements as interpreted by the architect, or core design professional. Subconsultants such as the structural, mechanical, and electrical engineers will use the architect's drawings to create the structural, mechanical, and electrical engineering portion of the plan set to make the space habitable and functional. This assumes the project requires these other disciplines.

All buildings are constructed with a definitive purpose and require professionals skilled in specific areas to make the design suitable. Just as most contractors develop an expertise in one market type of construction (residential, light commercial, etc.), design professionals often focus on a general area of expertise. Examples include a commercial kitchen designer for restaurant kitchens, or an architect that specializes in the hospitality or medical industry. Specialty drawings, included as part of the set, often require considerable coordination with the mechanical and electrical systems, as well as with the core drawings. The coordination aspect of the design between subconsultants cannot be understated. This ensures that structural columns are not placed in door openings, or that mechanical equipment is provided with power to operate.

Other drawings in the set include designs that are less concerned with the structure itself than with support services, such as utilities, that will be provided to the structure. These *civil* or *site drawings* include grading and drainage plans, which indicate how surface precipitation will be channeled away from the structure and collected onsite; landscaping and irrigation design; paving; and curbing layout. Ordinary site improvements, such as pedestrian walkways, fencing, patios, walks, flagpoles, and the like, are shown on a kind of catchall *site improvements drawing*. In short, the civil drawings help integrate the building with the site it is built on.

Some drawings are crossovers and show items of work or systems that may also be found in another set. For example, site electrical drawings indicating site lighting, power distribution, and low-voltage wiring (cable TV, telephone, and data) may also be shown in the electrical portion of the set.

In addition to drawings that show actual architectural or engineering improvements, there are drawings that provide data only or show compliance with codes and ordinances. These will be explored later in this chapter.

ORGANIZATION OF THE WORKING DRAWINGS

There is a distinct organizational structure to the working drawings, which is almost universally accepted and is as follows:

- **Architectural drawings:** Core drawings showing the layout of the building and its use of space. They convey the aesthetic value of the structure and show the dimensions and placement of all key features. The first architectural drawings in a set generally show large areas in less detail. As one progresses through the architectural set, the level of detail increases. These drawings are prefixed by the letter "A" and sequentially numbered. The architectural drawings typically follow the civil drawings in the actual set.
- **Structural drawings:** Illustrate how the various load-carrying systems will transmit live and dead loads of the structure to the earth. Structural design is based on the architectural features and is designed around the core drawings. (For example, columns and beams are designed to avoid interrupting a space.) Structural drawings are prefixed by the letter "S" and are sequentially numbered. They often follow the architectural drawings in the set.
- **Mechanical drawings:** Illustrate the physical systems of a structure, such as plumbing, fire suppression/protection, and HVAC (heating, ventilating, and air-conditioning) systems. These drawings may be prefixed by the letter "M" for mechanical or "H" for heating. Plumbing drawings use the

letter "P" and fire suppression drawings use "FP"(fire protection), "SP" (sprinkler system), or "F" (fire). All drawings are sequentially numbered and shown mainly in plan view with supporting details. Typically, the mechanical set follows the structural drawings in the set.

- **Electrical drawings:** Illustrate the electrical requirements of the project, including power distribution, lighting, and low-voltage specialty wiring, such as for fire alarms, telephone/data, and technology wiring. They often show the provision for power wiring of equipment illustrated on other types of drawings. They are prefixed by the letter "E" and are sequentially numbered. Electrical drawings follow the mechanical drawings in a set.

- **Specialty drawings:** Illustrate the unique requirements of various spaces' special uses (such as kitchens, libraries, retail spaces, and home theater systems). They define the coordination among other building systems, most commonly the mechanical and electrical systems. The drawings are sequentially numbered and named according to the type of drawings. For example, "K" might be used for kitchen drawings, "F" for fixture drawings, and so forth. Given the complexity of today's structures it is not uncommon to see more than one set of specialty drawings.

- **Site or civil drawings:** Illustrate the structure's relationship to the property, including various engineering improvements to the site, such as the sanitary system, utilities, paving, pedestrian walks, curbing, and so forth. They are sequentially numbered but have a less formal naming convention, open to the interpretation of the design engineer. They are easily recognized from the core drawings since they only deal with the site. They are often the beginning discipline in the full set of plans.

Drawings for each of these categories will show only the work of the particular discipline. All lines and symbols that are not specifically related to that discipline are grayed out or shown in a lighter line weight. This helps coordinate or locate the work of a specific drawing with other drawings that indicate adjacent but unrelated work. Since the overwhelming majority of plans are drawn today on computer, the graying-out of a line or feature is fairly simple and routine.

There are some common, basic elements in a set of contract drawings, which will be discussed in the following sections. These include a cover sheet, title block, revisions, and, to a lesser degree, a code analysis page. (Some revisions will be encountered on every drawing.)

Cover Sheet

The cover sheet, although very basic in nature, is one of the most important pages in a set of drawings. It lists information, such as the name of the project; the location; and the names of the architects, engineers, owners, and other consultants involved in the design. The cover sheet also lists the drawings that constitute the set in the order they will appear. The drawing list is organized by the number of each drawing and the title of the page on which it appears. The cover sheet may also list information specifically required by the building code having jurisdiction over the design of the project, including the total square-foot area of the structure, the building code use group the structure will fall under, and the type of construction. For larger, more complex projects, the code analysis sheet may be a dedicated sheet at the beginning of the set that identifies the specific code requirement and how this particular project complies. It is intended to aid the reviewer in determining compliance with a particular section of the code or ordinance.

Another important element on the cover sheet is a list of abbreviations or graphic symbols used in the drawing set. There is often a section that contains general notes, such as, "All dimensions shall be verified in the field" or "All dimensions are to face of masonry." These notes help set the standards for background information that you will encounter throughout the drawings. In the absence of a separate set of bound specifications (most common in the residential market, where separate specs are not often written), the cover sheet may list the general technical specifications that govern the quality of materials used in the work. Optional information, such as a locus plan locating the project with respect to local landmarks or roadways or an architectural rendering of the structure, may be included in the cover sheet.

Many cover sheets define energy conservation compliance calculations or adherence to a specific sustainable standard or program. The reader is advised to review these criteria carefully, as they can have a tremendous impact on project costs. The cover sheet might actually be several sheets long to provide the required details.

TITLE BLOCK

The title block is located along the right side, on the bottom, or in the lower right-hand corner of the drawing. Locations can vary by firm and even region. The title block should include the following information:

- The prefixed number of the sheet (to identify the discipline and order in the set) such as A1.0
- Name of the drawing (e.g., "First Floor Plan")
- Date of the drawing
- The architect's project number, or CAD file number, if applicable
- Initials of the draftsperson
- Any revision numbers and date to the final set of drawings
- Reference to north via a directional arrow
- Scale of the drawing if applicable
- The design professional, architect/engineer's seal/stamp (if appropriate for a record or permit set)

The date and scope of the revisions should be noted within the title block. If there is not enough space available, the revisions should be noted close to it. The title block should specify whether the entire drawing is one scale or whether the scale varies per detail, as in the case of a sheet of details. Sets of drawings for commercial projects, and some residential projects, require a stamp (and usually a signature) of the architect or engineer responsible for the design. This individual is referred to as the "design professional of record" and with the placement of the seal or stamp legally takes responsibility for compliance with the codes or ordinances that the project has been designed under.

Revisions

Often, after the set of working drawings has been completed, recommendations are made for correction or clarification of a particular detail, plan, or elevation. While major changes may require redrafting an entire sheet, smaller changes are shown as a revision of the original. All changes must be clearly recognizable. They are indicated with a *revision marker*, which encloses the revised detail within a scalloped line that resembles a cloud. Tied to the revision marker is a triangle that encloses the number of the revision. Revisions are noted in the title block, or close to it, by date and number. This procedure provides a mechanism for identifying the latest version of drawings. This helps ensure that the estimator is using the most up-to-date version of the plans as the basis of the estimate.

GRAPHIC FORMATS USED IN DRAWINGS

There are accepted standards or methods that architects and engineers use to present graphic information on drawings. Different views ensure that all required information is available on the drawings. There are six main graphic formats:

- Plan views
- Elevations
- Sections
- Details
- Schedules
- Diagrams

Each method illustrates the various aspects of a project from a different viewpoint. The information is most effectively presented when multiple views are used together. Showing the same item in different views helps confirm and adds to the information that can be seen in a single view. It is intended to be a check and balance feature for the information provided.

Plan Views

The most common graphic view, the plan view, is presented as if the viewer is looking down on the space. Plan views form the basis of the project and often provide the most complete view. The most common plan view is the *architectural floor plan*, which shows doors, windows, walls, and partitions. It provides the "big picture" view of the space.

Variations of plan views include *structural, fire suppression, plumbing, HVAC,* and *electrical plans*. Each shows the work of the respective trades in plan view as they fit into the architectural floor plan. Other types of plan views include *reflected ceiling plans*, which illustrate the ceiling as it would appear in a mirror, and *partial plan views,* which illustrate a particular area and enlarge it for clarity. Partial views are most often used in areas of high congestion or detail. *Demolition plans* show proposed changes to the existing floor plan. *Roof plans* show the roof layout as would be seen from overhead.

Plan views provide dimensions, which help the estimator calculate areas. Dimensions should be accurate, clear, and complete, showing both exterior and interior measurements of the space. It can also be used for counting features such as light fixtures on a reflected ceiling plan, or to confirm the number of windows shown on the exterior elevations. Plan views are also a starting point from which the architect directs the reader to other drawings for more information. The plan view on the core architectural drawings can often be used to verify information on the subconsultants plans. For mechanical and electrical drawings, the core drawings are shown as lighter line quality or shade to identify the location of electrical or mechanical features with respect to the architectural drawings.

Elevations

Elevations provide a pictorial view of the walls of the structure, similar to a photograph of a wall taken perpendicular to both the vertical and the horizontal planes. Exterior elevations may be titled based on their location with respect to the headings of a compass (north, south, east, or west elevation) or their physical location (front, rear, right side, or left side elevation). The scale of the elevation should be noted either in the title block or under the title of the elevation.

Interior elevations provide views of the walls of the inside of a room. They illustrate architectural features, such as casework, standing and running trims, fixtures, doors, and windows. Exterior elevations provide a clear depiction of doors and windows, often using numbers or letters in circles to show types that correspond to information provided in the door and window schedule. In addition, elevations show the surface materials of walls, and any changes within the plane of the elevation or facade. While the floor plan shows measurements in a horizontal plane, elevations provide measurements in a vertical plane with respect to a horizontal plane. These dimensions provide a vertical measure of floor-to-floor heights, windowsill or head heights, floor-to-plate heights, roof heights, ceiling heights, or a variety of dimensions from a fixed horizontal surface. The dimensions are provided for use in calculating measurements, areas, and volumes for specific tasks. They can also be used to confirm quantities of features such as windows or doors with other views such as plan view or schedules.

Building Sections

The building section, commonly referred to as the *section,* is a "vertical slice" or cut through a particular part of the building. It offers a view through a part of the structure not found on other drawings. Several different sections may be incorporated into the drawings. Sections taken from a plan view are called *cross-sections;* those taken from an elevation are referred to as *longitudinal sections* or simply *wall sections.* Wall sections provide an exposed view of the building components and their arrangement within the wall itself. By referring to sections, in conjunction with floor plans and elevations, an estimator can visualize the composition of the building component along with its necessary dimensions.

Details

For greater clarification and understanding, certain areas of a floor plan, elevation, or a particular part of the drawing may need to be enlarged. This enlargement provides information that is critical to a part of the building item that may otherwise not be available in another view. Enlargements are drawn to a larger scale and are referred to as *details.* Details can be found either on the sheet where they are first referenced or grouped together on a separate detail sheet included in the various disciplines they reference. The detail is shown in larger scale to provide additional space for dimensions and notes. Details are not limited to architectural drawings but can be used in structural and site plans and, to a lesser extent, in mechanical, electrical, or civil plans.

Schedules

In an effort to keep drawings from becoming cluttered with too much printed information or too many details, design professionals have devised a system to organize all types of repetitive information in an easy-to-read table, known as a *schedule.* Schedules list information pertaining to a similar group of items, such as doors, windows, room finishes, columns, trusses, and light or plumbing fixtures. The most common schedules are door, window, and room finish schedules. However, information on any repetitive type of item can be assembled into a table and incorporated in a set of drawings.

Schedules are not limited to architectural drawings but can be found in any discipline included within the set. A typical door schedule lists each door by number, or *mark,* and provides information on size and type, thickness, frame material, composition, and hardware. In addition, the door schedule provides specific instructions or requirements for an individual door, such as fire ratings, undercutting, weatherstripping, or vision panels. In the "remarks" portion of the schedule, the design professional lists any nonstandard requirements or special notes to the installer.

Diagrams

Some of the information presented in the set of drawings is more diagrammatical than pictorial. A *diagram* illustrates how the various components of a system are configured and is often provided for purposes of coordination. Diagrams are commonly used for mechanical and electrical drawings, because of the complex nature of the work. Common examples include diagrams for fire alarm systems, waste and vent piping risers, and fire protection. They are rarely, if ever, drawn to scale, and the symbols used are not intended to resemble the individual components. The lines connecting the symbols indicate how the components are assembled to become a system. Diagrams are typically additional information provided as a supplement or confirmation to another view.

DRAWING CONVENTIONS

Certain conventions have been adopted to provide a standard for drawings – from one design firm to another. The most common graphic features are lines, in-fill techniques, and shading, which can often contain subtle but very important information relative to the detail shown. While most of these conventions are widely accepted and practiced, there will always be minor deviations based on local practices. This is most apparent in the use of abbreviations and symbols. In many cases, any unfamiliar symbols and abbreviations will usually become clear by studying the drawings or cover sheet for the particular discipline.

Lines

Drawings must convey a great deal of information in a relatively small space, where there is no room for a lot of wording. Consequently, different types of lines are used to communicate information. The most common ones are discussed as follows:

- **Main object line:** A thick, heavy, unbroken line that defines the outline of the structure or object. Used for the main outlines of walls, floors, elevations, details, or sections.
- **Dimension line:** A light, fine line with arrowheads or tick marks at each end, used to show the measurements of the main object lines. The arrowheads fall between extension lines that extend from the main object lines to show the limits of the item drawn. The number that appears within the break in the dimension line is the required measurement between extension lines.
- **Extension line:** A light line that extends from the edge or end of the main object line, touching the arrowheads. Used together with dimension lines to helps one determine the limits of a particular feature.
- **Hidden or invisible line:** A light dashed line of equal segments that indicates the outlines of an object hidden from view, under or behind

some other part of the structure, such as a foundation shown in elevation that would be below grade.

- **Centerline:** A light line of alternating long and short segments that indicates the center of a particular object. Frequently labeled with the letter "C" superimposed over the letter "L."
- **Shaded or grayed-out line:** A lighter or gray-shaded line used to indicate features not part of the main view but provided for overall location reference. Features drawn as grayed-out lines can be assumed as existing to remain.

Material Indication Symbols and Shading

In-filling certain graphic features on a drawing help convey their content or composition. In-filling can indicate whether the feature is solid, as in the case of cast-in-place concrete, or hollow, as with concrete masonry block. In-fills are called *material indication symbols*. Because of the different views used on drawings, various materials must be recognizable at each view, from plan to section to elevation. As with abbreviations, material indications symbols are subject to change based on specific materials used in various regional locations.

Shading

Architects and engineers can convey information in a subtler manner by changing the intensity of a particular feature. This effect, called *shading*, increases or decreases the focus on the item, merely by its intensity. Items in the foreground or focus are often drawn darker or thicker. Objects in the background are lighter in color and drawn less sharply. Shading is often used to differentiate between proposed and existing work on renovation projects. It is also a means of determining the location or relationship of the main object to other background information or features.

Graphic Symbols

Graphic symbols are another means of providing a standardized way to recognize information and depict repetitive information on drawings. *Section markers* indicate where a section is cut through an object and can be directional or nondirectional. *Elevation symbols* direct the reader to the drawing that contains a noted elevation. They indicate differences in vertical height, such as the distance between floors, and provide a reference point to use in calculating the height of components in walls or partitions.

Frequently, the design professional draws a feature and, to save space on the page, uses a *break in a continuous line*. This symbol conveys that the feature is not drawn to scale and that there is more of the same feature that is not shown. Geometric shapes with letters, numbers, or dimensions within the shape define certain features or main objects. This graphic symbol is frequently used to name windows, doors, rooms, partition types, and ceiling heights. The important information is within the shape, not the shape itself. The shape used is often based on the preference of the individual design professional or the local accepted practice.

Trade-Specific Symbols

Like graphic symbols, trade-specific symbols depict items that are common to the various trades. Because of the highly diagrammatic nature of mechanical and electrical drawings, there is an abundance of unique, trade-specific symbols used on these drawings. Engineers typically provide legends that define the symbols used. Some symbols, such as for a water closet or sink, are highly recognizable because they mirror the real-life feature. Others have no relationship to the real-life feature and adopted by the particular discipline or industry and have become standards. Many electrical symbols illustrate this particular point.

Abbreviations

Abbreviations are used to save design professionals time, as well as space on drawings. There is a wide and varied selection of abbreviations used in daily practice. It is not necessary to memorize each abbreviation. Standard practice is to list the abbreviations on the cover sheet or legend sheet of the set of drawings. This compilation of abbreviations saves time by locating the meaning of each abbreviation in a central location or legend. For those not listed in a legend the estimator may be best served by investing in a construction dictionary or take their search to the internet. Many abbreviations are regionally based or even colloquial.

Occasionally cover sheets or subsequent sheets will provide definitions of words used in the documents. The construction industry often attaches very specific meanings to common words. For example: *furnish only* means to deliver a product onsite for installation by others. *Install only* means to install a product furnished by others and *provide* means to furnish and install a product. While there is a general acceptance of these meanings, it is suggested that the plans, and sometimes the specifications be searched for confirmation of the meanings and usage.

Scale

Since there are various physical limitations to drawing a building's actual size on a piece of paper, the drawings retain their relationship to the actual size of the building using a ratio, or *scale*, between full size and what is seen on the drawings. There are two major types of scales: the *architect's scale* and the *engineer's scale*. The estimator should become familiar with each and be able to recognize the different use or application.

Architect's Scale

The architect's scale is used for building drawings, as well as the mechanical, electrical, and structural engineering disciplines. The actual architect's scale may be flat, like a ruler, or three-sided. The three-sided architect's scale has 10 separate scales: ⅛" and ¼", 1" and ½", ¾" and ⅜", 3/16" and 3/32", and 1½" and 3". The one remaining side is in inches, similar to a ruler. For example, when used on a floor plan that is ¼" scale, each ¼" delineation represents 1'-0". The same rules apply for ⅛" scale, in that each ⅛" segment on the drawing represents 1'-0" of actual size. The same approach applies to each of the other scales. There is no strict convention that states which scale should be used on which drawings. In general, as the area of detail being drawn becomes smaller, or more focused, the scale often increases. For example, a floor plan may be fine at ¼" = 1'-0", yet the detail of an element within that floor plan would be better illustrated in ½" or ¾" = 1'-0" for clarity.

Engineer's Scales

The engineer's scale is similar to the architect's scale and is typically (though not exclusively) used to prepare or scale civil drawings. The difference is the size of the increments on the sides of the scale. The engineer's scale has six scales: 10, 20, 30, 40, 50, and 60. For example, the 10 scale refers to 10' per inch; the 20 scale is 20' per inch, and so on. Other specialty scales are divided into even smaller increments, such as 100 feet per inch.

The engineer's scale is used to measure distance on site plans when it is greater than would be encountered in the plans of the building. Occasionally, architects and engineers include a detail strictly for visual clarification. These details are labeled NTS, meaning "Not to Scale." This lets the estimator know that the details are not for determining quantities and measurements but for illustrating a feature that would otherwise be unclear. Diagrams are also typically not drawn to scale.

The use of scales in hand drafting has all but disappeared with the mainstream use of computer-aided design (CAD). Scales are now most often used by field personnel determining dimensions. Many estimators today employ take-off software that eliminates the need for scales. For those using scales in estimating, it should be noted that the profession recommends the use of printed dimensions even if it requires adding a string of dimensions in lieu of scaling. In addition to the triangular and flat scales previously mentioned there are rotometers, digital scale wheels, and tape scales available for the estimating professional.

Civil Drawings

Commercial and custom residential projects typically include a *site plan*, which illustrates the relationship of the proposed structure to the building's lot, as well as the various site improvements needed to accommodate the new building. The grouping of different types of site drawings, such as utility and drainage, grading, site improvement, and landscaping plans, is known under the general classification of *civil drawings*. Civil drawings encompass all work that pertains to a project, other than the structure itself. They have some unique conventions and nomenclature that merit a separate review. The most obvious difference between civil drawings and architectural drawings is the use of the engineer's scale. (As mentioned earlier, smaller scales are used on site drawings to indicate much larger areas.) It is important to note the scale in order to avoid errors in measuring during the takeoff. To avoid confusion, it is best to use the title block to clarify the type of drawing and scale.

The following sections review the most common terms and symbols associated with the various civil drawings.

Plot Plan

The main purpose of the plot plan is to locate the structure within the confines of the building lot. The plot plan is often the legal location of the lot within specific boundaries. This plan shows the property line with *bearings* and *distances* indicated for each line enclosing the lot. These are usually drawn and then each line is annotated with bearing and distance. An example of bearing and distance might be N 65° 10' 0" E and 545.01". This can be read more descriptively as "north 65 degrees 10 minutes and 0 seconds east for a distance of 545.01 feet." The bearing or direction is relative to the 360 degrees of the compass. The distance is the feet and inches dimensions converted to feet and portions of a foot. For example, the architectural dimension of 22'-6" would be 22.50' on a plot plan. This decimal system is used because it is the basis of measurement for the land surveyor, the engineer predominantly responsible for laying out the site. The plot plan is often a separate sheet and is stamped by a registered civil engineer or land surveyor dependent on the regulations of the state in which the lot is located.

Site Plan

Even the most basic site plans clearly establish the building's dimensions, usually by the foundation's size and the distance to property lines. The latter, called the *setback* dimensions, are shown in feet and hundredths of a foot.

As a starting point for the site design, a site survey is performed by a registered land surveyor, who also records special conditions. These may include existing natural features, such as trees or water, as well as manmade improvements, such as walks, paving, fences, or other structures. The new site plan shows how the existing features will be maintained, modified, or removed to accommodate the new design.

Another chief purpose of the site plan is to show the unique surface conditions, or *topography*, of the lot. Changes in the elevation of the lot, such as slopes, hills, valleys, and other variations in the surface, are shown on a site plan by means of a *contour,* which is a line connecting points of equal elevation. This convention is used to show three dimensions (length, width, and height) on a two-dimensional medium – paper. An *elevation* is a distance above or below a known point of reference, called a *datum.* The datum could be sea level, or it could be an arbitrary plane of reference established for the particular building.

Grading Plan

For projects in which the topography must be shown separately for clarity, a *grading plan* is used. Grading is necessary to direct water from precipitation away from the structure to a collection system on site. The grading plan typically shows the existing and proposed contours as dashed and solid lines, respectively.

A known elevation on the site for use as a reference point during construction is called a *benchmark.* The benchmark is established in reference to the datum and is commonly noted on the site drawing with a physical description and its elevation relative to the datum. For example: "Northeast corner of catch basin rim – Elev. 102.3" might be a typical benchmark found on a site plan. When individual elevations, or *spot grades,* are required for other site features, they are noted with a + and then the grade. For example: *+123.45* would designate a spot grade for a particular feature such as a catch basin frame or the elevation of an exterior step. Spot grades are accurate to two decimal places, whereas contours are expressed as whole numbers. Some site plans include a small map, called a *locus,* showing the general location of the property in respect to local highways, roads, and adjacent pieces of property.

Grades are used to calculate differences in vertical elevation and are extrapolated over the area of the site to determine quantities of excavation and backfill involved on a site. (This will be discussed in Chapter 23, "Earthwork.")

Drainage and Utility Plans

Larger projects have several site plans showing different scopes of related or similar work, such as drainage and utility plans. Drainage and utility work are typically shown on separate drawings. Utility drawings show locations of water, gas, sanitary sewer or septic systems, telecommunications, and electric utilities that will service the building. Drainage plans detail how surface water will be collected, channeled, and dispersed on- or off-site. Both plans illustrate, in plan view, the size, length, and type of pipes and special connections or terminations of the various piping. Because the effluent in certain types of pipes moves by gravity, the elevation of each end of the pipe must be different to ensure a flow.

Certain site plans require clarification in the form of a detail, similar to the architectural detail. Classic examples are sections through paving, precast structures, pipe trenches, and curbing. Details are not limited to typical views, but occasionally appear in the form of perspective drawings, which are not drawn to scale and are used as a means of clarification only.

Landscaping and Irrigation Plans

Landscaping plans show the locations of various species of plantings, as well as lawns and garden areas. The plantings are noted with an abbreviation, typically three letters, along with the quantity of the particular species. This designation corresponds to a planting schedule, which is a complete listing of plantings by common name, Latin or species name, and quantity and size frequently located on the drawing. Notes describing planting procedures or handling specifications accompany the schedule.

Irrigation drawings may be included, which illustrate how the landscaping elements will be watered: including sleeves under pavements, piping from water sources, controllers, valve boxes and both low- and line-voltage electrical needs. It is not uncommon to find low-voltage landscape lighting on landscape drawings. This should be coordinated with the electrical drawings in the set.

Paving and Curbing Drawings

To accurately show the layout of parking lots and driveways, a *paving/curbing drawing* is needed. This plan shows the various types of bituminous, concrete, and brick paving and curbing, and the limits of each – helpful for calculating areas and measurements. Again, it is important to review the legend symbols in order to clearly delineate where one material ends and another begins. Details showing sections through the surface are used to differentiate between thickness and the substrate below. In addition to the vehicular and pedestrian surfaces themselves these drawings can often depict the aggregate subbase required for each and detail sections through curbs to illustrate the trench and the backing material.

Site Improvement Drawings

When the project warrants, separate drawings may be needed to clarify various site improvements, such as walks, retaining walls, patio paving, fences, steps, benches, play areas, and flagpoles. Site improvement drawings are often used as a catchall to show the miscellaneous items that do not fall neatly into one of the aforementioned classifications of work. The estimator is advised to carefully review this drawing for miscellaneous items that are commonly left out of the estimate as well as the associated work of another section that may be required such as excavation and backfill, concrete and formwork for a retaining wall.

Sedimentation and Erosion Control Drawings

The excavation and backfilling of soils on a site can be extremely detrimental to existing surrounding environments and habitats. Codes and ordinances mandate measures to protect the environment. Many projects are required to illustrate via a drawing how the contractor will mitigate or manage erosion and sedimentation control during construction and keep it on site. To this end, many sets of civil drawings contain Sedimentation and Erosion Control plans. These drawings illustrate unique features, materials, and techniques for controlling erosion and the sediment it produces. Failure to comply with these requirements can result in fines.

Existing Conditions Site Drawings

For projects with existing drainage, utilities, and structures, an *Existing Conditions plan* is provided, which is invaluable for understanding and calculating the difference between actual conditions and proposed work. The existing conditions are shown in the background grayed out or lightly shaded, and the new work is shown darker in the foreground. This allows the estimator to determine the quantities of connection points of new work to existing work. Other methods include showing existing conditions as dotted or broken lines and proposed conditions as solid, darker lines. Occasionally *Existing Condition* drawings will show existing contours and new contours on the same sheet.

Sometimes *test boring* logs are provided, which document engineering tests to determine the load-bearing and general quality of the soil at the site. These can be borings or excavated test pits. Existing Condition drawings can often show or indicate features buried beneath the surface such as existing foundations, rock or ledge features, utility lines, and similar information.

For projects of a more complex nature, for sites with suspect soils, or high water tables, a geotechnical report can be included in the project manual. Individual borings are numbered and located on a plan in the civil set with the test results provided in a table format in the project manual. Each boring indicates the composition and vertical limitation of the various soil strata below the surface. The height of the water table is indicated as a dimension below the surface or as an elevation from a known datum. A word of caution for the estimator is advised. Geotechnical reports often come with a disclaimer advising that the soil conditions identified in the boring are for that boring only and that there is no guarantee, written or implied, that the quality of the soil is the same outside of the limitation of the boring casing. The estimator should consider this information when relying on the data in the geotechnical report. These reports can be accompanied by a narrative explaining the requirement for soil replacement or treatment.

Site Visits

It is becoming increasingly common for owners or awarding authorities to provide access and a tour of the site prior to bid. This site visit is called the *prebid walkthrough*. Many projects are requiring mandatory attendance at the prebid walkthrough as a way of reducing change requests resulting from not visiting the site. While site visits are not specifically part of the plans, they have the benefit of resolving questions that most often arise from the plans.

It is essential to become familiar with the drawings *prior* to the site inspection and start of the quantity takeoff. One recommended procedure is to review the plans with a pad of paper nearby. As questions arise, jot them down. As the review progresses, many answers to the questions become readily apparent and can be removed from the list. Any remaining questions may be answered during the specification review or during the prebid walkthrough. It is also common practice to have one's major subcontractors accompany the prime contractor on prebid walkthrough. This achieves the benefit of allowing subs to answer their own questions and secondly to show both the owner/awarding authority and the competition the serious intent of the bidder.

If allowed, a camera or smartphone can be used to take photos, which can help in recalling information during the preparation of the estimate later on. Most seasoned estimators recognize that there are specifics about the site that can influence the estimate, in a positive or negative way. Not everything can be obtained from the plans and specifications. Quite often things such as access to the site, storage availability on site, parking for trades, and even overhead clearance for cranes may benefit from a site visit.

SCOPE OF WORK

As mentioned in the opening paragraph of this chapter, not all projects are lucky enough to have a fully developed set of plans and specifications. In all fairness, not every project in need of an estimate warrants the full set of plans and specifications. Many projects of limited scope can provide sufficient information and direction to potential bidders so that an estimate and ultimately a bid can be submitted.

While there is no exact organizational structure to a Scope of Work, sometimes called a Statement of Work, there is a general consensus as to what needs to be included. Project owners may weigh each category below differently. However, a SOW should include as a minimum:

- Owner's information
- Invitation to bid with dates
- Project description or overview
- Project timeline/schedule
- Project deliverables

- Technical specifications
- Payment terms and conditions
- Contract form and insurance certificates
- Bid form or submission requirements
- Project signoff or closeout

SOWs are often used for projects of limited scopes such as the replacement of a rooftop HVAC unit or the painting of an office suite. The beginning and end of the work are clear and distinct so there is little confusion as to what is included. With a SOW there is more burden on the bidder's estimator to determine the list of tasks to be priced. Photographs are an integral part of the site visit(s). Qualifications to a bid may be allowed if stated in the bid form, whereas with a full set of plans and specifications qualifications may render a bid nonresponsive.

With SOWs there is typically a fixed time period for questions in the form of RFIs. In the pre-award phase, there can be a detailed review of the scope to ensure compliance with the deliverables.

SUMMARY

This chapter reviewed the types of plans and drawing elements that together constitute a full set of working drawings. The working drawings, along with the specifications (discussed in Chapter 2), are the bid documents. The bid documents, along with a site visit, are the basis of the estimate and will become the contract documents upon award.

Working drawings are only part of the contract documents. They comprise the graphic representation of the design professional's intent. Plans illustrate the project in a format that allows the estimator to determine quantities as part of the takeoff process.

A thorough review of the drawings often reveals discrepancies, conflicting information, or even omissions and helps determine whether to proceed with the next step in bidding the job. Note that the various views should be used together. Information located on one drawing can often be corroborated on another. This is part of a checks-and-balances process that is fundamental in estimating.

Some smaller projects of limited scope do not require the development of a full set of plans and specifications and opt for the use of a SOW that acts as the basis of the estimate. While SOWs are a viable option under the right conditions, the work and deliverables must still be clearly identified in the document.

It is essential that the first step in preparing any estimate is to become familiar with the bid documents.

QUESTIONS/ PROBLEMS

1. What role do the plans play in the estimating process?
2. Identify the six (6) main graphic formats that a reader would expect to see in a set of plans.
3. Name five (5) things a reader would expect to see in the title block of a drawing.
4. Define extension lines and dimension lines. Please ensure that the definition includes how they are used together.
5. What is the purpose of a revision marker on a set of plans?
6. Explain why design professionals use abbreviations on a set of plans.
7. What is a contour? Include in your explanation what information can be derived from a contour.
8. What can a reader expect to see on an *Existing Conditions* plan?
9. Define the term *benchmark* and explain its purpose for the estimator.
10. Explain the limitations of a geotechnical report for an estimator.

2 | Understanding the Specifications

Project owners and awarding authorities have continually sought to improve the definition of *industry standard* when defining the quality of materials or workmanship to be included in their projects. As a result, design professionals in all segments of the industry are relying on technical specifications to establish the quality level for owner's expectations and as a metric for contractors performing the work. During the last half-century, in fact, technical specifications have become increasingly popular as *the* standard of measurement for quality. When plans and technical specifications are used together, it allows the estimator to define the quantity and the quality of both the material and the labor so that the estimated value of the work is easily determined.

ROLE OF THE SPECIFICATIONS

The technical specifications, or *specs,* as they are commonly referred to, are part of the bid documents, along with the working set of drawings (discussed in Chapter 1); they define in detail the processes and materials for the project. Technical information about the quality of materials and workmanship is not always incorporated on the drawings themselves, because of a lack of space and the need to maintain clarity. For many projects, working drawings are issued with a separate set of specifications in a bound *project manual.* Even the simplest projects have some specifications, whether incorporated on the drawings or issued as a separate document to guide the contractor and subcontractors.

The specs perform a variety of functions, including:

- Serving as the legal basis for the Contract for Construction
- Defining the quality or grade of materials to be used in the project
- Defining the acceptable workmanship or providing standards to judge workmanship through tolerances

Estimating Building Costs for the Residential and Light Commercial Construction Professional, Third Edition. Wayne J. Del Pico.
© 2023 John Wiley & Sons, Inc. Published 2023 by John Wiley & Sons, Inc.

- Providing guidelines for resolving disputes between parties to the contract
- Providing a basis for accurately estimating cost
- Complementing the graphic portion of the project, the drawings

The specs are intended to be used in conjunction with the drawings. If the drawings are the *quantitative* representation of the project shown in a graphic format, then the technical specifications are the *qualitative* requirements of the project described in a written document.

Technologies, processes, and products are continually evolving in the construction industry, and architects and engineers incorporate these advancements more frequently into their designs. As a result, highly technical information is needed in the specifications. The materials and processes are described in such detail that the intent of the designer, as well as the product or system, can be upheld in case of a dispute or if products are installed incorrectly.

The specifications serve as a basis for bidding and performing the work. The person preparing the specifications, called a *specification* or *technical writer,* makes every effort to cover all of the items or segments of work shown on the working drawings. In the past, if there was a discrepancy between the specifications and the drawings, the specifications generally took precedence. A more thorough explanation is presented later in this chapter.

ORGANIZING SPECIFICATIONS BY CSI MASTERFORMAT®

Throughout this book, we will refer to the CSI MasterFormat®, which is the most widely accepted system for organizing construction specifications and estimates. Developed by the Construction Specifications Institute, the MasterFormat® system is also used for classifying data and organizing manufacturers' literature for construction products and services. CSI's multi-level system has allocated an eight-digit code and topic descriptions to all components of the specifications. MasterFormat® as it would appear in a project manual groups the information into four major categories:

- Bidding Requirements
- Contract Forms
- General Conditions
- Technical Specification Sections

MasterFormat® 2020, the latest version of the organizational system, consists of 50 construction divisions within the technical specifications section. MasterFormat® has been divided into five subgroups that loosely groups related divisions.

Each division is a compilation of similar or related work numerically organized into subsections called *levels.* Each level represents a further breakdown of the CSI division classification. With the publication of the 2004 edition, the MasterFormat® numbers and titles were revised to allow them to cover construction industry subject matter more adequately and to provide ample space for the addition of new sections. The titles that make up MasterFormat® were also revised, reflecting the new edition's renewed focus on work results. As a part of this process, the numbering system was extensively revised, meaning that all section numbers and many section titles have changed from the 1995 edition. The five-digit numbers used in the 1995 edition were expanded to allow room for more subjects at each level of classification. The old numbers were limited at levels 2 through 4 to only nine subdivisions. Because of this limited number of available spaces at each level, many divisions of MasterFormat® simply ran out of room to properly address topics. This lack of room often resulted in inconsistent classification. These limitations were solved by making the new MasterFormat® numbers six digits in length and arranging the digits into three sets of paired numbers, one pair per level. These pairs of numbers allow for many more subdivisions at each level. Meanwhile, the main six-digit number still represents three levels of subordination, as the numbers in previous editions of MasterFormat® have done.

For example, consider the CSI section number 03 30 53.40. The first group of two digits, 03, indicates MasterFormat® Level 1 and designates the division the work belongs to, in this case Division 3—Concrete. The second group of digits, 30, indicates MasterFormat® Level 2, which designates the subsection Cast-in-Place Concrete within Division 3. The third group of two digits, 53, indicates MasterFormat® Level 3, a further breakdown of the Cast-in-Place subsection, Miscellaneous Cast-in-Place Concrete. The last group of two digits, 40, is MasterFormat® Level 4 and deals with one component within the previous Level 3 section.

The 50 MasterFormat® divisions were determined based on relationships of activities in the actual construction process, and they roughly follow the natural order of the construction of a building. The specification divisions and a general summary of their contents are as follows:

Division 00—Procurement and Contracting Requirements: Advertisement for bids, procurement, solicitation, invitation to bid, instruction to bidders, prebid meetings, bid forms, wage rates, bond forms, conditions of the contract for construction, and related certifications.

Division 1—General Requirements: A summary of the work, as well as the definitions and standards for the project and project coordination, price and payment procedures, meetings, schedules, reports, testing, samples, submittals, shop drawings, closeout, cleanup, quality control, and temporary facilities. In addition, this division addresses pricing issues, such as unit prices, alternates, and allowances, as well was closeout requirements.

Division 2—Existing Conditions: Existing conditions of the site or structure; surveys; geotechnical reports and subsurface investigations; salvage of materials; lead, asbestos, and mold remediation; contaminated site material removal, structure moving, and the removal of underground storage tanks.

Division 3—Concrete: Formwork, reinforcing, precast and cast-in-place concrete, concrete curing, underlayment, grouting, concrete cutting and boring, and cementitious decks.

Division 4—Masonry: Brick, block, stone, mortar, anchors, reinforcement, corrosion resistant masonry, and masonry restoration and cleaning.

Division 5—Metals: Structural steel, metal joists, metal decking, light-gauge structural metal framing, and ornamental and miscellaneous metals.

Division 6—Wood, Plastics, Composites: Rough and finish carpentry, architectural millwork, casework, composite lumber products, structural composites, and plastic fabrications.

Division 7—Thermal and Moisture Protection: Waterproofing, dampproofing, insulation, roofing, air and vapor barriers, siding, fire and smoke protection, caulking, and sealants.

Division 8—Openings: Metal and wood doors and frames, windows, glass and glazing, skylights, roof windows, mirrors, finish hardware, and louvers and vents.

Division 9—Finishes: Gypsum wallboard systems, board and plaster systems, painting, coatings, and wall coverings, flooring, carpeting, acoustical ceiling systems, acoustical treatments, and ceramic and quarry tile.

Division 10—Specialties: Demountable partitions, toilet partitions and accessories, fire extinguishers, postal specialties, flagpoles, fireplaces, safety specialties, storage specialties, lockers, signage, and retractable partitions.

Division 11—Equipment: Specialized equipment for homes, banks, gymnasiums, schools, churches, laboratories, food service, prisons, libraries, hospitals, and entertainment.

Division 12—Furnishings: Cabinetry, rugs, tables, seating, artwork, fabricated casework and cabinetry, and window treatments.

Division 13—Special Construction: Greenhouses, swimming pools, integrated ceilings, incinerators, sound vibration controls, and clean rooms.

Division 14—Conveying Systems: Elevators, lifts, dumbwaiters, escalators, moving walkways, cranes, and hoists.

Divisions 15 through 20: Reserve divisions for future expansion.

Division 21—Fire Suppression: Fire suppression and protection systems.

Division 22—Plumbing: Plumbing piping, gas piping, special services piping, plumbing fixtures, pipe insulation, water heaters and circulating pumps.

Division 23—Heating, Ventilating, and Air Conditioning: heating, air-conditioning, ventilating, ductwork, controls, insulation, HVAC equipment, solar energy heating equipment, fuel systems, and humidity controls.

Division 24—Reserve division for future expansion.

Division 25—Integrated Automation: Network servers, integrated automation of HVAC, fire protection, electrical systems, communications, and terminal devices.

Division 26—Electrical: Electrical service and distribution, wiring, wiring devices, fixtures, site lighting, and power.

Division 27—Communications: Communication services, cabling and cable trays for communication, adapters, and software.

Division 28—Electronic Safety and Security: Fire alarm systems, closed-circuit TV, security alarm systems, access control, leak detection, and video surveillance

Divisions 29 and 30: Reserve divisions for future expansion.

Division 31—Earthwork: Clearing of the site, earthwork; bulk and general excavation, backfill and compaction, grading, soil treatments and stabilization, and heavy site work, such as shoring, pile driving, and caissons.

Division 32—Exterior Improvements: Paving, curbing, base courses, unit paving, parking specialties, fences and gates, landscaping, plantings, and irrigation.

Division 33—Utilities: Piping for water, sewer, drainage and related structures, fuel distribution utilities, and electrical utilities.

Division 34—Transportation: Railways and track, cable transport, monorails, transport signaling and control.

Division 35—Waterway and Marine Construction: Coastal and waterway construction, dams, marine signaling, and dredging.

Divisions 36 through 39: Reserve divisions for future expansion.

Division 40—Process Integration: Specialty gas and liquid process piping, chemical process piping, and measurement and control devices.

Division 41—Material Processing and Handling Equipment: Bulk materials handling and conveying equipment, feeders, lifting devices, dies and molds, and storage equipment.

Division 42—Process Heating, Cooling and Drying Equipment: Industrial furnaces and process cooling and drying equipment.

Division 43—Process Gas and Liquid Handling, Purification, and Storage Equipment: Liquid and gas handling and storage equipment and gas and liquid purification equipment.

Division 44—Pollution and Waste Control Equipment: Air, noise, odor, and water pollution control, and solid waste collection and containment.

Division 45—Industry-Specific Manufacturing Equipment: Oil and gas extraction equipment, mining machinery, food and beverage manufacturing equipment, textile, plastic, and a variety of other types of manufacturing equipment.

Division 46—Water and Wastewater Equipment: Package water and wastewater treatment equipment.

Division 47—Reserve division for future expansion.

Division 48—Electrical Power Generation: Fossil fuel, nuclear, hydroelectric, solar, wind, and geothermal electric power generation equipment.

Division 49—Reserve division for future expansion.

CSI MasterFormat® is categorized into five subgroups:

1. General Requirements Subgroup—Division 1
2. Facilities Construction Subgroup—Divisions 2 through 19
3. Facilities Services Subgroup—Divisions 20 through 29
4. Site and Infrastructure Subgroup—Division 30 through 39
5. Process Equipment Subgroup—Divisions 40 through 49

THE PROJECT MANUAL

Successful communication of the architect's or engineer's design intent to the contractor depends heavily on how well the *project manual* is written and organized. The project manual is the bound document that contains all of the four major categories of CSI MasterFormat®. Information in the manual must be written clearly and presented logically. It should be easy to follow and comprehensive in order to prevent delays as a result of the need for constant clarification.

Preparation of the project manual is a substantial task, primarily the responsibility of the architect. Individual disciplines, such as mechanical, electrical, and structural engineers, review, edit, and contribute to their individual sections of the technical specifications. Problems tend to occur when the various disciplines fail to coordinate their part of the work with each other and with the core language of the General and Supplemental Conditions, as well as with the drawings. The technical writer must create a complete document using very specific language that will guide the contractor in the bidding and building processes and will also serve as a powerful tool to enforce the contract. It takes a skillful use of language, a high level of proficiency in understanding and coordinating technical information, and the ability to process that data into usable information. A review of the structure of the project manual and its four main categories is essential prior to starting the estimate and is a significant part of the document review process.

Over time, the term *specifications has come to be synonymous with the project manual. In actuality, the specifications most often refer to the technical specifications contained in the 50 divisions, whereas the project manual is the entire bound document that includes:*

- Bidding Requirements
- Contract Forms
- General Conditions
- Technical Specifications

Bidding Requirements

The Bidding Requirements are composed of the following sections:

- Bid Solicitation
- Instructions to Bidders
- Information Available to Bidders
- Bid Forms and Supplements

Bid Solicitation

The Bidding Requirements begin with a solicitation for bids or proposals. This solicitation can be in the form of an *Invitation for Bid, Request for Proposals (RFP),* or, in the case of public work, an *Advertisement for Bid.* In the private sector, bid solicitations can also be offered as an Invitation to Bid to selected firms only. All are similar in that they request bids from contractors. The RFP invites qualified general contractors and subcontractors to submit proposals for a particular project. It identifies the name and location of the project, along with a brief summary of the work involved. It clearly defines the date, time, and location for bids to be submitted. The RFP should name the owner or authority responsible for the bid award, whether the bids will be publicly or privately opened. In the case of taxpayer-funded projects, the bids are usually opened publicly and made available for inspection by the general public. The RFP typically identifies the architect and key engineering firms contributing to the design. For publicly funded projects, the statute governing such considerations as bidding and payments is also identified, along with any established budget for the work.

Some publicly funded projects require that certain subcontractors submit their proposals separately prior to the general contractor's bid date. This practice is called *filed subbidding* and is the law in one form or another in many states. The trades required to be filed sub-bidders are listed according to their MasterFormat® division and section numbers. The RFP states the date, time, location, and manner for the submission of filed subbids. The bids are held in a subbid depository and reviewed by the awarding authority. Filed subbids are either approved or rejected based on the governing laws and responsiveness. The results are issued to the general contract bidders through an addendum.

Instructions to Bidders

The Instructions to Bidders contain any required prequalification or eligibility criteria to eliminate bidders who could later be considered unacceptable. In the case of private bidding, the invitation to bid may be all that is required. In some states, publicly bid projects require format qualification forms and a summary of the contractor's performance record. If a prebid conference or site inspection is scheduled, the date, time, and location are also stated. Additionally, the Instructions to Bidders define the various forms and amount of bid security or bid bond that will be required. It states any liquidated damages that may be part of the contract, times for the commencement and completion of the work, and addenda or rules governing interpretation of the documents.

The Instructions to Bidders portion of the Bidding Requirements indicate the date, time, and location for procuring a set of contract documents and the cost, if any, to bidders. Other pertinent information, such as the time frame for award or rejection, special wage rates, tax-exempt status, or legal rights of the awarding authority to accept or reject proposals, is also provided.

Information Available to Bidders

The Information Available to Bidders provides locations where bidders can obtain copies of additional documents helpful in the bidding process. These documents could include geotechnical reports or subsurface investigation, property surveys and record drawings, conservation commission reports or directives, and hazardous materials management reports. Much of this information can be hosted on an ftp site for ease of access. Bidders can be provided secure access and the owner can track the frequency and length of time spent on the site.

Bid Forms and Supplements

This section contains the forms developed by the architect or awarding authority for use by the contractors submitting bids, as well as bid security forms. Bid forms are used to keep proposals uniform in appearance and content. They provide the owner and architect with a mechanism for comparing "apples to apples." Bid forms provide the language of the proposal with blanks for the contractor to fill in. Space is provided to acknowledge addenda; add or deduct alternates; list unit prices; state the name, address, and signatory party of the bidding contractor; and, naturally, indicate the dollar amount.

Nonresponsive is the term applied to a bidder who has incorrectly filled out or inadvertently left out information on the bid form. It can also be applied to bidders that have qualified their bid when no qualifications are allowed. For privately funded projects the error or omission on the bid form may be deemed insignificant or in the case of publicly funded projects it may render that bidder ineligible for award. In any event, the estimator should pay careful attention to the bid form requirements.

Supplements, or supplemental forms, include a certificate of compliance with tax laws form that requires bidders to certify under the penalties of perjury that they have complied with the tax laws regulating the state in which the work is to be done. A *Non-Collusion Affidavit* attests that bidders have not colluded or conspired with any entity, including other bidders, to defraud the owner or awarding authority. Other forms, although less common, include *Conflict of Interest* and *Power of Attorney* statements.

The Bidding Requirements often contain sufficient information for contractors to decide whether the project is right for their firm and, essentially, worth bidding.

Contract Forms

The most important contract form is the agreement between the owner and the contractor, more commonly referred to as the *Contract for Construction*. This is a legal instrument supported by all of the contract documents. The Contract for Construction must contain the following basic items in order for it to be considered a functional document:

- Clear identification of the parties to the agreement
- Clear identification of the project
- Rights and responsibilities of each party
- Basis and terms of compensation

This agreement is incorporated within the project manual so that prospective bidders can carefully review the contract that will be executed when the project is awarded. A careful review of the proposed contract is often the first act in determining whether to bid the project. Language governing the terms for payment, penalties, or damage assessments, as well as any unfavorable or exculpatory clauses should be reviewed for acceptance. It may also be advisable to seek legal counsel for terms and conditions that may not be fully understood.

The American Institute of Architects (AIA) publishes a family of contract documents that are frequently used by owners and architects and have become the recognized standard. It should come as no surprise that contracts written by architects and/or owners tend to favor those parties. Some contracts are written in such a manner as to impose a disproportionate share of the risk on the contractor or to indemnify the owner and architect with significant exculpatory clauses. The contract should be analyzed carefully and if the risks greatly outweigh the chance for success or profit, the bidders may want to decline to bid the project. Bidders should be aware of this and make every effort to understand the contract and any related information prior to making a commitment of company resources to bid. In fact, bidders should either accept the contract language as written or decline to bid the project. Changes to unfavorable contract language after the contract has been awarded (or the bidder selected) are unlikely.

Performance and Payment Bonds

Other contract forms include Performance and Payment (P&P) bond forms. Many owners issue a standardized form to prevent the surety from including exculpatory language of their own. The issuing of a P&P bond on a project has serious financial implications to the company as well as its principals. Default and/or termination while under the protection of a Performance and Payment bond can be devastating. It is important to review the terms and conditions of the contract carefully with the surety's representative and even legal counsel. Securing a reputable surety to provide bonding to a contractor can be a long and detailed process. It is important to understand all of the liabilities that can be associated with bonding; do not venture into this process uninformed. Since the P&P bond benefits the owner or awarding authority the cost of the bonds must be captured in the estimate.

Certificates

The last section in the contract forms section of the specifications contains forms for insurance required for the project. This document defines the dollar limits for the various policies required. Most contractors understand basic insurance requirements included with policies, such as General Liability and Workers' Compensation. However, many projects today are required to carry more unique forms of insurance, such as *Owner's Protective* and *Completed Operations* policies, which have potential impact long after the project has been completed. Again, seek professional guidance from your insurance agents when new policies or limits are required. Insurance is a highly specialized industry and beyond the expertise of most contractors. Your insurance agency should be able to price specific policies so that the contractors can add the cost of the policy to their bid if they do not currently have the insurance. Also note that for projects that have a duration longer than 12 months, the insurance premium for the additional time must be added.

General Conditions of the Contract

Although the Contract for Construction is the primary legal instrument in the project manual, it is insufficient on its own. Because of its complexity, a separate set of guidelines is necessary, called the *General Conditions of the Contract for Construction*. The General Conditions are meant to complement the Contract for Construction, defining the complex relationships between the owner, architect, and contractor and the mutual responsibilities and rights of the signatory parties. They are included in the Contract for Construction by reference. The General Conditions include the definitions of key terms and provide procedures and mechanisms for resolving disputes or clarifying information provided on the drawings or specifications.

Many owners in both the private and the public sector have elected the AIA General Conditions of the Contract as the document of choice. It should be noted that there are other entities that produce similar documents in addition to the AIA. Some owners have drafted versions of their own General Conditions using the AIA A201 as a model. In short, the General Conditions of the Contract are the administrative ground rules for executing the contract and the work. All are similar in content and address the 15 basic articles found in AIA A201, which address specific relationships made as a result of the agreement and the unique situations that are created during the construction process. The 15 basic articles of AIA A201 are:

1. General Provisions
2. Owner
3. Contractor
4. Administration of the Contract
5. Subcontractors
6. Construction by Owner/Separate Contractors
7. Changes in the Work
8. Time
9. Payments and Completion
10. Protection of Persons and Property
11. Insurance and Bonds
12. Uncovering and Correction of Work
13. Miscellaneous Provisions
14. Termination or Suspension of the Contract
15. Claims and Disputes

The General Conditions establish the legal requirements of the project in general terms:

- **Articles 1–6:** Provide definitions of terms and relationships and define the responsibilities of the various parties to the contract. They also establish procedures for resolving disputes during the construction process. Article 6 covers the owner's right to contract separately with other independent contractors to perform work concurrently with the prime contractor.
- **Article 7:** Defines procedures for handling changes to the work, including when there is disagreement in price, scope, or time.
- **Article 8:** Explains time and its impact on the schedule and defines the remedies for delay, as well as the procedure for requesting an extension of time.
- **Articles 9–12:** Sets forth terms for payment and justification for withholding funds. Article 9 defines the project's completion in contractual terms. Article 10 assigns responsibility for the protection of persons and property, as well as safety programs and responsibilities governing hazardous materials. Article 11 deals with loss and insurance or bonds required to make the owner whole. Article 12 assigns responsibility for the correction of defective and nonconforming work.
- **Articles 13 and 14:** Explains the legal provisions for the assignment of rights or termination of the contract by either signatory party.
- **Article 15:** Explains the procedure for perfecting claims and the resolution of disputed work. It also addresses the limitations of the architect as the initial decision maker in the response to claims for an equitable adjustment to money or time.

Supplements to the General Conditions of the Contract

As noted previously, the General Conditions of the Contract address specific issues (in a general format) that could be considered applicable to the industry as a whole. Often, projects have specific needs or unique conditions that require an amendment to the General Conditions. Because the AIA A201 is intended to interface with an entire family of other AIA documents, any modifications to this document or any AIA document can have serious legal ramifications. For this reason, many architects leave the General Conditions intact as written and modify them by adding a separate document, called the *Supplemental Conditions of the Contract*, frequently referred to as the *Supplementary General Conditions*. This custom-tailoring process allows the author of the project manual great flexibility in meeting the specific needs of the individual client or project without risking the loss of continuity that the General Conditions provide.

The estimator must review and analyze the specific impact that the Supplementary General Conditions have on the General Conditions and the project in its entirety. The importance of this review cannot be overstated. The Supplementary General Conditions often are used to modify already stringent or restrictive contract language focused on the contractor. They are often presented in a way that a dollar value can be established against its impact. A classic example is insurance requirements. While the General Conditions describe the type and extent of insurance coverage, the Supplementary Conditions establish its limits. Using this information, you can establish the amount of increased insurance policy dollars and thereby include the difference in the appropriate category of the estimate.

The three-part format of the technical sections provides a consistent organizational system for locating pertinent information quickly and efficiently:

Part 1—General
Part 2—Products
Part 3—Execution

Technical Specifications

The last of the four categories of the project manual is called the *Technical Specifications*, or *Technical Sections*, which define the scope, products, and execution of the work. Many estimators refer to this as the "meat and potatoes" section, providing the estimator with the necessary information, in a highly organized and industry-accepted format, to accurately price and build the structure. The technical sections provide the following information for each activity:

- Administrative requirements
- Quality or governing industry standards
- Products and accessories
- Installation or application procedures
- Workmanship requirements
- Each technical section is organized into three distinct parts, which provide a consistent organizational system for locating pertinent information quickly.

Part 1, General

Part 1, the General section of the specifications, provides a summary of the work included within that particular section. It ties the technical section to the General Conditions and Supplementary General Conditions of the contract, an essential feature in maintaining continuity between the general contractor and the subcontractors. This is sometimes called the *flow-down* provision. Roughly translated, it allows the general contractor to assign responsibility to a subcontractor. Part 1 identifies the applicable agencies or organizations by which quality assurance will be measured. It defines the scope of work (SOW) that will be governed by this technical section, including, but not limited to, items to be furnished by this section only or furnished by others and installed under this section. It also identifies other technical sections that have potential coordination requirements with this section and defines the required submittals or shop drawings for the scope of work described in this section. Part 1 also establishes critical procedures for the care, handling, and protection of work within this section, including such ambient conditions as temperature and humidity. If applicable, it addresses inspection or testing services required for this scope of work.

Part 2, Products

Part 2 deals exclusively with the products and materials to be incorporated within this technical section of the work. It can also identify components that make up an assembly. In general, Part 2 can identify acceptable manufacturers as well as product lines as a guide to what is acceptable.

For products that are directly purchased by the contractor from a manufacturer or supplier, the items can be identified using one of four methods:

- Proprietary specification
- Performance specification
- Descriptive specification
- Specification compliance number

PROPRIETARY SPECIFICATIONS

These specifications spell out a product by name and model number. Proprietary specifications have the unique advantage of allowing architects or owners to select a product they desire or have used successfully on prior projects. The advantage of requiring specific products is the level of reliability they provide. The disadvantage is that they eliminate open competition. To help reduce the exclusivity of the proprietary spec, the architect often adds phrasing called the "or equal" clause, which allows limited competition. While the "or equal" clause opens the door to some competition, it can be risky, as it puts the burden of equality on the proposing party (the contractor or the subcontractor, or even the vendor) who proposes the substitution. If you are pricing alternate products that are not specifically listed, research substitutions carefully. What may appear as a comparable product might not pass muster under closer scrutiny by the architect during review of the submittals or shop drawings. If the proposed substitution is not acceptable, the contractor is responsible for providing the specified product originally named in the specification, even if the bid was based on the proposed substitution.

PERFORMANCE SPECIFICATIONS

An alternate method of specifying products and materials is based less on makes and models and more on the ability to satisfy a design requirement or perform a specific function. This type of specifying is called a *performance specification*. In lieu of specifying a particular product by name, the architect or engineer opens competition to all products or materials that can perform the specific functions required to complete the design. This approach allows healthy competition among various manufacturers that have a similar line of products. It ensures competitive pricing and more aggressive delivery schedules.

Performance specs can identify products by characteristics, such as size, shape, color, durability, longevity, resistivity, and an entire host of other requirements. Some products that are not specified by name can be identified generically by reference to a particular ASTM testing number or a Federal Specification number. Use caution when pricing materials or products by their conformance with an ASTM number, however, as there could be several different grades of one product with vastly different prices. Also, remember that the architect or engineer makes the final decision as to whether a product has satisfied the performance criteria. The contractor proposing the substitution should be able to prove performance compliance with comprehensive facts and evidence, such as copies of pertinent tests and their results, and manufacturers' data. For a specification section that involves custom-fabricated work, the language might be a mixture of proprietary and performance specifications.

DESCRIPTIVE SPECIFICATIONS

The third method of specifying a product or process is by using *descriptive specifications,* which are written instructions or details for assembling various components to make up a system or assembly. It is analogous to a recipe. Most often, descriptive specifying is used for generic products such as mortar, grout, or concrete. Frequently, no manufacturers' or proprietary names are mentioned or needed. It can also be used for specifying soils by their gradation or size that passes through a sieve.

SPECIFICATION COMPLIANCE NUMBER OR REFERENCE STANDARD

Another method of product specification is by the use of a federal, ASTM, or other testing agency reference number. While less popular in the general marketplace, this method tends to be more common with some federal agencies. It holds the product to a very strict set of guidelines or tolerances established through testing and a formal acceptance process. In lieu of a product name or performance spec, the compliance number identifies a product by a number that confirms it meets the standard or level or performance necessary. The estimator is advised to use caution with this means of specifying. A single digit difference can be significantly different in price.

Part 3, Execution

Part 3, called the *Execution*, deals exclusively with the method, techniques, and quality of the workmanship. This section makes clear the allowable tolerances of the workmanship. The term *tolerances* refer to very definitive descriptors such as: plumb, square, straight, flat, round, level, or true. Part 3 identifies what is considered acceptable quality of the installation with respect to these descriptors. For example, it may describe an acceptable masonry installation as "No more than ¼" out of plumb in a 10' vertical brick wall." Clearly this standard can be applied in the field as a means of determining compliance. The Execution section should also describe any required preparation to the existing surfaces in order to accommodate the new work, as well as a particular technique or method for executing the work. Take this method or technique into account while deriving the quantities of the task, as other methods may render the work unacceptable during review by the architect. In addition, verify the conditions as a precursor to performing the work, such as temperature, humidity, surface prep, and testing. Part 3 also addresses issues such as fine-tuning or adjustments to the work after initial installation, general cleanup of the debris generated, final cleaning, and protection of the work once it is in place. Some sections of Part 3 may identify any ancillary equipment or special tools required to perform the work, such as staging or scaffolding. In the absence of any detailed installation instructions in Part 3, most architects and engineers will specify that the installation must comply with the "manufacturer's recommendations." This is a common practice to avoid the assumption of liability when conflicting direction is provided. It also has the benefit of being verified in the field by a technical representative of the manufacturer.

Part 3, Execution, often contains phrasing or language, in more generic terms, that indicates the type of workmanship required for the task. These terms can be industry-accepted or can be project specific. A common example is *first-class workmanship*. First-class workmanship is defined as a standard of workmanship that leaves a construction task free from any defect materially affecting appearance or serviceability. It is applied to tasks that have an aesthetic value and that are considered a craft in comparison to a trade. For example: finish carpentry, plastering, painting, and exposed masonry. These all have an aesthetic value to the project.

A step down from first-class workmanship is referred to as *ordinary workmanship*. Ordinary workmanship is workmanship that meets a standard of quality that is uniform to the task and is functional, safe, and usable in the ordinary sense. This is often applied to trades. For example: excavation and backfill, piping and conduits, wood framing, formwork for concrete that will not be exposed. It often relies on the function the work will perform rather than its form or appearance. In simpler terms, hammer marks on wood framing are acceptable since its purpose is structural, but hammer marks on wood trim of a window or door are not acceptable.

CONFLICTS BETWEEN DRAWINGS AND SPECIFICATIONS

Occasionally, there are conflicts between what is shown on the plans and what is specified in the technical specifications. The intention of the design professional is to coordinate the plans with the technical writer creating the specifications. Information on the plans is not intended to be repeated in the specifications, and vice versa. When there is a conflict or contradiction between plans and specs, it is quite simply an error requiring clarification.

Discrepancies between the contract drawings and the specifications should be addressed, in writing, to the architect or owner immediately. When the drawings and specifications differ as to the characteristics of a particular product, a correction or clarification needs to be issued by the architect. One of the main objectives of the specifications is to equalize the bidding process. Clarifying any discrepancies helps to maintain a fair and equal process.

Specifications were once considered to supersede the plans. This was because plans are lines and symbols and can be subject to interpretation based on one's viewpoint. Words in the English language, however, have specific established meanings, not subject to interpretation. As an example, consider the words *may* and *shall*. Substituting one for the other has a profound change on the meaning of the phrase. The word *may* implies that something is optional and subject to an individual's choice, whereas the word *shall* leaves no doubt that it is a requirement. The dominance of the specifications is no longer always the case. Some specifications outline a hierarchy of precedence in the contract language or state that when there is a discrepancy between the plans and specifications, whichever results in the greater quantity, is more expensive, or is of greater benefit to the project will supersede. There is new contract language that roughly states: In the event of a discrepancy between the plans and specifications, the greater quality or greater quantity shall prevail.

The estimator should endeavor to ensure that each page of a particular specification section is within the document and numbered, including the CSI MasterFormat® section identification number. Much like checking that all plans are included within a bid set, this helps one verify that all the pages are intact and the complete information for each section is included.

MODIFICATIONS TO THE CONTRACT DOCUMENTS

Addenda

The bidding process often produces questions that require answers or clarification from the architect or engineer. Any changes to the contract documents made during the bidding period (the time period beginning on the date the drawings are issued and ending on bid day) in the form of modifications, clarifications, or revisions, for any reason, are called *addenda*. Most often, the architect will issue addenda, or an *addendum* (singular); however, on occasion they may be issued by the owner. Addenda must be issued in writing and will automatically become part of the contract documents, complete with all of the benefits of the Contract for Construction and the General Conditions of the Contract. Addenda should, at a minimum, contain the following information:

- Number of the addendum and date of issue
- Name and address of the architect and/or engineer
- Project name and location
- Names of bidders to whom the addendum is addressed
- Contract documents that are to be modified
- Explanation of the addendum's purpose

Bid forms include an area for bidders to acknowledge addenda, and failure to do so could render the bid nonresponsive. Be sure to evaluate how each addendum will affect the bid price, not only for the individual scope of the change, but also consider the impact on the entire project and its schedule. As addenda can affect the bids of all parties involved, subcontractors and material suppliers should be made aware of any addenda, so they can adjust their bids accordingly.

Bidders should take note of projects with an excessive number of addenda. This is frequently a result of underdeveloped drawings being released to the bidders too soon. It can also indicate that the design professional is relying on the bidders to sort out the conflicts, errors, or omissions in the bid phase. This has the benefit of reducing contract modifications once in the construction phase. Excessive numbers of addenda often have impacts beyond the obvious and may not be readily apparent until the work is executed.

Bid Alternates

Often, owners and awarding authorities want to see how a change in materials, method of construction, or addition or subtraction of work will affect the project's price. This information is presented in the form of additions or deletions to the base bid, called *alternates*. Typically, the alternate is listed at the end of the specification section that is affected by it and also in Division 1 under the section Alternates. Be sure to include the increase or decrease in cost for all work, including all taxes, labor burden, and overhead – both direct and indirect costs and profit. (See Chapter 26 for more on calculating direct and indirect costs and profit.)

For example, the total consequence of an alternate might look like this:

Alternate 1: Delete door, frame, and hardware including labor to install for Door#3 in its entirety

DEDUCT $1,500

In this example, not only would the estimator delete the cost of the door, frame, and hardware, and installation but the estimator must also include the additional wall materials, wood, gypsum board, paint, and so on, to fill in the area Door #3 originally occupied in the base bid. The actual price of the alternate is the difference between the two, or the net cost. In some cases, the addition or deletion of large scopes of work by alternates can have a tremendous effect on the project's duration, thereby increasing or decreasing overhead and other time-sensitive costs of the project. Projects with limited budgets often include a series of alternates as a way of choosing how to use the budget most effectively. It can also indicate an owner or awarding authority that is trying to match the contract amount to the funding amount.

Once again, projects with an excessive number of alternates present a challenge to the estimator. The addition or deletion of too many alternates can impact the project in other ways as well. It should come as no surprise that contractors structure the pricing of alternates either based on an order of award provided in the documents or as the contractor prefers to do them. For example, a general contractor that self-performs masonry work might price a masonry alternate more attractively. To reduce this tactic, many owners and awarding authorities state an order of acceptance of the alternates. For example, in Division 1—General Requirements it may state that the alternates will be accepted in the order in which they are offered (numbered). In this manner, an owner puts the value of Alternate #1 higher than Alternate #2 and as a result, before selecting Alternate #2, the owner must select Alternate #1.

Bid alternates provide the advantage of identifying the add or deduct cost of a particular feature while still in a competitive pricing phase. It also allows the owner to evaluate the cost of a feature based on its perceived value to the project. For the bidder, a structured order of acceptance prevents the owner from awarding the alternates in such a way as to manipulate the award.

Allowances

Frequently, as the contract documents are ready to be issued, certain items have yet to be finalized and are not ready for inclusion in the bid set. Rather than leaving the item out altogether, the architect or engineer includes a *cash allowance*. The allowance is a fixed lump sum such as: *$10,000 for the purchase and delivery of plantings*. The allowance can also be in the form of a unit price, such as: *an allowance of $750 per M* (where M = thousand) *for* *brick, including delivery to the job site.* Typically, it is clearly stated what the allowance is for: materials, furnished and delivered only; materials and labor; or the entire scope of work. If there is any doubt, request clarification. At the completion of the project, the actual cost is computed for items included as allowances, savings are returned to the owner, and overages are added to the contract price.

For the estimator, understanding what the allowance includes is often the most challenging part. Delivery costs may be insignificant if the product is shipped from 20-miles away. However, if the selected product is shipping from overseas, the delivery costs can consume much of the allowance.

Bidding by Unit Prices

In the course of design for some projects, architects or engineers are sometimes unable to provide sufficient detail to the drawings so that the estimator can determine an exact quantity of a certain task. An example of this is excavation of rock or removal of unsuitable soil. The architect or engineer may be aware of what needs to be done and the techniques or quality required but is unable to establish the exact amount of rock or materials to be removed. This prevents the estimator from determining a quantity of the work to be performed. In an effort to establish the cost of this work for postbid purposes, unit prices are requested and submitted as part of the bid form or proposal. Unit prices are included on the bid for each item by a unit of measure, such as excavation and removal of unsuitable soils at $100.00 per cubic yard (CY).

The unit price should always include markups for taxes, insurance, overhead, profit, and bonds if appropriate. It is intended to be a complete price for a unit of the work. The estimator is again cautioned to understand what is included in the unit price.

Whenever possible, the estimator should attempt to calculate the approximate quantity of the task, as unit prices tend to decrease as the quantity increases, a concept referred to as *economies of scale*.

Frequently, the unit price bid may be tiered based on stipulated quantities. For example:

- Excavation of unsuitable materials $100 per CY for 1 to 50 CY quantities
- Excavation of unsuitable materials $90 per CY for 51 to 200 CY quantities
- Excavation of unsuitable materials $80 per CY for 201 and over quantities

This example is referred to as a *typical unit price* or a *unit price without base quantities*. Under most contract delivery methods, the unit price for an item is established in the bid phase. Once the item is discovered and quantified on the project, the unit price is multiplied by the approved quantity and a change order, or other contract modification, is issued. As many contractors are aware, the change order process can take time – valuable contract time.

To avoid the delay of contract time, there is an alternate method for exercising the owner's unilateral right to change under the unit price. The second method involves the inclusion of predetermined quantities for each unit price task – not necessarily exact but approximate quantities, The quantities are offered as part of the bid phase. The bidder bids a unit price as shown here, then multiplies the unit price by the approximate quantity and carries the resultant dollar amount as a lump-sum amount in the bid. This provides several benefits:

- It allows the bidder to use the approximate quantity as a way of determining the unit price by considering the economy of scale.
- It allows the architect to adjust the contract sum up or down by the unit price if the quantities change.
- It avoids delays in issuing a change order as the lump-sum value of the unit price and quantity is already in the base contract.
- It allows the contractor to earn the value of the work as it occurs.
- It establishes the budget value of the work at the bid and reduces unknown risk from the bid process.
- It can also offer some disadvantages to both parties. Consider a unit price bid with a stipulated quantity of 500 CY for ledge (rock) removal. A contractor that is familiar with a particular site may gamble as to whether or not the site has ledge. In doing so the contractor underbids the unit price significantly since it will be added to the base bid as a lump sum item. For example: instead of a realistic unit price of $500.00 per CY, the contractor bids $10.00 per CY, assuming they will never be required to perform the work because there is no ledge on the site. As a result, $5,000 is added to the base bid instead of the appropriate $250,000. This would give the contractor a significant edge in their bid. Should ledge be encountered, the contractor is still contractually bound to remove the ledge at $10.00 per CY, thereby presenting a financial challenge to the contractor. This practice is called *penny bidding,* and aside from being dangerous, it is almost always prohibited. It can be grounds for a bid being deemed nonresponsive.

SUMMARY

Once the plans have been reviewed and studied, the estimator's attention should focus on the specifications, and any addenda by a detailed review of the project manual. If appropriate, once the review of the plans and specifications have been completed, the estimator should conduct a site visit. Upon completion of the site visit and familiarization with the bid documents, the quantity survey or takeoff can begin. (This process is discussed in Chapter 4.)

Remember that specifications are to be used in conjunction with the plans. The plans are the quantitative representation of the project, and the specifications are the qualitative representation of the project. Combined, there should be adequate information both in quantity and in quality for the estimator to determine the value of an item of work or the entire project. They should be studied carefully to ensure an understanding of the project.

Prior to beginning the takeoff, it may be helpful to review basic area and volume calculations, covered in Chapter 3.

QUESTIONS/ PROBLEMS

1. Identify the key functions of the specifications. Include their relationship to the plans in the bidding process.
2. Name the four categories of CSI MasterFormat® including a brief description of what can be found in each.
3. Explain what information is contained in Division 1 of the CSI MasterFormat®.
4. Define the term *nonresponsive* as it applies to the submission of a bid.
5. Provide a detailed explanation as to why the contract form is included in the bid documents.
6. Explain why the General Conditions of the Contract for Construction is included in the contract by reference. Explain its purpose.
7. What does Article 7 of the General Conditions of the Contract for Construction govern?
8. Identify the three parts of a technical specification section. Provide a brief description of what information is contained in each part.
9. Explain in detail how conflicts between plans and specifications should be handled.
10. Explain the purpose of Bid Alternates. Include in the explanation why they are used and how they are selected.

3 | Calculating Linear Measure, Area, and Volume

To perform even the most basic quantity takeoff, estimators should be well-versed in the calculations of linear measurements, area, and volume. This chapter reviews the basic formulas and relationships needed to perform these calculations. The formulas are fairly simple, and it is not necessary to memorize them. However, as they are used repetitively, they often get committed to memory. It is essential, though, to know which formula to use in the proper application, and where it can be found.

A large portion of estimators today use some type of quantity survey software (see Chapter 27 for a more thorough discussion of software) and as a result rely on the computer to do the correct calculation of the units of measure and quantity. Takeoff software contributes much to the estimating process. It can also be seen as a disadvantage to the estimator by reducing the estimator's skill set from overreliance on the software. This author is a firm believer that calculating quantities is a crucial part of the estimator's stock and trade of skills. Professional proficiency requires personal skills. As such, this chapter will focus on the manual skills of doing takeoff.

UNITS OF MEASURE

The most fundamental rule is to use the correct units of measure for area and volume. Area is always expressed in square units, most often square feet (SF) or square yards (SY). Volume is always in cubic units, the most common of which are cubic feet (CF) or cubic yards (CY). Another important point to remember is to be consistent and keep the units the same. For example, feet multiplied by feet results in square feet, yards multiplied by yards results in square yards, and so on. Multiplying a dimension in feet by a dimension in inches leads to an erroneous value. It is common to find different dimensions used in various parts of the drawings. Be sure to convert the dimensions given into the same units. Often, a dimension on the drawings is given in both feet and inches and its up to the estimator to convert this to its decimal equivalent.

DECIMAL EQUIVALENTS

For the majority of projects designed in the United States, architects and engineers use the *US Standard system* of measurement. This is the feet and

Estimating Building Costs for the Residential and Light Commercial Construction Professional, Third Edition. Wayne J. Del Pico.
© 2023 John Wiley & Sons, Inc. Published 2023 by John Wiley & Sons, Inc.

inches measurement that are seen on drawings. To calculate the area of a space that is 24'-6" × 20'-3", the dimensions must be converted to their decimal equivalents. The feet-and-inches dimensions are changed to feet in order to arrive at a measurement in square feet for the area. This is a fairly simple process of converting the inches portion of the dimension to its decimal form and then adding it to the whole number. Once this has been done, the values can easily be entered into a calculator. Following are the decimal equivalents of the 12" in a foot:

1" = 0,08'	7" = 0.58'
2" = 0.17'	8" = 0.67'
3" = 0.25'	9" = 0.75'
4" = 0.33'	10" = 0.83'
5" = 0.42'	11" = 0.92'
6" = 0.50'	12" = 1.00'

Note that the decimal equivalent of 2" in the preceding table is not precisely two times the decimal equivalent of 1". This is because two-place accuracy after the decimal point is sufficient for estimating purposes in most cases. (While the actual value of 2" in decimal form is .0166666', the value is rounded up to 0.17.) The decimal value is limited to two places after the decimal point and is always rounded up. For example: If the total quantity of concrete in a takeoff is 34.22 CY, round to 35 CY for use in the cost analysis portion of the estimate.

According to the preceding table, it should hold true that if:

1" = 0.08', then ⅛" = 0.01'.

Using this simple mathematical analogy, it is easy to convert any feet-and-inches dimension to decimals to an accuracy of ⅛". For example:

To convert a dimension of 20'-3⅝" to its decimal equivalent, start by converting the 3" to 0.25', then ⅝" to 0.05'. When these two are added together, 0.25" + 0.05", the result is 0.30". This can be added to the 20', with the result of 20.30'.

To calculate the area of a room that is 24'-6" × 20'-3⅝", first determine that the decimal equivalent of each measurement and multiply 20.50' × 20.30', which equals 497.35 SF. Remember that feet multiplied by feet equals square feet.

Decimal equivalents can be found using a calculator. Decimal equivalents of fractions of an inch are as easy as pushing a few buttons. For example:

3½" = 3.5"; 3.5" divided by 12" per foot = 0.29'.

The remainder of this chapter is divided into three sections:

- **Linear measurement:** The measurement of lines in a single dimension.
- **Area:** The measurement of surfaces in two dimensions – length by width.
- **Volume:** The measurement in three dimensions – length, width, and height or thickness.

LINEAR MEASUREMENT

Perimeter

Imagine the floor of a building as a simple planar surface with no depth, the sum of the sides of that planar surface is called its *perimeter*. Linear measurement is the distance between two points. Since perimeter is a linear measurement and the dimensions of the sides are added, its units are also linear, most often linear feet (LF). The perimeter of a surface can be found by adding the length to the width and multiplying it by 2, or by adding the length and width of all the sides.

Knowing how to find the perimeter is helpful in determining the length of numerous items found in a building, such as baseboard and other running trims within various rooms.

ANGLES   If a rectangle with four 90° corners was divided in half by a line connecting two opposite corners, the resulting shape would be a *right triangle*. A triangle has three angles that total 180°. It also has three sides: a base, an altitude, and a *hypotenuse*, the diagonal line connecting the endpoint of the base to the endpoint of the altitude.

Calculating the length of the hypotenuse is helpful in determining the length of rafters, stair stringers, grade slopes, and the like. To do so, the Pythagorean theorem or the right triangle law is used. The right triangle law is the same as the "3-4-5 triangle" that framers and contractors use to "square up" work in the field. It states:

> *The square of the hypotenuse of a right triangle is equal to the sum of the squares of the other two sides.* (See Figure 3.1.)

Converting this to a formula:

$$C^2 = A^2 + B^2,$$

where C = length of the hypotenuse, A = length of the altitude, and B = length of the base.

Using this formula, we can determine the length of a side of a right triangle, provided the lengths of the other two sides are known.

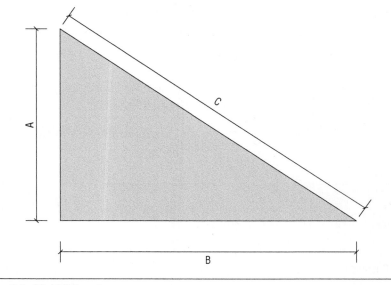

Figure 3.1 Right Triangle

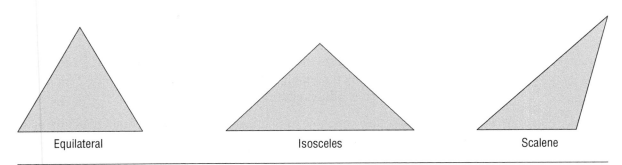

Figure 3.2 Equilateral, Isosceles, and Scalene Triangles

There are three types of triangles that frequently occur in a set of construction drawings. An *equilateral* triangle has three sides that are equal in length. An *isosceles* triangle has two sides that are of equal length. In a *scalene* triangle, none of the sides are equal in length. (See Figure 3.2.)

AREA AND SQUARE MEASURE

Possibly the most common calculation performed by estimators is determining the area of a shape. The most common shape in the construction business is some variation of the rectangle. A rectangle, by definition, has four sides, with all angles equal at 90°. If all four sides are equal in length, it is a *square*. A rectangle with only opposite sides that are equal in length and parallel is a *parallelogram* (the angles are not 90°). A *trapezoid* has two opposing sides that are parallel, but not of equal length. When no sides are equal in length, and none are parallel, the shape is a *trapezium*. (See Figure 3.3.)

Area calculations consider only the surface, not the depth. Areas are expressed in square units – most commonly square feet, square yards, or square inches. The area of a rectangle or square is defined as the product of its length and width. The formula for area is:

$A_R = L \times W,$

where A_R = area of a rectangle, L = length, and W = width.

Since a triangle is essentially a bisected rectangle, the formula for the area of a rectangle could be modified for a triangle:

$A_T = (\tfrac{1}{2}b) \times a,$

where A_T = area of a triangle, b = length of the base, and a = length of the altitude.

This formula requires that the angle between the base and the altitude be 90°.

Rectangle

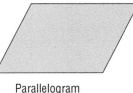

Parallelogram

Trapezoid

Trapezium

Figure 3.3 Rectangle, Parallelogram, Trapezoid, and Trapezium

Irregular Shapes

Estimators may be required to calculate the areas of more complex polygons and other irregular shapes. Determining the area of a construction feature is not always one simple area calculation; sometimes it requires additional calculations. Frequently, the area to be quantified is calculated by dividing it into several smaller areas, calculating each, then adding the results back together to arrive at the total area. This requires that odd-shaped features be broken into recognizable rectangles and triangles. The same holds true when determining a smaller portion of the whole. In this case, deduct all unwanted areas until the desired area is achieved. For most construction applications, a close approximation of the area of an irregular shape is sufficient.

Area of a Circular Shape

Many construction elements are circular, such as brick patios and concrete-filled cylindrical forms. Before reviewing the formula for the area of a circle, it is helpful to define various parts and some constant relationships.

The *circumference* is the perimeter of the circle. The *diameter* is a line drawn through the center of the circle, beginning, and ending on the circumference. Any number of possible diameters drawn on a circle should render the two halves of that circle equal. All diameters of the same circle are also equal.

The *radius* of a circle is a line from the center point within a circle to a point on the circumference. All radii of the same circle are equal in length. The radius, by definition, is equal to one-half of the diameter. (See Figure 3.4.) Knowing this relationship, we can establish the following formulas:

$D = 2 \times R \text{ or } R = \frac{1}{2} \times D,$

where D = diameter and R = radius.

A *constant* is a unitless number that expresses a relationship in a mathematical formula. The circumference of a circle has a constant relationship to the diameter of the same circle. That constant is the number 3.1416, or *pi*, which has a corresponding symbol, π. For most calculations in construction, π can be truncated at two places after the decimal point, or 3.14. If the area to be calculated is very large, π can be extended to a third decimal place, 3.142. The formula for the relationship between the circumference and the diameter is as follows:

$C = \pi \times D \text{ or } C = 2 \times \pi \times R,$

where C = circumference, π = 3.14, D = diameter and R = radius.

A *chord* is a straight line connecting two points on the circumference, without passing through the center of the circle. An *arc* is any portion of the circumference of the circle. A circle has 360°. Therefore, if the radius and

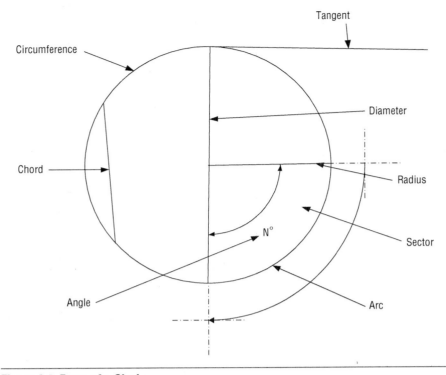

Figure 3.4 Parts of a Circle

the interior angle between two radii are known, the length of the arc between them can be calculated using this formula:

$$L_A = N/360 \times 2 \times \pi \times R,$$

where N is the central angle, $\pi = 3.14$, and R = radius.

The *tangent* of a circle is a straight line touching only one point on the circumference. A radius drawn to this point is at 90° to the tangent.

The area of a circle is the radius multiplied by the radius, then multiplied by π. As a formula:

$$A_c = \pi \times (R \times R), \text{ or } A_c = \pi \times R^2$$

An alternative method for calculating the area of a circle is to multiply the diameter by itself, then multiply the resultant area by the constant 0.7854. Expressed as a formula:

$$A_c = (D \times D) \times 0.7854, \text{ or } A_c = D^2 \times 0.7854,$$

where D = diameter and 0.7854 is a constant.

It is also possible to calculate the area of a portion of a circle. If one were to cut a pie-shaped piece out of a circle – two radii with an angle in between, with a known radius and known angle between – we can calculate that area, which is called a *sector*. The length of the arc is a fraction of the total circumference. A similar deduction can be used to devise a formula for the area of the sector:

$$A_s = N/360 \times \pi \times R^2,$$

where A_s = area of a sector, N = angle between the radii in degrees, π = 3.14, and R = radius.

Surface Area of Cylinder, Pyramid, and Conical Shapes

Estimators often need to calculate the *surface area* of a three-dimensional shape, such as a cylinder, pyramid, or cone. A good example is when estimating painting for one of these shapes. The outside surface area of a three-dimensional shape is referred to as its *lateral area*.

The formula for calculating the lateral area of a cylinder is:

$$A_L = C \times h,$$

where C = circumference of the cylinder base and h = height of the cylinder.

The lateral area of a pyramid can be expressed as:

$$A_L = p \times \tfrac{1}{2} \times h_s,$$

where p = perimeter of the base of the pyramid where the base is a regular polygon and h_s = slant height of the pyramid.

The slant height of the pyramid is a line drawn from the vertex, or converging point at the top of the pyramid, to the center of any one side of the base. The lateral area of a cone is the area of its tapering side. It can be expressed as:

$$A_L = C_b \times \tfrac{1}{2} \times h_s,$$

where C_b = circumference of the base of the cone and h_s = slant height of the cone.

The slant height of the cone is a straight line drawn from the vertex of the cone to the circumference. It can be calculated by solving for the hypotenuse in the Pythagorean theorem, where the altitude of the cone is its height perpendicular to the base, and the radius of the base of the cone is its base.

VOLUME AND CUBIC MEASURE

In contrast to area, which has only two dimensions, volume has a third dimension, depth. The depth of a shape can also be called its *thickness* or *height*. Once a shape takes on this third dimension, it is no longer planar, but becomes a *solid*. The term *cubic* refers to the volume of a solid, whereas *square* accounts only for its area. The standard units of cubic measure are cubic inches, cubic feet, and cubic yards, with the abbreviations CI, CF, and CY, respectively. There are numerous tasks encountered in construction estimating that require volume calculations. A few examples include excavation, backfill, and concrete for a form.

Volume of a Prism

If one were to visualize a rectangle as a surface area with a height, it would be called a *prism*. Prisms in construction are virtually everywhere. Examples of volume calculations of prisms that might be required include pilecaps, footings, and excavations. If a prism's dimensions of length, width, and height are the same, it is a *cube*. This is defined in the conversion of cubic feet (CF) to cubic yards (CY), where 1 CY = 27 CF (3' × 3' × 3').

If we modify the formula for the area of the planar surface by adding the new dimension, the result is the formula for the volume of a prism.

$$V = A \times h,$$

where V = volume of the prism, A = area of the base, and h = height.

This formula applies only to shapes whose ends and opposite sides are parallel. To further expand this formula:

$$V = l \times w \times h,$$

where l = length, w = width, and h = height.

Contractors may encounter an endless number of shapes that are variations of a prism. A triangular prism has a triangular surface area and a height. The rule of base area multiplied by height still applies:

$$V = \frac{1}{2} \times l \times w \times h,$$

where again l = length, w = width, and h = height.

In short, it shares the same relationship that the area of a triangle has with the area of a rectangle. The volume of a triangular shape is one-half of the volume of a prism.

Volume of a Cylinder

A common, yet more sophisticated, shape is the *cylinder*. The formula for its volume is essential in calculating the amount of concrete to fill a round column form. The volume is the area of its circular base multiplied by its height (see Figure 3.5). Expressed as a formula:

$$V_c = \pi \times R^2 \times h,$$

where V_c = volume of a cylinder, π = 3.14, R = radius, and h = height.

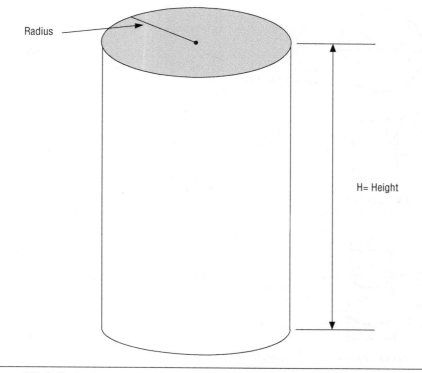

Radius

H= Height

Figure 3.5 Cylinder

Volume of a Pyramid

The volume of a pyramid is calculated as:

$$V_p = \tfrac{1}{3} \times A_b \times a,$$

where A_b = area of the base of the pyramid provided that the base is a regular polygon, and a = the altitude of the pyramid as defined by a line drawn at 90° from the base to the vertex of the pyramid.

Volume of a Cone

The volume of a cone is very similar to that of a pyramid. It is:

$$V_c = \tfrac{1}{3} \times A_c \times a,$$

where A_c = area of the round base of the cone and a = the altitude of the cone as defined by a line drawn at 90° to the base to the vertex of the cone.

SUMMARY

The shapes in this chapter are not intended to be an exhaustive array of all the shapes that may be encountered in a quantity takeoff, but they are the most common ones. Despite the extensive use of computers, it is still important that an estimator is capable of performing basic calculations of length, area, and volume. Calculating quantities accurately is essential to

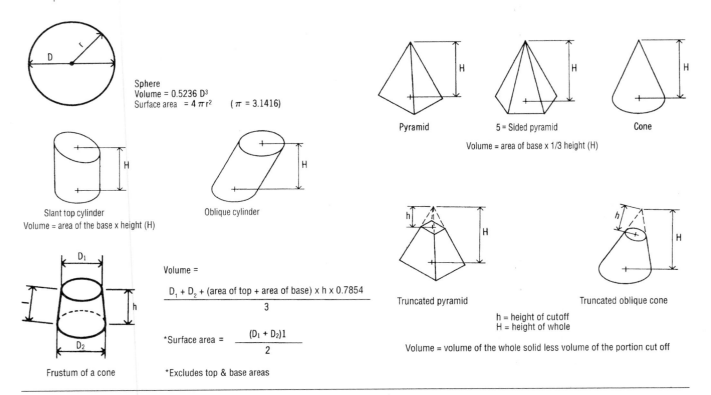

Sphere
Volume = 0.5236 D³
Surface area = $4\pi r^2$ (π = 3.1416)

Slant top cylinder
Oblique cylinder
Volume = area of the base x height (H)

Pyramid
5 = Sided pyramid
Cone
Volume = area of base x 1/3 height (H)

Volume =

$$\frac{D_1 + D_2 + (\text{area of top} + \text{area of base}) \times h \times 0.7854}{3}$$

*Surface area = $\dfrac{(D_1 + D_2)1}{2}$

Frustum of a cone
*Excludes top & base areas

Truncated pyramid
Truncated oblique cone
h = height of cutoff
H = height of whole
Volume = volume of the whole solid less volume of the portion cut off

Figure 3.6 Formulas for Less Common Shapes

formulating a solid base from which to start the estimate. If the quantities are incorrect, the estimate will be inaccurate. It is not a requirement for the estimator to commit the formulas to memory, but they should know where to find the formulas. To input dimensions into a calculator or computer, the estimator must convert the feet and inches of the US Standard system to their decimal equivalent. A key point to remember is that all dimensions must be the same units of measure such as feet times feet. Multiplying feet by inches will provide an erroneous value. For formulas and conversions not included here, refer to the internet. The table in Figure 3.6 lists additional formulas for less common shapes. Figure 3.7 is a table of conversions for linear, square, and cubic measure.

Now that linear, area, and volume calculations have been reviewed, the next step is to begin the takeoff.

QUESTIONS/ PROBLEMS

1. Convert the following dimensions to their decimal equivalent.
   a. 19'-10"
   b. 0'-3½"
   c. 33'-7¾"
2. Calculate the perimeter of a hexagonal shape if each side is 14'-7". Provide the answer in decimals.
3. What is the hypotenuse of a right triangle with a base of 12'-7" and an altitude of 8'-5"? What is the area of the same triangle?

Units and Equivalents	
1 cu. ft. of water at 39.1°F	= 62.425 lbs.
1 U.S. gallon	= 231 cu. in.
1 imperial gallon	= 277.274 cu. in.
1 cubic foot of water	= 1728 cu. in.
	= 7.480519 U.S. gallons
	= 6.232103 imperial gallons
1 cubic yard	= 27 cu. ft. = 46,656 cu. in.
1 quart	= 2 pints
1 gallon	= 4 quarts
1 U.S. gallon	= 231 cu. in.
	= 0.133681 cu. ft.
	= 0.83311 imperial gallons
	= 8.345 lbs.
1 barrel	= 31.5 gallons = 4.21 cu. ft.
1 U.S. bushel	= 1.2445 cu. ft.
1 fluid ounce	= 1.8047 cu. in.
1 acre-foot	= 43,560 cu. ft.
	= 1,613.3 cu. yds.
1 acre-inch	= 3,630 cu. ft.
1 million U.S. gallons	= 133,681 cu. ft.
	= 3.0689 acre-ft.
1 ft. depth on 1 sq. mi.	= 27,878,400 cu. ft.
	= 640 acre-ft.
1 cord	= 128 cu. ft.

Linear Measure		Square Measure	
1000 mils =	1 inch	144 square inches =	1 square foot
12 inches =	1 foot	9 square feet =	1 square yard
3 feet =	1 yard		
2 yards =	{ 1 fathom / 6 feet	30-1/4 square yds. =	{ 1 square rod / 272-1/4 square feet
5-1/2 yards =	{ 1 rod / 16-1/2 feet	160 square rods =	{ 1 acre / 43,560 square feet
40 rods =	{ 1 furlong / 660 feet	640 acres =	{ 1 square mile / 27,878,400 / square feet
8 furlongs =	{ 1 mile / 5280 feet	A circular mil is the area of a circle 1 mil, or 0.001 inch in diameter.	
1.15156 miles =	{ 1 nautical mile, / or knot / 6080.26 feet	1 square inch =	1,273,239 circular mils
3 nautical miles =	{ 1 league / 18,240.78 feet	A circular inch is the area of a circle 1 inch in diameter	= 0.7854 square inches
		1 square inch =	1.2732 circular inches

Dry Measure		Weight—Avoirdupois or Commercial	
2 pints =	1 quart	437.5 grains =	1 ounce
8 quarts =	1 peck	16 ounces =	1 pound
4 pecks =	{ 1 bushel / 2150.42 cubic in. / 1.2445 cubic feet	112 pounds =	1 hundredweight
		20 hundredweight =	{ 1 gross, or long ton / 2240 pounds
		2000 pounds =	1 net, or short ton
		2204.6 pounds =	1 metric ton
		1 lb. of water (39.1°F) =	27.681217 cu. in.
		=	0.016019 cu. ft.
		=	0.119832 U.S. gallon
		=	0.453617 liter

Figure 3.7 Standard Weights and Measures

4. What is the circumference of a round patio that has a 12'-8" radius? Provide the answer to two places after the decimal.

5. What is the area of a sector that has a radius of 9'-5" and an interior angle (between two radii) of 67°?

6. A concrete form is 36' long, 2'-4" deep, and 3'-0" wide. Calculate the amount of concrete needed to fill the form. Round the answer up to the nearest cubic yard (CY).

7. There are seven (7) concrete columns that require their surface be painted. The dimensions of each are 20'-6" high and 8'-0" in diameter. What is the total surface area requiring paint? Round the answer up to the nearest square foot (SF).

8. Calculate the net area of brick for an exterior wall that is 70'-0" long by 20'-8" high and has four (4) windows that are each 8'-0" × 6'-0". If there are 6.55 brick required per SF of wall, how many brick are required?

9. A parking lot is being repaved, the area of the parking lot is 7,680 SF and the thickness of the new surface is 2½". How many tons of paving are required if 1 CY of asphalt paving weighs 2.025 tons?

10. What is the volume of soil to be excavated from the trench that is 4'-0" deep, 100' long, and a width of 4'-0" at the bottom of the trench? A 45° (1:1) slope must be maintained at the sides of the excavation for safety. Round up to the nearest whole cubic yard (CY).

4 The Quantity Takeoff and Pricing Phases

A precise and thorough quantity takeoff is the basis for a sound estimate. Errors or inaccuracies in this portion of the estimate can be compounded during the pricing phase, regardless of how reliable the unit prices are. This chapter explores the most common practices of manually taking off quantities for a construction project and offers suggestions for developing routine procedures that help ensure accuracy. While the author acknowledges that the popularity of takeoff software has increased significantly since the second edition of this text, it is still an essential part of the estimator's skill set to both understand and develop a proficiency in doing manual takeoff. Understanding the process of takeoff can often enhance any software experience and prevent errors or omissions.

In the last half of this chapter, the application and sources of prices will be explored.

STUDYING THE DOCUMENTS

Before beginning an estimate, it is important to thoroughly review or, more appropriately, study the contract documents, which usually include:

- A set of plans
- A set of specifications
- Any related addenda or bulletins
- Additional relevant documents, such as geotechnical reports and any special documentation or unique requirements from local authorities. Examples include conservation commission directives or conditions and/ or amendments made by local inspection officials during the review of the plans, or for project without a set of plans and specifications,
- A Scope of Work or Statement of Work
- A visit to the location of the work

Estimating Building Costs for the Residential and Light Commercial Construction Professional,
Third Edition. Wayne J. Del Pico.
© 2023 John Wiley & Sons, Inc. Published 2023 by John Wiley & Sons, Inc.

In the absence of a formal set of specifications (bound and separate from the plans in a project manual), the plans should contain, at the least, a minimal amount of general information to be used as guidelines by the individual estimating the work. As discussed in Chapters 1 and 2, the plans and specifications each contain equally important, but distinctly different, kinds of information. Both are necessary to prepare an accurate defendable estimate. Information illustrated on the plans should be supported by the written language in the specifications. For example, if the plans depict a reinforced concrete footing, the specifications should define the strength and any special requirements of the concrete material and reinforcing in such a way that the estimator is able to price the material and labor necessary to complete the task.

Reviewing the documents is essential not only to understand the work that will be estimated but also to become familiar with the location of information within the documents. This preliminary review often raises questions as the result of insufficient, missing, or contradictory information. This is a key step in the process.

A reasonable amount of missing or contradictory information, however, is normal. The best architects, engineers, and designers are subject to human error, just as contractors are. Even a constructability review by a third party, such as a peer review or a hired consultant, does not always reveal all of the conflicts. A responsible estimator will document any discrepancies and, after a thorough search of the documents for an answer, will ask the architect, engineer, designer, or owner to clarify conflicts or provide missing information. This is conducted through a formal written process. The design professional responds to all bidders in writing in an attempt to keep the playing field level.

If the estimator elects not to seek clarification in writing, that bidder may have to bear the financial responsibility for any problems that arise as a result of unclear, missing, or erroneous information. It is recommended that the estimator make notes of any questions or problems while reviewing the documents, many of which will be answered as the review process proceeds and the documents become more familiar. A time-tested method is to keep an estimator's log or journal. This can be a simple spiral-bound notebook that accompanies the estimator during the review. It provides the estimator with a mechanism for jotting down not only questions but also reminders.

The study of the documents takes time – uninterrupted time. To complete the process correctly, it is recommended that the estimator set aside sufficient time for the review. Bidding a project is a commitment of both time and money, often with zero return. However, to do it correctly and increase the chances for award, time must be devoted to each phase in the process. Short-changing time spent on one phase will impact all phases of the estimate.

The estimating process can be broken down into three distinct phases: *document review, quantity takeoff,* and *pricing.* After a detailed review of the contract documents and all of the steps above are completed, the takeoff can begin. Ideally, it is best, to have all questions answered, although not essential.

THE TAKEOFF PHASE

Whether the takeoff is done manually or using software, there are similar steps. The quantity takeoff, sometimes called the *quantity survey*, breaks the project down into its elemental parts, called *tasks* or *activities*. (For the purpose of estimating, these terms are interchangeable.) Tasks are actual units of work to be performed, such as *place and finish concrete slab, form footings, frame exterior walls, or paint interior doors and frames*. They are described in detail in the estimate. Most tasks should have descriptors that identify an action, a location, and characteristics. They are quantified by details and dimensions provided on the drawings. Tasks are also identified in terms of the quality of materials and labor and possibly equipment required. This information is in the specifications. Establishing the quantity and quality of a task is an essential part of accurate estimating. Without them, the estimate is nothing more than a guess, which could cause conflict if the architect or owner requires different methods or materials when the construction work is performed. An example of a properly defined task is:

> *3000 psi concrete placed by direct chute in frost wall column line A.*

This line is accurate, concise, and sufficient to understand the task, the quality of the material, placement method, and location of the work.

Tasks or activities are made up of various components, called *elements*. Elements include material, labor, equipment, subcontractors, and occasionally non–production-related expenses, such as building permits, bond costs, and direct overhead costs. Not every task includes all of these components. For example, painting interior doors and frames would have a material and a labor component if the work is self-performed. If a subcontractor performs the work, the sub's cost might be the only component in the task. If the work is self-performed using spray equipment, the task would have material, labor, and equipment components. Determining which components apply is a combination of what the documents specify, along with judgment, experience, and strategy. How an estimator categorizes the components and whether the work will be self-performed or subcontracted is part of the bidding strategy the estimator uses. It is part of the estimator's – or more aptly, the contractor's – *means and methods*. Means and methods can be defined as how the work on the plans and specifications will be accomplished while complying with the terms and conditions of the contract. Simply put, there are sometimes multiple ways to perform a task – for example, spray paint the door frame or paint it by hand with a brush. Whether an estimator decides to subcontract the finish carpentry or self-perform it is also a part of the strategy called means and methods. Deciding on a particular means and methods is part of the responsibility of the estimator and will need to be decided prior to doing the takeoff. The means and methods employed in accomplishing a particular task or the project itself can have a profound impact on the cost of the work and should be decided in advance of the takeoff and pricing. It can also impact how the estimate for that task is performed. This will be explored more later in this chapter.

Some tasks in construction have physical relationships that create an order that cannot be changed. For example, the process of reinforcing concrete requires that the rebar be set in the forms prior to placement of the concrete. Another example might be that the trench for the footing is excavated before the forms are constructed and set. These are physical relationships that cannot (practically or economically) be changed. This is not what is meant be means and methods.

Lastly, the actual process of taking off quantities can be time-consuming and even tedious work, but it should not be rushed. Takeoff must be done methodically and thoroughly to make sure nothing is missed. Take breaks as needed and provide sufficient time to do the work as with the document review phase. Remember that 100% of items missed in takeoff do not get estimated in the pricing phase.

RULES FOR ACCURATE TAKEOFFS

The following takeoff rules are based on commonsense established practices that help prevent, or at least minimize, errors. They can make the takeoff better organized, more efficient, and more accurate. By following a standardized set of rules or practice the estimator develops competency and confidence in their skills.

Rule 1: Write Clear Task Descriptions

Descriptions should be clear and legible and should indicate the work needed and the part of the structure involved – or the location on the drawing where the quantity originates. The description should contain a verb to describe the action taking place, such as form, place, install, and so on. When taking off quantities, descriptions should be written according to the guidelines of the individual who will apply the unit prices. In smaller companies, one individual may do both. The following is an example of a thorough task description:

> *Form 24" × 12" footing including keyway along B-line as per Detail 5 on drawing S-2.*

The description should be sufficiently detailed so the estimator can accurately price the work in the pricing phase.

Rule 2: Use Industry-Accepted Units

A takeoff is not a list of materials for use in placing an order, but rather, a descriptive list of tasks with quantities derived from the dimensions on the documents. These quantities must then be extended into accepted units for pricing. For example, concrete is estimated in cubic yards (CY), because that is how it is sold. Cubic yards are also the accepted industry standard unit for its placement. The mathematical computations to change cubic feet to cubic yards are done in the takeoff phase, not in the estimating phase. Whenever possible, conversion factors should be included. For example:

3,782 LF of #5 rebar × *1.043 lbs./ft.* = *3,945 lbs. or 1.98 ton*

This provides an audit trail to review how one unit of measure resulted in the final unit of measure.

Rule 3: Follow a Logical Order

The takeoff should be logical and organized. The best approach is to proceed in roughly the same order as that in which the structure is built—from the ground up or as the work would be performed. This allows you to visualize the process while performing the takeoff. The logical thought process is to consider "What is the next step?" and organize the takeoff according to the CSI MasterFormat® classification system. (See Chapter 2 for more on MasterFormat®.) For example, all of the work in Division 3—Concrete should be complete before moving on to Division 4—Masonry.

Occasionally, the MasterFormat® divisions present a challenge in that items that would naturally be included as part of a construction task may have components in different divisions, and one element of work may be omitted or forgotten about by the time you reach the later division, where the element should be listed. For example, although the vapor barrier under a slab-on-grade is placed during concrete work (Division 3), it is classified as part of Division 7—Thermal and Moisture Protection. As with every rule, there are exceptions. In the estimate, place the task where it makes the most sense. In this example, it might be best to include the vapor barrier under Division 3 work, with a simple note in Division 7 that it has been accounted for in Division 3. For those who insist on adhering to divisional protocols, a line complete with the quantity can be included in one division with a note that it belongs to another division. Considering the previous example, the correct time to do the takeoff of the vapor barrier is while calculating the area of the slab over it.

Rule 4: Review Scales, Notes, Abbreviations, and Definitions

Review drawings and details carefully for notes and scale. Scale can change from drawing to drawing, and it is generally acknowledged that as the detail becomes smaller in focus, the scale becomes larger. It should also be noted that multiple scales can be on a single drawing.

It is necessary to become familiar with symbols and abbreviations that typically appear on the drawings. Frequently, drawings contain legends that define material and graphic symbols (as discussed in Chapter 1). Abbreviations, such as NTS (not to scale) and TYP (typical), may be used throughout the entire set of plans. Some words in the construction industry have unique meanings, such as *provide,* which is defined as *to furnish and install.* Carefully review any specification sections that include references or definitions.

Rule 5: Verify Dimensions

Wherever possible, use the dimensions exactly as they appear on the drawings. Add intermediate dimensions to arrive at total dimensions. Scale dimensions should be used only as a last resort. Develop the habit of occasionally checking printed dimensions against scaled dimensions. Discrepancies should be brought to the attention of the design team or the owner.

Always express dimensions in the same order, such as length × width × depth when calculating volume. This method prevents errors when referring to the size of certain sections of work. With items such as windows or doors, use the correct industry nomenclature of width × height, which is how the dimensions appear on the door or window schedules.

Convert dimensions to the same units for calculation purposes: feet × feet × feet. For example, when calculating the volume of concrete in a slab with the dimensions of 10' long × 10' wide × 4" thick, convert the 4" to 0.33' so that all units are the same. Converting all dimensions to the same unit prevents one from arriving at erroneous values that do not represent the work.

Rule 6: Be Consistent

Develop a systematic approach when working with the drawings. For instance, take measurements in a clockwise direction around a floor plan. Begin counting similar features, such as light fixtures, from left to right and from top to bottom. Whatever the procedure, it should become a standard, systematic approach. This helps the estimator develop confidence and competency.

Rule 7: Number Takeoff Sheets

Always number each takeoff sheet and keep them in order, even if the takeoff is in a notebook. Whenever possible, tasks, groups of similar tasks, or entire sheets should be identified by their location on the drawings (e.g., *Phase 2-B-Building Foundation Footings*). Numbering takeoff sheets or pages is a way to ensure that all sheets are included and are in order. The use of a notebook or log allows space for the estimator to provide reminders for the pricing phase.

Rule 8: Define Units for Material, Work, and Assembly Items

Items or tasks that have no labor component are called *material items.* These are furnished only and will be installed under another scope of work. Good examples are lintels for masonry openings. Typically, steel lintels are furnished under Division 5—Metals and are installed by a masonry contractor as the brick or block is laid up. The labor to install would be in Division 4—Masonry.

Items that have no material component and require labor only are referred to as *work items*. Examples include fine grading gravel or finishing of concrete. All tasks or items should be labeled with a unit of measure, which will be extended to the final price. For example:

Calculating the volume of concrete in a footing that is 27' long × 3' wide × 2' deep results in 162 CF (cubic feet). However, concrete is priced in cubic yards, so the calculated quantity of 162 CF must be converted to 6 CY by dividing 162 by 27. (1 CY = 27 CF.)

Calculating the area of a rectangular shape results in a square area unit, most often square feet or square yards (refer to Chapter 3 for more on calculations). Even the area of a circle is expressed in square feet or square yards. For example:

The area of a room 12' long × 17' wide = 204 SF.

When calculating the volume of a shape, the results should be expressed in cubic units, most often cubic yards. For example:

The volume of a prism that is 12' long × 13' wide × 14' deep is 2,184 CF, or 80.88 CY.

All items that have a cost value in the estimate should be assigned a unit of measure. Most items or tasks have defined units of measure. Others are less clear and are sometimes assigned a more arbitrary unit of measure, called a *lump sum* (LS), most often applied to work items that are not measured or expressed in more conventional terms. A good example is cutting an opening in a masonry wall. The actual work item includes four separate tasks:

1. Cut the opening with a masonry saw.
2. Clean up and dispose of debris.
3. Patch in the masonry at the jambs.
4. Install the lintel at the head.

A compilation, or *assembly*, of work items often uses a lump-sum unit. When tasks are repetitive, sometimes it is easier to group them into one lump-sum item. If the lump-sum (LS) unit is used to incorporate multiple tasks into one item, it is important to accurately and adequately define what tasks are included even if it requires individual quantities for each task in the assembly.

Rule 9: Use Decimals

Decimals are preferable in the quantity takeoff in lieu of fractions. Decimals are faster, more precise, and easier to use with a calculator or computer. Dimensions on drawings should be converted to their decimal equivalents. For example:

A dimension of 24'-6" should be converted to 24.5'.

In calculating the area of a room that is 24'-6" × 24'-6", converting both dimensions to decimals and performing the multiplication results in an area of 600.25 SF. Always check the final units of the dimension. Adding linear dimensions results in linear feet (LF) or linear yards (LY). For example:

The perimeter of a rectangle that is 12' long × 13' wide is 50 LF.

Rule 10: Verify the Appropriate Level of Accuracy

While accuracy is important, overaccuracy wastes time. There is an old adage in the estimating profession that warns, "Don't spend ten dollars of estimating time figuring a one-dollar item . . . unless there are hundreds of them." Accuracy is relative to the task being taken off or estimated. Rarely is it necessary for a number to be calculated to more than two places after the decimal.

There are acceptable parameters for rounding, depending on the particular task being calculated. Most items can be rounded to the nearest full unit. In some cases, it is necessary to round to the nearest sales unit, if the balance of the sales unit has no inventory or future value. (Waste factors are discussed in Rule 11 and in its own section following these rules.) Rounding quantities should be done when appropriate. For example:

If the volume of concrete for each of 10 footings is 34.56 CF, then the total would be 345.60 CF, or 12.80 CY for all of the footings. The total concrete yardage would be rounded to 13 CY for the total amount of concrete, rather than rounding each individual footing to the nearest CY.

Rule 11: Calculate Net Versus Gross Quantities

Some materials require an added allowance for waste. Waste is applicable to materials only and should not be applied to labor or confused with productivity. Waste can be loosely defined as the difference between what one must purchase and what one needs. Before waste is added, quantities are referred to as *net quantities*. After an allowance for waste has been added, quantities are referred to as *gross quantities*. (See the "Accounting for Waste" section later in this chapter.)

Rule 12: Check the Takeoff

The quantity takeoff should be checked by another individual for accuracy. Ideally, another completely separate takeoff and estimate of a task should be completed as a means of checking the first. However, this may not be practical or cost effective. As computer estimating and takeoff has become mainstream, there has been a reduction, even elimination, of checking one's work. Having one's work checked by another had a two-fold benefit: uncovering errors and building competency.

Quantities derived by hand (without the use of computer software) should be randomly checked. Select several work items or tasks throughout the estimate and recalculate their quantities. A reliable clerical staff person who can use a calculator can even check extensions from the takeoff quantities to the final pricing units. The extension of quantities involves calculations that can be checked by anyone with an understanding of simple mathematics and a calculator.

Rule 13: Mark Up the Drawings as Bid Documents

Mark the drawings using checkmarks, colored pencils, or highlighters to indicate the work that has already been taken off. These documents serve as the estimator's work papers and should be kept as a record of how quantities were derived. If original drawings must be returned, make copies and file them as records. This aspect of the takeoff is critical for projects that are bid and then go to contract. Once the project has been awarded, it will be turned over to the project manager, and the bid documents become crucial for relaying information about how the project was bid. The information learned during the estimating process, such as the means and methods strategy, and which subcontractor or vendor was carried in the bid is invaluable to the assigned project manager.

Rule 14: Focus on the Task

Those who perform the takeoff and pricing require a high level of concentration in order to accurately do the job and should not be subject to interruptions from the normal distractions of the construction office. Phone calls, frequent drop-ins by coworkers, and any type of concentration breakers are detrimental to accurate performance. Distractions and attempts to multitask are often the greatest sources of error. When the takeoff must be interrupted, select a natural stopping point and mark it clearly so that when work is resumed there is no doubt about the starting point. Again, it is wise to stress that the entire process of estimating is an investment of time; there is no way to get around it.

Rule 15: Organize the Documentation

Careful organization and neatness of work papers and takeoff sheets are crucial. If supporting work papers are needed (including sketches or details of how unusual features were estimated), they should be retained and attached to the pertinent quantity sheet. Even if the takeoff and estimate is done on a computer, there will still be work papers and notes.

All calculations should follow a logical and sequential process. Preprinted takeoff sheets and forms, such as those shown in Figure 4.1, are a good way

			SHEET NO.
PROJECT			ESTIMATE NO.
LOCATION		ARCHITECT	DATE
TAKEOFF BY		EXTENSIONS BY	CHECKED BY

DESCRIPTION	NO.	DIMENSIONS		UNIT		UNIT		UNIT	

Figure 4.1 Quantity Sheet

to track the takeoff. This is also true if the takeoff is done in a spiral bound notebook. Erasures should be neat and clean. Work papers, quantity sheets, and all components of the estimate should be maintained for a minimum of one year. The estimator should also make use of the contemporaneous data that the estimate contains. The estimated values and quotes from subs and vendor can often lend insight to future estimates.

Projects are sometimes abandoned for a number of reasons. Often, those same projects are restarted at a later point in time, due to changes in the economy, ownership, or need. Retaining the estimate and its various components is a reliable way to check for what has changed or remained the same over time.

ACCOUNTING FOR WASTE

Quantities derived during the takeoff process are often not the same quantities that are purchased when the work is actually in process. For example, it may be determined that the area of a floor to be covered with plywood is 300 SF. However, plywood is sold in full sheets, which are each 32 SF. When the 300 SF required to cover the floor deck is divided by 32 SF per sheet, the result is 9.375 sheets. Since only full sheets are sold, 10 sheets must be included in the takeoff. The difference between what to include in the takeoff and estimate and what is actually installed is called *waste*. As mentioned earlier, the material quantities before waste is added are called *net* quantities; they are called *gross* quantities after waste has been added. Material pricing is done at the gross quantities level, not on the basis of net quantities.

Waste may need to be added for any of three primary reasons:

1. To adjust to the standard sales unit
2. As anticipated waste resulting from handling
3. To achieve a specific assembly lap, as in shingles or rebar

Adjusting for Standard Sales Units

Materials often go through some on-site modification. The classic example is wood framing. Framing lumber is purchased in standard lengths, usually in 2' increments, delivered to the site and then cut to exact lengths for the specific component of the frame. It is purchased in lengths as close to the in-place length as possible, to minimize waste. For example, if a partition has a single top and bottom plate with a plate-to-plate height of 8'-0", the estimator should include studs to be purchased at 8' long in the takeoff. The studs would then be cut down to 7'-9" for installation between the plates. The remaining 3", sometimes referred to as *falloff*, is the waste. It has no real value but still needs to be accounted for in the estimate, because it is paid for at the time of purchase.

It is important to be attentive to other types of materials with similar waste requirements. Any material with a standard sales unit larger than needed for the task qualifies as having a waste component. Construction materials sold in lengths, rolls, bundles, boxes, or sheets, and fluids sold in gallons, drums, or barrels, should be reviewed for waste.

Waste Resulting from Handling

Waste can occur as a result of handling or placement, which is fairly common. Even with careful planning and execution of a task, waste occurs. One example is concrete with specific type of placement. Concrete placed by a chute has minimal waste, as the concrete slides directly into its final resting place, the form. There is no real handling of the concrete using this method. Concrete placed by pump, however, results in a significant loss of the total amount, for the quantity in the hose and the hopper. This needs to be accounted for in the estimate as waste.

Other examples of common materials with waste include soil, gravel, and stone delivered by the truckload. Often, these materials are distributed by equipment. Due to the inaccurate nature of placing earthen materials, waste can be significant. Generally, the more the materials are handled, the more waste can be expected.

Waste Required for Lap

Often, additional materials are required to satisfy a specific lap in order to maintain continuity of a particular feature. Examples include concrete reinforcement, siding materials such as clapboard and shingles, and roofing underlayment such as bituthene membranes and felt paper. Allowing for lap does not meet the strict definition of waste, since the material is actually used in the project, but lap requires additional materials, so the same principle applies.

Suppliers may be able to provide price break points for certain materials. Brick, for instance, is sold by the pallet, varying in quantity based on the size and type. For example, assume that there are 500 bricks per pallet. If it is determined that 63,485 bricks, including normal waste, are needed for a particular project, it might be more economical to order a full pallet rather than having loose bricks. This would mean ordering 63,500 bricks, or 127 full pallets.

Consider the placement of welded wire fabric (WWF) for temperature reinforcement in a slab, which is required to be continuous by design. For the WWF to be effective, there must be no break in the continuity. It is sold in specific sales units, most commonly as a sheet measuring 5' × 10', or 50 SF. The specifications define the amount of lap required based on the design. In this particular example, Section 03 30 00 of MasterFormat®, Cast-in-Place Concrete, might define the lap as a minimum of 12" on side laps and 12" on end laps. If the *effective area*, or the net area that one sheet will cover, is compared with the individual sales unit, a significant loss for lap is evident. If a 12" lap is maintained on the end and side of a single sheet, the effective area is reduced from 50 SF to 36 SF, or 72% of its original area. This represents a 28% waste as a result of the lap required. While not meeting the strict definition of waste, it still requires that estimator increase the net quantity to the gross quantity to account for the lap.

OTHER FACTORS THAT IMPACT QUANTITIES

Economies of Scale

In addition to the specific examples already noted, there are other considerations that, while not specifically considered waste, have an impact on the quantity of materials included within the takeoff. Price breaks based on total quantities, referred to as *economies of scale*, should also be considered. This is a simple economic principle that can be defined for our purposes as securing a better unit price for a large quantity of a material purchased. Distributors and vendors often charge a fee to remove quantities from a pallet of brick or a lift of plywood. As a result, estimating the entire pallet or lift can result in a discounted overall price.

Compaction and Swell

Other types of tasks require additional materials, though they do not fit the standard definition of waste. Consider for example, soil placement. When soils are imported to a site, placed in an excavation, and compacted, there is a portion of the *in-truck*, or loose volume, that is "lost" due to compaction. This is expected and must be accounted for in the takeoff process. The same is true in reverse. As soils are excavated from their natural in-ground maximum density (*in situ*) they swell thereby increasing the volume. Imagine an in-ground volume of 100 CY is excavated and loaded onto 20 CY trailers for export. If the soil swells 20%, then the loose volume is 120 CY and would require an additional trailer for export.

Swell and compaction are accounted for by adding a percentage to the actual calculated volume. Different soils have different swell and compaction factors based on their characteristics and gradation.

Finally, much of the leftovers from waste not only have no real value to the project but add a further expense for disposal. Consider the falloff from the earlier framing example. It has no appreciable value to the project that can be acknowledged in the estimate and will cost money to dispose of. Most wood frame projects, when completed, have a pile of lumber scraps that need to be disposed of. Associated costs might include dumpster and disposal fees, along with the labor to put the scraps in a dumpster. Other materials may require recycling or separate containers for disposal. The estimator must consider this when dealing with waste.

THE PRICING PHASE

Once the takeoff has been completed, the next step is to start the pricing portion of the estimate. In order to accurately apply unit prices to quantities, one must understand the different types of costs associated with the unit price system.

There are two main classifications of costs associated with the unit price estimate:

- Production costs
- Nonproduction costs

Production Costs

Production costs are part of the actual physical project, including materials, labor, tools, equipment, and subcontractor costs for each task or activity incorporated in the final work. If one analyzes the unit price of an individual task, such as a concrete slab that is placed by a contractor's own employees, the individual components are concrete material itself, labor to place and finish it, and equipment rental/ownership costs for the power trowels. If the slab was placed and finished by a subcontractor, the costs might be estimated as a single cost (lump sum) that includes all of these components. Productions costs consume resources: material, labor, and time on the schedule. Their performance can and are measured and tracked.

On any given workday a crew's performance can be measured by subtracting the end point from the starting point. This can be considered the crew's production for the day. For the same crew over extended periods the production may vary slightly or even significantly; however, over time an average is established. This is the historical production for the crew and can be reproduced for similar work in the future.

Nonproduction Costs

Nonproduction costs are still part of the project and cost money, but often do not consume schedule time. They are *overhead*, or the cost of supporting the production activities. Overhead costs are divided into two categories:

• Direct overhead costs
• Indirect overhead costs

Direct Overhead Costs

Direct overhead costs, sometimes referred to as *project overhead*, *general requirements*, or *general conditions*, are directly related to one individual project and only that project. These include a wide variety of costs such as temporary facilities, trailer rental, supervision, telephone usage, dumpsters, permits, and electrical power usage. All are costs that are directly related to one specific project. They can be broken down further for estimating purposes into:

• Time-sensitive costs
• Fixed costs
• Variable costs

TIME-SENSITIVE COSTS

Time-sensitive costs are costs whose price is driven by time. The longer the particular item is on-site, the higher the cost. A good example is a project trailer. Most contractors rent trailers, and since there is a monthly fee for rental the cost is a function of time or is considered *time-sensitive*. The longer the project goes on, the higher the accrued cost for the trailer rental. There are numerous other examples of time-sensitive costs that can be found in project overhead, such as cell phone costs, electrical usage, and the superintendent's wages. To accurately assign a dollar value to these costs, a schedule should be developed to determine how and when they apply during the term of the project. Initial schedules for determining time-sensitive costs tend to be rudimentary and develop with more information as the estimate proceeds. It is not uncommon for a project schedule to evolve through multiple generations before it is considered sufficient for use. (The development and evolution of schedules for estimating purposes are discussed in detail in Chapter 6 as well as a more thorough explanation of overhead.)

FIXED OVERHEAD COSTS

Direct overhead costs that are not affected by time are classified as *fixed* project overhead costs. Examples include building permit fees, project sign, registered site layout, engineering fees, access roads or ramps, and so forth. In most circumstances, there is a single occurrence for each of these costs, independent of the project schedule.

VARIABLE OVERHEAD COSTS

Variable overhead costs are more challenging for the estimator to calculate. They include costs such as fuel usage for heat during winter conditions, or snow plowing and shoveling to clear a site. It is difficult, if not impossible, to pinpoint a cost for these due to their variable nature. Accurately forecasting the amount and frequency of snowfall or the future cost of heating fuels is a guess at best, even when using statistical or historical information. Some projects use an allowance to cover variable costs.

Indirect Overhead Costs

Indirect overhead costs are referred to as *main office overhead*, which is any cost of a general nature that is not unique to a specific project but must be included as part of the estimate. Indirect overhead costs are associated with being in business. Some examples include main office rent or mortgage, salaries and benefits of staff, base insurance policies, company vehicle costs, and so forth. The costs associated with indirect overhead are accumulated while work is being performed and must be recovered in each estimate when bidding on future work. Each project bid and performed is expected to absorb a percentage of these costs. These costs are recovered by assigning an incremental piece of the indirect overhead to each project that is bid. The most common method is by a percentage of the estimated cost of the work.

All costs associated with indirect overhead are tabulated for a specific review period, most often quarterly (covering three-month periods) for the established company. (Companies experiencing rapid growth or new companies might review indirect overhead costs as frequently as once a month.) Indirect overhead costs are compared to the dollar amount of work completed in the same review period. The idea is that there is a relationship between the dollars spent and the dollars of revenue generated. For example:

If the total indirect overhead costs for one calendar year were $500,000, during which time $5,000,000 in construction work was completed, the percentage of indirect overhead could be calculated as 10% of the work completed.

$$\frac{Indirect\ Overhead\ Cost}{Construction\ Work\ Completed} \times 100 = Indirect\ Overhead\ Percentage$$

$$\frac{\$500,000}{5,000,000} \times 100 = 10\%$$

This is a very simple example of the method used to calculate indirect overhead costs for estimating purposes. It shows that the indirect overhead costs are about 10%, or $1 in indirect overhead costs for every $10 of construction costs.

Establishing the home or main office overhead cost as a percentage of sales is often a task that is arrived at after careful review of overhead costs and a clear understanding of the accounting procedures of the individual company. Traditionally, upper management and the company's accounting professional do this.

There are alternate methods used to determine indirect overhead costs. Certain types of contractors (such as wood framing contractors) provide labor as their primary product. They allocate indirect overhead using a different method. Their overhead costs are still tracked and reviewed periodically, and adjustments are made based on sound financial practices. The difference is that overhead costs are allocated by *billable hours of work,* sometimes referred to as *service billing allocation.* A framing contractor may have ten billable employees, each working an average of 2,000 hours per year (50 weeks at 40 hours per week), less adjustments for vacations, sick time, and downtime for weather and other potential lost hours. This translates to 20,000 billable hours per year. If the same company's indirect overhead costs for the same period were $200,000, it would make sense that each billable hour would need to support that $10-per-hour overhead expense. In mathematical terms:

$200,000 ÷ 20,000 hours = $10.00 per hour

In addition to the hourly wage of the employee, the labor burden, and any direct benefit program costs, such as medical, dental, or retirement, the contractor would have to add $10.00 per hour to cover the cost of the company's indirect overhead expenses. The sum of all these costs, plus an additional markup for profit, would be the billable rate for the individual employee of the framing contractor.

While each method has distinct advantages depending on the type of work, each also has potential flaws. The downside is the risk associated with the difficulty predicting future business. As long as business is as good as predicted or better, both formulas are successful. Any downward trend in the business plan or reduced volume of work under contract will require adjustments, such as cutting back on unnecessary overhead costs. Failure to make adjustments in a timely manner results in lost profits or worse.

Bear in mind the above explanation has been overly simplified as the process can vary from company to company. Most companies keep their internal overhead and gross profit margins a closely guarded secret. This is the construction equivalent of the "secret sauce recipe" for a fast-food chain. Often, senior management adds the overhead and profit to the estimate to maintain the secrecy.

Variances in Accounting Procedures for Overhead Costs

Some nonproduction costs are less clear in terms of whether they should be classified as direct or indirect, and they can vary depending on the accounting practices of the contractor or even the individual project. In either case they must be captured for estimating purposes. A classic example is project management costs. If the project manager is assigned to one project and only one project, the project management costs would be considered a direct cost. If that same project manager's time was split between several projects, it might be better considered an indirect overhead cost.

Other unique considerations include insurance premiums. Some premiums have an annual base cost for the policy that fluctuates in conjunction with the volume of work under contract or the direct labor costs. In this circumstance, the base policy is considered an indirect overhead cost, and the incremental increases relative to specific projects would be classified as direct overhead costs.

A simple, reliable way to determine if a cost is part of direct overhead or indirect overhead is to ask the following question:

> *If the contractor did not have the project under contract, would the cost still exist?*

If the answer is yes, then the item is considered an indirect cost. If the answer is no, it is a direct cost. If the answer is both yes and no, then the cost can be shared as both an indirect and a direct overhead cost.

In order to be successful, the professional estimator must:

- *Be highly organized and efficient*
- *Understand the construction process and its materials*
- *Read and fully understand plans and specifications*
- *Be able to visualize the project being built*
- *Have a working knowledge of basic mathematics and geometry*

Predicting Overhead Costs

Contractors with a history of success have a distinct advantage over new, growing companies. Over time and with the completion of multiple projects, experienced contractors have the ability to regulate and establish indirect overhead costs through a series of reviews and adjustments until they have determined their typical overhead costs. This allows them the opportunity to forecast future indirect costs based on trends and a business plan for future revenues. While the formula is not always exact and can change as personnel and resources come and go, it is a fairly good indicator of what a company's overhead will be and the volume of work the company needs to maintain to be financially sound and profitable. The need to predict and control costs explains why so many contractors develop a market niche and tend to establish a reputation in one type of construction. Few contractors can compete in a wide variety of construction markets. Successful contractors quickly learn the limitations and capabilities of their companies.

That having been said, many successful companies desire to grow. They analyze business and economic: micro and macro conditions and look toward the next market. Many successful contractors use the profits made in one market to expand into other markets by hiring employees with the required expertise.

UNIT PRICE ESTIMATE

While there are various reasons for business failures in the construction industry, a major contributor is the inability to produce an accurate, profitable, and defendable estimate. At the root of this chronic problem are poor or unprofessional practices on the part of the estimator.

Estimating is a labor-intensive and costly operation that does not always result in success. At best, an estimate only *approximates* the cost of a construction project as seen through the experience and judgment of the professional estimator while employing specific means and methods. Every element within the estimate, ranging from quantities to unit prices, must be substantiated in factual terms in order to be considered professional work. Each cost line should be able to be trailed back to the plans or specifications. This is the basis for a defendable, detailed estimate.

The work product of the professional estimator is the detailed, or *unit price*, estimate. By definition, the unit price estimate consists of breaking the project down into tasks, quantifying those tasks, and then applying a price based on the units of each task. As mentioned previously, each task can have various and multiple components, such as materials, labor, tools, equipment, and subcontractors that make up the unit price. For example:

Task Description	Quantity	Unit	Unit Price	Total
Place and finish concrete slab	10,000	SF @	$1.00	=$10,000

The unit price of $1.00 per SF in this example consists of the labor and equipment used in the placing and finishing of the concrete slab. The example illustrates that the $1.00 unit cost is based on the SF unit. Once the unit price has been multiplied by the quantity, the resultant cost is called the *extended cost*, or total cost for the item.

Other types of estimates, such as conceptual and square-foot estimates, are acceptable under certain circumstances, but they are in the realm of the seasoned professional estimator. Both are less accurate than the unit price estimate and are used as tools for budgeting and feasibility studies during various phases of the design development.

The accuracy of an estimate can often be impacted by the contract delivery method the project will use. Firm fixed price or *stipulated sum* contracts that are the results of *hard* or *competitive bidding* require spot-on accuracy, whereas guaranteed maximum price contracts resulting from negotiated bids can offer grant "a second bite of the apple" to ensure everything is included. This is by no means an advocacy for sloppy or weak estimating, but merely advice that it is difficult to produce a unit price estimate with documents that do not have the detail or level of completion.

A correct unit price estimate has a quantity and a price for each item of work or expense identified in the bid documents. When the estimate is summarized and profit is added, it is submitted as the offer, or the *bid*. Should the contractor be successful in attaining the contract, this same estimate now becomes the basis for the cost control system that will determine whether the work has been performed and the costs incurred as estimated – in short, if the project made or lost money.

The unit price estimate serves as the guideline for awarding subcontracts and purchasing materials and the standard for measuring performance once the work commences. It is valuable in assigning a dollar value to each category in the schedule of values that will become the basis for requisitioning payment. In essence, the estimate becomes the foundation of every project. This is why it is of such importance—it is part of virtually every aspect of the construction process. The most proficient contracting firm cannot overcome flawed estimates that represent work taken below cost or that do not allow for an appropriate profit.

At the heart of the successful estimate is the unit price. The term *unit price* can be defined as the incremental cost per unit of work.

$$\frac{\text{Cost}}{\text{Unit of Work}} = \text{Unit Price}$$

The following are parameters that define the unit price:

- Unit prices are based on dollars, or portions of a dollar, per unit.
- The pricing unit or unit cost should match the extended unit of each task quantity.
- Most unit prices are based on a specific time frame as a means of measuring productivity.
- Labor unit prices can be based on an individual's production or on the production of a crew.
- Unit prices can be based on historical cost data or published cost data.
- Unit prices can include all, one, or a combination of cost types (material, labor, equipment, subcontractors, etc.).

Sources of Unit Price Cost Data

Collecting, organizing, and analyzing the data for the estimate can be daunting. Being able to use the data efficiently is a result of how well it has been organized. There are several sources for pricing data, but most fall into these five categories:

1. Written quotations or published prices from suppliers and vendors
2. Written quotations from subcontractors, including materials and labor
3. Written quotations from equipment suppliers or rental agencies
4. Historical cost data from your company's own previously completed projects (similar to the one currently being estimated)
5. Cost data from published sources, such as RSMeans data from Gordian

Categories 1 through 3 are referred to as *contemporaneous* pricing. They are based on a review of the project documents, with full understanding of the project's unique or special conditions. These prices can usually be assumed to be an offer or bid from a source looking to do business. Pricing by interested parties doesn't just happen. It is typically the result of time (and money) invested in soliciting pricing from various sources. Failure to establish contacts, gain the interest of the bidder, and follow up as bid day approaches can often result in gaps in the estimate on bid day. Quotes from bidders, reviewed and qualified, are usually the best source of costs.

Current pricing should be accepted only from individuals or firms that will provide prices in writing. Frequently, pricing may be verbal, with written confirmation to follow. Such quotations should be written by the person receiving the quotation on a standardized form, called a *telephone quotation sheet*. Another more common practice is to send quotes electronically via email. At the time the quotation is received, it should be *qualified*. This involves asking the person providing the quote what is included. An intimate knowledge of the work involved in each task is necessary in order to ask the right questions. Sometimes all that is required is to confirm that the quoting party has included what was required or requested. The qualification process can also help prevent duplication errors. For example, if a material quote includes sales tax, this should be noted so that tax is not added a second time.

Any notes or pertinent information discussed during the qualification process should be written down on the telephone quotation sheet, so that they can be compared against the written follow-up quote. At a minimum, the telephone quote must document the name of the individual delivering the quote, the date and time, and how long the price is valid. In the absence of a written quote, follow up after the bid with a "confirmation of price" fax or email. This procedure has the benefit of documenting the price while it is fresh in both parties' minds. While most people in the construction business know that the quote is only as valid as the integrity of the firm or individual that makes it, a written quote goes a long way

toward ensuring compliance. A well-defined price is a big help when comparing various competing numbers.

Category 4 sources are records of actual costs of similar work previously performed by your own company. Historical costs are recorded – they are actual costs to perform a quantified, specific task. That means that the costs can be analyzed and compared with the project now being bid to obtain unit prices. If the total cost is divided by the total quantity, the result is the unit price. Consider the following hypothetical example:

Over the last several jobs, production was tracked, analyzed, and it was determined that our in-house carpentry crew of 4 hung an average of 5,500 SF of ⅝" drywall on wood stud partitions per 8-hr day. For that same 8-hr day each carpenter had a daily cost of $475.00, which translates to a crew cost of $1,900 per day. One could conclude that:

$1,900. ÷ 5,500 SF = $0.345 or $0.35 per SF

If the conditions were similar and the cost per day for the crew remained constant, the estimator could use the $0.35 per SF for the current labor to hang drywall on the project being estimated.

Historical data shows not only actual project costs for previously completed projects, but also the accuracy of those projects' estimates.

Historical costs are typically taken from the company's records for self-performed work. However, it is not uncommon for some general contractors to study and record the time spent and methods employed by other contractors or subcontractors to perform a defined quantity of a specific task in the hopes of gaining better estimating insight for their own future projects.

While historical costs are actual, it is rare that the unit price can be applied without making at least minor modifications to allow for different circumstances between the previously completed project and the project currently being bid. At the very least, there are considerations for cost escalations of wages and materials over time. There may be differing site or weather conditions, varying productivities, and supervision, learning curves, and the degree of difficulty that might not be readily apparent from empirical data alone.

Historical data must be analyzed carefully, with some understanding of the unique set of circumstances under which the work was performed. All work papers from mathematical computations from the analysis should be retained for the record. Whenever possible, another individual should check the accuracy of the computations to uncover possible errors before they are incorporated into the estimate and a bid is offered.

Category 5 costs are from published sources, such as the numerous RSMeans cost data books and software. As with any published construction cost data, there is an applied set of parameters. It is essential to understand

all of the conditions that have been applied to the data (normally averages), so that the estimator can make their own adjustments for factors such as location, skill of the crew, climate, availability of resources, and supervision.

Published costs must be adjusted for these considerations. Cost data can be presented in two ways: as *bare costs* without overhead and profit and as *total costs*, which include a markup for overhead and profit. It is important to understand whether the work will be self-performed or subcontracted. If it is self-performed, bare costs are preferable so that the markup can be applied at the end of the estimate. If the work is to be subcontracted, make sure that an allowance for the overhead and profit of the *installing contractor* is carried in the estimate before recapitulation.

Published data is based on normal working conditions, during regular business hours, and under average conditions. There is no accounting for unpredictable costs associated with labor resource shortages, supply-and-demand cycles, or travel and per diem costs. It is also important to understand price and crew size in published cost data to make adjustments based on experience and judgment. Published cost data is often best used in the development of a budget. It is rarely used as data in a hard bid except under unique types of contracting methods. Published cost data can be used for comparison and analysis of subcontractor quotes.

APPLYING PRICES TO THE QUANTITIES

After the quantity takeoff for each individual task or activity has been assembled, organized, extended into its final units, and checked, prices must be applied. As discussed previously, unit prices can be obtained from a variety of sources and should be noted within the estimate. In the case of published cost data, such as the RSMeans cost data books and software, a 12-digit individual line item can be cited. In the case of historical or contemporaneous prices, the specific source should be noted and documented within the estimate. This helps establish the credibility of the unit price.

Cost Analysis Sheets and the Estimate Summary

For estimators who have yet to make the transition to estimating-by-computer, the cost analysis portion of the estimate can be recorded on preprinted forms. (See Figure 27.1 for a sample Cost Analysis form.)

Cost Analysis sheets are different from quantity takeoff sheets in that they provide space for unit prices. Unit prices can be broken into component parts, such as materials, labor, and equipment. Since each component has a different modifier, each component must be calculated and summarized separately. The format of the form allows you to apply unit prices for each component in a columnar sequence.

As each task or activity is priced, its total estimated cost is tabulated on the right-hand side of the sheet. Once each task line has been priced and

totaled, the columns are totaled from the top to the bottom of the sheet. Finally, they are totaled for each CSI MasterFormat® section or subsection number.

Tasks can be grouped together for work of a similar nature. For example, the estimator might summarize and total all of the carpentry framing tasks or the entire masonry scope of work for a project. This is done so that the totals for each MasterFormat® division can be brought to an *Estimate Summary* sheet (see Chapter 27 for an example of an Estimate Summary sheet). This represents an entire scope of work that might be subcontracted if the project bid is won. Totaling the cost of work by CSI MasterFormat® division is an easy and logical way to organize the estimate. The total cost for materials, labor, and equipment can be summarized and brought to the estimate summary sheet so that you can view a single number that represents a well-defined scope of work. This allows a price generated for a division or segment of work to be compared with a price that was submitted by a subcontractor bidding the work.

The format of the Estimate Summary sheet is straightforward. The description of the tasks is in column format along the left side of the page, with parallel columns for materials, labor, equipment, and subcontractor costs that start adjacent to the description and move from left to right. At the far right is the total column, which contains the total value of the various components of each grouping of tasks or CSI MasterFormat® division. The values of each component are added along the individual rows, which, when totaled, represent the total value for that scope of work.

As a check-and-balance system, each column is totaled. Then the total of each column is added and compared with the last number in the lower right-hand corner of the sheet. Totals for each column are carried over to the reverse side of the sheet and entered into the "Totals Brought Forward" row, which again allows for an accuracy check. When all costs have been transposed from the Cost Analysis sheets to the Estimate Summary sheet, costs for project overhead are added. At this level of costs, the totals are referred to as *raw costs* or *unburdened costs*. Modifiers such as sales tax are added to each category, and the indirect overhead is applied to the subtotals. This is referred to as the *burdened*, or real, cost. In theory, if each task was performed and the costs realized exactly as detailed in the estimate, the real cost would be the break-even point with no profit being realized. The last steps would be to add any contingencies deemed justified, profit, and, finally, any costs related to performance and payment bonds, if required. (See Chapter 26 for more on contingencies and profit.) This final number rounded to the whole dollar represents the bid amount for a lump-sum bid.

For estimators that have made the transition to estimating software, much of the mathematical computations and summaries are done by the software, which saves time and reduces error. For many, this is sufficient reason alone to use software.

For those estimators that are still on the fence, that have yet to fully transition to software but are beyond paper and pencil estimating there is an intermediate option. Many estimators create their own spreadsheet using Microsoft Excel™. The cells are populated with the data and the software does the calculation and tabulation (Chapter 27 explores estimating by computer in more detail).

SUBMITTING A BID

The bid is a byproduct of the estimate – a proposal to do the work estimated. In short, a bid is an offer to perform work for a certain price. It is the natural progression of the estimate. In simple terms, a bid is composed of a minimum of:

- An offer to perform a defined scope of work
- A stipulated compensation for performing the work
- Name of the party making the offer
- Name of the party to which the offer is made

It may be best at this point to define a few terms. Four main terms require definition:

- An *estimate* is an in-house work product. It is never intended to go beyond the estimating department or the management of the firm. It contains insight into the company's estimating practice.
- A *bid* is a formal offer, almost exclusively on a bid form provided by the owner or design professional. It has limited space for things such as price, company name, signatory party, and acknowledgment of addenda, for example. It assumes that all bidders have the same scope: the plans and specifications, thereby making a comparison of bids rather simple – price. It is intended to prevent the bidder from qualifying their bid.
- A *proposal*, while less formal, is still a method for submitting a bid. It allows bidders to *propose* what they will provide. Since there is no form regulating what can be proposed, a bidder is free to qualify what is and is not included in the bid. This leaves the recipient of the proposal to determine best value or price.
- A *budget* has multiple definitions. It is an estimate done at early stages of design to maintain cost control. It can have a number of names: conceptual estimate, preliminary estimate, ROM estimate, for example. A budget can also be the result of a detailed estimate that is used for project cost-control purposes. It uses the line items in the estimate as a cost breakdown structure to compare actual cost performance against estimated cost performance after the work has commenced.

Some offers are submitted from a firm on letterhead, while some are on generic preprinted proposal forms, and others are on bid forms provided by the architect or owner. In the absence of a bid form, the party making the offer has to quantify, and sometimes qualify, the exact scope of work being provided within the proposal. The bidder must identify the documents used in preparing the bid by listing the plans and specifications, the date of the

documents, and any addenda issued that modified the documents. The offer should include any assumptions or qualifications on which the proposal is based. The bid should be executed by a responsible individual having the authority to do so, assigned by the company making the offer. The person who signs the bid should include his or her title. The offer should have an expiration date and should provide a space for the receiving party to sign, date, and accept the offer if this step is appropriate.

Using letterhead or preprinted forms in lieu of a bid form has some distinct advantages to the contractor and disadvantages to the owner or architect. The contractor can include or exclude any item or scope of work chosen and can qualify or quantify any item of work within the scope without rendering the proposal null and void. The disadvantage to the owner is that it becomes difficult to compare multiple proposals from different contractors if each has noted some qualifications that affect the price.

When a bid form is provided, it alleviates some of the problems encountered with bids that are submitted on letterhead. Bid forms are printed forms specific to the project with the language of the proposal already typed in the body of the form. They are provided by the architect or owner as part of the documents and contain blank spaces for the contractor to fill in only the information required and offer no space for the contractor to qualify a bid. Using a standard bid form ensures a level of conformity so that each bidder's proposal includes the same scope, where the only difference is in the offered amount. The advantage to the architect and owner is an immediate acknowledgment of where each contractor's price lies in comparison to the prices of the other bidders.

SUMMARY

A reliable and comprehensive estimate is a combination of quantities derived from the plans and specifications during the takeoff and the application of correct unit prices. Unit prices can be categorized into materials, labor, and equipment costs for self-performed work or subcontractor costs for work performed by those individuals who are not direct employees. There are also overhead or nonproduction costs that must be applied. An offer can be a bid or a proposal. It is the point in the estimate when all costs are summarized, profit added, and the price is offered in the hope of securing work.

The estimate is the result of three distinct phases: the document review, the takeoff or derivation of quantities, and the pricing and summarizing of those prices. Sources of pricing can vary depending on the use of the estimate. Pricing derived from historical and contemporaneous sources is often best for competitive bidding but may not be required for budget or preliminary estimates.

A proper estimate cannot be undervalued. It is the foundation of a company, and if done incorrectly it can damage a company beyond repair. Adequate time should be allotted for each phase of the estimate after a definitive commitment has been made to bid the project.

**QUESTIONS/
PROBLEMS**

1. Why is it essential to use industry-accepted units of measure when doing the takeoff?
2. What is an element of a task? Explain your answer with an example.
3. What are the three reasons for adding waste in the estimate?
4. Why is it necessary for an estimator to adjust the quantities to the standard sales unit?
5. Explain the premise behind compaction and swell of soils and why additional materials may be required.
6. What is the difference between production and nonproduction costs?
7. Explain in detail the difference between direct and indirect overhead costs.
8. Explain the various sources of cost data and when each is appropriate to use.
9. Provide an explanation of how indirect overhead percentages used in an estimate are derived.
10. What is the purpose of using a bid form when requesting an offer from a contractor?

5 | Understanding the Cost Elements of a Unit Price

Before finalizing materials, labor, and equipment costs as unit prices within an estimate, they must first be modified. Some of the modifications, such as for materials, are rather simple. Others are more complex, such as modifiers to labor or depreciation of equipment. We will explore the process of modifying these costs in this chapter. While modifiers vary from state to state and city to city, the process is the same.

MATERIALS

Two modifiers are added to the costs of materials to arrive at an accurate unit price: sales tax and waste. Some taxpayer- and government-funded projects are exempt from sales tax, so there may be no need to add it. However, most residential and commercial projects are subject to a sales tax on materials. Some states also apply sales tax to labor and equipment rental.

Waste is applicable to most materials for several different reasons. (See Chapter 4, "The Quantity Takeoff and Pricing Phases.") Waste is project specific; different projects require different waste factors. Properly accounting for waste is often a matter of judgment based on experience. If the estimator were to prepare a detailed materials list for all lumber and sheathing materials needed on a project, then solicit pricing from local lumber suppliers, in return the estimator would receive quotations for materials and delivery. Often, the quote is a lump sum derived by multiplying a series of quantities by unit prices. Based on experience, the estimator would know that these are net quantities and must be adjusted for waste.

It would be impossible to predict which materials will incur more waste, so the most practical approach is to add a percentage in dollars to the lump-sum quote. For example:

There are some unpredictable and uncontrollable reasons to add waste, including theft, damage from weather, damage by other trades, and human error. This waste can be accounted for by adding a percentage to the lump-sum quote provided by the supplier.

Estimating Building Costs for the Residential and Light Commercial Construction Professional, Third Edition. Wayne J. Del Pico.
© 2023 John Wiley & Sons, Inc. Published 2023 by John Wiley & Sons, Inc.

> *The quote from a lumber supplier is $76,890. It was determined based on experience and/or historical cost data that an appropriate waste factor for the particular project is 8%. This translates to an approximate added cost of $6,152, which must be included in the estimate. The two costs added together now equal $83,042. Since the state in which the project is being built has a 5% sales tax on materials, the gross amount of the subquote for the lumber needs to be increased by 5%: $83,042 × 0.05 = $4,152. When combined and rounded, the total is $87,194.*

Traditionally, waste is added to the quantities and not the unit price. However, on occasion waste can be accounted for in the unit price itself. The following example illustrates the calculation of carpet material costs for an office space based on a waste factor incorporated into the unit price. To determine a unit price cost per square yard (SY) of carpet to cover an office floor, the necessary quantities should first be reviewed.

> *The net area to be covered with carpet is 2,000 SY. However, the carpet is sold in 15'-0"-wide rolls, and, when layout is considered, the actual quantity needed is 2,160 SY, or an additional 8%. Applicable sales tax must be added (in this example, 5%). These factors are then reviewed, and their impact assessed in the following calculation:*

Carpet (as specified)	*$38.00*	*per SY*
Waste factor of 8% applied	*3.04*	*per SY*
Subtotal	*$41.04*	*per SY*
Sales tax of 5% applied	*2.05*	*per SY*
Total	*$43.09*	*per SY*

If the $43.09 is extended over the 2,000 SY of the net area, the cost is $86,180. Since the actual quantity to be installed is 2,000 SY, it would be incorrect to use the 2,160 SY to represent the quantity to be installed, since this would add an 8% error in the amount to be installed to the labor portion of the estimate. In addition to the costs, profit would be added to the unit price of $43.09 per SY. If a 10% profit were added, the billing unit price would be $47.40 per SY.

This approach is sometimes used when the unit price is the basis of payment. In the above example, the contractor is paid by the unit price of $43.09 for the net area only, so the waste must be included in the unit price.

A more recent modifier to materials is the expediting costs as a result of supply chain failures. Of late, materials are more difficult to secure and mobilize on site and require that the estimator capture the costs associated with the expediting and shipping of the product. While this is not expected to last forever, it is proving to be the new normal for the time being. The estimator would be prudent to ask the question when evaluating quotes from suppliers and vendors.

LABOR

Calculating the cost of labor unit prices is more difficult than for material unit prices. First, the correct wage rate must be calculated. There are several modifiers that must be added to the cost of the actual wage rate. These include state and federal taxes on wages, insurances, and benefits. It might be helpful to start by defining some basic terms.

Billing Rate

The term *wage rate* for this discussion refers to the "in the envelope pay" or the agreed-upon wage between the employee and employer. Once the wage rate has been modified by all taxes, insurance, benefits, and overhead components, the rate is referred to as the *burdened labor cost*. After profit has been added, the result is the full value of the labor-hour, referred to as the *billing rate*. This is the hourly rate that would be charged for labor in a time and material application. The following is a review of the components that make up the billing rate and how those costs are calculated.

Wage Rate

The wage rate, or the hourly rate of pay on which the employee's paycheck is calculated, is regulated by the agreement between the employee and the employer, a collective bargaining agreement, or by a contractually obligated prevailing wage scale. For example, a carpenter hired at $35 per hour would gross $1,400 in wages for a 40-hour workweek. From this amount, taxes would be deducted, and the employee would receive a net paycheck.

Benefits

Many employers provide benefits in the form of medical and dental policies, vacation pay, paid sick days, retirement package contributions, annuities, or a variety of other compensatory benefits. The benefits can be part of a collective bargaining agreement which the employer has signed-on to or it can be a matter of company policy. The costs of these benefits are typically included within the calculation of the burdened rate based on the wage. These benefits represent a cost to the employer that must be recovered, which can be broken down into a percentage of the hourly rate and then extended to a dollar amount per hour. This dollar amount is then added to the wage rate.

State and Federal Taxes and Insurance

All states and the federal government apply a tax to wages to provide a source for unemployment benefits. The Federal Unemployment Tax Act (FUTA) and State Unemployment Tax Act (SUTA) are taxes paid by the employer based on the dollars earned by the employee. They are levied on the employee's gross taxable wages up to a maximum called a *cap*. Once the employee earns more than the cap, the tax stops. The SUTA tax varies from state to state and fluctuates depending on the employer's experience, which is measured by the number of employees that the employer has laid off. The fewer employees claiming benefits, the less the employer's contribution. As the number of employees claiming benefits increases, the amount of the employer's contribution also increases. The SUTA tax can vary with a number of factors based on the employer's experience in the workforce, number of workers, and the type of work. There is typically a cap on the wages subject to SUTA, as with the FUTA tax.

The Federal Insurance Contribution Act (FICA), or Social Security and Medicare taxes are provided by the federal government, though more appropriately considered insurances than taxes. The employee is taxed at the rate of 7.65% of gross taxable wages, deducted from the employee's paycheck. Additionally, the employer makes a matching contribution of the same percentage up to a certain cap. It is the employer's percentage that is recouped in the wage calculation.

Worker's Compensation Insurance

Another type of insurance that protects employees is *workers' compensation*, provided by the employer in the event of injury, disability, or death occurring in the workplace. In most states, this is a compulsory insurance. Workers' comp, as it is commonly referred to, is different for each classification of worker or trade, and is based on a percentage of the gross nonpremium wages of the employee. (In other words, overtime or premium wages do not affect the insurance premium.) Insurance premiums for individual companies performing the same work will vary by experience. Premium rates are based on the employer's number of days without a serious accident on the job site. A safe work environment translates to lower workers' compensation premiums based on the *experience modification rate* or *EMR*. This is a rating system that insurers use to analyze a contractor's safety record. Thus, the insurer rewards the employer for promoting a safe work environment with reduced insurance premiums. Lower premiums are reflected in lower labor costs, thereby enabling the contractor to be more competitive in bidding.

General Liability Insurance

General liability insurance protects the project and any adjacent property from damage. This insurance compensates the contractor and, ultimately, the owner for damage by parties under the control of the general contractor, including employees and subcontractors. General liability premiums are again rated on the contractor's success in reducing risk. They are calculated based on two factors: employee wages and subcontracted work. For most general contractors who subcontract more than 70% of their contract work, the largest portion of the premium is based on the costs incurred by those parties under subcontract. However, there is still a portion of the premiums that is based on gross employee wages. Percentages will vary with the limits of the policy, size of the work, performance history of the contractor, and any regulations mandated by the state or insurer.

Public Liability Insurance

Public liability insurance protects against actions brought on by the general public. For contractors that perform public-access oriented work a public liability policy may be a necessity. It can be covered as part of home office overhead or as a function of the hourly wage of the employees. This is a matter of internal policy for the company.

Labor Productivity

Productivity can best be described as the rate at which work is produced by an individual or crew per unit of time. The unit of time most often used is the day, or *workday*. (The hour is too short of a period to provide an accurate measurement.) A workday is considered an eight-hour period, over which productivity will fluctuate up or down, depending on breaks, learning curves, starts and stops, instruction, and material handling. Productivity on the same task can vary when performed by two different crews, or during different times of the day. It can also vary with extremes in weather or temperature. When the work produced is measured at the end of the workday, it represents a daily output accounting for all of the normal fluctuations of the day. It is fairly easy to take a "snapshot" at the beginning of the day and again at the end of the day and compare the difference. If the productivity of the crew or individual is tracked over extended periods of time it begins to represent an average productivity for the task and for that individual or crew. This productivity becomes the historical record for the task and can be used for future estimates accounting for any differences in the task or means and methods.

There are also many factors to consider when determining the expected productivity on a task. For example, consider two separate roofs with the same number of units (squares), height from the ground, asphalt shingles, and configuration (style). The only difference is the pitch of the roof; one is a 6:12 pitch and the other 12:12. Most construction professionals, even novices, would agree that the steeper-pitched roof reduces mobility, thereby making the roofing process slower and more difficult. This is translated into reduced productivity and, thus, a higher unit cost for labor.

As noted previously, the best measure of a crew's productivity is its own historical data. Job records for payroll and labor-hours to complete a specific task offer the best guidelines for predicting future performance. Experienced estimators know that while this is no guarantee of future performance, it provides a logical model. Specific adjustments can be made based on the needs of the individual project.

Individual and Crew Tasks

All tasks that have a labor component can be categorized for estimating purposes into either an individual task or a crew task. An *individual task* is one performed by a single individual. Performance and productivity are measured by the sole output of the individual doing the work. As an example, consider a carpenter trimming a modestly sized interior window with wood casing. This is clearly an individual task. The performance of the individual carpenter can be easily measured by counting the quantity of windows trimmed in a single day, provided the windows are of similar size. After several days of trimming windows, an average productivity could be established, and a model could be created for future window-trimming work. The cost of the carpenter per day could then be divided by the number of windows trimmed to establish a baseline cost model per window. Adjustments could be made to this cost model to predict the productivity for trimming windows of sizes different than the one used to create the cost model, as well as adjustments for the labor cost as wages increased.

> *Here's an example of a cost model for the carpenter trimming the windows: 8 carpenter hours × $54.00 per hour = $432.00. For multiple days, the carpenter will trim 10.5 windows per day. The cost per window will be $41.15. It could also be concluded that approximately 0.76 carpenter-hours are needed to trim a window.*

Crew tasks are more challenging to estimate in terms of productivity. They combine the performance of multiple individuals, each of whom performs a specific function to complete the overall task. The measurement of productivity is a byproduct of how well the team members perform together. Crew tasks have a different dynamic: efficiency as a team. For example, a crew installing brickwork might consist of two bricklayer tenders and three bricklayers, one of whom is a foreman. Each member of the crew has a specific function that contributes to the overall performance of the crew. The foreman directs the crew, lays out the work, and aspires to a steady, predictable productivity. The bricklayers, including the foreman, lay up the brick. The foreman produces less as a result of other duties, but still contributes. The bricklayer tenders, or laborers tending the bricklayers, mix mortar, stock the scaffolding with materials, and provide support services. While they do not install the brick, their duties are essential to successful production.

Each crew member's performance relies on the other members in order for the crew to perform their work efficiently. Adding or removing a single individual changes the crew and the productivity. Selecting the most efficient crew is critical to successfully estimating. This is where skill and experience play major roles. Contractors typically consult historical data for determining the most effective crew size to maximize efficiency. This can sometimes be by trial and error.

The productivity of a crew is measured as an entire crew, this is especially true with *blended* crews or crews with different trades. A crew of four carpenters trimming windows could be analyzed to determine the productivity of a single carpenter, assuming they all had similar skill sets and experience levels; however, the same could not be done with the bricklayers and tenders because their production is a composite.

Subcontractors

Subcontractors are, by definition, independent contractors. They are an indispensable part of the construction process and perform an increasingly larger percentage of work on construction projects. Subcontractors "hire and fire" their own labor, purchase their own materials, run their own equipment, and, quite often, define their success entirely on their own performance. Subcontractors provide their own insurances: both general liability and workers' compensation. They are held responsible for their actions and are required to pay their own taxes.

By virtually every test, subcontractors are considered independent of the general or prime contractor, yet why are general or prime contractors held contractually responsible for the errors, losses, and overall performance of their subcontractors? This can be explained in two ways. As a result of ever-advancing technology in the construction industry, projects are becoming more complex. Construction requires a higher level of subcontractor specialization, which in turn demands more training and brings more risk for error or loss to the general/prime contractor.

Subcontractors are bound by agreement to general or prime contractors to execute a "portion of the work." This portion of work is most often well defined but of limited scope, with a stipulated sum as the basis of compensation. This sum is a piece of the overall or total amount that comprises the contract value between the general/prime contractor and the owner.

Typical agreements between owners and general contractors do not explicitly recognize subcontractors. The agreement assigns responsibility for subcontractors to the general or prime contractor, who is the only signatory party with the owner. This translates to poor scope evaluation or errors and omissions from a subcontractor's bid becoming the responsibility of the general or prime contractor. General or prime contractors realize that their firms must be compensated for the coordination, supervision, and risk assumed when hiring subcontractors.

Since subcontractors are separate entities, there is no need to apply markups for taxes and insurance on wages, or taxes and waste on materials. These are implied to be included by the subcontractor, as is the subcontractor's own markup for indirect overhead costs and profit. In order to apply markup to prices received from subcontractors, the general contractor should consider these factors:

- Dollar value of subcontractor quote. How large of a percentage of the overall total contract is the work of this subcontract?
- How often has your company done business with this subcontractor, and how successful has the relationship been?
- How financially stable is the subcontractor? Does the sub require more frequent payments than the terms of your contract with the owner?
- Can the subcontractor provide a performance and payment bond?
- Will the subcontractor require more management time than other subcontractors?
- What is the project duration?

These are all criteria for determining the markup that should be applied to the subcontractor's price. There is no standard or acceptable range; it depends on what the market will bear. If the general contractor adds too much, however, the bid will not be competitive. If the amount is too little, the bid could be too low, which would result in substandard return or worse – performing the work for free. Neither of these is the hallmark of a successful contractor.

The estimator's responsibility when using subcontractors is threefold. First, it is the estimator's responsibility to ensure that there is adequate competent coverage for subbids. Adequate competent coverage can be defined as a minimum of three legitimate and competent bidders for each major subcontract scope of work. This may require the estimator or estimating team contact the subcontractor multiple times during the bid phase to gauge interest in the project and ensure they will quote the work. It may also require due diligence in checking references of new bidders.

The second responsibility is to generate a *parametric* or a budget estimate for the work. This allows the estimator to approximate what the expected bid price will be. It also has the benefit of forcing detailed thinking about the scope of work of that subcontractor so that they can ask intelligent and comprehensive questions during the qualification process. The parametric estimate can be based on historical data from previous bids or completed projects. Consider as an example three bids from a previous project that can be distilled to a cost per square foot. More than likely there would be a low, high, and intermediate cost per square foot. This could be used to generate an expected range for the subcontractor's quotes on a similar project being bid.

The third and last responsibility is the review or qualification process, which is sometimes referred to as "scoping" the subcontractor bid. This is a crucial part of the process – some would say the most important part of the process. The estimator must review the scope of work contained in the bid to ascertain whether the subcontractor has priced what was requested or the complete scope of work required. This is done with the subcontractor submitting the bid over the phone or in-person before it can be added to the GC's bid.

One final note concerning subcontractors: The Internal Revenue Service (IRS) has very specific criteria for defining a subcontractor. These criteria have survived many a challenge. All parties should be aware of applying the term *subcontractors* to what otherwise could be considered a thinly veiled attempt at avoiding the payment of insurance and taxes on wages. The common practice in the residential industry of hiring individuals, paying them as subcontractors by issuing a Form 1099 at the end of the year, and not insuring them with workers' compensation and liability insurance will not survive the challenge and could result in serious fines or penalties being assessed to both parties.

Wage Scales

In the United States today there are four main wages scales for construction labor. They are open shop, union, Davis Bacon Prevailing Wage, and residential. The following is a brief description of each.

Residential

Residential wage scale applies to residential construction: home building and residential renovations. It has most all the same trades as the commercial wage scales. It is highly affected by location and both the macro and micro economic conditions of the area being considered. Residential wages change based on the abundance of work in an area. They are volatile and can go up or down with availability of the tradespersons or work. They are not regulated or predictable like commercial wages. Residential tradespersons are less likely to have completed a formal apprenticeship program. The residential sector of the industry is the largest.

Union

Union wages are based on a *collective bargaining agreement*. The rank and file agree to forgo individual negotiations in lieu of the negotiation strength in the numbers of members. The union management negotiates a wage and benefit package approved by the membership for a specific period of time. The agreement dictates the wages and benefit package its signatory contractors must pay the employees. The agreement also defines jurisdictional work for the trade. Unions provide apprenticeship training so that each member upon reaching journeyman status has the same training. This is intended to portray a homogenous workforce whose members can produce similar daily outputs. Union wages scales are generally higher than their open shop counterparts and have a benefit package. Unions represent a small percentage of the commercial construction workforce.

Open Shop

Open shop wages are based of the typical wages for a trade for the demographic and typically have no organization or formal management structure. Wages are reflective of the economic conditions of an area and have no established increases. Membership in any regulatory organization, other than the statutory licensing, is not required to work. There is often no established benefit package other than those required by law or provided by the company. There is no jurisdictional outline of what work is considered the work of that trade other than typical practice. Training of apprentices can be by trade schools or from on-the-job training. Wage scales tend to be lower than union wages and reflect an "at will" work environment. Open shop wages represent the overwhelming majority of the domestic commercial workforce.

Prevailing Wage Scale

The third and perhaps least familiar wage scale is called *prevailing wages*. It is a wage scale modeled after the federal Davis Bacon Act of 1931, which requires private contractors and subcontractors to pay mandatory predetermined wage scales to employees on all taxpayer-funded construction projects over $2,000. Most often the prevailing wage of the area corresponds to the union wage. The Davis Bacon Act was intended to help close the gap between the higher union wage and benefit and lower open shop wages when competing for federally funded projects. Awarding Authorities are issued a predetermined minimum wage scale by the Department of Labor and Industries to include as part of the project manual. Contractors and subcontractors can pay higher wages but must meet these minimum requirements. Failure to comply can result in debarment, fines, and/or imprisonment.

EQUIPMENT AND OPERATING EXPENSES

Many of the tasks performed on a construction project require *equipment*. Equipment helps perform a variety of tasks and has a cost that must be captured accurately in the estimate. It can be loosely defined as something used by the labor to install the material and is considered a production cost. Equipment is a temporary cost measured by the unit of time it is required to perform the work on site. The units of time are calculated by the day, week, or month. When it is no longer needed it can be removed from the project and returned to the yard or rental store. It should be noted that the rental term *day* is a 24-hour period, the *week* is a 5-day week, and the *month* is typically a 21-day month. Some pieces of equipment have costs associated with delivering them to the site and removing them from the site after the work has been completed. These costs are called *mobilization* and *demobilization*, respectively. Whether a piece of equipment is owned or rented, it is estimated at the "going rate." In other words, the over-the-counter rental rate that can be charged for that particular piece of equipment. The only time the rate is discounted is if the contractor owns the piece of equipment, its cost is significantly less than the rental rate, and the discount provides the contractor with a competitive bidding advantage.

Determining the cost of a piece of equipment can be as simple as obtaining a quote from the rental company for the term required including a delivery and pickup fees. With the exception of fuel consumption and an operator, (if applicable) the costs of ownership, depreciation, and insurance is included in the rental. For equipment owner by the contractor the rate to carry in the estimate may be more challenging. It can include the cost of purchase or lease, depreciation, maintenance, repair, expected use per year, and insurance, to name a few. For contractor-owned equipment the cost per day, week or month is usually calculated by the company's accounting professional.

Deciding on renting or owning is often a matter of expected usage. For contractors that use the piece of equipment every day it makes more sense economically to own equipment, and possible multiples of the same piece. This is especially true if it is foundational to the performance of their work. For general contractors that subcontract large portions of the work only the most common equipment is owned. Generators, scaffolding and planks, and lighting, are fairly routine equipment owned. Specialty equipment may be leased, purchased, or rented on a per project basis. For many general contractors, it is more about the liability posed by the equipment than the cost. This often results in hiring specialty subcontractors to perform the scope that includes the equipment. Consider scaffolding the exterior of the building for repairs. Many general contractors will hire a subcontractor to erect, maintain, insure, and dismantle the scaffolding as a means of reducing the liability.

Also note that rental or owned equipment may be subject to a sales or use tax depending on the laws governing the project.

INDIRECT OVERHEAD

Costs related to the operation of the company's main office and its staff are called *indirect overhead* expenses. These costs are not solely attributable to a specific project but are associated with maintaining a construction business. Indirect overhead is discussed in greater detail in Chapter 6.

PROFIT

Profit is the reason for doing business. It is the end result of a project done right and/or the reward for risks taken. While profit is not specifically a cost, it does need to be captured in the estimate or the calculation of the billing rate. Without it, the project would be considered a failure. Profit is most often assigned as a percentage of all costs of the work. However, it can be assigned as a lump sum or stipulated fee if appropriate. Calculating profit is based on careful consideration of a host of factors – some less tangible than others. A complete discussion of the assignment of profit is covered in Chapter 26, "Profit and Contingencies."

Following is a sample calculation to illustrate deriving the billing rate for a carpenter from the wage rate, using hypothetical rates for insurance, taxes, and other markups.[1]

Carpenter wage rate	*$35.00*	*per hour*
Benefit package	*15.50*	
FICA (7.65% on wage)	*2.67*	
FUTA (0.8% on wage)	*0.28*	
SUTA (7% on wage)	*2.45*	
Workers' comp at carpenter rate of 13.05% (on wage)	*4.57*	
General liability insurance at a rate of 0.82% (on wage)	*0.29*	
Subtotal	*$60.76*	*Raw labor cost*
Indirect overhead at 18% of all costs	*10.94*	
Subtotal	*$71.70*	*Burdened labor cost*
Profit at 10% of all costs	*7.17*	
Billing rate	***$78.87***	***per hour***

---

[1] The rates used in this example are for illustration purposes only and should not be considered appropriate for all conditions.

SUMMARY

Understanding the various modifiers that apply to the material, labor, and equipment components of the unit price is essential to good estimating practice. Modifiers on materials are typically statutory sales tax in the area, however not all projects are subject to sales tax. This can also be true of equipment costs. Additionally, the estimator would be prudent to investigate any special delivery or supply chain charges for materials. The estimator must take special care to ensure that all of the modifiers for labor are accounted for when calculating the hourly billing rate. Frequently, the actual percentage rates for modifiers such as FICA, FUTA, and SUTA or the current EMR for the Worker's Comp rate may require the assistance of the accounting department. Same is true for the cost per day for owned equipment and its operating expense. Some equipment requires the estimator include the cost of mobilization and demobilization.

Productivity is the amount of work produced by a crew or individual per workday. Productivities for repetitive or unfamiliar tasks should be tracked over extended periods to develop historical data that can be applied to future estimates. Deviations for the typically daily output should be analyzed for causes. Labor costs and benefits fluctuate based on the wage scale of the project. The estimator must understand what labor scale the project is governed by to determine an accurate labor cost.

Subcontractors form an integral part of the construction industry by performing large portions of the general contract. The estimator must review and qualify a subcontractor's bid before inclusion in the general contractor's bid. Failure to do so may result in errors or omissions of scope that the general contractor may be responsible for under the contract with the owner. Subcontractors are independent entities that are responsible for their own materials, labor, equipment and occasionally subcontractors. Their bids should include all taxes, insurances, overhead costs, and profit.

Most estimators calculate the hourly billing rate for trades that self-perform their work. This may include the assistance of the payroll or accounting staff to be accurate. Frequent review is also required as changes occur with time or performance.

QUESTIONS/ PROBLEMS

1. The cost of an element in an estimate has modifiers. Identify and explain each modifier for labor.
2. Explain the term *productivity*. Why is it important in estimating?
3. Opine as to how subcontractors are accounted for in an estimate. What is the estimator's responsibility with regards to a subcontractor's scope of work?
4. What was the purpose of the Davis Bacon Act with regards to labor on a project?
5. Why is direct or project overhead not included as part of a billable hourly rate.

6. Calculate the billing rate for a bricklayer with the following: wage of $37.50 per hr., $15.90/hr. benefit package, FICA at 7.65%, FUTA at 0.87%, SUTA at 7.2%, workers' compensation at 31.77%, General Liability at 0.91%, main office overhead of 16.5% and a profit of 15%.

7. Calculate the billing rate for a bricklayer tender (helper) with the following: wage of $30.50 per hr., $12.45/hr. benefit package, FICA at 7.65%, FUTA at 0.87%, SUTA at 7.2%, workers' compensation at 38.5%, General Liability at 0.91%, main office overhead of 16.5% and a profit of 15%.

8. Using the results from questions #6 and #7, calculate the daily billing rate for a crew of three bricklayers and two bricklayer helpers.

9. If the daily output for the crew in #8 is 1,655 bricks, what is the labor cost to install single brick?

10. Using a factor of 6.55 brick per SF and a material cost of $0.87 per brick, and 5% sales tax, calculate the material and labor cost for laying up 40,000 SF of brick veneer.

6 | Division 1—General Requirements

As discussed in Chapter 02, the technical portion of the specifications is based on the 50 divisions of the CSI MasterFormat®. The first division is appropriately named Division 1—General Requirements and is the only topic within the General Requirements subgroup. This division deals primarily with:

- Project overhead requirements
- The basis for administering the project, as defined by the General Conditions of the Contract for Construction
- Administrative requirements of the project

While the General Conditions of the Contract and the General Requirements of Division 1 stand alone as two separate documents, they are related. General Requirements are important to the estimating process because they provide the information needed to assign a monetary value to the project overhead items.

General Requirements include items such as:

- Temporary facilities and controls
- Project meetings
- Reference standards and definitions
- Submittals, mockups, and samples
- Testing requirements
- Project closeout
- Project record documents
- Quality control
- Commissioning
- Project phasing
- Progress schedule (critical path method)

Estimating Building Costs for the Residential and Light Commercial Construction Professional, Third Edition. Wayne J. Del Pico.

Division 1 also identifies the contractor's special contractual obligations that have an associated cost and must be accounted for in the bidding process. This category includes development and provision of:

- Unit Prices
- Alternates
- Allowances

While the General Requirements are the first division of the CSI MasterFormat®, they are often the last to be estimated (priced). This is because most General Requirements items require a complete understanding of the entire project, which is not possible until after the estimator has become familiar with the contract documents, performed the takeoff of each division, made a site visit (if applicable), and estimated the majority of the project. Many of the General Requirements can be highly subjective which requires that the estimator have a complete picture of the project.

It should also be noted that the General Requirements are known by three other names as well: Project Overhead, Direct Overhead, and General Conditions. The name General Conditions should not be confused with the contract document of the same name. Division 1 also includes other information that does not require pricing such as payment terms, contract modification procedures, and reference standards.

CLASSIFYING DIRECT OVERHEAD COSTS

Most items in the General Requirements can be classified for estimating purposes into two main categories: fixed costs and variable costs. Both classifications are considered nonproduction costs as they are more about contract compliance and project necessities than moving the project toward completion. Many estimators use a checklist when reviewing the General Requirements to establish what is needed.

Fixed Costs

Fixed costs are associated with one-time project requirements. Building permit fees are a good example; these may be calculated by a formula that is cost driven (e.g., $12 per $1,000 of building cost) or by some other fixed means, such as the square footage of the floor plan. Either way, the estimator can determine a fixed cost for these fees to incorporate into the estimate. Other examples of fixed costs include street opening fees, water tap permits, water and sewer betterment fees, trailer furnishings, prepaid insurance premiums, project sign, utility connection fees, and project mobilization/demobilization costs. The essence of a fixed General Requirement cost is that it is incurred once on the project, despite the project schedule, so it can be estimated as a fixed sum.

Variable: Time-Sensitive Costs

Variable costs are schedule driven and are determined based on the length of time the items or services are needed on the project. Examples of time-sensitive costs include trailer rental, telephone and electrical power usage, supervision costs, and temporary toilets, office, and storage trailers. Variable costs can also be thought of as more subjective in the sense that they are frequently defined based on experience and judgment. Variable, or time-sensitive, costs require that a "means and method" technique be used in development of the estimate. For example:

If scaffolding is required along the exterior façade of a building, the estimator will need to develop a method for staging the structure to perform the work. The options may be numerous and vary greatly in cost. Staging could be included for only one portion of the building at a time and then dismantled, relocated, and re-erected as many times as necessary to complete the work. Each move would have specific costs relative to how long the staging is required on-site. Another alternative might be to use a power lift, which may improve mobility but might limit the number of workers that can be carried. A third option might be a roof-based swing platform, which is less mobile than a lift, yet can provide more work area for personnel. Although all three methods satisfy the requirements for scaffolding, one approach must be selected. Once a strategy is selected, the duration and cost implications can be determined.

Other time-sensitive costs that are less subjective but still require experience and judgment include supervision, temporary utilities (such as job site electricity and water usage), and construction controls (such as barricades or temporary fencing). Many of the utility, cell phone, and water usage costs are the result of the tracking of historical costs much like what is done for production costs. All of these costs are clearly tied to a specific length of time they are needed for the project. Time-sensitive costs have a direct relationship to the project progress schedule and potentially the project phasing plan.

The first step in accurately determining the duration of time-sensitive costs, is to create a schedule. The estimator, being most familiar with the project and the means and methods employed in estimating, is the likely candidate to draft the schedule. While this preliminary schedule often requires multiple generations, it has the benefit of being less detailed than a working CPM schedule. Drafting and revising a preliminary schedule is essential for accurately predicting time-sensitive costs.

Other Variable Costs

Other variable costs are less straightforward to calculate. While their costs can vary, they are not impacted by time. Consider the category of *winter conditions*. For projects impacted by cold weather and snowfall, mitigating winter conditions during construction is often a contract requirement.

They can include snowplowing, hand shoveling of work areas, spreading ice melt on frozen surfaces, and providing temporary heat for a structure under construction. This consumes material, labor, and equipment resources. Calculating the amount needed, the frequency, and cost is dubious at best, but required in the project overhead. Accepted methods for determining snowfall include consulting websites that track and forecast weather months in advance. The site most often used for climatic predictions is the *National Oceanic and Atmospheric Administration,* or NOAA. It tracks past years' temperature, snowfall amount, and frequency and uses the data to predict future winter weather conditions. While NOAA can provide the most scientific prediction, it is still only that – a prediction. In addition, trying to forecast the cost of heating oil or natural gas a year or more in the future is not the domain of the construction estimator.

An option is to include a dollar allowance for winter conditions in the estimate. When this is unacceptable, the estimator is advised to document their strategy clearly and carefully for pricing winter conditions, including retaining quotes and NOAA data. This information will be helpful should a claim for additional costs develop.

Schedules for Time-Sensitive Costs

Depending on the complexity of the project, several generations of schedules may be required as the estimator becomes more familiar with the project or as input is added from other members of the project team, such as the superintendent or the project manager. It should be noted that the schedule used for estimating purposes is not the same version that might be required after the job is awarded. The schedule used to estimate general requirements is a timeline showing where an activity occurs in the project and its duration. This initial schedule should be comprehensive, but it should be noted that it is for use only as a tool for preliminarily estimating project durations and seasons in which tasks occur.

A second but less desirable method of determining time would be to use the overall time for performance provided by the owner or awarding authority. This is language in the bid documents that stipulates a project's duration. While this gives an overall time allotted for the project, it fails to identify when certain weather-sensitive tasks occur.

BAR GRAPHS AND GANTT CHARTS

The most common form of schedule used for residential and light commercial projects is the *bar graph,* or *Gantt chart.* The bar graph or Gantt chart is a simple graphic representation of the project timeline. It is easy to understand for the professional, novice, and nonconstruction personnel. The bar graph roughly follows the way the work will actually be executed. Each activity or major activity group is represented graphically by a bar. Each has a start date, a duration, and a finish date. The bar graph implies a sequencing of activities, with one activity following another. However, each activity is shown independent of its succeeding or preceding activity. This implies that each activity is not affected by other activities, which, of course, is not the case on construction projects. There is, in fact, interdependence among activities on a construction project. If one task is delayed, it can have a "downstream effect," disrupting future tasks and, ultimately, delaying the project as a whole.

For estimating time-sensitive cost items on simple projects, a bar graph may be sufficient on a project-by-project basis, but less than ideal. It shows the overall duration of the project, as well as the duration of each activity, and allows the estimator to locate specific activities, such as weather-sensitive tasks, relative to calendar dates. This can be important in parts of the country where construction projects may be required to include winter conditions.

CRITICAL PATH METHOD (CPM) SCHEDULING

The *critical path method* or CPM schedule is useful for larger, more complex construction projects that require a more dynamic and sophisticated scheduling procedure. With computer scheduling now mainstream, there are many software applications that do the calculation part of scheduling. Like most computer applications, it is designed to reduce the time inputs of the professional and are ideal for creating schedules to accompany an estimate. The CPM format enables the user to create a schedule based on the means and methods adopted and to identify potential problems in advance. It also allows the advantage of viewing different sequences or scenarios called "what-ifs" that may aid in fine-tuning the schedule and ultimately the estimate.

CPM schedules are more detailed, and the activities are shown linked. This linking of activities creates the interdependence between activities, which ultimately identifies a critical path through the project. If one task on the critical path is delayed, it will cause the delay of all succeeding critical path tasks and the project as a whole, if corrective action is not taken. The CPM schedule also illustrates the fact that on a real construction project, multiple tasks can occur independently at the same time, while other tasks must occur in a certain order and are highly dependent on completion of the preceding task. Developing a CPM schedule for bidding purposes can be more time-consuming than a bar chart schedule, but it is also far more reliable for predicting the project schedule.

Regardless of the schedule method selected, it is often recommended that other team members review the schedule and be allowed to make suggestions. This "multiple sets of eyes" approach is extremely helpful in gaining insights from different viewpoints.

Schedule Limitations in the Contract

As mentioned previously, projects have a prescribed project duration or set "available time," called *time for contract performance*. Statements such as

"The Contractor will have 270 calendar days from the Notice to Proceed to achieve Substantial Completion" are fairly common. As a result, many contractors believe this to be the schedule. For any project with an owner-dictated time for performance, the estimator would be wise to still draft a schedule to ensure that the estimator's means and methods vision of the project fits within the allotted time. It is highly recommended that the project schedule be determined independently of any such statements within the documents.

The initial schedule should be developed with normal working times in mind. If the critical path needs to be compressed in order to achieve substantial completion on the required date, then the costs associated with the acceleration must be included as part of the bid. Acceleration options may include:

- Working extended daily hours
- Working weekends and holidays
- Adding labor resources
- Using multiple work shifts and crews
- Other, more creative options such as off-site prefabrication

The costs of acceleration that may be required to meet contract time for performance are sometimes overlooked. The contract-specified project duration may also tend to influence the estimator to shorten the schedule to match the time available. Using the contract time for performance as the construction schedule can be risky at best and should be avoided whenever possible.

IDENTIFYING AND QUANTIFYING GENERAL REQUIREMENTS COST ITEMS

As noted earlier in this chapter, the costs for the General Requirements of the project are often left to the end of the estimate, when the estimator is most familiar with the overall project. A common practice for developing a thorough and realistic cost for Division 1 involves maintaining a checklist or even a handwritten list of reminders as each section of work is estimated. As a particular task is taken off, related General Requirements activities are noted. These may include such items as overtime supervision, cranes or other such equipment, temporary protection, or electrical needs. When it is time to determine the Division 1 costs, the estimator can note the General Requirements items that were checked off or listed while estimating the other divisions. Another option is to develop a checklist of all possible General Requirements items to use for each project, selecting the items that apply. In any case, a system of checks and balances is necessary to ensure that all General Requirements costs are addressed. The Project Overhead Summary sheet in Chapter 27 of this text can be used as a starting point for a project overhead checklist.

The estimator should keep in mind that some General Requirements are the responsibility of individual subtrades and are specified within the subtrade's technical specification section by exact language or by reference. For example, it is not uncommon for the roofing subcontractor to be responsible for his/her own hoisting or for the electrical subcontractor to be responsible for installation and removal of temporary power and lighting on a project. The estimator should review and qualify a subcontractor's bid to avoid duplication or omission of Division 1 items.

Project overhead can be impacted by change orders that extend time specifically for tasks that have general requirements like lifts or temporary heat. Not all change orders impact project overhead, so the estimator (or project manager) should consider the impact of the change on the project as a whole.

Project overhead items are not always easily quantified as with production items. Many require the estimator to sort through the language and establish quantities based on their interpretation. For example, consider the statement: "The contractor will provide sufficient number of portable toilets to accommodate the tradespersons. The portable toilets will be maintained in a clean condition and shall be stocked adequately with toilet paper and hand sanitizer."

This statement clearly implies that there will be a sufficient number of clean portable toilets on site as needed. To calculate the cost of the toilets, the estimator would again consult the schedule for duration in weeks, multiplied by the weekly cost of the toilet, including stocking and cleaning. However, how does the estimator know exactly how many are needed. Again, the calculation may be subject to experience or historical data.

Many Division 1 requirements use broader, more general words than in other technical sections. Words such as *sufficient* and *adequate* leave room for interpretation and allow the contract administrator to adjust items without exposing themselves to a contract modification. Consider the interpretation as the technical writer has intended whenever possible or seek clarification.

PRICING GENERAL REQUIREMENTS ITEMS

Pricing fixed-cost items is fairly straightforward. Prices for time-sensitive or variable-cost items can be determined using the list of needed items, together with the schedule. General Requirements costs, such as supervision, are reasonably clear. The General Conditions will stipulate that the project is required to have "adequate and competent" supervision by the general contractor, defined as supervision licensed by the authority having jurisdiction, with a particular education and experience level in the type of project. The General Requirements might mandate that the supervisor be on-site during all times that the work is in progress, including weekends and holidays. These statements help to determine the level of supervision that is required and the associated salary and benefit costs. At that point, it becomes a simple unit price multiplied by the duration.

Other time-sensitive costs may be required only intermittently or for specific periods within the overall project duration. For example, the General Requirements may stipulate that the general contractor provide an office trailer on-site until the building is closed in and the office can be relocated inside. The estimator would then calculate by the schedule how many weeks or months until the building will be closed-in and an indoor office can be set up. The period (in weeks or months) from the start of the project until the date the building could house an office would be determined and then multiplied by the appropriate rate (per week or month).

Costs for General Requirements categories such as project meetings and project closeout are judgment calls and are almost exclusively experience based. It is difficult to predict accurately how long site meetings will take before the project starts. Any prediction is based purely on past experience, and that is still no guarantee. Possible considerations in determining costs for meetings include level of architect involvement (the more involved, the longer the meetings), level of document development (the more complete, the shorter the meeting time), and contractual requirements. Many contracts mandate a weekly meeting.

One last note on calculating project overhead costs. Some estimators apply a percentage of construction costs as the cost for project overhead. This should be discouraged except when the contractor is in full control of Division 1 requirements as in the case of most homebuilders. Repetitive building of the same model of home often smooths out the schedule and identifies the direct overhead needed, which can then be calculated as a percentage of construction costs.

PROJECT CLOSEOUT

Most commercial projects have a formal process at the end of the project called *project closeout*. Project closeout is the organized and orderly transfer of project control from the contractor to the owner or awarding authority. It involves contract and administrative review and reconciliation and frequently testing and supporting reports. Project closeout should start by developing a schedule of the documents and other items that need to be transferred at the end of the project. Since project closeout is tied to the release of retainage, it is best completed as expeditiously as possible.

Project closeout can involve tasks such as punchlist, commissioning, (a complex and expensive process), HVAC testing and balancing, the submission of rebate documents, training of maintenance personnel on new systems, warranties and guaranties, and as-built or record documents, just to name a few. Much of the project closeout work is handled by project managers or the administrative support staff working with project managers. The emphasis is that it takes time, resources, and ultimately money to complete; therefore, it should have a corresponding value in the estimate.

Many estimators assume the costs to be covered in the assignment of project management costs in the estimate as a direct or indirect cost. This is somewhat true, although not completely. It is recommended that the estimator or support staff compile a list of closeout requirements with an adjacent column that identifies who will complete the task. Using this method, the estimator can assign labor-hours and dollars to the tasks that are not covered under direct or indirect overhead costs. The cost of project closeout should not be trivialized or undervalued, as it can be expensive.

INDIRECT OVERHEAD

As noted in Chapter 5, costs related to the operation of the company's main office and its staff are called indirect overhead expenses. These costs are required to maintain a construction business but are not directly attributable to a specific project. Like project overhead, *indirect overhead* is known by other names. Home office, general administrative, or main office overhead are the same as indirect overhead. It includes, but is not limited to, expenses such as:

- Corporate officers' salaries and benefits
- Rent or mortgage for the main office
- Monthly telephone, fax, or internet charges
- Clerical and office associates' salaries, insurance, and taxes
- Corporate vehicles, insurance, and maintenance/operating costs
- Estimator's salary and benefits
- Accounting staff's salaries and benefits
- Legal, accounting, and technology service fees outside the firm
- Heating, electricity, and maintenance of the main office

Depending on the company's accounting policies, the cost of project managers can be classified as either a direct or indirect overhead. For a project manager dedicated to a single project, with no other responsibilities, the answer is clear. That individual would be a direct cost. For the project manager overseeing multiple projects, the cost could be proportionally shared among the direct overhead for each of the projects. Other companies prefer to include project managers as part of the indirect overhead.

The cost of home office overhead is intended to be absorbed by all projects under contract. To recover these costs, a percentage representing the main office overhead as a function of the total dollar value of work is added to each future bid. This percentage is derived by the company's accounting staff and senior management. As a result of its proprietary nature, the actual percentage and methodology under which it is determined are closely guarded secrets. To illustrate the calculation, consider the following scenario.

A construction company has established a successful market niche in building and renovating schools. Over the most recent calendar year (January 1 to December 31) the company has bid, contracted, executed, and successfully completed $20,000,000 worth of construction work, referred to as *work-in-place*. During the same calendar period, the accounting department has determined the indirect overhead for the company was

$2,000,000. The accounting department could calculate the indirect overhead as follows:

$$\text{Indirect overhead costs} \div \text{Contract value of work in place} = \text{Overhead \%}$$

$$\$2,000,000 \div \$20,000,000 = 0.10, \text{ or } 10\%$$

A reasonable assumption, based on the above information, would be that it costs the company $1 in overhead for each $10 worth of construction billings. The aforementioned example has been simplified to illustrate the general concept. There are clearly more considerations to developing an overhead percentage, but most are beyond the scope of this text or the responsibility of the estimator.

Home office costs are tracked biweekly or monthly and adjusted as dictated by the economy the company is operating within. Adjustments of the overhead from quarter-to-quarter or year-to-year are intended to accommodate seasonal changes to earnings and any desired growth of the company. In lean economic times, adjustments can include the downsizing of the cost or the outright elimination of the cost to maintain competitiveness and financial health. Well-established companies with firm market niches have little fluctuation in their overhead costs other than normal cost escalation. Other companies experiencing rapid growth may have trouble maintaining a stable overhead percentage.

One final note on indirect overhead: In the above example, the 10% overhead was as a result of a simple mathematical example. It is not intended to imply that the 10% was an industry-accepted percentage. In fact, companies in different markets can have significantly different overhead percentages. It is, however, a common assumption that competing companies in the same market have similar overhead.

SUMMARY

In summary, the General Requirements are the administrative costs associated with the execution of the contract and the project. They are the nonproduction or project overhead costs. Determining these costs can be a matter of interpretation of the Division 1 language and of prior experience with similar projects. Many contractors establish a market niche by doing the same types of projects repeatedly. This provides historical cost data that the estimator can reference for project overhead costs.

Project overhead has two classifications of costs: fixed and variable. Fixed costs are not impacted by time or weather. They are incurred once on the project. Variable costs are impacted by time on site or weather. Time-sensitive costs require a schedule to accurately predict. Schedules are often developed when bidding to ensure that the estimator knows not only the overall project duration, but also when weather-sensitive work will occur relative to the calendar. This allows the estimator to determine what tasks may be subject to Winter Conditions.

The costs of Division 1—General Requirements in an estimate cannot be relegated to a specific percentage of the cost of the work. The cost of the General Requirements for each project will vary and can be determined only by a careful review and pricing of the General Requirements dictated by the contract. Traditionally, projects with high security requirements or unique conditions in which the work is performed tend to have a higher cost of the General Requirements.

Indirect overhead is the cost of doing business. It includes office and administrative expenses that are absorbed by each project under contract. Indirect overhead is most often applied as a percentage of construction contract value. Larger projects, based on dollars, are expected to absorb more than smaller ones.

Chapter 27, "Estimating by Computer," provides a summary of overhead items that could be used as a checklist.

QUESTIONS/ PROBLEMS

1. Define project overhead. As part of the definition, explain how project overhead is calculated. Include within your answer the different types of overhead.
2. What are some of the advantages of using computer software to generate schedules?
3. Explain how indirect overhead is different from direct overhead.
4. Is project management a direct overhead cost or indirect overhead cost? Explain your answer.
5. Please explain why the critical path method of scheduling is uniquely suited to construction projects.
6. Explain what is meant by "means and methods." How does this impact an estimate?
7. Explain the difference between a fixed overhead cost and a variable overhead cost.
8. Explain why some Division 1 language is intentionally ambiguous.
9. Explain the relationship between the General Conditions of the Contract for Construction and the General Requirements of the project.
10. Explain how a change order can impact project overhead.

7 | Existing Conditions

The second division of CSI MasterFormat® is called *Existing Conditions*. It is the first division in the Facility Construction subgroup. As the title suggests, this division deals exclusively with providing information about the current condition of the project site and its structure. It includes information relative to subsurface investigations of the site, site surveys of structures and utilities, assessment of environmental conditions, location and species of hazardous materials, Selective Demolition, Site Demolition, and a wide variety of remediation topics. The quantifying and pricing of much of the work of Division 2 is highly specialized. It is one of the most difficult parts of an estimate due to the number of unpredictable factors and the highly specialized nature of the work. The estimator must carefully study all available contract documents, as well as any supplemental information provided. A site inspection should be conducted only after becoming familiar with the documents.

ASSESSMENT

Surveys and Assessments

Division 2 sets forth surveys for use by the estimator. Surveys include site surveys, boundary and marker surveys, archeological surveys, traffic and accessibility surveys, and acoustic assessments. Traditionally, the surveys included in Division 2 are for informational purposes only and don't necessarily constitute a task in the estimate. They do, however, have the potential to affect how the estimator views the work of other sections. These survey or assessments are included in the technical sections of the specifications but can also include drawings within the set (this will be discussed later in this chapter). The surveys can also include a legal description of the property boundaries in a plan within the set, so the project

Estimating Building Costs for the Residential and Light Commercial Construction Professional, Third Edition. Wayne J. Del Pico.
© 2023 John Wiley & Sons, Inc. Published 2023 by John Wiley & Sons, Inc.

can be laid out for *line and grade*. Line and grade, known as control, has to be established on site before work can begin in earnest. The cost of a registered engineer or land survey crew should be included in the estimate if a registered survey is required for control.

Another part of the assessment portion of Division 2 is the *Hazardous Material Assessment* section. It evaluates, in a report format, the levels of specific hazardous materials on the project. Subsections within the group include assessments for; asbestos, lead, mold, polychlorinated biphenyls (PCBs), and biological hazards. These reports often provide the estimator the background on the location and concentration of specific hazardous material that are part of the remediation or *abatement* process. The surveys and assessments are an invaluable resource for providing guidance in means and methods as well as pricing for subcontractors performing the work.

On some projects the abatement of hazardous material is done by the owners or awarding authority. It can be done to mitigate the risk of liability or to comply with law. The results of the abatement in the form of regulatory paperwork can be included in this section.

SUBSURFACE INVESTIGATION

Geophysical Investigations

Geophysical investigations include reports that provide background or studies on less routine topics such as seismic activity, gravity, magnetics, electromagnetics, electrical resistivity, and magnetotellurics. Geophysical reports are highly specialized reports that are beyond the scope and project types covered in this book.

Geotechnical Investigations

Architects often retain the services of geotechnical engineers, when the budget allows, to investigate the conditions below the surface of the soil. These investigations provide useful criteria for the designing of foundations on commercial and high-end residential structures. The purpose of the investigation is to define the conditions of the soil that will support the structure, and to provide bidders with an idea of the subsurface conditions that may be encountered during excavation. Included are conditions such as soil types, strata, rock locations, and levels of the water table, all of which can have a tremendous affect on the pricing of sitework. The detail and specificity of the geotechnical reports can run the gamut from test pits to extensive soil analysis and deep borings.

Several methods can be used to sample the soil beneath the surface. The simplest is the *test pit,* which allows a visual inspection of the soil. Information such as soil content, grading, stratification (layering of soil), water table height, and cohesiveness of the soil can be obtained through visual inspection. Unfortunately, the test pit is restricted by the reach of the excavating equipment, and most nonresidential structures require analysis of soils at greater depths.

A fairly common method for reaching greater depths is *test boring*, which provides an actual sample of the materials as they occur in place and identifies the location of the water table. The soil samples are analyzed, and the results are interpreted by a geotechnical engineer and distributed in the form of a report. These results are used by the structural engineer in the design of the foundation system for transferring loads to the ground. The report typically provides a plan that locates each boring with respect to the proposed structure. The locations of the borings can often be found on the Existing Conditions drawings, to be discussed later in this chapter. The actual report contained in the specifications is in the form of a simple chart providing the physical description (composition) of the soil sample as it occurs vertically and the corresponding depth. The location of water, if any, should be clearly noted, as should any major obstructions or nonsoil fill materials. The subsurface investigation is frequently accompanied by a narrative from the geotechnical engineer that interprets the information contained in the analysis. The geotechnical report can often serve as the basis for the removal limits of *unsuitable materials*. Unsuitable materials or soils are fill materials that are unsuitable or unstable as structural fill under a building foundation or paving. Many projects use this report as supporting documentation for the quantification, removal, and disposal of unsuitable soils in the bid either as a unit price or as a means of defining quantity.

When available and interpreted correctly, the subsurface investigation is a valuable tool. The geotechnical report can aid the estimator in pricing the excavation. The identification of soils by species and grading helps the estimator to determine the ease of excavation and the daily outputs. It can also establish the need for overdigging to maintain safe stable slopes.

Make careful note of the disclaimer that accompanies each report. It will explicitly state that the information contained in the boring reports is for the convenience of the contractor and that the geotechnical firm assumes no responsibility for the representation of the soil conditions of the site as a whole. This disclaimer is a way to mitigate risk and reduce claims for differing conditions.

The estimator can also find information on groundwater monitoring during the construction process. For projects near aquifers or wells that supply public drinking water, it is not uncommon for larger commercial projects to have regular monitoring of the water source to prevent contamination from construction.

DEMOLITION AND STRUCTURE MOVING

Demolition

Demolition refers to dismantling, removing, and disposing of unwanted existing work. It can be classified into two major groups for estimating purposes.

Full or Total Demolition

The first is *Full* or *Total Demolition*, which includes the removal of entire structures, both above and below grade, in preparation for a new structure or engineering improvement. Within the Total Demolition group are the following categories:

- **Whole Building Demolition:** The complete removal of a structure above grade without salvage concern for any specific building element, component, or material during the demolition. This work is performed with large pieces of construction equipment that break up the structure, load it onto trailers, and transport it to disposal sites. The limits of work end at grade (ground) level of the building. This type of demolition is equipment intensive, and labor is generally limited to the operators of the equipment, truck drivers, and supervisors. Since the advent of recycling, the demolition process can require a predemolition survey for recyclables such as metals or salvage items for resale.
- **Below-Grade or Foundation Demolition:** The removal and disposal of parts of the structure that are at or below grade, such as the slabs, piers, foundation walls, footings, or grade beams. This type of demolition is performed predominantly with large equipment but may also require general labor to saw-cut concrete, drill, or use pneumatic-powered equipment to separate components. Again, it would be wise to consider the sustainability requirements of the project or the community. This may include the separation of concrete from rebar on site and separate recycling containers.

Selective Demolition

The second classification, called *Selective Demolition*, involves a more coordinated and detailed removal of specific elements, without damaging adjacent work. Within this group are several classifications of selective demolition, as follows:

- *Full Interior Demolition*, or **Gutting:** The systematic removal of interior finishes, mechanical and electrical systems, and all nonstructural components of the building. No concern is shown to nonstructural elements in the removal. This is labor-intensive work and requires considerable use of hand tools, ranging from shovels to pneumatic or electric tools. It frequently requires scaffolding or lifts and may even require transportation within the building. Shoring or bracing of the structure is common when structural elements are part of the gut. Chutes attached to the exterior of the building to remove debris from upper levels may also be necessary. This type of demolition may also require the coordination of other trades for the removal of any toxic/hazardous materials such as coolants, refrigerants, flammable fluids, lead, asbestos, or any other regulated materials deemed hazardous. It may also include "make safe" operations by the electrician, plumber, or sprinkler contractor. These resultant materials must be selectively removed by the appropriately licensed trade and disposed of legally off-site prior to commencing with the demolition.

- *Selective Demolition* **is the careful demolition and removal of specific items, finishes, or elements of a structure:** Selective Demolition considers adjacent surfaces or elements that are existing and scheduled to remain. It often includes a layout component that is not necessarily required in projects with more extensive demolition scopes. Coordination with other trades, as noted, may also be required to complete the Selective Demolition process. This is labor-intensive work and requires considerable use of hand tools. It frequently requires scaffolding or lifts and may even require transportation within the building. Shoring or bracing of the structure is common when structural elements are modified. Chutes attached to the exterior of the building to remove debris from upper levels may also be necessary. This type of demolition may also require temporary protection between work and occupied spaces or temporary enclosure as might be the case for window or door replacement projects. It may also include providing control of the air between spaces occupied by the contractor and those being used by the owner. Selective Demolition may also include the cutout of openings in walls and floors that have to be coordinated with other trades for their use, such as duct chases or holes for pipe/conduits.

The cost of handling and loading demolition debris is in addition to the cost of the actual demolition, except for total demolition work. Hauling and disposal are typically added to the costs of demolition and loading for all demolition costs. Other costs such as site fencing for protection, permits, dust control, special insurance for damage to adjacent property, material separation for recycling, or sweeping should also be included. Expendables such as concrete saw blades, bits, or jackhammer points must also be addressed as the task demands. Exterior debris chutes are extra to the demolition costs and are common for evacuating debris from upper floors.

Selective Demolition classifications can include a wide variety of different tasks and are usually labor and equipment intensive. It often requires a plan for the execution of the work and how it will be supervised.

Selective Demolition can be applied to the site as well and includes such tasks as paving or utility removal, the removal or partial removal of site improvements, and removing and disposing of plantings and lawn. Site Demolition is often referred to as *Site Preparation* and may even have a dedicated drawing(s).

The work of Selective Demolition is detailed and often requires coordination with other drawings within the set. It may also require mechanical and electrical disconnections or "make safe" procedures to allow the demolition to proceed safely. These costs may be estimated under the individual trade sections performing the work.

The estimator is advised to make a site visit whenever possible prior to submitting a bid, to collect additional information that may not be discernible from the drawings. For example, if the drawings call for removing and disposing of VCT (vinyl composition tile) flooring, it is important to know whether it is in good condition and therefore difficult to remove, or in poor condition and potentially easier to remove. Other considerations would include the location of panels for electrical connections of equipment or water sources for duct control. One of the most challenging aspects of demolition is quantifying the debris generated for the purpose of determining trucks or dumpsters.

Dumpsters and Disposal

The cost for dumpsters and disposal can be a rather large part of the demolition process and the estimate. The cost for disposal should be a separate item so that estimator can analyze various options for pricing.

Dumpsters are steel containers delivered to the site to be loaded with debris from demolition or general construction work. Once full, they are hauled off-site to a legal disposal facility or recycler and an empty dumpster replaces the full one. The cost of a dumpster is contingent upon its capacity (size) and expected contents. Dumpsters can range in size from 6 CY to 40 CY. The capacity of each size is limited by volume and weight. Trash-hauling contractors charge a fixed fee per size, provided the base weight limit is not exceeded. Surcharges for excess weight are typically assessed by the ton. The fixed fee most often includes disposal costs and rent for the dumpster for a week or month but should be verified with individual contractors.

Hauling and disposal can be done by alternative methods such as trailers and dump trucks, especially in the case of total demolition. Trailer trucks are loaded as the debris is generated and trucked to the disposal site. Charges can be incurred by the hour for the truck and driver or by the load and haul cycle.

More creative methods may include the daily loading and disposal of a smaller dump vehicle as the work progresses. This is fairly common on smaller sites, where a dumpster may hamper access or space is limited.

Removal and Salvage

Another type of demolition included within Division 2 is called *Removal and Salvage*. This work requires the careful disassembly of specific aspects or items on a project for reuse, resale, or reincorporation in the work. Estimating removal and salvage requires that the estimator have a particular plan or method for the removal in mind when performing the estimate. Part of the scope of this section frequently includes the protection and storage of the work until its final disposition. Storage on urban projects with restricted space may require hauling to off-site storage and back to the site later for installation. Removal and salvage tasks are labor intensive and can often take between half and twice the normal installation time of the element.

For larger items that cannot be removed and salvaged by hand, equipment and rigging may be required, as well as work platforms and protection of both the item and the public. The estimator is advised to review the proposed method of removal and salvage with team members that will actually perform or oversee the work before committing to a specific method in the estimate. Removal and salvage is often the work of a specialty subcontractor with experience in the work. The estimator should consider soliciting pricing from multiple sources for all types of demolition.

Last, it is common for contract documents to hold the contractor financially responsible for items that are damaged or destroyed during the removal and salvage process or for any item demolished unintentionally. The estimator is directed to review the documents carefully for such language and investigate possible risk mitigation through insurance or some other method.

Structure Moving

Part of the site preparation may require the relocation of an existing structure. For all but the smallest structures, the relocation of an existing building is labor intensive and requires highly specialized estimating knowledge and experience. Ancillary costs such as surveys of roadways, fees for utility disconnection and reconnection, police details, road permits, public notification of utility outages, and even special insurance can contribute significantly to the cost of the move. The estimator is urged to seek the expertise of specialty contractors for bids on structure moving.

Taking-off and Pricing

Frequently, units of demolition are difficult to label. Demolition tasks can be quantified by the linear foot (LF), square foot (SF), cubic foot (CF), cubic yard (CY), or by the piece (EA). There are some tasks or segments of work that contain more than one operation but are better left as a whole for the purposes of estimating. This work is sometimes classified as a *lump sum* (LS). An example is removing a metal window and frame from an existing concrete block wall, cutting the jambs to the floor, and removing debris. It might be wise to consider a lump sum for this demolition job because of the variety of tasks to be performed.

Remember that Selective Demolition is a labor-intensive process. Calculate the number of labor-hours necessary to perform each segment of the lump-sum task. The hours would be summarized, priced, and added to the cost of any other items required to perform the work, such as special tools or protection. The total cost would then be entered into the estimate as a lump-sum amount. This method of estimating demolition can be extremely useful when there are numerous occurrences of the example cited here.

Sometimes demolition work requires erecting barriers, ramps, chutes, staging, and work platforms, or protecting surrounding work. Protection can be as simple as a polyethylene dust barrier or as complex as building temporary partitions to protect existing features. Be sure to include these costs. Include both the erection and the dismantling of this work as well as any short- or long-term rental fees that may be applicable.

Carefully review the documents for any special considerations about working within an occupied building. Certain portions of the work may have to be performed after normal business hours, thereby requiring shift work. Other costly means of protection may include stringent air quality provisions, monitoring, and periodic tests by certified agencies. The application of negative air machines and filtering systems to prevent the migration of dust or odors to the occupied adjacent areas can be expensive and require supervision and maintenance. Review these requirements carefully, as the cost can be significant.

The important consideration with demolition is to ensure that adequate labor-hours and supporting tools and equipment are included to complete the task.

Dumpster counts can be calculated by the volume of debris to be hauled. Keep in mind that an in-place volume of debris will expand as it is demolished and loaded into dumpsters. Alternate methods for determining dumpster counts include providing a dumpster per a specific square-foot area of the building scheduled for demolition. This is often arrived at based on experience and historical data.

Removal and salvage is calculated by the quantity of labor-hours required to perform a specific task. A general rule is that removal and salvage will take between one-third and twice the normal time to install the same item new. This is basic and should be evaluated for each task.

Almost any classification of demolition can have the unpleasant consequence of disturbing the natural habitat of rodents and pests. This is especially true of whole building demolition. Many projects require that the contractor hire competent professionals in the rodent/pest field to create, monitor, and control a plan to mitigate infestations of other adjacent structures. The estimator is encouraged to read the specifications, including Division 1 and boilerplate language, in addition to Division 2 for just such language. When required, the estimator should seek a price from a specialty contractor for such services, as they can be expensive.

REMEDIATION   Decontamination

One of the principal aspects of *Site Remediation* is the decontamination of the project prior to the start of construction. This includes a wide variety of decontamination processes, ranging from simple surface cleaning to radioactive decontamination. Topics in this subsection range from physical, chemical, thermal, and biological decontamination.

The estimator is directed to seek the guidance of a contractor who performs this work on a regular basis. This type of work is specialized and highly regulated, and it is typically beyond the scope of the general knowledge estimator's expertise. It requires an in-depth understanding of the laws and regulations that govern individual topics as well as specially trained individuals and equipment to perform the work. It can also require special licenses, permits, and insurance beyond those a general contractor can secure.

Removal and Disposal of Contaminated Soils

Many projects require the removal of contaminated soils and the replacement with clean fill (soil) material. The most common example of this work is the removal and replacement of fills beneath buried leaking gasoline storage tanks at gas stations. The actual process of excavation is performed in much the same manner as normal excavation is done in Division 31—Earthwork. However, the handling, removal, storage, and final disposal of the contaminated soils are much different.

For large quantities of excavation and removal, the estimator is directed to seek the guidance of a contractor who performs this work on a regular basis. Careful review of containment and shipping procedures is essential to ensure that all costs associated with the removal and disposal are covered. Disposal fees and long-term insurance can be costly as well and should be researched thoroughly.

For the takeoff scope of the work, the reader is directed to Division 31—Earthwork, discussed later in this text.

Underground Storage Tank Removal

Underground storage tank removal can be classified in two ways: leaking and nonleaking. Prior to proceeding with the removal, tests should be done in the presence of the appropriate authorities to determine if the tank is leaking, and if the surrounding soil is contaminated. The procedure for removing an underground storage tank is as follows:

- Remove covering over tank such as pavement or concrete.
- Excavate to the top of the tank.
- Disconnect and remove all piping and sensors.
- Open access ports and tank vents.
- Pump out all liquids and sediment/sludge.
- Purge the tank interior with an inert gas.

Component	Tasks (hours)				Component Total	Labor-hours/unit
	Disassembly	Processing	Prod. Support	Non-prod.		
Interior						
1. Interior doors, frames, trim	5.75	5.25	---	---	11.0	0.55/each
Baseboards	4.75	5.0			9.75	0.19/lf
2. Kitchen cabinets	2.75	0.5			3.25	0.27/each
Plumbing fixtures	7.75	1.75	---	---	9.5	0.59/each
Radiators	1.5	0.5			2.0	0.13/each
Appliances	0.25	2.75			3.0	0.60/each
3. Bathroom floor tile	2.50	0.50	---	---	3.0	0.038/sf
4. Oak strip flooring	19.25	27.0	0.25	---	46.50	0.038/sf
5. Plaster - upper level	34.25	10.0	5.50	---	49.75	0.012/sf(plaster area)
6. Plaster - lower level	23.75	10.75	2.0	---	36.50	0.009/sf(plaster area)
7. Piping and wiring	6.75	3.25	0.50	---	10.50	0.0072/lbs
8. Interior partition walls	6.25	24.75	3.0	---	34.0	0.18/lf
9. Windows and window trim	10.0	2.50	0.50	---	13.0	0.54 each
10. Ceiling joists	1.0	4.75	0.5	---	6.25	0.0075/lf
11. Interior load-bearing walls	2.75	15.5	1.75	---	20.0	0.027/lf
12. Second level sub-floor	16.0	6.0	1.25	---	23.25	0.023/sf
13. Second level joists	7.25	16.25	1.5	---	25.0	0.027/lf
14. First level sub-floor	7.75	8.0	---	---	15.75	0.016/sf
15. First level joists	7.0	10.0	---	---	17.0	0.020/lf
16. Stairs	2.5	0.75	0.75	---	4.0	0.3/riser
Exterior						
17. Gutters, fascias, rakes	2.25	1.0	---	---	3.25	0.014/lf
18. Chimney	33.25	40.5	4.75	---	78.5	0.16/cu.ft.
19. Gable ends	8.0	3.0	0.75	---	11.75	0.053/sf
20. Masonry walls - upper section	14.75	104.5	20.5	---	139.75	0.25/sf(brick area)
21. Masonry walls - lower section	15.75	84	5.25	---	105.0	0.078/sf(brick area)
Roof						
22. Roofing material	17.75	18.25	1.75	---	37.75	2.68/100 sf
23. Roof Sheathing boards	21.25	14.5	1.5	---	37.25	0.028/sf
24. Roof framing	7.25	9.75	7	---	24.0	0.021/lf
25. Shed roof framing at entry	1.25	2.25	---	---	3.5	0.036/lf
Building Subtotal	291.25	433.5	59	---	783.75	
26. Talk shop	---	---	29	29.5	58.5	NA
27. Supervision	---	---	9.5	---	9.5	
28. Meetings, paper work, daily roll-out and roll-in of tools, etc.	---	---	38	43.5	81.5	
29. Research monitoring	---	---	---	89.5	89.5	
30. Lunch, breaks, idle	---	---	---	118.75	118.75	
Business Subtotal	---	---	76.5	280.25	357.75	
Grand Total	291.25	433.5	135.5	280.25	1141.5	

Figure 7.1 Deconstruction-Labor Summary of Task Performed

Source RSMeans Estimating Handbook, Third Edition. © 2009 John Wiley & Sons, Inc.

- Provide access to the inside of the tank and clean out the interior using trained personnel and personal protective equipment (PPE).
- Excavate soil surrounding the tank using proper personal protective equipment (PPE) for workers.
- Pull (extricate) and properly transport and dispose of the tank.
- Clean up all contaminated materials (if applicable) from the site.
- Test soil beneath and adjacent to tank after removal.
- Install new tank, backfill the excavation, or proceed to contaminated soil remediation, depending on test results. (See Figure 7.1.)

Facility Remediation

Facility remediation generally means "providing remedy" for the toxins, contaminates, or pollutants within a facility or structure. The removal, handling, and, in some cases, long-term storage of the remediated material is part of the Facility Remediation process.

Scopes of work within the Facility Remediation section include contaminates or hazards such as asbestos, lead, PCBs, and mold. The reader should note that earlier in this chapter there was a discussion of the survey and assessment of these same topics. The survey and resulting assessment (report) significantly impact the method and scope of the remediation process. Everything from the size of the containment area, crew, and equipment required to disposal are affected. The report identifies the location, concentration, or scope of the contaminants and the recommendations for the individual remediation process. On the other hand, the Facility Remediation subsection for each of the topics defines in typical CSI format (Part 1, Part 2, and Part 3), the contract requirements for removal and disposal.

It is worth noting that the remediation of hazardous material is a specialized and highly regulated industry and is typically beyond the scope of the general knowledge estimator's expertise. It requires an in-depth understanding of the laws and regulations that govern the handling and disposal of the hazardous material. It is *strongly recommended* that the estimator seek the guidance of a contractor or estimator experienced in the respective remediation topic.

Asbestos Remediation

As noted, asbestos removal or *abatement* is a specialized and highly regulated industry. It should always be performed by a trained and licensed team with the necessary equipment and insurance.

Asbestos can be found in numerous construction renovations, especially if the project was built prior to the 1970s. Asbestos was used in adhesives, pipe, insulating materials, and fire-retardant applications, to name but a few.

A general outline of the steps in the abatement process is as follows:

- Obtain an asbestos abatement plan from an industrial hygienist if one is not provided in the contract documents.
- Monitor the air quality in and around the removal area and the path of travel from the removal and transport area to establish a background level for the contamination.
- Construct an approved two-part decontamination chamber at the entry and exit of the area to be abated.
- Install a HEPA[1] air filtration system to maintain a negative air pressure in the area to be abated.
- Construct wall, floor, and ceiling protection in accordance with the hygienist's plan.
- Have the hygienist inspect the protection to confirm compliance with the plan prior to starting the work.

---

[1] A HEPA filter: high-efficiency particulate air filter.

- Provide temporary supports for pipes/conduits that may be affected during the removal process (if applicable).
- Proceed with the removal and bagging process while monitoring air quality as noted. Cease operations when airborne contaminants exceed safe levels as determined by the air monitoring process.
- Use shot blasting or chemical removal and scraping procedures to lift adhesives.
- Document the legal disposal of all materials existing in the abatement area to ensure compliance with Environmental Protection Agency (EPA) requirements.
- Thoroughly clean all surfaces and crevices in the removal area.
- Have hygienist conduct reinspection to ensure compliance with the approved plan.
- Provide the required documentation from a licensed industrial hygienist certifying that the contaminant levels are within acceptable standards prior to returning the area to normal service.

Lead Paint Remediation

As with all forms of remediation, special training, licensing, and insurance are required to perform lead paint remediation. In addition to federal regulations governing lead paint many states have developed their own regulations as well. There are several acceptable methods to perform this operation, not all of which require the removal of the lead paint. The appropriate method is dependent on the specific circumstance and the intended result. These methods are detailed in the following list:

- **Abrasive blasting:** The sand or metallic particle blasting of the surface containing lead paint. Prior to the start, the work area is contained so that particles do not escape into the atmosphere. The work can be performed under negative pressure similar to the asbestos abatement process.
- **Chemical stripping:** Uses strong chemicals to remove lead-based paint on the surface. Areas under and adjacent to the work must be protected from the chemicals and to catch the chemicals and lead-based paint. After the chemicals have been applied, the work is covered with paper, the paper is removed, and the lead paint is removed along with it. Residual lead may remain, which requires multiple applications. The chemical stripping process usually requires a neutralizing agent and several wash-downs after the paint has been removed. Worker protection includes neoprene or other comparable clothing and respiration protection with a face shield. Inspection from an industrial hygienist is required intermittently throughout the process. (See Figure 7.2.)
- **Power tool cleaning:** This accomplished using shrouded needle blasting guns. The area is blasted with hardened needles, the shroud catches the lead with a HEPA filter vacuum, and it is deposited in a holding tank. An industrial hygienist monitors the project. Workers must wear protective

Hazardous Material Remediation	Crew	Daily Output	Labor-Hours	Unit
Removal Existing lead paint, by chemicals, per application				
Baseboard, to 6" wide	1 Painter	64	.125	LF
To 12" wide		32	.250	"
Balustrades, one side		28	.286	SF
Cabinets simple design		32	.250	
Ornate design		25	.320	
Cornice, simple design		60	.133	
Ornate design		20	.400	
Doors, one side, flush		84	.095	
Two panel		80	.100	
Four panel		45	.178	
For trim, one side, add		64	.125	LF
Fence, picket, one side		30	.267	SF
Grilles, one side, simple design		30	.267	
Ornate design		25	.320	
Pipes, to 4" diameter		90	.089	LF
To 8" diameter		50	.160	
To 12" diameter		36	.222	
To 16" diameter		20	.400	

Figure 7.2 Installation Time in Labor-Hours for Lead Paint Removal by Chemicals

Source RSMeans Estimating Handbook, Third Edition. © 2009 John Wiley & Sons, Inc.

clothing and gear until the air quality is determined to be safe. After the work has been completed, the removed lead is disposed of as hazardous waste.

- **Encapsulation:** A process that leaves the well-bonded paint in place after the loose and flaking paint has been removed. Before the work can begin the adjacent area is protected to catch the loose paint scrapings. The scraped surface is then prepared by washing with a detergent and thoroughly rinsing. The prepared surface is then painted with up to 10-mils of paint. A reinforcing fabric can be added for added protection. The scraped paint is disposed of as hazardous waste. Protective clothing and respirators are required. (See Figure 7.3.)

- **Remove and replace:** Another effective way to remove lead-painted materials such as wood, gypsum drywall, or masonry block. The painted materials are removed, and new materials are installed. Workers are required to wear protective clothing and respirators while in the removal phase. Removed materials must be disposed of as hazardous materials if they fail the TCLP test (Toxicity Characteristics Leaching Process).

- **Enclosure:** The process that permanently seals the lead paint in place. This process has many applications, such as covering the painted drywall with new drywall or covering exterior painted materials with an air barrier (Tyvek™) and residing. The seams on all enclosing materials must be securely sealed. An industrial hygienist monitors the project. Workers must wear protective clothing and gear until the air quality is determined to be safe.

All of these techniques require clearance monitoring and wipe testing by an industrial hygienist.

Description	Crew	Labor-Hours	Unit
LEAD PAINT ENCAPSULATION, water based polymer coating, 14 mil DFT, interior, brushwork, trim, under 6"	1 Painter	.033	LF
6" to 12" wide		.044	
Balustrades		.027	
Pipe to 4" diameter		.016	
To 8" diameter		.021	
To 12" diameter		.032	
To 16" diameter		.047	
Cabinets, ornate design		.040	SF
Simple design		.032	"
Doors, 3' × 7', both sides, incl. frame & trim			
Flush		1.333	EA
French, 10-15 lite		2.667	
Panel		2.000	
Louvered		2.909	
Windows, per interior side, per 15 S.F,			
1 to 6 lite		.571	EA
7 to 10 lite		1.067	
12 lite		1.391	
Radiators		1.000	
Grilles, vents		.029	SF
Walls, roller, drywall or plaster		.008	
With spunbonded reinforcing fabric		.011	
Wood		.010	

Figure 7.3 Installation Time in Labor-Hours for Lead Paint Encapsulation

Source *RSMeans Estimating Handbook, Third Edition.* © 2009 John Wiley & Sons, Inc.

Taking-off and Pricing

The appropriate takeoff quantities for Facility Remediation tasks will vary with the specific methodology for removal. Generally, asbestos is taken off by the square foot (SF) for surface area, by the linear foot (LF) for pipe insulation (based on pipe diameter), and by the individual piece (EA) for fittings or doors containing asbestos.

Lead remediation can include everything from paint removal or encapsulation to the removal and disposal of lead-contaminated soil. EPA legislation has routinely intensified the standards for removal and disposal of lead-based painted products even in residential construction. The estimator is strongly urged to seek the available training and education for the removal, handling, and disposal of the lead containing construction waste, which will provide excellent guidance for estimating the cost of the work.

Polychlorinated biphenyls were widely used as dielectric fluids in transformers and capacitors and as coolants especially in components of early fluorescent light fixtures and electrical transformers. Quantities for takeoff can be designated by the gallon of liquid (GL) or by the individual piece (EA), as in the case of ballasts in light fixtures. Many of the PCBs being abated on construction projects fall under this heading or as flexible additives to exterior applied caulking. Remediation calls for the removal and handling under very strict federal guidelines. Individual types of PCBs are managed differently, thereby requiring special procedures and techniques to be estimated. It is strongly recommended that the estimator seek the guidance of a contractor or estimator experienced in the respective remediation topic.

Specific topics in Division 2 have been intentionally vague or omitted, as they are beyond the scope and project types covered within this text. Again, it is stressed that the general knowledge estimator seek specialized education and training for estimating topics within Division 2.

EXISTING CONDITIONS DRAWINGS

Existing site details are shown on the appropriately named *Existing Conditions* plan. Existing Conditions drawing(s), frequently labeled XC or EX, are a land survey of the conditions of the site prior to the proposed engineering improvements that constitute the project. These plans illustrate:

- Property boundary markers and lot lines
- Existing elevations, both contours and spot grades
- Drainage structures and known existing drainage piping
- Known existing utilities: water, sanitary sewer, electrical, and similar features
- Existing paving, drives, walks, parking lots, and curbing
- Existing site improvements: benches, flagpoles, signs, and similar features
- Existing trees, shrubs, vegetation, and site irrigation
- Location of soil borings or test pits
- Setbacks for wetlands or environmental features

There is a high level of detail on Existing Conditions plans, since the aim of these plans is to provide an accurate accounting of what is currently there and the condition of the site. Estimators use plans showing the existing conditions and features of the site to calculate the amount of work that needs to be performed *in preparation for* the proposed improvements. Each drawing is meant to be complementary and used with the others to help determine accurate quantities for pricing. For example:

- The Existing Conditions site plan must be used in conjunction with the Site Grading drawing to determine the quantity of bulk cut or fill required to achieve the final grade.
- Site Preparation drawings should be used with Existing Conditions drawings to determine the depth of excavations for the removal of existing piping or structures.
- Existing Conditions plans can be used with any listing of Division 2 items scheduled for removal.

SUMMARY

Division 2—Existing Conditions includes information relative to the existing state of the site and the structure as a means of providing background information for the estimator. The background information is provided in the form of surveys or assessment reports. This information is used to more accurately approximate the costs of changing the current state of the site/structure to its proposed improvements defined by the contract documents.

Much of the work of Division 2 is to remediate the substandard conditions of the site as a precursor to the start of the new work. It may have been done by the owner or awarding authority prior to the start of the general construction work. This is done as a means of controlling and containing the risk resulting from these types of work. The results of the remediation work, in the form of a report, is provided for the contractor's use in bidding the project.

Another form of report is the geotechnical report and test borings. This allows the estimator to see a sampling of the contents and soil conditions in specific areas beneath the site. This can be helpful in pricing the excavation based on the type of soil that will be encountered.

Note that with the exception of demolition, much of the work of this division is specialized and highly regulated, requiring an in-depth knowledge and fluency with the rules and regulations governing this work. It is strongly recommended that the general knowledge estimator seek specialized education and training for estimating remediation topics within Division 2 and whenever possible a subcontractor quote for the performance of the work.

QUESTIONS/ PROBLEMS

1. Define the process of selective demolition. Explain the difference between selective demolition and removal and salvage from an estimating viewpoint.
2. Select one of the six processes identified for lead paint removal and explain it in detail.
3. How is removal and salvage calculated by the estimator?
4. How are Existing Conditions drawings helpful to the estimator?
5. Explain in detail why some selective tasks are quantified as a lump sum unit of measure.
6. Opine as to why it is important for an estimator to secure a subcontractor quote for much of the work or Division 2.
7. Explain in detail one of the challenges in calculating the number of dumpsters needed and the weight of the debris within it.
8. Explain the reasons and under what conditions negative air would be required.
9. Select one of the categories of demolition and explain what it includes.
10. What is the purpose of a test boring report? How can it be used by the estimator? What are the disadvantages of a test boring?

8 | Concrete

Concrete is a composite material consisting of sand, coarse aggregate, cement, and water. It forms a stonelike material when mixed and allowed to harden. Concrete has high compressive strength, durability, ability to withstand the weather, and relative ease in handling and shaping. It is a versatile material in widespread use in the construction industry due in part to its fire resistance and low maintenance costs. When used with steel reinforcing, it takes on the ability to withstand elongation, called *tensile strength,* along with its own compressive strength, making it an ideal product for use as a structural component. Because of its versatility, concrete is used in many forms in most construction projects.

To prepare an accurate takeoff for pricing, first review all drawings in the set for concrete work. CSI Division 3—Concrete work is found predominantly in the structural drawings, specifically the foundation plan and sections showing details of footings, walls, piers, slabs-on-grade, and elevated slabs. Site drawings should also be reviewed for items such as concrete walks, pads, and miscellaneous site concrete. Site concrete for exterior pads, walkways, or as paving surfaces is specified in Division 32—Exterior Improvements. To a lesser degree Division 33—Utilities can specify concrete as thrust blocks, encasement, or bedding for utilities for example: the concrete encasement of duct banks.

Other drawings, such as mechanical and electrical, should be studied for items like boxouts for piping and conduits, or "housekeeping" pads for specific pieces of floor-mounted equipment, such as furnaces and hot water heaters. Architectural drawings may include indications of surface finish treatments or architectural concrete with exposed surfaces. The specifications and drawings should be reviewed to determine the strength of the mix, or the particular additives specified, in order to properly

Estimating Building Costs for the Residential and Light Commercial Construction Professional,
Third Edition. Wayne J. Del Pico.
© 2023 John Wiley & Sons, Inc. Published 2023 by John Wiley & Sons, Inc.

separate the different mixes for accurate pricing. The price of concrete is impacted by its strength and additives.

Concrete work can be organized into the following categories for takeoff and estimating:

- Concrete materials
- Formwork for containment of concrete
- Concrete placement method
- Concrete finishing
- Concrete curing
- Concrete reinforcement
- Embedded items
- Precast concrete

CONCRETE MATERIALS

While concrete has many applications in construction, its main function is as a structural component. Concrete has an extremely high ability to resist a crushing force, usually imposed by the weight of the structure it supports. Concrete materials are rated by their compressive strength. Once thoroughly mixed, the ingredients undergo a chemical process called *hydration*, which produces heat as a by-product and results in a cured, rock-hard building material. With water as a key ingredient, mixed concrete is greatly affected by temperature and weather extremes. In cold weather, cast-in-place concrete must be protected against freezing. Concrete that freezes prior to curing will never achieve its full strength. Similarly, placing concrete on hot, dry days may cause evaporation of essential water used for hydration before it has a chance to set. Both conditions must be avoided if the concrete is to perform as designed. Evaluating the requirements for placing and curing concrete is covered later in this chapter.

Taking-off and Pricing

Concrete material in its plastic or wet state will assume the shape of the container it is placed in. It is measured in units of volume and is taken off by the cubic foot (CF) by calculating the volume of the container or formwork that will hold it. It is then converted to cubic yards (CY), the accepted units for both estimating and purchasing. One CY of concrete contains 27 CF. The quantities of different types and strengths of concrete should be listed separately in the takeoff and estimate, because a variety of factors affect pricing. Concrete components should also be segregated by application. For example, concrete for foundations should be separated from concrete for *flatwork*, concrete placed as a slab, pad, or walkway. Study the specifications carefully for type and strength, as well as any additives to the mix.

The following factors affect the price of concrete and should be acknowledged in the takeoff description so that pricing is accurate:

- The strength of the mix specified as the concrete's compressive strength per square inch after curing (e.g., 3,000 psi, 3,500 psi, or 4,000 psi).
- The use of additives that accelerate or retard the curing process, such as calcium chloride.

- The size and type of the coarse aggregate used in the batching, such as gravel, peastone, stone, and/or local aggregates, such as slag.
- The percentage of air incorporated into the mix in the form of tiny air bubbles, known as air entrainment, which adds to workability.
- Special cements, such as High Early Strength Portland Cement, that will achieve the same strength in 72 hours that other types normally achieve in seven days.
- Plasticizer for pump mixes, which increases workability and flow through a concrete pump, frequently used where normal access is not available.
- Hot water or ice chips to control the water temperature (offsetting the effect of high or low ambient temperature on the concrete mix). This is common practice in areas with extreme climatic changes.
- Fibers for reinforcement.

Calculating a waste factor for concrete is often a matter of experience. A reasonable waste factor of approximately 2% can be applied for an average concrete placement. Excessive handling or transporting of concrete after it has left the mixer may require a 5% waste factor, but this is more the exception than the rule. For example, a concrete placement using the direct chute method (from the truck to the form directly) done correctly will have less than 2% waste; however, a placement that requires the use of a concrete pump may have a waste factor of 5% or more due to the concrete lost in the pump hose and hopper. Waste factors for concrete placed as a slab-on-grade can be 5% or more depending on how well the substrate has been graded.

READY-MIXED CONCRETE

Other factors affecting the price may be more closely related to local batching practices and should be reviewed on an individual basis. As a result of significantly better quality control, the use of ready-mixed concrete has all but eliminated the job-site batching of concrete for most light commercial/residential projects. Some projects with the need for large quantities of concrete may set up a site-based batching plant. *Ready-mixed concrete* refers to concrete that is batched at an off-site location, then transported to the site in mixers.

When calculating the unit price of ready-mixed concrete, consider any small quantities, typically under 5 CY, as "short loads." Most ready-mixed concrete companies have a 5-CY minimum charge for short loads unless prior arrangements have been made. Difficulty in placing concrete may also affect the unit price in the form of a charge for "waiting time" accrued by a mixer that is delayed while unloading the concrete. Be sure to separately list the cost of concrete for short loads and placements that take extra time as this will impact cost for both materials and placement.

FORMWORK

Due to its plastic consistency during placement, all cast-in-place concrete must be contained in some type of *formwork*. Formwork varies in size and composition but is most often constructed from a wood facing applied over a steel or wood frame. Simpler forms, such as those used in forming a footing, may be no more than a plank anchored by stakes and straps. The cost of formwork should include the erection and bracing of forms until the concrete has cured and also the stripping and cleaning of the forms. Each of these tasks can be separately priced, or the work can be priced as a single process that includes all of the steps. This tends to be the more common practice. Contractors performing the work themselves may prefer to list the cost items separately. For developing a budget for work that will later be subcontracted, a single, overall cost may be sufficient. Formwork is typically listed in separate categories based on each application. Some of the more common types of formwork and their respective takeoff units are described in the following section. Bear in mind that not all formwork is modular and reusable. Occasionally forms are constructed from new lumber and sheathing and have a one-time use. In this circumstance the entire cost of the lumber and sheathing is included as a material cost of the formwork.

Many estimators solicit subcontractor quotes for the formwork scope of a project. Since the work requires special carpentry skills, an array of forms and bracing, as well as labor skilled in placement of the mix, all but the smallest formwork scope is subcontracted.

Footing Forms

There are two main types of footings: *continuous strip footings,* on which walls will be erected, and *isolated spread footings,* which are used for supporting interior columns or piers. Figure 8.1 illustrates both types.

Strip Footings

Continuous strip footings are designed to support a uniformly distributed load as would be applied at the perimeter of a structure. To evenly distribute the dead and live loads, strip footings are wider than the walls they support. They are typically formed on both sides and braced on the top with temporary wood braces at 2' to 4' intervals. The bottom of the footing is braced with a perforated metal strap at approximately 2' to 4' intervals on the earthen substrate. The forms used for footings are rough planking, similar to staging planks, approximately 2" × 12" in varying lengths. Common sizes for footings are 20" to 36" in width by 12" to 18" in depth, but can vary based on the structural engineer's design and the bearing capacity of the soil. The materials, with the exception of the perforated straps, are reusable. (See Figure 8.2.)

The formwork for continuous strip footings is taken off and priced by the linear foot (LF), because most strip footings remain constant in height. Changes in elevation at the bottom of the footings may be required, based on the design criteria. When this occurs, the footing bottom is *stepped.* Note any changes in elevation that may require stepping the footings. Figure 8.3 illustrates a stepped footing.

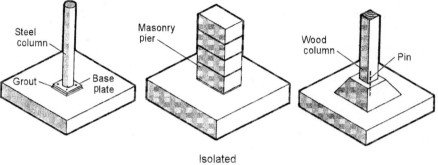

Figure 8.1 Continuous and Isolated Footings

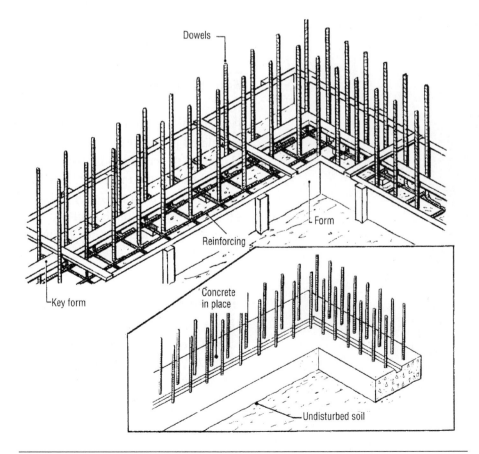

Figure 8.2 Strip Footings

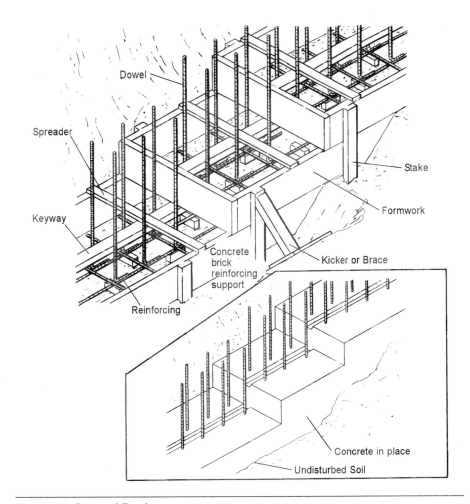

Figure 8.3 Stepped Footings

Productivity is reduced for stepped footings, so an additional cost is typically calculated per occurrence. To reduce the lateral movement of the wall to be placed on the continuous strip footing, a small trough, called a *keyway*, is formed by using a tapered 2" × 4" member embedded in the top surface of the wet concrete in the footing. Takeoff and pricing of the keyway can be separate or part of the completed formwork unit. If priced separately, it is taken off and priced by the LF. Figure 8.4 illustrates the keyway form and the resulting trough after the forms have been stripped.

Spread Footings

Spread footings, also known as *isolated footings,* are isolated masses of concrete, often square or rectangular in shape, with thicknesses varying from 12" to 24". Their main purpose is to support point loads of columns that will rest on them. Their actual size can vary, depending on the load carried and the soil's bearing capacity. A typical spread footing is shown in Figure 8.5.

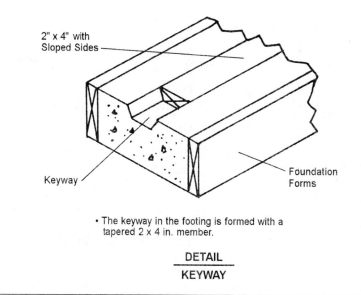

2" x 4" with
Sloped Sides

Keyway

Foundation
Forms

• The keyway in the footing is formed with a
tapered 2 x 4 in. member.

DETAIL

KEYWAY

Figure 8.4 Keyway Form

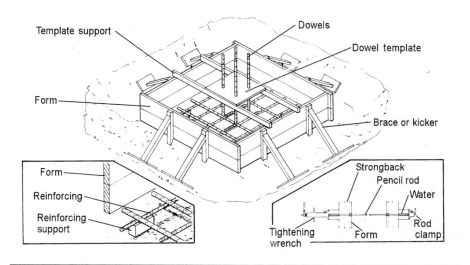

Template support

Dowels

Dowel template

Form

Brace or kicker

Form

Reinforcing

Reinforcing
support

Strongback

Pencil rod

Water

Tightening
wrench

Form

Rod
clamp

Figure 8.5 Spread Footings

Taking-off and Pricing

The form material may be planks or panels, depending on the thickness of the footing. Footing forms are erected and braced in a manner similar to that used for strip footings.

Continuous footings are taken off and priced by the LF of a complete form (two sides). An alternate method is to calculate the square footage of area for which the concrete is in contact with the form. This is referred to as *square foot of contact area* (SFCA) and mainly used in commercial formwork.

Spread footings are typically taken off and priced by the piece; footings of repeated sizes are labeled each (EA). Foundations with variably sized footings are typically listed in a footing schedule shown on the structural drawings. Spread footings may require a template for embedded anchor bolts to attach the column, which can be listed separately from the formwork, as it requires precise layout and, often, the services of a site engineer. (This topic is discussed in more detail at the end of this chapter.)

New footings attached to existing footings may require the drilling and embedment of rebar to continue the transfer of load. This typically has an additional labor to install and material cost for the epoxy.

Foundation Walls and Piers

Concrete foundation walls are cast-in-place below grade to support the structure above. They can be cast in a variety of heights and are supported by footings. In a structure with a basement, foundation walls act to retain or hold back the soil. A *pier* is a short column of plain or reinforced concrete used to support a concentrated load. Piers are used as components in foundation walls or as isolated, separate members.

Formwork for foundation walls is constructed of smooth wood sheathing applied to a 2" × 4" wood or steel frame. Formwork is built in modular sizes starting at approximately 8" in width and increasing to 16" widths in 2" increments. Larger panels for longer straight runs are in 24" and 48" widths. Standard panel heights are 48", 72", and 96". Foundation walls are formed by erecting and fastening modular panels side by side on top of the strip footing. Foundation wall formwork requires "doubling up" panels to create a narrow box to hold the concrete until it has cured. The panels are held apart at a predetermined space using metal ties. This space is ultimately the thickness of the wall. Figure 8.6 illustrates formwork for a typical wall.

Ties are usually spaced at 24" on center, both horizontally and vertically, though they may require closer spacing for greater loads imposed by the wet cast-in-place concrete or as dictated by panel size. In addition, the panels are braced on the exterior (against the hydrostatic pressure caused by the wet concrete) by a series of horizontal wood or metal braces known as *wales* or *walers*. Diagonal braces running from the wales to stakes in the ground at predetermined spacing complete the erection of the formwork.

Taking-off and Pricing

Formwork for foundation walls can be quantified and priced using one of two separate units. The first unit is the area of the form that comes in contact with the concrete. It is listed as SFCA. While this is the most accepted and long-standing method, it does not always represent the pricing structure in some segments of the industry.

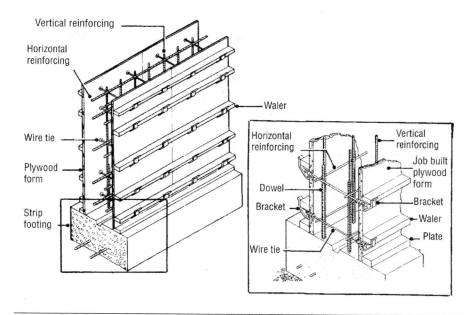

Figure 8.6 Foundation Wall

Another unit that can be used in the takeoff and pricing of foundation walls is the LF, most often used by residential and light-commercial contractors. This method is best where contact area is less critical than the size of the form panels used. For example, pricing formwork for a foundation wall that has concrete to 3'-10" versus 3'-6" in height is irrelevant since both would require a 4'-0" panel. In this application, the cost of 1 LF of formwork actually reflects two completed sides. Once the concrete placement exceeds the 4'-0" height of the panel, there is an added cost because larger panels are needed.

Because the price of forming a wall is affected more by the size (height) of the panel than the amount of concrete that will later be placed in it (and estimated separately), separately list formwork of different sizes (heights)—specifically 4', 6', and 8' heights. Walls to be formed in excess of 8' in height should be listed separately in vertical increments of 2', because of the premium cost for such formwork. Odd-shaped, custom-built, one-of-a-kind, or round forms should also be listed separately for pricing. The pricing of formwork based on SFCA still is used in one-of-a-kind applications, where the materials are used once, and the formwork is custom fabricated versus assembled. It can also be estimated by pricing the materials to be used and the labor-hours to fabricate, form, and strip the assembly as a lump sum cost.

As part of the cost of the formwork, the estimator should include costs for coating the forms with a release agent to break the bond between the wood and the concrete; for the ties that remain in the wall; and for stripping, cleaning, and reloading the forms on trucks after the work has been completed. Releasing agents are liquid and are converted from SF of area to gallons per SF, and, finally, to actual gallons needed, including 10% to 15% waste. Wall ties are by the piece (EA), converted to the common method of purchase – per hundred count.

Study the specifications thoroughly for incidentals such as breaking off ties, patching holes, or "honeycombing." These can be time consuming and costly if not estimated correctly or if missed altogether. Again, the formwork cost per LF proposed by a subcontractor is intended to be a complete product with ties, release agent, breaking ties, filling voids, and concrete placement included with the formwork. Qualify any subcontractor proposals to ensure completeness.

Other costs that may be necessary include the rubbing of "green" concrete. Exposed concrete walls, such as retaining walls may, in some cases, require a rubbed finish to make the work more visually appealing. In this process, the forms are stripped before the concrete has fully cured and then rubbed with a special abrasive float that gives the appearance of a smoother finish. The labor-intensive work is done by hand and may require mixing small quantities of mortar to patch holes or voids in the rubbed surface area. This process is taken off and priced by the square foot of surface area (SF) to be rubbed.

Grade Beams and Elevated Slabs

Grade Beams

These are horizontal beams supported at the ends, as opposed to foundation walls, which are supported by footings on the ground. The structure's load is carried along the grade beam and transmitted through the end supports (piers) to the soil below. Formwork for grade beams is taken off and priced by the SFCA. Grade beams differ from wall formwork in that they sometimes require forming the bottom of the grade beam as well as its sides. (See Figure 8.7.) For grade beams that are cast on grade using the earth as the bottom form follow the takeoff and pricing for walls. If custom-made or one-time-use forms are required for certain applications, list, and price this work separately.

Elevated Slabs

Elevated cast-in-place slabs are often integrated with concrete beams, similar to grade beams. Again, the unit of takeoff and pricing is SFCA. Formed horizontal areas should be listed separately, because they require considerably more bracing to support the weight of the concrete they contain. This process can also require specialized equipment and forms. Figure 8.8 illustrates an elevated cast-in-place slab.

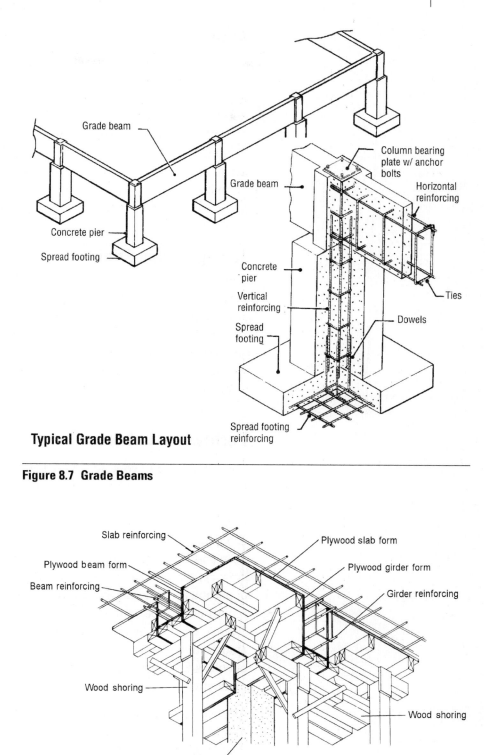

Grade beam

Grade beam

Concrete pier

Spread footing

Column bearing plate w/ anchor bolts

Horizontal reinforcing

Concrete pier

Vertical reinforcing

Spread footing

Ties

Dowels

Spread footing reinforcing

Typical Grade Beam Layout

Figure 8.7 Grade Beams

Slab reinforcing

Plywood beam form

Beam reinforcing

Plywood slab form

Plywood girder form

Girder reinforcing

Wood shoring

Wood shoring

Square Concrete Column

Figure 8.8 Elevated Slab

Edge Forms

The simplest type of form, called the *edge form*, is most commonly used to contain shallow pours of concrete (generally 12" or less) for slab-on-grade, walks, or pads. Edge-form materials are typically rough-grade lumber in the dimension required by the depth of the pour, such as 2" × 4", 2" × 6", and 2" × 8". Other types of edge forms include those more flexible to allow for bending to achieve radii on walks or patios. The process is the same. The actual edge form is held in place by wood or metal stakes, driven into the ground at spacing needed to support the work and prevent bowing. The full process includes the removal of the form once the concrete has cured.

Taking-off and Pricing

Edge forms are taken off and priced by the LF. Quantities of straight edge forms should be listed separately from curved edge forms because of the reduced productivity in installation. Curved forms are often strips cut from sheets of plywood and maybe doubled up for strength. Also note any vertical surfaces that may be used as edge forms. For example, using a foundation wall as an edge when placing a basement slab may eliminate most of the edge form requirement.

Formwork Labor

The labor for formwork, with very few exceptions, is based on the productivity of a multi-person crew. Crews are typically composed of carpenters and laborers. Carpenters erect, brace, and eventually disassemble formwork, while laborers distribute panels, bracing, and other related components of the system. The laborers often place the concrete within the form and then clean the form when it's stripped. The total labor cost for the crew per day is then divided by the production quantity in LF (or SF, if applicable) to arrive at the unit cost. Specific conditions of the individual project need to be addressed, as these can affect productivity dramatically. These conditions include access to the site, extreme climatic conditions, amount of direction changes in the wall, steps in the wall, and boxouts or openings for penetrations. All of these conditions have the potential to reduce productivity and subsequently increase costs.

As always, the best source of formwork labor is the historical data for projects of a similar nature. This is usually calculated as a function of the daily output of completed assembly and disassembly or as a cost per linear foot (LF) of completed wall.

Placement of Concrete

Concrete placement into forms is most often calculated separately from formwork. The placement of wet concrete in erected formwork (as opposed to flatwork) is taken off and priced by the CY. Concrete is ordered, sold, delivered, and handled by the CY. The cost of placement is determined by the number of yards, calculated in the concrete materials takeoff. Placement is most often done by a crew of two or more, depending on the size of the task.

Other considerations when pricing concrete placement include pumps, buckets, cranes, and provisions such as additional chutes. Small equipment, such as a handheld vibrator for concrete, is part of the normal hand tools of the trade. Again, the method of placement affects the cost significantly. Concrete placed directly from a chute off a ready-mix truck yields the lowest placement cost. To direct-chute concrete, the truck must have sufficient access to unload (place), and the formwork must be lower than the lowest point on the chute, as this method relies on gravity. Some specifications require the use of a vibrator to evenly distribute the mix within the form and to fill voids caused by air pockets. This is included as part of the placement cost. For placements that require a pump, costs are typically based on half-day and full-day rates regardless of the quantity being pumped. Mobilization and demobilization costs including cleaning times for equipment may also impact final in-place costs.

FLATWORK

Concrete is frequently used as a flat surface, such as a slab-on-grade, a composite slab, or a walkway. This type of concrete placement is called *flatwork*. The calculation of the concrete ready-mix material follows the same process as all concrete; calculate the volume contained within the limits of the form. The costs for placement of flatwork concrete are different from those for walls and footings. The labor component is a two-step process. The first step consists of the placement of the fresh concrete and the leveling or smoothing of the slab's surface. This is commonly referred to as *placing and screeding*. Placing and screeding of concrete slabs, elevated or at grade is labor intensive and may include a pump to access elevated areas.

Placement of concrete for flatwork can be by the same methods as concrete in formwork. Direct chute is the least expensive due to the minimal amount of handling. Concrete pumps are also common but add substantially to the cost. For any application of concrete where access is limited or height above the ground makes direct-chute or wheeled placement impractical, concrete pumping equipment may be necessary. Concrete pumping involves depositing wet concrete into a truck-mounted pump, which uses a piston-type action to push the mix. The cost of placement per unit by such means is typically high in comparison to direct chute. Another placement method involves the use of a bucket and crane. Concrete is deposited directly from the mixer to a bucket hoisted by a crane to the forms, thereby minimizing the amount of handling. This method is often more economical than the concrete pump, but it does require that the placement location be free from overhead obstructions. Analyze the placement to determine whether there is sufficient quantity to justify the use of a bucket and crane.

A fourth method of placement, mainly for ground-level flatwork, is the use of a power buggy or even a wheelbarrow. This method is less productive than the direct-chute method but may be the most economical for inaccessible areas and small quantities. The use of equipment of any type in the placement of concrete may warrant the separation of CY placement units from SF finishing units so that different options can be evaluated.

It is important to determine whether the quantity of concrete required warrants the use of pumping equipment. Most concrete pumping companies charge by the day or a minimum of one-half day, regardless of the quantity pumped. Concrete pumping is often used for difficult access situations, such as when the formwork takes up the entire excavated area and the access area, making it difficult to place by chute, power buggy, or crane and bucket. The cost of erecting, maintaining, and dismantling riser piping (temporarily attached vertically to the structure) should be listed and estimated separately.

Be sure to coordinate the sequence of concrete placement in such a way as to maximize cost-effectiveness. This means placing concrete only when sufficient forms are available or using concrete pumps or the bucket-and-crane method when there is a sufficient quantity to keep the equipment busy for the entire day. Frequently, concrete slab placement requires materials to separate flatwork from other surfaces. This separation is achieved by the use of an expansion joint.

The second step in the flatwork process is the finishing of the concrete surface. Depending on the desired surface finish this can be labor intensive. Concrete flatwork surfaces are finished using a hand or power trowel, a float, or even a broom. For pedestrian walkways or concrete driveways, a rough or slip resistant surface is typical. Once the surface has been floated and edged, a broom is dragged across the surface to create a rough finish. This is priced by the square area of surface (SF) to be finished. (See Chapter 24, "Exterior Improvements" for a more thorough discussion of pedestrian walkways and exposed aggregate surfaces.)

For interior slabs, not subject to weather conditions or that will receive flooring such as carpet or tile a more refined finish is required. The fresh concrete surface can be finished by hand with magnesium floats and steel trowels if the area is small enough. It may also require a power trowel or multiple power trowels if the placement is large enough. In both cases, the process is labor-intensive and time consuming. Some of the labor cost can be the wait time for the concrete to set and cure sufficiently so that it can support the weight of the crew and equipment. The cost of the finishing is dependent of the level of finish. Complying with flatness and smoothness criteria in the specifications, as in the case of warehouses and big box retailers, the process can require addition time and labor.

Concrete Polishing

Concrete polishing is the mechanical process of grinding or cutting a concrete surface to refine the surface to a desired finish. There are multiple levels of concrete polishing with different finishes:

- *Ground* concrete has a flat appearance with little or no reflection or sheen.
- *Honed* concrete has a matte appearance with slight reflection and varying levels of sheen.
- *Polished* concrete has a highly reflective appearance with a glasslike sheen.

Each of these processes is achieved with the use of an abrasive medium, typically diamond grit on an abrasive disc. The discs are attached to an upright machine that grinds the concrete to the desired level of finish by repeated passes with progressively finer diamond grit. Concrete polishing can be done wet or dry. It can be done as a finish to new concrete after sufficient curing or in renovation work when existing slabs are being repurposed.

Taking-off and Pricing

Concrete placement for slabs, walks, or drives is taken off and priced by the CY required. It is calculated by the surface area multiplied by the depth of the placement. Added costs may be incurred for placements in excess of 6" thick. Be sure to calculate the SF area of the slab to be placed based on areas shown on plan view drawings. These are simple length-by-width calculations of the surface area.

Finishing, or cement finishing as this task is referred to in the trade, requires more than one tradesperson, except when finishing small slabs or pads. The productivity rate is based on the individual production of each cement finisher combined to represent the output of the crew. Overtime costs as a result of waiting time for concrete slabs to "set up" should also be considered. Regardless of the finish requirements the finishing process is estimated by the square foot of surface area to be finished, as SF.

Concrete polishing is taken off and priced by the SF of area to be polished and the desired level of finish. Included within the cost of the work are the expendables such as abrasive discs, preparation, and cleanup. Large open areas are priced separately from smaller areas. Areas adjacent to vertical surfaces, such as walls or columns, may require considerable handwork to achieve the same finish. Quantities of edgework along walls or columns should be taken off separately by the SF or LF.

Figure 8.9 provides guidance on the installation time in labor hours for slab-on-grade flatwork.

Description	Labor-Hours	Unit
Slab-on-Grade		
Fine Grade	.010	SY
Gravel Under Floor Slab 6" Deep Compacted	.005	SF
Polyethylene Vapor Barrier	.216	SQ
Reinforcing WWF 6 × 6 (W1.4/W1.4)	.457	CSF
Place and Vibrate Concrete 4" Thick Direct Chute	.436	CY
Expansion Joint Premolded Bituminous Fiber 1/2" × 6"	.021	LF
Edge Forms in Place to 6" High 4 Uses on Grade	.053	LF
Curing w/Sprayed Membrane Curing Compound	.168	CSF
Finishing Floor		
Monolithic Screed Finish	.009	SF
Steel Trowel Finish	.015	SF

Figure 8.9 Installation Time in Labor Hours for Slab-on-Grade

Source RSMeans Estimating Handbook, Third Edition. © 2009 John Wiley & Sons, Inc.

EXPANSION JOINTS

Concrete, like other construction materials, expands and contracts with temperature changes. To allow for safe expansion and contraction without defects to the work, certain precautions must be taken during construction. A premolded joint filler (which compresses as the concrete expands) allows space for expansion. These asphalt-impregnated fibrous boards come in a variety of widths, the most common of which are 4" and 6". Thicknesses range from ½" to 1". Expansion joints typically occur at the perimeter of a concrete slab, where it terminates at a masonry or concrete wall or any change in substrate. Figure 8.10 shows the use of a premolded joint filler at the perimeter of a slab that meets a foundation wall.

More complex expansion joints that are embedded in the fresh concrete may be specified. These are typically a metal alloy and a multiple component

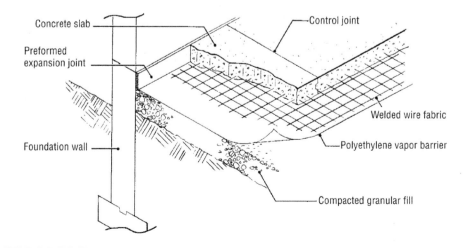

Figure 8.10 Use of Premolded Joint Filler

assembly. Individual products can vary significantly in price based on the application and performance requirements such as weather resistance. The estimator is recommended to solicit pricing for the individual product and any available guidance for installation. They are most often specified under Expansion Control in Division 7—Thermal and Moisture Protection. They are taken-off and priced in the same division.

Taking-off and Pricing

Expansion joints are taken off and priced by the LF. List each quantity separately by size and thickness. A modest allowance of 3% to 5% for waste should be included.

CONTROL JOINTS

It is not uncommon for concrete to crack, even when placed correctly. As noted earlier, concrete expands and contracts with temperature. Containing or controlling the cracking by predetermining where it will occur is the function of the control joint. A control joint is a formed, sawed, or tooled groove in a concrete surface. Its purpose is to create a weakened plane to regulate the location of cracking that results from the dimensional changes in large volumes of placed concrete.

Tooled joints are accomplished by the use of a handheld tool and are typically used for exterior walkways and driveways. The grooves are cut perpendicular to the length of the walk or drive at intervals of approximately 5' during the finishing process. *Saw-cutting* is done after the surface is hardened but before the final strength of the concrete has been achieved. Saw-cutting is done with a diamond blade set to a specific depth, commonly ½" to 1". Saw-cut control joints occur at column lines in large slabs. *Formed control joints* are created by edge-forming areas to be placed at different times. (The slab is edge-formed in a checkerboard fashion, and alternate squares are placed.) Formed joints are typically created at the intersection of a column's base and the surrounding slab. Figure 8.11 shows a control joint at the base of a column.

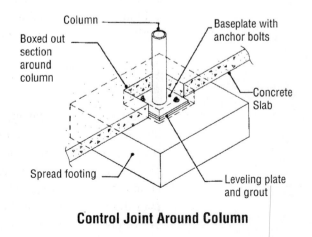

Control Joint Around Column

Figure 8.11 Control Joint at Base of Column

Taking-off and Pricing

Each method of providing control joints – formed, sawed, or tooled – involves taking off and pricing the work by the LF. It is not uncommon to have a combination of all three methods on the same project. The estimator should consider the quantity of sawcut joints as large quantities may require additional blades for the saw. Formed control joints require form material to establish the edge of the joint, as well as labor to form and remove the work after curing.

CURING AND SEALING

Concrete curing involves maintaining the proper moisture and temperature in the environment where concrete has been placed to ensure the proper hydration (chemical reaction) and hardening. Curing can be as simple as spraying water on the concrete surface after it has initially hardened and protecting it from evaporation. It can also be accomplished by spraying chemicals on freshly finished concrete to create a membrane on the slab surface. This membrane prevents the premature evaporation of water in the mix before the concrete has had adequate time to cure properly. Applications are as required based on weather conditions. Repeated applications may be required in dry or hot-weather conditions. The unit of takeoff and pricing for this method is SF and is regarded mainly as a labor cost since most projects have an ample supply of water. The duration of curing may be as long as 24 hours, requiring overtime or shift work.

A similar method calls for covering the moistened surface of the finished slab with a vapor barrier, such as polyethylene, to prevent evaporation of the surface moisture.

During winter, in cold-weather climates, additional protection, such as temporary portable heaters and insulating materials, may be required to ensure proper curing. Sometimes, the heat produced during the chemical reaction of hydration is sufficient to maintain a temperature above freezing and covering the work with straw and polyethylene or insulating blankets is adequate.

Portable kerosene heaters, called *salamanders,* or similar heaters fueled by propane, are also used to maintain temperature. Both require supervision while in use, because of the danger of fire and the continued requirement for adequate ventilation. Constant supervision and the consumption of fuel can be costly. Therefore, portable heaters are employed only when no other means will suffice.

Concrete sealing is the application of a chemical coating over the finished surface of the concrete flatwork. It is used predominantly to prevent the dusting of the concrete surface or the absorption of unwanted materials such as oil into the concrete's surface. Sealing can be performed while the concrete is still curing so that it can be absorbed into the surface or applied over the cured concrete similar to painting. Concrete sealing is traditionally performed by applying a liquid sealer, either by spraying or roller over the surface of the concrete.

Taking-off and Pricing

The area to be sprayed for curing is calculated in SF. This figure is converted to gallons (GL), based on the individual product's coverage per manufacturer recommendations. For example, if the product has a coverage of 400 SF per GL and the area to be covered is 4,000 SF, 10 GL would be required. Quantities should be rounded up to the nearest gallon and may need to be rounded to the nearest common sales unit, such as a 5-gallon container.

Curing by polyethylene/burlap cover is also calculated and priced by the SF. The area is then converted to the required size roll. Sufficient allowances for overlap and waste are necessary for full coverage. A maximum of 10% waste and overlap is usually adequate.

The cost for temporary heat is computed based on the time required, usually by the day, and the fuel consumption of the individual heating apparatus. Equipment rental or costs associated with ownership should also be included. Allow for the setup and dismantling of any necessary enclosures to contain the heat, including both materials and labor-hours. The cost of this work is often based on historical data. This is a difficult cost to accurately predict since it involves predicting future weather conditions and the fuel commodities market (refer to Chapter 6, "General Requirements" for additional information). The experienced estimator uses available historical data to predict this cost and carefully checks the specifications for any language preventing the use of portable heaters.

The area to be sprayed for sealing is calculated in SF and follows the procedure for curing. This figure is converted to gallons, GL, based on the individual product's coverage per manufacturer recommendations and can be extended to the most economical sales unit. Multiple applications may be required. It should also be noted that many of the products may require adequate ventilation to disperse the fumes as well as personal protective equipment, (PPE) for the workers applying the sealants. These costs should also be included within the estimate as part of the cost of the work. Labor costs can be calculated based on the area to be sealed.

REINFORCEMENT

Concrete reinforcement refers to placing steel bars or wire fabric within the formwork prior to placing concrete. Reinforcement in concrete is either structural reinforcement or temperature reinforcement. The concrete and steel reinforcing are designed to act as a single unit when placed together, providing both compression and tensile strength in resisting the forces caused by the weight and mass of the structure. The addition of reinforcing to concrete allows the finished product to act as a flexural member, thereby resisting failure in loading/unloading cycles or in a seismic event. Temperature steel reinforcing is added to slabs to counteract the forces of expansion and contraction due to temperature. There are two basic types of reinforcing covered in this section: welded wire fabric for temperature reinforcing and steel reinforcing bars called *rebar* for structural reinforcing.

Welded Wire Fabric

Welded wire fabric (WWF), also called *welded wire mesh*, is a series of longitudinal and transverse wires of various gauges arranged at right angles to each other and welded at all points of intersection. Welded wire fabric is used in concrete slabs, both to provide reinforcement against thermal expansion and contraction and to reduce cracking. In addition to WWF, reinforcing to prevent shrinkage can also be accomplished by the use of fibers embedded in the mix. This is accounted for in the price per CY of the batched concrete.

Taking-off and Pricing

WWF is taken off and priced by the SF. It is manufactured in sheets and rolls and is sold based on the price per SF. Rolls are 250' × 5' wide, and flat sheets are 5' × 10'. Figure 8.12 is a table of specifications for common styles of WWF and an illustration of its parts.

	New designation	Old designation	Steel area per foot				Approximate weight per 100 SF	
	Cross sectional area (in.) – (100 in.²)	Space wire gauge (in.) – (AS & W)	Longitudinal		Transverse		lbs.	kg
			in.	cm	in.	cm		
Rolls	6 × 6 – W1.4 × W1.4	6 × 6 – 10 × 10	0.028	0.071	0.028	0.071	21	9.53
	6 × 6 – W2.0 × W2.0	6 × 6 – 8 × 8(1)	0.040	0.102	0.040	0.102	29	13.15
	6 × 6 – W2.9 × W2.9	6 × 6 – 6 × 6	0.058	0.147	0.058	0.147	42	19.05
	6 × 6 – W4.0 × W4.0	6 × 6 – 4 × 4	0.080	0.203	0.080	0.203	58	26.31
	4 × 4 – W1.4 × W1.4	4 × 4 – 10 × 10	0.042	0.107	0.042	0.107	31	14.06
	4 × 4 – W2.0 × W2.0	4 × 4 – 8 × 8(1)	0.060	0.152	0.060	0.152	43	19.50
	4 × 4 – W2.9 × W2.9	4 × 4 – 6 × 6	0.087	0.221	0.087	0.221	62	28.12
	4 × 4 – W4.0 × W4.0	4 × 4 – 4 × 4	0.120	0.305	0.120	0.305	85	28.56
Sheets	6 × 6 – W1.4 × W1.4	6 × 6 – 6 × 6	0.058	0.147	0.058	0.147	42	19.05
	6 × 6 – W4.0 × W4.0	6 × 6 – 4 × 4	0.080	0.203	0.080	0.203	58	26.31
	6 × 6 – W5.5 × W5.5	6 × 6 – 2 × 2(2)	0.110	0.279	0.110	0.279	80	36.29
	6 × 6 – W1.4 × W1.4	6 × 6 – 4 × 4	0.120	0.305	0.120	0.305	85	38.56

Notes: (1) Exact W-number size for 8 gauge is W2.1
              (2) W-number size for 2 gauge is W5.4.

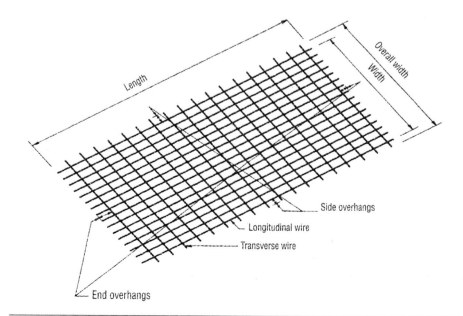

Figure 8.12 Welded Wire Fabric

The estimator should list the different sizes of WWF separately. Although waste is minimal, overlap can be substantial. (See Chapter 4.) Review the specifications or structural drawing details for specified overlap at the sides and ends of the sheet. In the absence of a specified overlap, add a minimum 15% for overlap.

The specifications should also be consulted for specified coatings on the WWF, which can have a dramatic effect on the cost of the material. The most common coating is epoxy, which reduces the deterioration caused by alkalines in the concrete mix or salts from ambient conditions, thereby extending the life of the WWF. Uncoated WWF is the least expensive with stainless steel being the most expensive. The labor costs for all types are reasonable similar.

Reinforcing Bars

Steel reinforcing bars, commonly referred to as *rebar*, are deformed or knurled round bars of high-grade steel used to provide tensile strength. Rebar is available in stock lengths of 20', or it can be cut, formed, or bent into any required shape. Bars are designated by a number that refers to the nominal diameter of the bar in eighths of an inch. Standard bar designation numbers are 3, 4, 5, 6, 7, 8, 9, 10, 11, 14, and 18. Therefore, the diameter of a #3 bar is ³⁄₈". Figure 8.13 is a table of standard rebar weights and measures. Rebar placement in walls or footings is horizontal and/or vertical at specific spacing defined by the structural engineer. Continuation of horizontal bars and the intersection of walls or corners require the bar be lapped to maintain reinforcing. These bars are connected with tie wire.

Bar Designation No.**	Nominal Weight, lb./ft.	U.S. Customary Units			Nominal Weight, kg/m	SI Units		
		Nominal Dimensions*				Nominal Dimensions*		
		Diameter, in.	Cross-Sectional Area, in.²	Perimeter, in.		Diameter, mm	Cross-Sectional Area, cm²	Perimeter, mm
3	0.376	0.375	0.11	1.178	0.560	9.52	0.71	29.9
4	0.668	0.500	0.20	1.571	0.994	12.70	1.29	39.9
5	1.043	0.625	0.31	1.963	1.552	15.88	2.00	49.9
6	1.502	0.750	0.44	2.356	2.235	19.05	2.84	59.8
7	2.044	0.875	0.60	2.749	3.042	22.22	3.87	69.8
8	2.670	1.000	0.79	3.142	3.973	25.40	5.10	79.8
9	3.400	1.128	1.00	3.544	5.059	28.65	6.45	90.0
10	4.303	1.270	1.27	3.990	6.403	32.26	8.19	101.4
11	5.313	1.410	1.56	4.430	7.906	35.81	10.06	112.5
14	7.65	1.693	2.25	5.32	11.384	43.00	14.52	135.1
18	13.60	2.257	4.00	7.09	20.238	57.33	25.81	180.1

Figure 8.13 Reinforcing Steel Weights and Measure

*The nominal dimensions of a deformed bar are equivalent to those of a plain round bar having the same weight per foot as the deformed bar.

**Bar numbers are based on the number of eighths of an inch included in the nominal diameter of the bars.

Similar to temperature reinforcing, the coating on the bar impacts the price. Uncoated or plain bar is the least expensive with stainless steel bars being the higher priced bars. As with most steel, pricing is a function of weight.

Taking-off and Pricing

Horizontal bars are taken off by total length measured in LF multiplied by the number of bars shown. Vertical bars are taken off by dividing the total length of wall or footing by the spacing and multiplying by their height or length, again measured in LF. For accuracy an additional bar is added at the beginning of the count. All rebar is converted from LF to weight. (See Figure 8.13.) Rebar is priced by weight, for both material cost and installation cost. For total weights in excess of 2,000 lb., the quantity is reported in tons (TNS). For quantities less than 2,000 lb., the quantity can be reported in pounds (LBS).

Figure 8.14 lists symbols and abbreviations commonly found on structural drawings referring to reinforcing steel.

List each size separately (by bar designation; #3, #4, #5, etc.) and shape in order to calculate weights for correct pricing of rebar. Coated bars should also be listed separately. When the length of continuous reinforcing exceeds the length of stock bars (20'-0"), overlap becomes necessary to maintain the structural integrity of the member. Most specifications or structural drawings indicate the amount of overlap as a multiple of the bar's diameter:

For example:

#4 bars will be overlapped by 20d, where 20d refers to 20 times the bar diameter (in inches).

In the case of #4 bar: 20 × ½" = 10" so each lap of a number 4 bar will be 10".

To determine the amount of overlap required for continuous horizontal bars, calculate the overlap as a function of the maximum length of the continuous

#	Indicates size of deformed bar number
Ø	Round, used mainly for plain round bars
@	Spacing, center to center, each at
→	Direction in which bars extend, span
↔	Limits of area covered by bars

PL	Plain bar	OF	Outside face	₵	Centerline
Bt	Bent	NF	Near face	∠	Angle
Str	Straight	FF	Far Face	WWF	Welded wire fabric
Stir	Stirrup	EF	Each face		
Sp	Spiral	Bot	Bottom		
Ct	Column tie	EW	Each way		
IF	Inside face	T	Top		

Figure 8.14 Symbols and Abbreviations Used on Reinforced Steel Drawings

To calculate quantities of rebar in slabs, the length and width of the slab are divided by the longitudinal and transverse spacing, respectively. Multiply the quantity in pieces by the length of each piece to determine a total linear footage. For example:

Calculate the LF of #4 bar in a 20' × 25' slab where the bars are spaced at 12" o.c. (on center) each way.
Longitudinal (length): 25'/12" o.c. = 26' pcs × 20' = 520 LF
Transverse (width): 20'/12" o.c. = 21' pcs × 25' = 525 LF
Total of #4 bar = 1,045 LF
Using Figure 8.12, the total linear footage of 1,045 can be extended to pounds (LBS) by multiplying the 1,045 LF × 0.668 LBS per LF, which equals 698 LBS.

---

[1]Note that the final quantity of pieces includes one to start.

bar being used. This could be converted to a percentage added to the total length. For example:

Calculate the total amount of #5 rebar required, including the specified overlap of 20d, if the footing is 1,000 LF with 3 continuous bars.

Start by calculating the amount of overlap required on one bar, assuming the maximum bar length is 20'-0".

$20 × 5/8$" = 12½", or approximately 1' for every 20' length of rebar. This translates to 1 divided by 20, or 0.05. This can be converted to 5%. Therefore, the overlap can be calculated at 5% of the total length of horizontal bar.

1,000 LF × 3 pieces × 1.043 LBS/LF = 3,129 LBS. If this number is now increased by 5%, the overlap can be included: 3,129 × 1.05 = 3,285.45 LBS.

Rebar may have additional costs associated with the production, submittal, and review of rebar shop drawings, required by most commercial projects. Costs for shop drawings may also be included as part of the costs per ton. Also, be sure to include costs associated with the transportation, delivery, unloading, storage, and protection of rebar. Accessories, such as slab bolsters or joist chairs for the support of the reinforcement, should be included in the cost. Specifications should identify any required accessories and their spacing and coatings.

Labor to Set Reinforcement

Labor costs for setting rebar are calculated per ton. Productivity is based on an individual ironworker for smaller-diameter bars (#3 up to and including #6). For bar designations of #7 or higher, or longer lengths of smaller diameter bars, productivity will be reduced due to higher bar weight per LF. A multi-person crew is needed to set the long-length bars. In some instances, cranes are needed for setting larger rebar configurations, such as spiral reinforcement used in round columns or mats in footings and pile caps. Equipment costs are typically included in the per-ton setting price. Rough estimates for setting rebar ranges 8 to 11 labor-hours per ton depending on the size of the bar and application. The estimator would be advised to add time to the labor costs for "shake out of the steel," which is loosely translated to laying the rebar out to determine where each type and quantity is before installation begins.

PRECAST CONCRETE

Precast concrete includes reinforced structural concrete components formed, placed, finished, and cured in a location other than their final position in the structure. Precast components are most often fabricated off-site and must be transported to the site. Items such as bulkhead enclosures, precast steps, wheel stops, lintels, wall panels, and concrete planking (also known as hollow-core planking) are common examples of precast concrete members. Occasionally, concrete structural elements are cast into molds and cured at the jobsite and lifted or tilted into place by crane. Job precast units are not restricted to size limitations of those that require trucking to the site.

Taking-off and Pricing

Precast items are typically shown and identified on structural drawings in plan and sectional views, but they can also be noted on site drawings, in the form of items such as precast steps or wheel stops. The units for estimating precast concrete items vary. Quantities should be listed according to size, length, or any other characteristics that would allow accurate pricing. Most are priced by the individual piece, or each (EA). Other products, such as wall panels and concrete planking, are priced by the SF of area. Lintels are priced by the LF, and precast steps can be priced by the riser (RSR). As most precast items require equipment for handling and installation, include sufficient equipment and labor-hours based on standard production rates for the individual items. On-site cutting of precast plank or caulking of joints between the planks should be included as part of the installation cost. Both are quantified and priced by the LF. Other costs associated with the installation of precast wall panels, such as welding or bolting, are taken off and priced by the piece (EA). It should be noted that precast concrete structures for use in sewer and drainage work and precast concrete curbs are specified under Division 32— Exterior Improvements and/or Division 33—Utilities.

CEMENTITIOUS DECKS

Cementitious decks are lightweight, noncombustible panels predominantly used in roof construction with steel framing. In place of concrete, a gypsum core is used, as it offers a lightweight, fire-resistant construction. The panels can be combined with insulation to provide an energy efficient assembly.

Taking-off and Pricing

Cementitious decks are taken off and priced by the square area covered by the material. Cementitious decks are taken off by either the individual piece (EA) or by the square foot (SF). Note that cementitious deck materials are available in varying thicknesses, with varying prices. Hoisting the panels to the roof is usually accomplished with a crane, but the precise placement and attachment are done manually. Installation costs should include the welding of clips or metal edging to the steel support framing, and these costs are typically calculated per labor-hour of a specific crew. Costs for welding are also based on labor-hours per individual welder. All costs can be combined to represent a cost for the completed assembly by the SF.

GYPSUM CEMENT UNDERLAYMENT

An underlayment is a material placed on the floor substrate, beneath the finish floor materials. It is used to stabilize the substrate and provide a level, or near level surface for new flooring. The most common type of self-leveling underlayment is *gypsum cement underlayment*. It is lightweight and cost less than cement-based underlayments. In addition, it has sound deadening capabilities and can be used as part of a fire-rated assembly. The disadvantage is it are not as strong as the cement-based version and is subject to damage caused by moisture. These products are batched with gypsum cement, fine aggregate, binders, and water. The product is pumped into place similar to pumping concrete but with a smaller diameter hose. It is considered self-leveling as it has a more liquid consistency but is often floated by labor with hand tools. After the underlayment cures it can be sealed depending on the manufacturer's recommendations.

Taking-off and Pricing

Gypsum cement underlayment is taken off and priced by the square area, in SF, covered by the material and its depth of thickness. The material component can be converted to cubic measure, either in cubic yards (CY) or cubic feet (CF). For large areas it can be labor intensive and requires equipment to place the mix for any size project. Typically, the labor component of the price is calculated by crew size, based on the required output per day. This can be translated to cost per SF. The crew most often consists of cement finishers and laborers, with at least one laborer dedicated to filling, mixing, and maintaining the pump full of product. The cost of the equipment is priced per day for each application. Even for the smallest quantity of underlayment there can be a setup, breakdown, and cleaning cost for the pump.

EMBEDDED ITEMS

Embedded items are encased or cast within the formwork for connecting future work or supporting reinforcing within the formwork. The most common embedded items are described as follows. Accessories for rebar and wire fabric can also be required but are typically taken off and priced with the reinforcing.

Sleeves

Sleeves are block-outs within the formwork that hold back concrete, so that when the formwork is stripped, there is a hole in the concrete wall or slab. Sleeves are specifically located to facilitate the passage of piping through the wall after the concrete has cured. When the sleeve is removed, it should allow clear passage of the piping. Sleeves can be made of PVC pipe, wood, or rigid insulation. A sleeve is fitted between formwork panels and fastened by nails or other means. Sleeves are taken off and priced by the individual piece and listed according to size. Included in the price should be the cost of patching around the piping with mortar or concrete after the pipe or conduit has been inserted through the sleeve.

Sleeves are not always shown on the drawings, but they may be required by the specifications where utility piping passes through the foundation wall.

Coordinate with the mechanical and electrical drawings for the locations, size, and quantity of sleeves. The specifications typically require that the formwork contractor provide and install sleeves, if needed.

Anchor Bolts

An *anchor bolt* is a threaded rod or bolt with a right-angle bend that is embedded in cast-in-place concrete. Anchor bolts are used to connect steel columns, beams, or wood sill plates to the concrete structure. The size, quantity, spacing, and location of anchor bolts are noted on the structural drawings. Anchor bolts are taken off and priced by the piece and are priced as each, (EA). A template (usually made from wood) may be needed to hold the bolts in position during the curing. The template is stripped along with the formwork. Figure 8.15 illustrates a template supporting anchor bolts on an isolated footing.

Other Embedded Items

Steel window frames for basement window sashes and doors in residential construction are installed in a manner similar to sleeves. They are taken off and priced by the piece and separated according to size. After the forms are stripped, the steel frames remain, and the wood or steel sash is installed within the frame. The locations are noted in both plan and elevation views on the architectural drawings.

Another item commonly embedded in concrete are waterstops. Waterstops are rubber or vinyl inserts placed in concrete joints to prevent water from penetrating the joint.

Taking-off and Pricing

Quantities of embedded items are taken off and priced by the individual piece (EA). Labor to install all embedded items including anchor bolts is calculated by the individual piece (EA) or by the labor-hour for tasks that include a multi-person crew. The estimator should consider additional costs for layout time and even the fabrication of templates for the setting of anchor bolts for columns, if the application requires it. There are numerous configurations of waterstops, but all are taken-off and priced by the LF.

VAPOR BARRIERS AND RIGID INSULATION

Vapor barriers are frequently specified between the compacted subbase and the cast-in-place slab-on-grade to prevent the transmission of moisture though the slab. Also used under the slab or foundation level is rigid insulation. Rigid insulation is adhered to the interior side of a vertical foundation wall or placed on grade under the concrete slab. The insulation is used to prevent the loss of heat through the structure into the ground. Although they are formally part of

Description	Labor-Hours	Unit
Formwork	.105	SFCA
Reinforcing		
#4 to #7	15.239	ton
#8 to #14	8.889	ton
Placing Concrete under 1 C.Y.		
Direct-Chute	.873	CY
Pumped	1.280	CY
Crane and Bucket	1.422	CY
Over 5 C.Y.		
Direct-Chute	.436	CY
Pumped	.610	CY
Crane and Bucket	.640	CY
Anchor bolt or dowel templates	1.000	EA

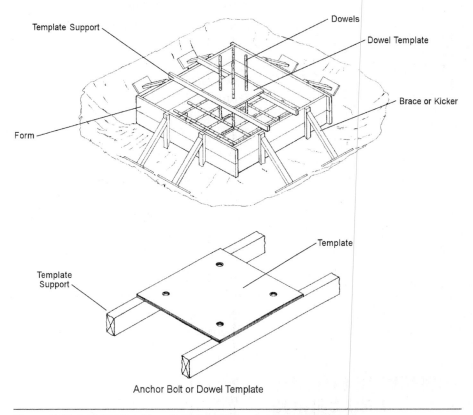

Figure 8.15 Installation Time in Labor-Hours for Spread Footings

Division 7—Thermal and Moisture Protection, vapor barriers and the rigid insulation found under the slab are sometimes included in Division 3—Concrete.

The thickness of the polyethylene vapor barrier is noted on the drawings in mils (with 1 mil equal to 0.001 inch). Polyethylene is commonly specified in 4- or 6-mil thicknesses. The vapor barrier is installed by spreading the polyethylene on the subbase just prior to placing the concrete. More sophisticated vapor barriers can be as thick as 15-mils and sold in sheets. The manufacturer's recommendations can require that the seams be taped with compatible tape as well as taping the vapor barrier at the penetrations. This clearly adds to the labor component for the installation.

Rigid insulation is designated by *R-value*, or its thermal resistivity, which can be translated to inches of thickness of the insulation. Architectural plans, specifically foundation wall or slab sections, show the insulation and possibly the thickness. The specification Section 07 20 00—Thermal Insulation should be consulted, if available, for the specified type of insulation and R-value. Labor may also be required to caulk any penetrations through the insulation with sealant.

Taking-off and Pricing

Vapor barrier is taken off and priced by the SF and listed separately by thickness, defined in mils and type. A reasonable allowance of 7% to 10% is usually sufficient for the overlapping of seams and waste. The estimator is also directed to review the specification Section 07 27 00—Vapor Barriers to determine if the seams are required to be taped, as this can add considerable time to the installation. The estimator should allow additional time for setting and finishing concrete slabs with vapor barriers, as water is retained longer as a result of the inability of the water to pass directly to the subbase. Often there is a 2" to 3" sand bed between the vapor barrier and the slab.

Rigid insulation is taken off and priced by the SF. Different types and thicknesses (R-values) should be kept separate in the estimate. Insulation area can be converted to typical sales units of 2' × 8' or 4' × 8' sheets and rounded to the full sheet, adding 3% to 4% for waste, as damage during shipping and handling is common.

As noted previously, vapor barriers and rigid insulation are specified in Division 7—Thermal and Moisture Protection but can often be taken off and priced in Division 3—Concrete. It is prudent for the estimator to clearly identify where the work is in the estimate to avoid duplication.

CONCRETE CUTTING AND BORING

Saw cutting and boring of concrete is specified in Section 03 80 00 and can be required in both new and renovation work. It requires the use of equipment; electric and gas powered to cut or core through a concrete wall or slab. The process is different than that of cutting expansion joints as it requires the cut to be the full depth of the slab or wall. The work requires the use of equipment and an operator to run the equipment as well as water for cooling the blade and dust control. There is also a layout component which may or may not be part of the work. Saw blades and coring bits are considered expendables and for large quantities of each type of work, the estimator should include additional based on the manufacturer's recommended life expectancy.

These tasks often include the removal of the resulting pieces, called *spoils* to an on-site dumpster. Depending on the size and quantity of the spoils it may require a small piece of equipment like a skid steer loader or it can be hand loaded into wheelbarrow. When cutting trenches in concrete slabs, the estimator should consider a layout of the work to calculate the quantity. In addition to the sides of the trench, the lengths will need to be cut at intervals of 3' to 5' for ease of handling. This is appropriately called dicing. The same process applies to vertical sections of concrete walls once the main cut has been made. The aforementioned processes are similar for concrete masonry block and brick veneer applications.

Labor to clean up the slurry generated by the blade or bit and the water is often part of the work. This can require a wet vacuum or in larger applications shovels and wheelbarrows. This work can be time consuming and labor intensive.

Taking-off and Pricing

Saw cutting of concrete walls and slabs are taken and priced by the linear foot of (LF) cut, per inch of thickness of the material with 4" thickness being a minimum. Reinforced and nonreinforced concrete should be separated as they have different costs. The same is required for vertical (walls) and horizontal (slabs) as the saw cutting as the vertical work will require that piece of equipment be bolted to the wall for stability. Coring holes through walls and slabs are based on the vertical (walls) and horizontal (slabs) application and whether the work is reinforced or plain concrete. Coring is priced based on the diameter of the core, with small diameters holes being less expensive. The unit of measure for the individual core is EA.

It should be noted that while CSI classifies this work as Division 3— Concrete, it can frequently be part of a demolition scope in Division 2— Existing Conditions. Again, the estimator is advised to clearly identify where the work is in the estimate to avoid duplication.

SUMMARY   Accurately estimating the cost of concrete work is critical to a successful overall project estimate. Concrete work is not limited to foundation and slab work alone, but an integral part of the entire project. There are numerous other applications in which concrete is used separately or as part of another trades work. In any application, a thorough understanding of concrete installation practices, as well as the jobsite conditions, are essential to a reliable estimate. Special considerations for seasonal costs in temperate climates and cold/hot weather practices should also be considered based on the project schedule.

Concrete material cost will vary by mix with additives and the compressive strength of the mix being the main determining factors. Most concrete requires some type of formwork as containment for the concrete in its plastic state. Formwork can be custom built or may have multiple uses. Placing concrete whether in forms or as a slab requires labor and possibly equipment. The estimator must consider means and methods for placement.

To improve its tensile strength concrete is often used with rebar. This allows the composite to act as a flexural member and broadens its use as a structural component. Other steel reinforcement in the form of a mesh that reduces damage due to temperature changes may also be required.

Other forms of concrete are cast off site and are delivered for use on the project. Precast items used on a project are numerous and diverse. The estimator should study multiple plans; structure and site to ensure all are accounted for.

Cementitious products can be used part of a lightweight roof system or as an underlayment for residential and light commercial floor systems.

Many cast-in-place concrete tasks require that items be embedded with the concrete. Most embedded items are estimated by the piece or by length. Vapor barriers and insulation while part of Division 7 can often be estimated in Division 3. The estimator should note which division the work is in to avoid duplication.

QUESTIONS/ PROBLEMS

1. A building has a perimeter of 300' and a footing size of 32" × 18". Assuming the bottom of the footing is at the same elevation along the perimeter, calculate the quantity of cubic yards required to fill the form. Round your answer up to the nearest whole yard.

2. If a crew of two laborers will place 50 CY per day, how many labor hours should be allotted to place the concrete from the answer in the previous question?

3. On top of the footing in question #1, there is a 4'-0" foundation wall form that will be filled with concrete to 3'-6". Calculate the amount of concrete required if the wall is 10" thick. Round your answer up to the nearest whole yard.

4. Calculate the placement cost per CY of a 20,000 SF, 5" thick slab-on-grade if the crew cost $3,650 per day and the crew will place 165 CY per day. Assume 5% waste and round to the whole day.

5. How many days would a pump be needed to place the concrete in problem #4? Round your answer up to the nearest whole day.

6. A footing for a retaining wall is 95' long × 7' wide and has 6 pieces of continuous #5 rebar in the footing and a lap of 20d. Using *Figure 8.13 Reinforcing Steels Weights and Measure* calculate the weight of the rebar in tons.

7. If an ironworker will place 157 lbs. per hr. how many labor hours should be estimated to set the rebar in problem #6?

8. A 12,000 SF slab requires WWF. The sheets are 5' × 10', and the specifications require a side and end lap of 6". How many sheets will be required and what is the percentage lost to lap?

9. A subfloor needs to be leveled with gypsum cement underlayment. The subfloor is 25' wide and 100' long. From side to side (across the 25' dimension) the floor has a ¾" difference in elevation. If the specifications call for a minimum thickness of ½", what is the total quantity in cubic feet (CF) required to level the floor while maintaining the minimum thickness?

10. If the underlayment in problem #9 is estimated at $85.00 per CF in place what would the price be to level the floor? Assume a 5% waste factor.

9 | Masonry

The use of masonry in construction is appealing for a number of reasons. It is fireproof and durable, it requires little or no maintenance, and it can be configured to satisfy a wide variety of structural requirements. Masonry units are available in an enormous selection of colors, shapes, textures, and sizes that can be installed for an aesthetically pleasing appearance. CSI Division 4—Masonry, includes brick, block (CMU), glazed block, glass block, natural products such as fieldstone and cut stone, as well as the labor and equipment required to install these materials.

Brick and block are the two most common types of masonry units, and the process of installing them is referred to as *unit masonry*. Brick is modular and adaptable, and it can be installed in a variety of patterns, called *bonds*. Concrete masonry units, commonly referred to as CMU, consist of concrete blocks. CMU is primarily used for walls and partitions that will support a structural load or as the backup for a face brick veneer. It is used extensively in commercial and industrial building applications and as a foundation material in residential construction. All masonry work is installed with mortar, the "adhesive" that holds the units together. Mortar is spread between the joints of individual masonry units and allowed to harden, which bonds the units together.

In reviewing the contract documents, there are several factors to consider that affect the cost of masonry work:

- What type of unit masonry will be used? Are there multiple types?
- What is the bonding pattern of the masonry and the size of the mortar joint?
- Is the masonry reinforced vertically and laterally?
- Will scaffolding be required?

Estimating Building Costs for the Residential and Light Commercial Construction Professional,
Third Edition. Wayne J. Del Pico.
© 2023 John Wiley & Sons, Inc. Published 2023 by John Wiley & Sons, Inc.

- What are the seasonal conditions that will affect productivity when the work is ongoing?
- Will enclosures or temporary heat be required?
- Are there incidentals that need to be considered, such as precast lintels or sills, through-wall flashings, steel lintels, embedded items, chimney components, and so on?
- Will other trades that have items embedded in the masonry work need coordination?

MORTAR

Mortar is a composition of water, fine aggregates (such as sand), cement (Portland, hydraulic, or masonry), and lime. Mortar requirements are specified in the products section of the masonry specifications. The mortar, or its components, is often listed as complying with American Society for Testing and Materials (ASTM), one of the primary agencies that sets standards for masonry products and procedures. Mortar is placed between the joints, and butt ends of the masonry units. It seals the spaces between the units and compensates for the variations between the units. Mortar is spread by hand by bricklayers or masons using trowels. It is placed while plastic and hardens or cures and bonds the individual units together. There are different strengths of mortar based of the application.

Types of Mortar

The compressive strength of mortar varies with the proportions of the ingredients and the amount of cement in the mix. The air temperature and relative humidity during the mortar's curing period can affect its bond strength. The four basic types of mortar and their uses are based on mixing proportions and strengths:

- **Type M:** A high-strength mortar used primarily in foundation masonry, retaining walls, walkways, sewers, and manholes. In general, Type M mortar is used when maximum compressive strength is required.
- **Type S:** A relatively high strength mortar that develops maximum bonding strength between masonry units. It is recommended for use where lateral and flexural strength are required.
- **Type N:** Medium-strength, general-use mortar for above-grade exposed applications.
- **Type O:** A low-strength mortar for interior non–load-bearing applications.

Figure 9.1 lists the mixing proportions and strengths of the four most commonly used types of mortar.

Taking-off and Pricing

Since the mortar quantity required is directly related to the number of bricks or blocks, first calculate the quantity of masonry units in order to determine the quantity of mortar needed. The size of the joints must also

Calculate the quantity of mortar required for laying up 50,000 standard bricks with a ½" joint thickness, where M is a unit of measure for a thousand bricks:

*50,000 bricks/1,000 bricks = 50 M-bricks ***

50 M-bricks × 11.7 CF per M-bricks = 585 CF of mortar

If 585 CF is increased by 15% for waste:

585 CF × 1.15 = 672.75 CF, rounded to 673 CF

To convert to CY, divide by 27:

673 CF / 27 CF per CY = 24.93, or 25 CY

Brick Mortar Type and Mix Proportions by Parts

Type	Portland Cement	Hydrated Lime	Sand	Water	Strength	Compressive Strength at 28-days	Application
M	1	0.25	3.75	*	High	2500 psi	Used where high strength is required. High compressive strength. Used in above or below grade applications.
N	1	0.5	4.5	*	High	1800 psi	Used in general applications where high lateral strength is required
S	1	1	6	*	Medium	750 psi	Used for masonry above or below grade where high compressive and lateral strengths are not required.
O	1	2	9	*	Low	350 psi	Used in interior applications of non-load bearing capacity. Not acceptable for masonry exposed to severe weather or below grade.

* Potable water of sufficient volume to make a paste-like consistency.

Figure 9.1 Brick Mortar Mix Proportions and Compressive Strengths

be specified or determined in the absence of a specified size. Mortar quantities are typically in cubic feet (CF) or can be converted to cubic yards (CY; 27 CF = 1 CY). Waste factors should be added at the quantity take-off stage and can range from 15% to 25% depending on the type of brick or block. Quantities are frequently determined from established tables, such as in Figure 9.2.

The quantity of mortar for setting stone varies with the size of the stone and the joint. However, a general rule is that 4 to 5 CF of mortar is required per 100 CF of stone. Quantities for various strengths of mortar should be listed separately for accurate pricing. Higher-strength mortar is more expensive due to the cement content as well as the type of cement. Mortar that requires special coloring should also be listed separately, as the pigment required to color mortar and the labor to batch will affect the cost. Review the specifications carefully for additives that will affect the price of the mortar or its longevity after it has been batched. Labor costs may vary for preparing mortar in the proper strength and color, as each mixture requires some experimentation. There are many additives that can be specified in the mix. Mortar is mixed on-site by laborers and a mixing machine. Only in the rarest of circumstances is the mortar mixed by hand, such as the refractory mortar for fireplace brick. Hand mixing is reserved for small repairs when machine setup is not practical. The labor cost of batching the mortar is frequently included as part of the labor costs for setting the brick or CMU or it can be part of the cost of the mortar itself. Typically, mason sand, bags of

Unit Masonry Type	Actual Unit Size (W x H x L)	Joint Thickness (inches)	Mortar Requirement (in CF) with 15% waste per 1000 units
Standard Brick	3-5/8" x 2-1/4" x 8"	3/8"	10.1
Standard Brick	3-1/2" x 2-1/4" x 7-1/2"	1/2"	12.5
Modular Brick	3-5/8" x 2-1/4" x 7-5/8"	3/8"	9.4
Modular Brick	3-1/2" x 2-1/4" x 7-1/2"	1/2"	10.1
Roman Brick	3-5/8" x 1-5/8" x 11-5/8"	3/8"	11.2
Roman Brick	3-1/2" x 2-1/4" x 11-1/2"	1/2"	15.0
Norman Brick	3-5/8" x 2-1/4" x 11-5/8"	3/8"	11.7
Norman Brick	3-1/2" x 2-1/4" x 11-1/2"	1/2"	15.7
CMU 12" x 8" x 16" 3-core	12" x 7-5/8" x 15-5/8"	3/8"	104.2
CMU 12" x 8" x 16" 2-core	12" x 7-5/8" x 15-5/8"	3/8"	89.7
CMU 8" x 8" x 16" 3-core	8" x 7-5/8" x 15-5/8"	3/8"	81.0
CMU 8" x 8" x 16" 2-core	8" x 7-5/8" x 15-5/8"	3/8"	72.4
CMU 6" x 8" x 16" 3-core	6" x 7-5/8" x 15-5/8"	3/8"	69.5
CMU 6" x 8" x 16" 2-core	6" x 7-5/8" x 15-5/8"	3/8"	63.7
CMU 4" x 8" x 16" 3-core	4" x 7-5/8" x 15-5/8"	3/8"	57.9
CMU 4" x 8" x 16" solid	4" x 7-5/8" x 15-5/8"	3/8"	88.9

Figure 9.2 Mortar Quantities for Brick and Concrete Block Masonry

cement, lime, color, and more are shipped in bulk to the job and mixed by laborers in quantities that can be used before it sets.

BRICK

Brick is a solid masonry unit made of clay or shale and formed into a rectangular prism while soft, then burned or fired in an oven, called a *kiln*, until hard. There is a wide and ever-changing variety of bricks available to the industry. The estimator is advised to obtain a material price delivered to the job site for the brick. This is especially true for large quantities or nonlocal brick. Due to the weight, shipping costs can be significant. For estimating purposes, most brick can be classified into one of the following groups.

Classifications of Brick

- **Face brick:** Used where appearance is important (e.g., veneer walls). Its manufacture is closely controlled so that color, size, hardness, strength, and texture are uniform.
- **Common brick:** Used in applications where performance is more important than appearance (e.g., below-grade masonry, as a backup for face brick, or in manholes).
- **Glazed brick:** Fired with a ceramic or other type of glazing material on the exposed surfaces, glazed brick is typically used in applications where durability and cleanliness are essential (e.g., restrooms, kitchens, and hospitals).

- **Firebrick:** Used in areas of high temperatures, such as furnaces and fireplaces.
- **Brick pavers:** Used as a wearing surface for floors, walks, and patios. Pavers are typically hard and durable, with a high resistance to damage from both freeze-thaw cycles and the corrosive salts used to melt snow.
- **SCR brick:** A patented type of brick developed by the Structural Clay Products Institute (SCR). SCR brick has a nominal width of 6" in contrast to the nominal 4" associated with modular brick. It can therefore be used structurally in a single width. It is most commonly used in the residential market for single-story applications.

To thoroughly understand the plans and specifications for masonry work and create an accurate takeoff, it is important to become familiar with brick nomenclature. (See Figure 9.3.)

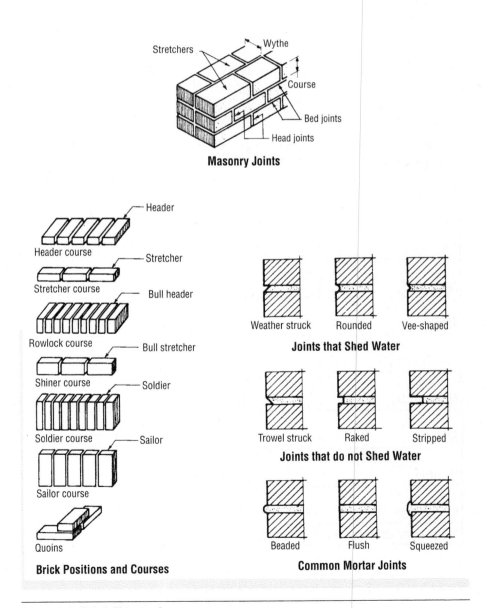

Figure 9.3 Brick Nomenclature

Most bricks are modular in design, and either the width, length, or coursing height is a multiple of 4". Brick sizes are designated as nominal, versus actual size. The difference between the actual and the nominal is made up by the thickness of the mortar joint. Figure 9.4 provides the bricks per square foot and waste factors for most brick types and their respective bonds used today.

Since brickwork is modular, the estimator must determine the width of brick for each application. Face bricks in veneer walls are one brick width, while other construction may be two, three, or four bricks in width. While the popularity of this type of multi-width masonry construction has declined in recent years, it still does occur in unique circumstances, especially renovation or historic preservation work. To accurately determine the quantity of brick in the takeoff, list the work separately by width of wall.

Brick masonry units are taken off by the square foot (SF) of wall in the case of single-width walls such as veneers. Multiple-width walls are taken off by the cubic foot (CF). In either case, the final conversion for pricing is to the quantity of brick. Determine the number of bricks per SF or CF based on the size of the brick, the size of the mortar joint, and the brick bond. The *brick bond* is the pattern of overlapping of one brick on another, either along the

Brick Type	Nominal Dimensions (inches)			Actual Dimensions (inches)			Coursing*
	W	H	L	W	H	L	(inches)
Standard Face	4	2-2/3	8	3-1/4	2-1/4	7-1/2	3C = 8
Engineer	4	3-1/5	8	3-1/2	2-11/16	7-1/3	5C = 16
Norman	4	2-2/3	12	3-1/2	2-1/4	11-1/2	3C = 8
Roman	4	2	12	3-1/2	1-1/2	11-1/2	2C = 4
Jumbo Utility	4	4	12	3-1/2	3-1/2	11-1/2	1C = 4
Norwegian	4	3-1/5	12	3-1/2	2-11/16	11-1/2	5C = 16
SCR	6	2-2/3	12	5-1/2	2-1/4	11-1/2	3C = 8
Double	4	5-1/3	8	3-1/2	4-13/16	7-1/2	3C = 16
8" Jumbo	8	4	12	7-1/2	3-1/2	11-1/2	1C = 4
* Number of courses and height in inches based on 1/2" mortar joint							

Figure 9.4 Brick Requirements and Waste Factors

length of the wall or through its thickness. Bricks are shifted so that the vertical mortar joints of successive courses do not line up, except in the case of stack bond. There are two basic arrangements of bricks in a bond. The frequency of these arrangements determines the bond. The first is the *stretcher,* where bricks are laid parallel to the face of the wall. The other arrangement, *header*, features bricks laid perpendicular to the face of the wall. In addition, there are ornamental courses, such as rowlock, soldier, sailor, and shiner, examples of which are illustrated in Figure 9.3.

The most common types of bonds and their respective coursing are illustrated in Figure 9.5. Running bond is the continuous use of stretcher courses, where each alternating course overlaps the preceding course by half

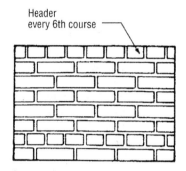

Common bond
(running bond if no headers)

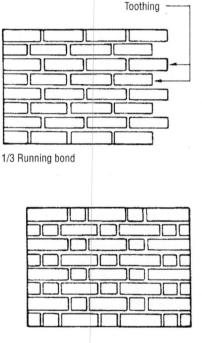

1/3 Running bond

Stack bond
(A pattern bond)

Flemish bond

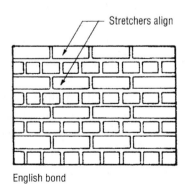

English bond

Cross bond
(english, flemish, or dutch)

Figure 9.5 Brick Bonding Patterns

of a brick. Modifications of the standard running bond are also used, such as one-third running bond.

Other bonds illustrated in Figure 9.5 are defined as follows:

- *Common bond*, or *American bond*, is similar to running bond, but has a course of headers at every fifth, sixth, or seventh course.
- *Stack bond* is created by stretchers laid up directly over one another, where all vertical joints line up. The bond is a pattern for appearances only and has relatively little structural value.
- *English bond* consists of alternating courses of stretchers and headers where the vertical joints of alternating courses align.
- *Dutch bond*, or *English Cross bond*, is similar to English bond except that the stretcher courses do not align but alternate by half a brick.
- *Flemish bond* consists of alternating headers and stretchers in each course. The header is centered over the stretcher in each consecutive course.

There are other types of bonds, but they are variations of the preceding types.

To calculate SF for veneer walls, multiply the length by the height. Always make deductions in full for areas greater than 2 SF. Show, as closely as possible, the actual number of bricks needed to do the work.

When deducting window or door openings from the wall, consider the depth of the jambs or returns, commonly called *reveals*. If the reveal is only the width of a brick (approximately 4" for most types of brick), the entire opening must be deducted. If the depth of the reveal varies, subtract the additional brick in the reveal from the opening size to be deducted.

For example, if the opening is 5'-4" × 6'-8" with a 16" reveal on either jamb, the calculation would be:

$$2 \text{ sides} \times 16" = 32", \text{ or } 2'\text{-}8".$$

Subtract 2'-8" from 5'-4", which equals 2'-8".

Therefore, the actual deduction for the opening will be 2'-8" × 6'-8", or 17.81 SF.

Quantities for waste for brickwork vary with the application, type of brick, bond, and quality of workmanship. On face brick veneers set in a running bond with no header courses, a standard acceptable waste would be between 3% and 5%. The estimator should bear in mind that for more accurate results, the brick added for waste should be identified in the estimate separately. While adding waste is essential, the labor for the brick waste is not required. Labor should be calculated on the net area or net amount of brick to be laid up, exclusive of the amount of brick added for waste. Consider an exterior elevation that has a gross area of 2,000 SF with 500 SF of window openings. The net area of brick that the bricklayers will lay up is 1,500 SF, despite adding 5% for the material waste, the bricklayers will only lay up 1,500 SF of brick. For more complex bonds, Figure 9.6 lists the most common brick bonds and the associated wastes.

Type of Pattern	% of Waste
Running or stretcher bond	The face brick are all stretchers and are tied to the backing by metal or reinforcing. Waste — 5%
Common or American bond	Every sixth course of stretcher bond is usually a header course. Waste — 4%.
Flemish bond	Each course has alternate headers and stretchers with the alternate headers centered over the stretcher. Waste — 3 to 5%.
English bond	Consists of alternate headers and stretchers with the vertical joints in the header and stretcher aligning or breaking over each other. Waste — 8 to 15%
Stack bond	Has no overlapping of units since all vertical joints are aligned. Usually this pattern is bonded to the backing with rigid steel ties. Waste — 3%.
English Cross or Dutch bond	Built up on interlocking crosses. This wall consists of two headers and a stretcher forming a cross. Waste — 8%.

Figure 9.6 Waste Allowances for Various Brick Bonding Patterns

Number of bricks per SF based on brick type and size of joints.

	Size of Joint (in inches)					
Brick Type & Size	**1/4**	**1/3**	**3/8**	**1/2**	**5/8**	**3/4**
Standard face brick (8" × 2¼")	6.98	6.70	6.55	6.16	5.81	5.49
Standard common brick (8" × 2¼")	6.98	6.70	6.55	6.16	5.81	5.49
Concrete brick (7⅝" × 2¼")	7.31	7.00	6.86	6.45	6.07	5.73
Modular brick (7½" × 2⅙")	7.68	7.35	7.19	6.73	6.35	5.98
Modular Roman brick (11⅝" × 1⅝")	6.47	6.15	6.00	5.59	5.22	4.90
Modular Norman brick (11⅝" × 2¼")	4.85	4.66	4.57	4.32	4.09	3.88

Note: Above constants are net, i.e., no waste is included.

Figure 9.7 Brick Quantities per SF

This table provides factors for determining the additional quantities of brick needed when specific bonding patterns are used.

For Other Bonds Standard Size Add to SF Quantities				
Bond Type	**Description**	**Factor**	**Description**	**Factor**
Common	Full header every fifth course Full header every sixth course	+20% +16.7%	Header = W x H exposed Rowlock = H x W exposed	+100% +100%
English	Full header every second course	+50%	Rowlock stretcher = L x W exposed Soldier = H x L exposed	+33.3% —
Flemish	Alternate headers every course every sixth course	+33.3% +5.6%	Sailor = W x L exposed	–33.3%

(See "Brick Quantities per SF" table in figure 9.7 for basic quantities of brick per SF.)

Figure 9.8 Adjustments to Brick Quantity Factors

Once the SF area of brick, the size of the mortar joint, and the brick bond are known, consult the tables in Figures 9.7 and 9.8 to determine the quantity of brick per SF and, finally, the total number of bricks. The final quantity of brick could be converted to thousands of bricks designated by the letter M, (where 1,000 brick = 1M-brick). Figure 9.7 refers to brick laid in a running bond pattern with various size joints. Figure 9.8 shows factors in determining the accurate quantity of brick materials in a particular bond.

Labor to Lay-up Brick

The labor task of installing brick is called *laying-up* brick. It is a crew task typically composed of bricklayers, or masons, and bricklayer-laborers called *tenders*. It is the laborer's job to tend to the bricklayers by stocking brick on scaffolding, mixing mortar and filling tubs, erecting and dismantling scaffolding and work platforms, washing new masonry, and general clean-up so that the bricklayers can focus on production. A typical crew size is three bricklayers and two tenders. For larger projects multiples of the above 3 : 2 ratio is applied. Crews typically start smaller and then increase as the learning curve expires.

Labor-Hours Required

The bond has a direct effect on productivity as well. Generally speaking, as the quantity of brick per square foot increases, the labor-hours per square foot to install the brick also increase. Figure 9.9 shows the labor-hours required per square foot, vertical linear foot, or linear foot of brick in various bonds.

There are many factors that affect the amount of brick a bricklayer will set in a day. If 10 bricklayers work side by side under identical conditions, producing the same class of workmanship, no two would lay the same number of bricks in a day. For this reason, averages are established. Extenuating circumstances, such as weather, temperature, access, mobility, and the specific application, affect productivity. Long, straight walls with few or no interruptions proceed more quickly than those with numerous window or door openings that require layout calculations or cutting jamb brick. The actual conditions must be studied carefully to ascertain an average productivity. It is not uncommon for productivity to change depending on the different applications around the building. The class of workmanship required also affects productivity. The estimator is wise to consider multiple daily outputs, or productivities for different applications and areas of the structure.

First-Class versus Ordinary Workmanship

Since most brickwork—and masonry in general—has an aesthetic value, the class of workmanship expected is normally considered first rate. This is referred to as *first-class workmanship* and is sometimes defined as such in the Execution portion of the specifications. However, there are times when first-class workmanship is not required. The next acceptable classification of workmanship is called *ordinary*. Both classifications are produced by skilled

Description	Labor-Hours	Unit
Brick wall		
Veneer		
4" thick		
Running bond		
Standard brick (6.75/SF)	.182	SF
Engineer brick (5.63/SF)	.154	SF
Economy brick (4.50/SF)	.129	SF
Roman brick (6.00/SF)	.160	SF
Norman brick (4.50/SF)	.125	SF
Norwegian brick (3.75/SF)	.107	SF
Utility brick (3.00/S.F.)	.089	SF
Common bond, standard brick (7.88/SF)	.216	SF
Flemish bond, standard brick (9.00/SF)	.267	SF
English bond, standard brick (10.13/SF)	.286	SF
Stack bond, standard brick (6.75/SF)	.200	SF
6" thick		
Running bond		
S.C.R. brick (4.50/SF)	.129	SF
Jumbo brick (3.00/SF)	.092	SF
Backup		
4" thick		
Running bond		
Standard brick (6.75/SF)	.167	SF
Solid, unreinforced		
8" thick running bond (13.50/SF)	.296	SF
12" thick, running bond (20.25/SF)	.421	SF
Solid, rod reinforced		
8" thick, running bond	.308	SF
12" thick, running bond	.444	SF
Cavity		
4" thick		
4" backup	.242	SF
6" backup	.276	SF
Brick chimney		
16" × 16", standard brick w/8" × 8" flue	.889	VLF
16" × 16", standard brick w/8" × 12" flue	1.000	VLF
20" × 20", standard brick w/12" × 12" flue	1.140	VLF
Brick column		
8" × 8", standard brick 9.0 VLF	.286	VLF
12" × 12", standard brick 20.3 VLF	.640	VLF
20" × 20", standard brick 56.3 VLF	1.780	VLF
Brick coping		
Precast, 10" wide, or limestone, 4" wide	.178	LF
Precast 14" wide, or limestone, 6" wide	.200	LF

Figure 9.9 Labor-Hours Required for the Installation of Brick Masonry

workers, trained, and experienced in the craft, and are not differentiated by experience (unlike the difference between a mechanic and an apprentice, for example). The difference lies in the average productivity that can be expected. To illustrate the difference between first-class workmanship and ordinary workmanship, consider the two different applications of brick. Assume that the cost of the materials, labor rate, and individuals laying up the brick are identical in both applications:

An interior partition 50' long × 10' high will feature a 4" brick veneer in a running bond. It will be a major focal point in the main corridor of the building, and it must be done as first-class workmanship.

Application #1—First-Class Workmanship

The masonry crew, consisting of two bricklayers and a tender, will take eight days to complete the work as first-class workmanship. This represents a productivity of approximately 410 bricks per day.

If this major focal point wall was suddenly changed by addendum to be furred out and drywalled over, the masonry crew could perform this work as ordinary workmanship.

Application #2—Ordinary Workmanship

The same masonry crew from Application #1 could install the same wall in six days as ordinary workmanship. This now represents an approximate productivity of 546 bricks per day, or a 33% increase in production.

The above hypothetical example is meant to illustrate a point and is not intended to represent real productivities or differences in productivity. Productivity is also affected by the type of mortar joint. A flush mortar joint is the most economical because it does not require a separate operation to "tool" the joint. It is struck flush by the bricklayer's trowel in the normal setting process. For rounded, beaded, V-shaped, or weathered joints, the process requires the additional operation after the joint has been struck flush and has started to cure. The bricklayer uses a trowel or a joint tool to achieve the final joint. Raked or stripped joints are even more labor intensive and require a limit on the height of the brick courses until the work has cured. This requires that the bricklayer stop the production of brick and finish the joints of the brick that has been already set.

Excessive heights also reduce productivity because of the effort and time required for the handling or hoisting of mortar and brick to the elevated work areas. Scaffolding and the cutting of masonry units are other factors in estimating masonry work. They are discussed in detail later in this chapter, along with cleaning of new masonry work.

The national average for an individual bricklayer ranges from a low of 400 bricks per day to a high of 800 bricks per day, with higher outputs for face and veneer brick. This includes finishing joints as well.

CONCRETE MASONRY UNITS (CMU)

Concrete masonry units have, over the years, come to be known as *concrete block* and *concrete brick*. They are used extensively as interior or exterior load-bearing walls, or as the backup for brick veneer walls. Because of their larger size, fewer units are required, and the setting labor is less than if the same backup wall was constructed from brick. CMU is composed of

Portland cement, water, and a variety of fine aggregates (sand, crushed stone, and shale) or lightweight aggregates (perlite, vermiculite, or pumice). They are modular in size to accommodate brick veneers and have a variety of compressive strengths.

Concrete Brick

Concrete bricks are solid, modular units of concrete used in much the same way as regular brick. They are typically manufactured to 2¼ "high × 3⅝" deep × 7⅝" long, although additional sizes may be available in some locations. They are also used as in-fill material in concrete block walls, where cut block is impractical. Concrete bricks are taken off and priced the same way as regular bricks. The total quantity of bricks required is a function of the number of bricks per square foot multiplied by the number of square feet. (Refer to Figure 9.7 for the quantity of concrete bricks per SF.) Include a 3% to 5% waste factor for exposed concrete brick. Labor to lay up concrete brick is much the same as for regular brick. However, concrete brick is rarely laid up in bonds other than running bonds.

Concrete Block

Concrete block consists of hollow-core, load-bearing masonry units. The standard nominal dimensions of concrete block are 8" high × 16" long × 4", 6", 8", 10", or 12" wide. The actual size is ⅜" smaller in length and height so that the addition of a ⅜" mortar joint produces an 8" × 16" finished unit. Solid versions of the hollow units are available, as well as half-block for corners, and specialty block such as bond beam and bull-nose block. Concrete block is manufactured using two basic weights of concrete: heavyweight concrete (weighing approximately 145 lb. per CF) and lightweight concrete (approximately 100 lb. per CF). This translates to heavyweight units that range from 40 to 50 lb. each, and lightweight units that vary in weight from 25 to 35 lb. each. Concrete block is manufactured and sold in a variety of designs, colors, shapes, and surface textures.

The most common types of CMU are:

- **Scored block:** Units scored across the face to give the appearance of smaller units.
- **Split-face block:** Units split lengthwise and installed with the split face exposed.
- **Split-rib block:** Units with a corrugated look achieved by molding the block with coarse, vertical ribbing.
- **Deep-groove block:** Units with deep vertical grooves scored at intervals along the face of the block.
- **Slump block:** Units manufactured so that the face of the block sags, giving a unique appearance.

- **Glazed concrete block:** Units with factory-applied glazing on one or more faces. These are used for applications that require a durable, washable surface.
- **Ground face block:** Units that have been ground on the face to provide a coarse, textured surface.

Clay masonry units follow the same procedure for estimating as CMU.

Taking-off and Pricing

The standard procedure for taking off concrete block is the same as that used for brick. Calculate the areas of walls and partitions by multiplying the length by the height of the various walls and partitions and computing a total area in SF. Separate the different types of blocks by size, weight, surface texture, shape, and color, as each affects pricing. Convert the SF area to the actual number of blocks by dividing the total area by the area of an individual block, based on its nominal 8" × 16" size (when using a standard ⅜" mortar joint). For example:

> *A standard hollow concrete block with the nominal dimensions of 8" × 16" has an individual area of 128 square inches, or 0.89 SF. If this calculation is applied to 100 SF, the result is approximately 112.5 blocks per 100 SF, or 1.125 blocks per SF.*

It is important to accurately calculate the actual quantity of CMU required. Make deductions for all openings over 2 SF to get an accurate count on the block. Corners should not be counted twice. Because concrete blocks are typically installed in a running bond pattern, include half-block at vertical terminations of the wall such as control joints and door or window jambs. These are taken off by counting the half-block at alternating courses at each location. For example:

> *Both sides of the control joint are 20 courses high. Therefore, 20 courses divided by 2 (for alternating courses) equals 10 pieces of half-block. Since there are two sides to the control joint, multiply the 10 pieces by 2. The result, 20, is the total number of half-blocks needed, not including waste.*

In addition, specialty blocks, such as *bond beam* blocks, must be included as part of the takeoff. Bond beam blocks are trough-shaped concrete blocks typically installed at the top course (or a story height) to provide horizontal reinforcing. Steel reinforcing bars are laid laterally within the trough and grouted solid to provide a continual lateral reinforced beam. (Both grout and reinforcing are discussed later in this chapter.) Bond beams are taken off by the LF and converted to the number of blocks by dividing the length by the nominal length of the block (16"). Figure 9.10 illustrates a bond beam and its use.

Include an allowance for waste on all CMU following the procedure identified in the previous section for brick. Under normal circumstances, an

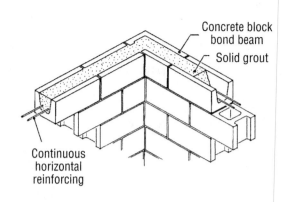

Figure 9.10 Bond Beam

average of 3% and a maximum of 6% is adequate; however, rough handling of block can increase waste factors. If considerable cutting is required for top-of-wall courses of CMU, the waste should be increased an additional 2% to 3%. Also, for custom-manufactured units, add an additional 2%. It is always better to have extra than to be short.

Labor to Lay-up Block

The labor cost of installing CMU is based on the crew's productivity, which varies depending on whether the CMU is lightweight or heavyweight block, the number of openings in the wall or how "cut up"the wall is, and the finishing of the joint. Most CMU is finished with a rounded or V-shaped tool. CMU below grade, as in the case of foundation work, proceeds faster than above grade. If the CMU is meant as the final finished surface on both or either side of the wall, calculate the costs for a tooled finish on one or both surfaces. Because of the size and weight of concrete block, some union regulations require that two masons set a block. This has a direct effect on productivity and should be reviewed. Figure 9.11 is a table of productivity for the installation of various types of CMU.

Other Cost Considerations

Other cost considerations include the cost of hoisting CMU to staging or above grade. Mechanical means of transporting CMU to and from the staging area are often required while the work is in progress. CMU like most masonry products is heavy and requires some type of equipment to handle. Masonry saw and diamond blades should also be considered. For large quantities of cutting, additional blades may be needed. For projects with large amounts of cutting, a bricklayer or laborer is dedicated to the task. Cleaning of completed CMU is typically figured separately, based on the surface area to be cleaned. Reinforcement, both lateral and vertical, and the masonry grout to fill the cells solid are typically estimated separately.

Description	Labor-Hours	Unit
Foundation walls, trowel cut joints, parged 1/2" thick, 1 side, 8" × 16" face		
Hollow		
8" thick	.093	SF
12" thick	.122	SF
Solid		
8" thick	.096	SF
12" thick	.126	SF
Backup walls, tooled joint 1 side, 8" × 16" face		
4" thick	.091	SF
8" thick	.100	SF
Partition walls, tooled joint 2 sides 8" × 16" face		
Hollow		
4" thick	.093	SF
8" thick	.107	SF
12" thick	.141	SF
Solid		
4" thick	.096	SF
8" thick	.111	SF
12" thick	.148	SF
Stud block walls, tooled joints 2 sides 8" × 16" face		
6" thick and 2", plain	.098	SF
Embossed	.103	SF
10" thick and 2", plain	.108	SF
Embossed	.114	SF
6" thick and 2" each side, plain	.114	SF
Acoustical slotted block walls Tooled 2 sides		
4" thick	.127	SF
8" thick	.151	SF
Glazed block walls, tooled joint 2 sides 8" × 16", glazed 1 face		
4" thick	.116	SF
8" thick	.129	SF
12" thick	.171	SF
8" × 16", glazed 2 faces		
4" thick	.129	SF
8" thick	.148	SF
8" × 16", corner		
4" thick	.140	EA

Figure 9.11a Installation Time in Labor-Hours for Block Walls, Partitions, and Accessories

Reinforcement and grout are discussed in detail later in this chapter. Some CMU are specified with insulated sleeves within the core of the block. This will increase the cost of the CMU material.

Mortar for laying up block is calculated similarly to brick. The square surface area of CMU to be laid up is determined and then converted to the number of blocks required. The quantity of mortar is then determined from

Description	Labor-Hours	Unit
Structural facing tile, tooled 2 sides		
5" × 12", glazed 1 face		
4" thick	.182	SF
8" thick	.222	SF
5" × 12", glazed 2 faces		
4" thick	.205	SF
8" thick	.246	SF
8" × 16", glazed 1 face		
4" thick	.116	SF
8" thick	.129	SF
8" × 16", glazed 2 faces		
4" thick	.123	SF
8" thick	.137	SF
Exterior walls, tooled joint 2 sides, insulated		
8" × 16" face, regular weight		
8" thick	.110	SF
12" thick	.145	SF
Lightweight		
8" thick	.104	SF
12" thick	.137	SF
Architectural block walls, tooled joint 2 sides		
8" × 16" face		
4" thick	.116	SF
8" thick	.138	SF
12" thick	.181	SF
Interlocking block walls, fully grouted		
vertical reinforcing		
8" thick	.131	SF
12" thick	.145	SF
16" thick	.173	SF
Bond beam, grouted, 2 horizontal rebars		
8" × 16" face, regular weight		
8" thick	.133	LF
12" thick	.192	LF
Lightweight		
8" thick	.131	LF
12" thick	.188	LF
Lintels, grouted, 2 horizontal rebars		
8" × 16" face, 8" thick	.119	LF
16" × 16" face, 8" thick	.131	LF
Control joint 4" wall	.013	LF
8" wall	.020	LF
Grouting bond beams and lintels		
8" deep pumped, 8" thick	.018	LF
12" thick	.025	LF
Concrete block cores solid		
4" thick by hand	.035	SF
8" thick pumped	.038	SF
Cavity walls 2" space pumped	.016	SF
6" space	.034	SF

Figure 9.11b *(Continued)*

a chart or table based on the width of the CMU. Again, waste can be considerable, in the 15% to 25% range.

Average daily outputs are less predictable due to the variety of applications. Consult Figure 9.11 for relative labor-hour outputs.

Description	Labor-Hours	Unit
Joint reinforcing		
Wire strips regular truss to 6" wide	.267	CLF
12" wide	.400	CLF
Cavity wall with drip section to 6" wide	.267	CLF
12" wide	.400	CLF
Lintels steel angles minimum	.008	lb.
Maximum	.016	lb.
Wall ties	.762	C
Coping for 12" wall stock units, aluminum	.200	LF
Precast concrete	.188	LF
Structural reinforcing, placed horizontal,		
#3 and #4 bars	.018	lb.
#5 and #6 bars	.010	lb.
Placed vertical, #3 and #4 bars	.023	lb.
#5 and #6 bars	.012	lb.
Acoustical slotted block		
4" thick	.127	SF
6" thick	.138	SF
8" thick	.151	SF
12" thick	.163	SF
Lightweight block		
4" thick	.090	SF
6" thick	.095	SF
8" thick	.100	SF
10" thick	.103	SF
12" thick	.130	SF
Regular block		
Hollow		
4" thick	.093	SF
6" thick	.100	SF
8" thick	.107	SF
10" thick	.111	SF
12" thick	.141	SF
Solid		
4" thick	.095	SF
6" thick	.105	SF
8" thick	.113	SF
12" thick	.150	SF
Glazed concrete block		
Single face 8" × 16"		
2" thick	.111	SF
4" thick	.116	SF
6" thick	.121	SF
8" thick	.129	SF
12" thick	.171	SF
Double face		
4" thick	.129	SF
6" thick	.138	SF
8" thick	.148	SF

Figure 9.11c *(Continued)*

STONE

Because it is a product of nature, stone varies dramatically in type, size, shape, and weight with each different species and geographic location. In general, stone for construction purposes can be classified in one of the following groups:

- **Rubble:** Irregularly shaped pieces broken from larger masses of rock, installed or "laid up" with little or no cutting or trimming.

Description	Labor-Hours	Unit
Joint reinforcing wire strips		
4" and 6" wall	.267	CLF
8" wall	.320	CLF
10" and 12" wall	.400	CLF
Steel bars horizontal		
#3 and #4	.018	lb.
#5 and #6	.010	lb.
Vertical		
#3 and #4	.023	lb.
#5 and #6	.012	lb.
Grout cores solid		
By hand 6" thick	.035	SF
Pumped 8" thick	.038	SF
10" thick	.039	SF
12" thick	.040	SF

Figure 9.11d *(Continued)*

- **Fieldstone:** Irregularly shaped rocks used as they are found in nature. Most commonly used in fireplaces and stone walls in landscaping.
- **Cut stone:** Stone that has been cut to specific shapes and sizes, with a uniform texture. Most commonly used for veneers.
- **Ashlar:** Characterized by saw-cut beds and joints, usually rectangular in shape, with flat or textured facing.

Taking-off and Pricing

Stonework prices vary by location and species of stone. In general, the following rules apply.

Most stone used in landscaping walls is taken off by the CF. This is done by multiplying length by height by width (or thickness) of the wall. Then the CF volume is converted to tons (TNS), based on the individual stone's volume per ton. The stone materials are priced by the ton for both purchase and setting. The volume in CF per ton of stone is often available from the distributor for each type of stone. The volume in CF per ton varies with the species of stone, as well as the individual size, shape, and density of the wall. Stone can range roughly from 100 lbs. to 160 lbs. per CF for this application. Stone in veneer wall and fireplace applications is taken off by the SF and converted to tons based on the average thickness in much the same manner. Stone conversion factors are often listed on a distributor's website, along with waste and other helpful information.

It is recommended that the estimator obtain a direct quote for each species of stone delivered to the job site. Prices vary with availability and popularity. Because most types of stone are extremely heavy, costs for shipping and handling can be significant. Allowances for waste are based on the individual type of stone and its application. A general rule is that for stone, ashlar, or regular-shaped stone, the waste will vary between 4% and 10%. Irregularly shaped stones, such as rubble or fieldstone, often have higher wastes because of "unusable"stone in a delivered batch. Waste factors for lower-grade stone can approach 20%. Remember that some walls are *dryset,* or set with a minimum amount of mortar or none. They are typically used in landscape applications. Styles vary with the type of stone and local practices. Stone is sometimes *dressed* at the site, which is the practice of shaping or cutting stone on-site both for visual appearance and to fit the application. This can be extremely time consuming and difficult to estimate.

Labor to Set Stone

The labor cost of setting stone is based on the crew's productivity, which varies depending on the shape of the product and the desired look of the finished surface. Cutting stone into random shapes to fit gaps in the wall can be labor intensive, especially with a mallet and chisel. Crews tend to be smaller, such as two stonemasons and one laborer, again as a tender. Depending on the size and type of stone, it may require a piece of equipment to set the stone or to stage it near the application. Other smaller stones may be one- or two- person handwork. For specific guidance as to the daily output of a particular type of stone, the company selling the product should be consulted.

MASONRY REINFORCEMENT

The term *masonry reinforcement* refers to the use of steel reinforcing bars and wire mesh–type lateral reinforcing installed in the coursing of masonry units, and the vertical rebar grouted into the voids in concrete block. Reinforcing steel in masonry work adds tensile strength to the wall or partition in much the same way as in reinforced concrete. It also allows the reinforced masonry to move during a seismic event with minor damage in comparison to the unreinforced masonry, which collapses. Masonry reinforcing is classified in two groups: lateral (horizontal) and vertical.

Lateral Masonry Reinforcement

Lateral reinforcing can be further divided into two types: *wire mesh,* or *strip joint reinforcing,* and *horizontal steel rebar.* Rebar is frequently the same type and grade used in concrete applications. Both types provide reinforcement against lateral stresses and horizontal movement. Wire mesh or strip joint reinforcing is accomplished by the use of wire mesh–type strips installed between the courses of masonry units. The most common types of wire strip joint reinforcement are the truss type and the cavity-wall ladder type. Figure 9.12 illustrates the major types of joint reinforcement.

Taking-off and Pricing

Horizontal joint reinforcement is taken off and priced by the LF. Most plans and specifications are specific about the location of horizontal joint reinforcement, which is typically noted by the course. To arrive at the total LF

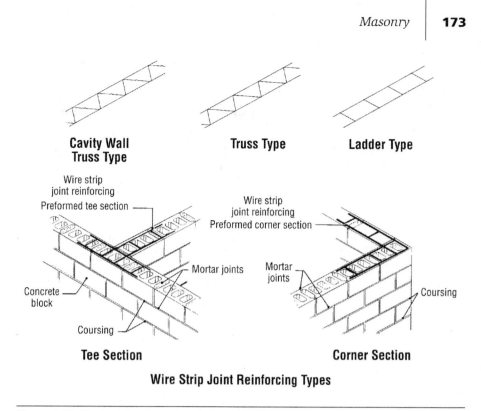

Cavity Wall Truss Type **Truss Type** **Ladder Type**

Tee Section **Corner Section**

Wire Strip Joint Reinforcing Types

Figure 9.12 Typical Concrete Block Reinforcing Types

of joint reinforcement, count the number of courses that require joint reinforcement and multiply by the LF of reinforcement for each course. Materials and labor are priced per LF, which can be extended to 100 LF quantities by dividing the total linear footage by 100 and labeling the result with the unit CLF (100 LF).

Horizontal reinforcing in the form of rebar is typically used for bond beam applications. Steel reinforcing bars are laid within the trough of the bond beam and are grouted solid to form a continuous ring of lateral reinforcement at specific locations within the masonry wall or partition. Horizontal rebar in masonry is taken off by the LF. Plans and specifications designate the location of horizontal reinforcing, as well as the size of the bar, the number of bars to be used, and the lap of the bars.

Rebar quantities are determined by adding all linear footage at each location to arrive at a total LF. This total is then multiplied by the weight per foot of the specific bar designation for pricing. Figure 9.13 lists bar designations.

Calculate the weight of all bent bars for corners and intersections separately, and then add the weight to the total for that bar designation. Additional weight must be added for overlap of continuous bars. Include costs relative to storage and handling of the rebar, in addition to setting costs. Special grades of steel or coatings should also be noted for accurate pricing. Procedures for takeoff and pricing follow the same rules as when used with concrete. Consult Chapter 8 for additional information on determining waste factors.

Bar Desig- nation No.**	Nominal Weight, Lb./Ft.	U.S. Customary Units			Nominal Weight kg/m	SI Units		
		Nominal Dimensions*				Nominal Dimensions*		
		Diameter in.	Cross Sectional Area, in.²	Perimeter in.		Diameter, mm	Cross Sectional Area, cm²	Perimeter mm
3	0.376	0.375	0.11	1.178	0.560	9.52	0.71	29.9
4	0.668	0.500	0.20	1.571	0.994	12.70	1.29	39.9
5	1.043	0.625	0.31	1.963	1.552	15.88	2.00	49.9
6	1.502	0.750	0.44	2.356	2.235	19.05	2.84	59.8
7	2.044	0.875	0.60	2.749	3.042	22.22	3.87	69.8
8	2.670	1.000	0.79	3.142	3.973	25.40	5.10	79.8
9	3.400	1.128	1.00	3.544	5.059	28.65	6.45	90.0
10	4.303	1.270	1.27	3.990	6.403	32.26	8.19	101.4
11	5.313	1.410	1.56	4.430	7.906	35.81	10.06	112.5
14	7.65	1.693	2.25	5.32	11.384	43.00	14.52	135.1
18	13.60	2.257	4.00	7.09	20.238	57.33	25.81	180.1

*The nominal dimensions of a deformed bar are equivalent to those of a plain round bar having the same weight per foot as the deformed bar.
**Bar numbers are based on the number of eighths of an inch included in the nominal diameter of the bars.

Figure 9.13 Reinforcing Steel Weights and Measures

Calculating the amount of grout for the bond beam and vertical cells is discussed later in this chapter.

Vertical Masonry Reinforcement

Vertical masonry reinforcement refers to steel rebar installed within the cells of hollow concrete block (CMU) walls and grouted in place to form a single unit. When installed within an engineered design, the reinforced masonry walls resist stresses exerted by wind, seismic, and other forces. In the typical application, the vertical bars are spaced at predetermined on-center spacing, noted on the plans. Figure 9.14 illustrates a reinforced concrete block wall.

Taking-off and Pricing

Vertical rebar is the same type and grade of steel used in the horizontal masonry rebar applications. The takeoff is performed in a manner similar to that used for vertical rebar in a cast-in-place concrete wall. The total length of the reinforced walls is divided by the on-center spacing to determine the number of bars. Once the quantity has been determined, it is multiplied by the length of the individual bars. Additional bars may need to be added per specification requirements at the corners or jambs of openings, or at intersecting partitions. These should be added to the total. Calculate bar overlap based on the specific grouting conditions (covered later in this chapter). Convert it to weight and add this to the total weight of rebar.

In order for the rebar to act as a monolithic unit and provide the tensile strength to the compressive strength of the CMU, the two must be bonded together by grout, discussed in the next section.

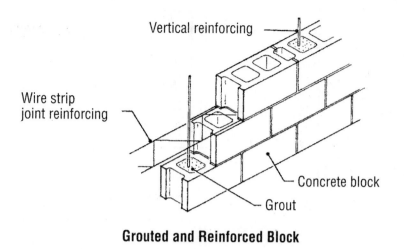

Grouted and Reinforced Block

Figure 9.14 Reinforced Concrete Block Wall

GROUT

Grout is a composition of Portland cement, sand, lime, and water mixed in similar proportions and strengths to mortar. Additional water is used to bring the consistency to a more fluid or plastic state for flowability. Grout is then pumped or poured into the cells of set concrete block that contain vertical or horizontal reinforcing bars. Vibrating or tamping may be required to ensure complete embedment of the rebar. Once the grout has cured, it forms a single solid unit of grout, block, and rebar. Grouting has some minimum requirements; the masonry has cured for at least 4 hours, the grout slump is between 10" and 11", and there are no intermediate bond beams between the top and bottom of the pour height.

Taking-off and Pricing

To determine the quantity of grout needed, calculate the volume of the cells to be filled. This is accomplished by determining the location and quantity of vertical columns (cells) to be filled and multiplying that amount by the volume of each column of cells. This figure should be computed in CF, the typical unit of pricing. The estimator may choose to convert the CF quantity to CY (1 CY = 27 CF.) depending on pricing units preferred. To expedite the calculation of the grout quantity, most estimators use predetermined grout charts or factors. *To simplify the calculation of grout quantities, use Figure* 9.15. The quantity of grout specified in this table is listed according to specific spacing and various wall thicknesses and is based on SF of wall area.

Grout quantities for bond beams vary by type and size of the bond beam block. Volume can be calculated by multiplying the length of the bond beam by the width and depth of the trough. All units should be in feet and converted to CY for pricing. An 8" wide bond beam will require approximately 0.31 CF of grout per linear foot (LF) of bond beam, not including waste.

Center-to-Center Spacing	6" C.M.U. Per SF		8" C.M.U. Per SF		12" C.M.U. Per SF	
	Volume in CF		Volume in CF		Volume in CF	
Grouted Cores	40% Solid	75% Solid	40% Solid	75% Solid	40% Solid	75% Solid
All cores grouted solid	.27	.11	.36	.15	.55	.23
cores grouted 16" o.c.	.14	.06	.18	.08	.28	.12
cores grouted 24" o.c.	.09	.04	.12	.05	.18	.08
cores grouted 32" o.c.	.07	.03	.09	.04	.14	.06
cores grouted 40" o.c.	.05	.02	.07	.03	.11	.05
cores grouted 48% o.c.	.04	.02	.06	.03	.09	.04

Note: Costs are based on high-lift grouting method.

Low-lift grouting is used when the wall is built to a maximum height of 5'. The grout is pumped or poured into the cores of the concrete block. The operation is repeated after each five additional feet of wall height has been completed. High-lift grouting is used when the wall has been built to the full story height. Some of the advantages are: the vertical reinforcing steel can be placed after the wall is completed, and the grout can be supplied by a ready-mix concrete supplier so that it may be pumped in a continuous operation.

Figure 9.15 Volume of Grout Fill for Concrete Block Walls

There are two different methods of grout placement: high-lift and low-lift. Each has particular advantages and disadvantages. Most project specifications define which method is acceptable. In the absence of a specified method, consult the building code that has jurisdiction. Some building codes do not allow the use of high-lift grouting.

Low-Lift Grouting

Low-lift grouting calls for the placement of grout within the cells of an erected wall at a maximum of 5'. Grout is poured to within 1½" of the top of the top block to allow for a keyway when additional lifts are poured. Low-lift grouting has the advantage of ensuring that cells are filled solid down to the previously poured lift. It also allows grout to be mixed in smaller quantities on-site. The disadvantage is that setting production is limited to a maximum of 60". The setting is then stopped and the CMU is grouted. This is translated to an additional cost for delayed production in the erection of the wall.

High-Lift Grouting

High-lift grouting is the placement of grout in the cells after completion of the top course of the masonry work (greater than 5'). This method requires that cleaning/inspection holes be left out at the bottom of each cell. This allows access for cleaning mortar droppings out from within the cell to be grouted. This method requires that the grouting procedure be continuous to the top of each cell. It has the advantage of allowing the grouting operation to be performed in one application, in contrast to low-lift grouting, which is done in smaller portions and at various times. It also allows the vertical reinforcing to be placed after the wall has been completed. The major disadvantage is that it is difficult, if not impossible, to ensure that all rebar has been adequately surrounded by grout through the entire height of the cell. Some structural failures have been attributed to weak spots in the wall where voids have occurred and the rebar has deteriorated. Regardless of the specified method of grouting, the calculation for volume is the same. Newly developed self-consolidated grouts are more fluid, assuring that grout makes it to the bottom of the pour and surrounds the vertical reinforcement.

Labor to Grout

The labor to perform grouting can be calculated per labor-hour per CY or per labor-hour per SF of wall surface area. High-lift grouting most often requires the use of a grout pump. This is similar to the pump used in the placement of concrete, as discussed in Chapter 8 but with smaller-diameter hoses. Many larger masonry contractors own their own grout pumps. If the pump is to be rented, include this as part of the cost of grout placement, plus the normal operating costs for the day. Rental can be calculated by the day or by the half day. Grouting can often require labor to clean the equipment when the work has been completed. Waste should also be allowed for the grout remaining in the hose and the pump. Waste in general should be included as the process can be difficult to avoid waste.

MASONRY ANCHORS AND TIES

To join multiple-width masonry walls together or a nonstructural veneer brick wall to a structural backup wall, an anchor or tie must be used. Anchors or ties are manufactured of coated metals or synthetics that will not deteriorate when in contact with the corrosive elements in mortar. They are available in a wide variety of shapes, sizes, and methods for fastening. Some of the more common types of anchors and ties are listed as follows:

- **Dovetail anchors:** Used to tie masonry veneers to cast-in-place concrete backup walls. The dovetail slot is poured within the forms and, when stripped, provides a vertical slot in which to attach the dovetail anchor.
- **Corrugated wall ties:** Hot-dip galvanized strips of corrugated metal for tying brick veneers to wood-framed or concrete block backup walls, often the most economical.
- **Box-type cavity wall anchors:** Loop-shaped metal wires of various gauges used to tie multiple-width masonry walls together.
- **Welded anchors:** Used to tie masonry veneers to structural steel columns and beams.

Anchors are embedded in the horizontal courses of masonry work and fastened to the backup wall by screws, welding, or embedment (in the case of dovetail slots). Ties for multiple-width masonry walls are embedded in the respective courses of both walls. The spacing of anchors and ties is noted in the specifications. They are typically described in terms of both horizontal and vertical spacing. For example: "Ties and anchors will be spaced at 24" o.c., both horizontally and vertically."

Review the masonry specification and the documents carefully to determine the specified tie and its method of installation/attachment. The cost of material varies dramatically among different ties. The cost of the method of attachment also varies with the actual application. For example, corrugated ties can be installed in the CMU backup wall as the block is laid up. However, the same tie attached to a metal stud and sheathed wall will require considerably more labor, as it is a completely separate application.

Labor is typically calculated based on a specific quantity that is expected to be installed per hour of labor.

Taking-off and Pricing

Anchors and ties are taken off and priced by the piece and listed per each (EA). Frequently, this quantity can be extended to sets of 100 (C) or per 1,000 (M). This requires that the area of wall to be anchored be calculated by the SF and then divided by the spacing specified. For example:

Calculate the ties required for 1,000 SF of masonry wall, with the ties spaced at 2' o.c. each way.

If the spacing is 2' each way, approximately one tie will be required for every 4 SF of wall.

1,000 SF/4 SF per tie = 250 ties

Allow about 5–7% for waste resulting from handling. Also consider jurisdiction issues on union projects. Some collective bargaining agreements call for the installation of the ties to substrates other than masonry to be by trades other than bricklayers, depending on the method of attachment. Labor can be based on productivity of the trade installing it or as an ancillary cost to the bricklayer, reflected as a small reduction in daily output.

MASONRY RESTORATION

Masonry restoration refers to restorative work on masonry that is already in place. Most of this type of work includes cutting out old mortar joints and *repointing* or refilling the existing joints with fresh mortar. This process is sometimes called *tuckpointing*. Masonry restoration may also include the removal and replacement of damaged or deteriorated individual masonry units, most often brick.

Cutting out old mortar joints is typically done by an electric saw or a grinder with a diamond or carborundum blade. The saw or grinder is set to a specific depth, as required in the specifications, and is passed along the existing mortar joint until it is free of old mortar. Removing damaged or deteriorated brick is done in much the same manner.

Additional chipping by means of a hand mallet and chisel may be required to remove individual bricks. The newly cut and cleaned joint is then filled with mortar and tooled to achieve the desired joint. Replacement bricks are "buttered" with mortar and fitted into place. The replacement of individual bricks in areas subject to the most damage, such as outside corners, is most common. Removing old courses to create a bond for the new work is referred to as *toothing* or *toothing-in*. Restoration work may also include sandblast cleaning of the surface area—the process of forcing fine sand or slag through a hose at high pressure. The abrasive force removes debris, graffiti, and the old surface. Excessive sandblasting will damage the surface and remove the mortar from the joint. Some repointing of damaged joints may be necessary as a result.

Taking-off and Pricing

Sandblasting is taken off and priced by the square foot area to be cleaned (SF). Openings for windows and doors are not deducted. The cost of sandblasting is affected by several factors. Costs can be reduced if the sand or slag can be salvaged. Often, up to 60% of the material can be salvaged and reused. Brick with normal dirt and grime has higher production rates and success after the first pass. Bricks with multiple coats of paint or waterproof coating may require multiple passes. Sandblasting, like most masonry restoration work, is a crew task in which the production is based on a multiperson crew. The estimator must also consider costs associated with sandblasting such as protection, staging, and cleanup, all of which are in addition to the actual costs of sandblasting.

The standard takeoff unit for both cutting out the old joint and repointing is the SF. The estimator may choose to separate the takeoff for each process for estimating purposes. Cutting out and repointing the old joint have different unit costs. The area is calculated by multiplying the length or width by the height of the surface to be tuckpointed. Labor to cut out old joints is based on production per hour and varies with the age and condition of the brick and the mortar.

Labor for Masonry Restoration

Labor costs to repoint old work have fewer variables. Consider the width of the mortar joint when calculating production and materials costs. Also include the cost of the carborundum wheels and possibly even the costs of the grinders, as cutting out old mortar generates a lot of wear on tools and equipment. Removal and replacement of old brick can be taken off and priced by the individual piece (EA) or by the SF. Individual brick to be replaced in random locations may be best taken off and priced by the piece. Larger quantities grouped together with definable dimensions should be taken off and estimated by the SF. In either case, the quantities of each should be noted separately on the takeoff. Be sure to include an allowance for removing and replacing adjacent bricks that may be damaged in the process. Matching existing brick and mortar colors often involves additional labor required for the multiple attempts to get it correct.

The cost of restoration work is affected by a variety of factors, including the type of staging to access the work. Movable staging, such as swing staging or motorized platforms, may best accommodate the frequent moves associated with tuckpointing. Typically, the cost for renting both types of staging exceed the cost of conventional scaffolding. Evaluate and price the most efficient method. Another factor to consider is the need for temporary bracing when large quantities of damaged or deteriorated brick are removed. The possibility of collapse of old brickwork already weakened by the removal of old mortar is a realistic concern. Bracing should be evaluated by the individual work area. Ground labor to support the workers on the staging is usually required and must be accounted for in the estimate. Ground support provides mortar, materials, tools, and equipment for the crew above. Restoration work is labor intensive and should not be underestimated. In both processes it is difficult to achieve the production normally associated with masonry due to small areas and random locations typically found in this type of work.

Protection requirements for sandblasting procedures vary with the size of the area being cleaned and the location of the work. Protection of surface features, such as windows, doors, or ornaments, can slow productivity. Because the use of sandblasting equipment is environmentally regulated, check the requirements for individual areas. Work locations with dense populations generally require airborne particles to be contained. Enclosure of the work should be included in the estimate and may consist of dismantling and re-erecting the containment enclosure several times as the work progresses. Cleanup of the spent sand or slag should also be included as part of the work. Maintenance and repair of the enclosures should also be considered for extended projects.

Most often, although not always, masonry restoration crews are smaller than production crews. For sand blasting there is an operator for each piece of equipment and a maintenance crew to keep them running. Larger projects may have support labor to help the sandblaster.

MASONRY CLEANING

Some form of cleaning is required on all new masonry work, to remove splattering of mortar and dust. Masonry cleaning is usually done after the work has set but before the mortar has reached its full strength. Several methods can be used, including mild detergents and a coarse hand brush to scrub the surface. Stronger solutions containing chemicals, such as muriatic acid, may also be required. The basic process is the same, although personal protective gear (PPE) may be required. An alternate method involves high-pressure washing. Water is forced through a nozzle at high pressure to remove the surface debris. Check the specifications to ensure that pressure washing of new masonry is not prohibited.

Taking-off and Pricing

Regardless of the method used, masonry cleaning is quantified and priced by the SF of surface area to be cleaned. Different methods for the same project should be listed separately, as each will have a different cost. Solutions used in hand cleaning and power washing are usually diluted to a specified strength. Base the quantity of solution on the area to be covered. The quantity of solution often varies with the product used, and most often it is a matter of judgment or may be stated in the specifications. Chemical cleaners, such as those used for paint removal, are often expensive and may require multiple coats applied by hand with rollers or brushes. Many of these chemicals are allowed to set before they are removed. This also requires protection laid over the ground so that the residual chemicals do not contaminate the surrounding soil.

Like most masonry restoration, cleaning often requires scaffolding for access. On new work with standard pipe scaffolding, the work is usually cleaned before the staging is dismantled and moved. A ground crew may be needed for larger cleaning jobs to support the personnel doing the cleaning. Staging is discussed in detail later in this chapter.

Toxic chemicals or solutions may also be governed by environmental regulations and may require protection or special cleanup. Protective gear for workers can be costly and may reduce productivity. Disposal of the residual byproduct is costly and should be investigated beforehand. Consult the specifications for the approved method of cleaning, and price the work accordingly. Substitutions of less expensive methods may not be acceptable. This can frequently be the work of a specialty contractor. It is recommended that the estimator solicit subcontractor quotes for the work whenever the opportunity allows.

Labor for masonry cleaning is most often a crew task and not that of an individual. However productivity is measured by the individual laborer cleaning the work. It is measured as a function of the laborer's daily output. Water is required to wash and rinse most masonry cleaners. Cleaning may be relegated to specific seasons in more temperate climates.

MASONRY INSULATION

Masonry insulation consists of installing rigid insulation boards between the veneer brick and the CMU backup wall. Another common method of insulating concrete block is to use masonry fill insulation, a granular material composed of water-repellent vermiculite or silicon-treated perlite. Both materials are poured from bags or blown-in from trucks into the voids in cavity walls or the empty cells in concrete block. Review the specifications carefully for the methods required. Specifications referencing code requirements should be investigated carefully for materials and procedures.

Taking-off and Pricing

Masonry insulation is taken off and priced based on the SF area to be insulated. The procedure for determining the quantity of masonry fill insulation is the same as that used for quantifying grout for reinforced cells. (See "Grout" earlier in this chapter.) The takeoff units are CF. Figure 9.15 can be used to calculate the volume of masonry fill insulation required, which can be extended into the typical sales quantities (bags) of the various products.

Calculate the number of bags required and allow a waste factor of 10–12%. The labor to install masonry insulation, regardless of the material employed, can be significant. Masonry fill insulation costs are based on the quantity of CF that can be placed in a day.

With blown-in insulation, the cost of equipment is added and someone to oversee and fill the equipment is also required. Insulation does not flow like grout so daily production can be less. It also requires labor to clean up after the work is complete.

Rigid insulation is applied to the cavity wall by several methods, the most common of which is using the masonry tie to hold the insulation in place. Sometimes this requires the use of an adhesive to temporarily hold the sheet in place until the ties are installed. The labor costs include cutting the insulation as required to fit around openings in the exterior walls, which is included as part of the cost of the work. Exterior walls with many openings will reduce productivity, and unit costs should reflect this. Again, union jurisdictional issues may require the installation be installed by a trade other than a bricklayer.

FLASHINGS

Flashings are impervious sheets of material commonly installed at the base of the exterior masonry walls to deflect water from going into the structure. Flashings are fabricated from a variety of different metals, including copper, lead-coated copper, asphalt-coated copper, steel, and some metals with proprietary coatings. The thickness of the metal sheet will affect the unit cost of the material as well as the cost of fabrication. Review the specification carefully for ownership of the flashings and its fabrication. Many specifications indicate that metal flashings are fabricated by other trades, typically the roofing or sheetmetal subcontractor, and provided to the masonry subcontractor for installation. This is often because the flashings are intrinsic to weatherproofing the roof system. Other types of flashings can be specified under the masonry section of the specifications.

Taking-off and Pricing

Flashings are taken off and listed for pricing by the LF. Flashings of different compositions, thicknesses, and species should be listed separately, as this will affect the costs of the materials and possibly the labor. The price of flashing materials varies dramatically based on thickness and species. To determine the base costs of the materials, calculate the number of sheets of metal required. (Sheet metal flashings are discussed in Chapter 12, "Thermal & Moisture Protection.") This is done by determining how many pieces can be produced from a single sheet of metal. An allowance for the overlap of the individual pieces must also be included. Labor costs for installation are reflective of the type of flashing. Some flashings are set in a bed of asphalt mastic and are mechanically fastened to the nonmasonry backup wall. Others are "coursed" into the masonry CMU backup as the CMU is laid up. In either instance, the cost for installation is determined by the LF of flashings to be installed. Labor costs can be a single individual or a crew of two for large pieces. Flashings that are bent into a particular shape off site will have a fabrication component to the price as well. The estimator is advised to secure a price for the metal and the fabrication whenever possible. Flashings made of copper are subject to daily price changes on an international commodities market. Figure 9.16 shows a typical masonry cavity wall with insulation and flashings.

CUTTING MASONRY UNITS

It is common for brick or block to be cut on-site. This can be time consuming if there is a large quantity and can even require a dedicated individual to do the cutting. Be sure to include the cost of cutting masonry units, such as

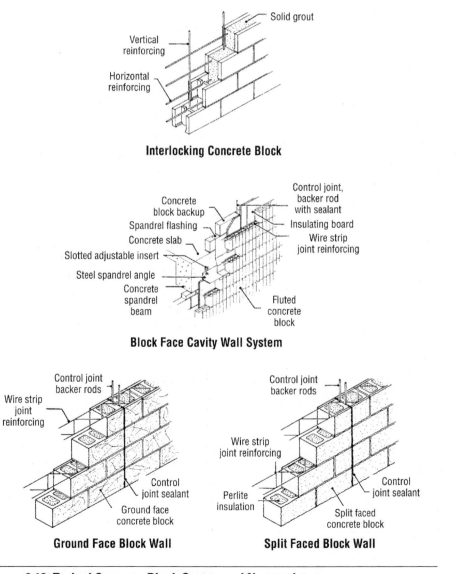

Figure 9.16 Typical Concrete Block System and Nomenclature

block and brick. Some block and brick materials are cut with a brick hammer and chisel, but when exposed surfaces require cuts, it may necessitate the use of a gas or electric masonry saw with a diamond or carborundum blade. Larger pieces of cutting equipment that resemble a table or radial arm saw may also be required, as well as water to cool the blade during cutting.

Taking-off and Pricing

Cuts are taken off and priced by the piece (EA) and can be listed as a total quantity for each type of masonry unit. The cost for minor cutting, especially with a hammer and chisel, may be negligible in the overall estimate. This is frequently considered part of the cost of laying up the brick. Projects that require extensive cutting must be evaluated accordingly. The cost of cutting masonry has two main components: the labor to cut and the carborundum saw blades. The cost of the labor is based on the number of pieces that can be cut per day or per hour. This depends on the density/hardness of the material being cut and the sophistication of the cut. Naturally, multiple passes through the saw with the same piece entail more time. All masonry blades have a limited life, which varies with the quality of the blade. That is to say that the blades eventually become ineffective after so many cuts. Determine how many blade changes are required and include that as part of the cost. A masonry saw is basic equipment for the serious masonry subcontractor, and its cost is calculated as either part of indirect overhead or direct overhead of the company. Occasionally, the saw may be rented, and then this cost must be included as part of the bid.

Formwork for Masonry

Some kinds of masonry work require the use of forms or braces to support installed work until it has cured and can support itself. Classic examples are the forms needed to hold brick or stone over half-round windows, or formwork to brace the bottom of a bond beam poured as a lintel for an opening.

Taking-off and Pricing

This work can be quantified by the opening or piece (EA). Different types and applications should be listed separately in the estimate, as they are likely to have different costs. Material costs in most applications have multiple uses and should be estimated accordingly. Labor costs can be calculated several ways, most typically by the labor-hour per form. Remember that forms are temporary and also are required to be removed and dismantled as part of the cost. Review the specifications carefully to make sure that the formwork portion of the masonry is the work of the masonry subcontractor. It is not uncommon to include this work within another section, such as rough carpentry. It should also be noted that after most formwork is dismantled, there is a minor amount of remedial work to bring the masonry covered by the formwork up to the same standard as adjacent work. This can include cleaning and filling joints with mortar.

ITEMS FURNISHED BY OTHER TRADES

It is common for masonry contractors to install items furnished by other trades as part of the masonry scope of work. Typical examples are lintels (structural steel members that support the weight of masonry over openings in masonry walls), joist bearing plates (steel plates with anchors embedded in bond beam courses to tie bar joists to masonry walls), hollow metal door frames, sleeves for conduits and piping, and electrical boxes and conduits enclosed in masonry walls for receptacles and switches. These items must be installed as the various courses of masonry are laid. The specifications should clearly enumerate the items furnished by other trades and installed under the masonry scope of work. However, this is frequently done in a general statement that requires a careful review of all drawings for implied work that is not itemized.

Taking-off and Pricing

While this work is often incidental, it still must be calculated and included within the estimate. Larger lintels that must be installed at overhead doors have considerable weight and may require a piece of equipment to set in place. This is frequently done with a rubber tire telescoping forklift vehicle used for stocking brick, block, and mortar to staging. Since most materials are provided to the mason for installation, the cost of materials is negligible. There is, however, a labor and, potentially, an equipment component that must be acknowledged. Labor is based on the quantity of pieces installed per hour by the bricklayer. Again, this varies with the size and weight of the item to be installed and should be considered on an individual basis.

CONTROL JOINTS

Masonry walls of considerable straight length require the installation of a vertical control joint to allow for expansion and contraction of the wall. Control joints require the use of a compressible premolded joint filler (similar to the type used in concrete construction, discussed in Chapter 8) to break the bond between adjacent units in courses of brick or block masonry. The location of control joints should be clearly shown on both plan view and elevations of masonry walls on the drawings.

Taking-off and Pricing

Control joints are taken off and priced by the vertical linear foot (VLF). Control joints of varying thickness (for different wall thicknesses) should be listed separately. While the cost of the actual joint filler is minimal, the labor costs are not. If control joints are placed in locations that do not fall on coursing, the block or brick must be cut at each course (both sides). This has a cost impact for both labor and equipment, as noted previously in this

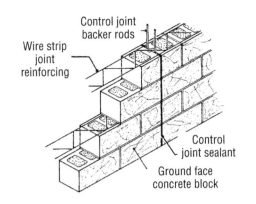

Figure 9.17 Control Joint in Block Wall

chapter. Figure 9.17 illustrates a control joint in a masonry wall. Also note that the control joints require backer rod and caulking as a finished product. This is typically not within the scope of the masonry work and is addressed in Division 7—Thermal and Moisture Protection.

INCIDENTALS FOR FIREPLACE AND CHIMNEY CONSTRUCTION

Fireplace construction requires the masonry contractor to furnish and install some special masonry items, such as dampers, flues, clean-out doors, and firebrick. *Dampers* are cast-iron operable traps that regulate the draft of the fireplace. *Clean-out doors* are steel access doors and frames in varying sizes to allow the removal of ashes and debris. They are installed within the courses of masonry as the work proceeds and are located at the lowest point in the fireplace. Masonry flues are noncombustible, heat-resistant, rectangular, or round-shaped tubes made of fireclay to allow the passage of smoke from the fireplace, boiler, furnace, or solid fuel stove through the chimney. They vary in size as required by the design of the system.

Taking-off and Pricing

Clean-out doors are taken off and priced by the piece (EA). Different sizes should be listed separately, as this will affect the cost. Flue lining is taken off and priced by the LF, measured from the top of the firebox to the very top of the chimney. Each size and shape should be listed separately for accurate pricing. Typically, there is only one damper per firebox, which should be listed by the piece (EA). Any special characteristics (such as manual or rotary operation) should be noted, as these can affect the cost.

Labor to Install Fireplace Incidentals

Labor for the various components of a fireplace is often calculated on a labor-hour basis. Note that many of the tasks in the construction of a fireplace and chimney require multiple personnel to complete efficiently. For others, the installation consists of nothing more than setting the item in place. Cutting of the masonry units in the construction of a fireplace is typical and should be considered in the estimate. Staging or scaffolding for chimney construction is also a concern. Figure 9.18 provides a checklist for fireplace and chimney construction.

Component	
Foundation footing	Footing at lowest level of fireplace: basement or under slab on grade. Reinforcing steel bar and formwork. Maybe by others.
Block foundation	CMU foundation from top of footing to underside of hearth. May be reinforced or non reinforced. Cells typically filled with grout.
Ash dump	Steel or cast iron door. Clean out trap door for access to ash dump in foundation block.
Hearth and fire box brick	Fire box and hearth brick on bottom, rear and sides of firebox, including refractory cement.
Hearth extension	Hearth extension. Brick or other masonry/stone for extension of hearth.
Concrete support	Cast in place concrete extension under the hearth (if required)
Masonry veneer	Brick, stone, etc. masonry products that will be the exposed faces of the fireplace and chimney, including mortar.
Damper	Damper at top of fire box including damper control
Lintel	Steel or masonry lintel over firebox opening. Add length for required bearing. Decorative stone or masonry lintel at veneer if required.
CMU chimney	CMU for chimney construction including mortar. Grout for vertical spaces between flue and CMU.
Chimney construction miscel. items	Ties on CMU for masonry veneers, rebar for vertical reinforcing, miscellaneous masonry or stone to support mortar/grout when wet. Bond beams and grout at floor levels. Lead or lead-coated copper for step flashing and continuous at roof connection.
Flashings and counterflashings	Lead or lead coated copper for step flashing and continuous flashing at roof connection. Flashing and counterflashing at masonry veneer rising walls adjacent to chimney.
Access	Scaffolding and/or roof brackets and planks for chimney construction at exterior. Scaffolding for access at interior.
Flue and flue extension	Flue lining and refractory cement for mortaring of sections. Flue extension above chimney.
Masonry cap	Masonry or cast in place concrete cap at top of chimney. Spark arrester at top of flue.
Cleanup	General clean up of masonry fragments and accumulated debris. Breakdown of scaffolding and washing of exposed masonry.

Figure 9.18 Fireplace and Chimney Estimating Checklist

FREIGHT OF MASONRY UNITS

Because most masonry units and stone materials are quite heavy, the costs for freight and delivery to the job site can be substantial. The estimator is advised to secure a price from the supplier to include all shipping and handling costs to the site.

Freight costs can be taken off and priced either per ton (TN) or for the entire load as a lump sum (LS). Quantities listed by weight require the approximate weight per 1,000 units (brick, block). Weight of stone may be by the pallet or shipping container or even by the dump truck load. Both should be available from the supplier. In addition, the supplier may charge for the unloading and distribution of pallets of brick or stone once on site.

STAGING AND SCAFFOLDING FOR MASONRY WORK

Staging and *scaffolding,* for the purpose of this discussion, are considered interchangeable terms. Staging consists of a temporary elevated platform and a supporting structure erected against, within, or around the work to support workers, materials, and equipment. The most common type is conventional pipe scaffolding and planking, which are erected from the ground level and built or added to as required to maintain access to the work. Other options are available and should be considered based on the specific application.

Taking-off and Pricing

The takeoff and pricing units for staging is SF of surface area of the masonry. The estimator may elect to convert the SF area to actual sections of scaffolding and planks needed. This conversion is based on the actual size of the available staging units. This required the conversion of the area to quantities of frame sections, cross braces, outriggers (brackets attached to the side of vertical sections of staging to provide support for planking), and planks. As most masonry contractors own some quantity of staging and planking, the cost of materials may be negligible. If insufficient staging is available, rental must be included along with loading, delivery, and pickup charges. Because staging is an ongoing process as the work progresses and requires dismantling, moving, and re-erection, the estimator must consider the number of moves based on productivity of the crew. This is a matter of judgment based on experience. Other types of staging may prove more efficient for some jobs. (Refer to the "Masonry Restoration" section earlier in this chapter.) Figure 9.19 is a table that provides various productivities for staging erection and dismantling.

Power lifts or extended boom lifts may be more effective in executing the work, due to access and flexibility. While these types of lifts are often more expensive to rent and operate, they can provide more efficient means of executing the work, thereby saving labor costs.

Labor to Erect, Dismantle, and Relocate Scaffolding

The labor to erect, dismantle, and move staging is calculated on a labor-hour basis. Simple staging applications can often be done by a crew of one- or

This chart is a summary of the labor involved in assembling scaffolding and staging. It shows typical crews, expected daily outputs, and the expected labor-hours per unit for various areas of scaffolding erection.

Scaffolding	Crew Makeup	Daily Output	Labor-Hours	Unit
SCAFFOLD, Steel tubular, rented, no plank, 1 use per month				
Building exterior 2 stories	3 Carpenters	17.72	1.350	CSF
4 stories	"	17.72	1.350	CSF
6 stories	4 Carpenters	22.60	1.420	CSF
8 stories		20.25	1.580	CSF
10 stories		19.10	1.680	CSF
12 stories	↓	17.70	1.810	CSF
One tier 3' high × 7' long x 5' wide 1 use per month	1 Carpenter	14.90	.537	CSF
2 uses per month		14.90	.537	CSF
4 uses per month		14.90	.537	CSF
8 uses per month		14.90	.537	CSF
5' high × 7' long × 5' wide 1 use per month		20.65	.387	CSF
2 uses per month		20.65	.387	CSF
4 uses per month		20.65	.387	CSF
8 uses per month		20.65	.387	CSF
6'-6" high × 7' long × 5' wide 1 use per month		26.85	.298	CSF
2 uses per month		26.85	.298	CSF
4 uses per month		26.85	.298	CSF
8 uses per month	↓	26.85	.298	CSF
Scaffold steel tubular, suspended slab form supports to 8'-2" high				
1 use per month	4 Carpenters	31	1.030	CSF
2 uses per month		43	.744	CSF
3 uses per month	↓	43	.744	CSF
Steel tubular, suspended slab form supports to 14'-8" high				
1 use per month	4 Carpenters	16	2.000	CSF
2 uses per month		22	1.450	CSF
3 uses per month	↓	22	1.450	CSF
SCAFFOLDING SPECIALTIES				
Sidewalk bridge, heavy duty steel posts & beams, including parapet protection & waterproofing				
8' to 10' wide, 2 posts	3 Carpenters	15	1.600	LF
3 posts	"	10	2.400	LF
Sidewalk bridge using tubular steel scaffold frames, including planking	3 Carpenters	45	.533	LF
Stair unit, interior, for scaffolding, buy				EA
Rent per month				EA
SWING STAGING for masonry, 5' wide × 7', hand operated				
Cable type with 150' cables, rent & installation, per week	1 Struc. Steel Foreman 3 Struc. Steel Workers 1 Gas Welding Machine	18	1.780	LF
Catwalks, no handrails, 3 joists, 2" × 4"	2 Carpenters	55	.291	LF
3 joists, 3" × 6"	"	40	.400	LF
Move swing staging	1 Struc. Steel Foreman 3 Struc. Steel Workers 1 Gas Welding Machine	37	.865	LF

Figure 9.19 Productivity in Scaffolding Assembly

two laborers, larger assemblies, or vertical building of staging require a large crew. The costs of loading and unloading trucks to and from the shop must also be included. For larger projects the masonry estimator should consider soliciting quotes from subcontractors who erect, dismantle, rent, and maintain scaffolding assemblies. On projects where trades other than the bricklayer will use scaffolding, this may be the responsibility of the prime contractor.

Telescoping power lifts are not always part of the basic equipment of the masonry contractor. Estimators should consider rental and operating costs as part of the estimate, as well as mobilization and demobilization costs associated with delivering and removing the equipment.

CLEANUP

Most masonry contracts provide for cleanup of debris generated by the work. Occasionally, the specifications will require that the masonry subcontractor is also responsible for removal and disposal of masonry debris. The amount of cleanup varies to some degree based on the contract or specifications and the stage of the building progress when the work is done. Cleanup typically consists of picking up broken or discarded brick, block, or other such masonry units, as well as ties and forms, and scraping or sweeping mortar droppings. Laborers generally perform masonry cleanup most often accomplished by hand although a backhoe might be used to clean up larger amounts of debris. Use judgment, based on experience, to determine the labor-hours necessary for the level of cleanup required. The cost for disposal in the form of dumpster rentals or landfill fees should be included. Consult Chapter 07, "Existing Conditions," for costs associated with dumpsters.

MASONRY SEALING

Masonry sealing is the application of a coating to the exterior surface of the masonry work. The coating can be silicone based, aimed at eliminating the penetration of water or moisture through the building's exterior walls. Other coatings can be applied to prevent graffiti from adhering to the building. Both applications fill the pores of the masonry against permanent damage or penetration of the surface.

Products are applied to the surface of the masonry by spraying, roller, or even brushing. Larger projects may use power spray equipment much like that used by painters.

Taking-off and Pricing

Masonry sealers are taken off and priced by the square foot of surface area to be sealed (SF) and the number of applications. Some require the use of primers to prep the surface. The specific method of application will affect the price and should be noted as the means of application. Protection of areas not scheduled to be sealed, such as windows, doors, and non-masonry surfaces, must also be included in the work. Also, protection of adjacent surfaces to the spray application is also a cost to be considered. As with all masonry work, the cost of accessing (staging) should be included based on the application. Labor costs are determined by the production and cost of an individual or crew, depending on the process.

SUMMARY

Masonry work is most often a crew task requiring that incremental pieces be assembled to create the structural element or the veneer facing. In addition to the specific considerations discussed in this chapter, the estimator should consider the impact temperature extremes and the use of scaffolding or specialized equipment may have on the project. Masonry work on the exterior of the building requires insulation, which can be applied to the surface of the substrate or included with in the cells of the CMU.

Similar to concrete, vertical and horizontal reinforcing is required in CMU to counteract the effects of seismic events and allow slight movement without damage. Rebar is calculated in the same manner as rebar in cast-in-place concrete both horizontal and vertical are taken off by length as a function of the spacing and converted to weight for pricing. Other types of horizontal reinforcing called lateral reinforcing may be required as well.

Productivity is often affected by a variety of factors, including weather, location of the work, type of masonry unit, and application. Cooperation and coordination of the work of other trades with items embedded in the masonry should also be included in the estimate.

For many masonry projects, especially with brick veneer exteriors or glazed block partitions, the architect will frequently require *mock-up* panels as part of the submittal process. A mock-up panel is a sample of the assembled wall or application. Mock-up panels allow the architect and owners to view the finished product before it goes on the building. These can be small (4' × 4') or large (10' × 10') with various shades of the specified mortar and joint. They can be built with a CMU backup complete with flashings, insulation, and vapor barrier. The point is that they can be costly and time-consuming. The estimator should consider this when reviewing masonry specs and subcontractor proposals.

In addition to the masonry work itself, costs such as shipping of stone or brick, scaffolding for access to the work, cleaning of masonry, and masonry sealing may also be part of this Division.

QUESTIONS/ PROBLEMS

1. A CMU building is 77'-2" × 40'-0" with a height of 22'-0" to the top course of the 8" CMU. There are three (3) 12' × 12' overhead doors, three (3) 3' × 7'-2" passage doors and eight (8) 6' × 4' windows. What is the net area (SF) of the block walls? With a 5% waste factor included how many blocks are required?

2. Using the net area derived in question #1, how many cubic yards of grout are required to fill the cells solid if the cores are grouted at 32" o.c. and the block is 40% solid. Use the appropriate table from Chapter 9 and round your answer up to the nearest whole cubic yard. Cite the table used.

3. Calculate the amount of lateral reinforcement in LF, needed if the wall is reinforced at every other course. Do not deduct for the openings.

4. Using the appropriate table from Chapter 9, what is the labor cost in dollars if the cores are grouted at 32" o.c. and it takes 0.038 labor-hours per SF of wall? The cost per labor-hour is $96.50.

5. For the building in question #1, there is a continuous bond beam at 10'-0" and the top course of the building. Calculate the total material cost of the grout to fill just the bond beam courses if there 0.34 CF per LF and the cost per CF is $11.50.

6. Using the appropriate figure in Chapter 9, calculate the amount of brick required to cover (veneer) the front elevation (40') only, full height with a standard face brick (8" × 2¼") with a ⅜" mortar joint, set in a running bond. The front elevation has one (1) 3' × 7'-2" passage doors and three (3) 6' × 4' windows. Do not include bricks for the window-sills. Cite the table used.

7. If the crew will lay up 1,500 bricks per day complete, how long will it take to lay up this wall? Round your answer up to nearest whole day.

8. Using the appropriate table in Chapter 9, what would be the difference in the number of bricks needed if the wall was changed to a Modular Norman brick with the same joint thickness? Cite the table used.

9. Using the appropriate table in Chapter 9, calculate the labor hours required to lay up the Modular Norman brick in question #8. Round up to the nearest whole labor hour.

10. Using the appropriate table from Chapter 9, how many labor-hours are required to wash the face of the brick? Round your answer up to the nearest whole labor-hour.

10 Metals

CSI MasterFormat® Division 5—Metals includes one of the most versatile materials used in the construction industry – structural steel. Structural steel can best be defined as the steel members that make up the frame of a building, which transmit the load to the foundation. Steel has the capacity to support large loads of a relatively compact size, in both compression and tension, making it an ideal material for flexural components. The variety of steel shapes and sizes available provides engineers with an economical solution to many structural design problems. Figure 10.1 lists the most common shapes and their respective designations. Structural steel is also available in different strengths or grades for particular loading or stress conditions. When estimating Division 5 work in general, the following categories must be considered:

- Structural steel shapes
- Miscellaneous steel items
- Open-web bar joists
- Metal decking
- Erection crew and cranes
- Field welding
- Items furnished for installation by other trades (such as steel lintels set by a mason)
- Shop and field priming/coatings
- Shop drawings

This chapter is limited to a discussion of the more common shapes of structural steel, open-web bar joists, and metal decking.

STRUCTURAL STEEL

Each steel shape, or *section*, is prefixed by a letter and numbers. These designations are more than simple identifications; they provide important information about the individual section. The letter stands for the

Estimating Building Costs for the Residential and Light Commercial Construction Professional,
Third Edition. Wayne J. Del Pico.
© 2023 John Wiley & Sons, Inc. Published 2023 by John Wiley & Sons, Inc.

The upper section of this table shows the name, shape, common designation, and basic characteristics of commonly used steel. The lower portion explains how to read the designations used for the illustrated sections above.

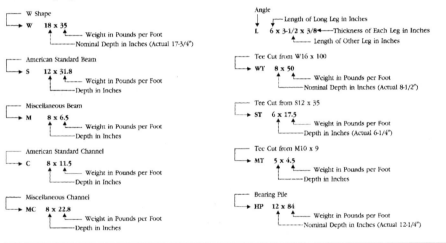

Shape & Designation	Name & Characteristics	Shape & Designation	Name & Characteristics
W	W Shape / Parallel flange surfaces	MC	Miscellaneous Channel / Infrequently rolled by some producers
S	American Standard Beam (I Beam) / Sloped inner flange	L	Angle / Equal or unequal legs, constant thickness
M	Miscellaneous Beams / Cannot be classified as W, HP or S; infrequently rolled by some producers	T	Structural Tee / Cut from W, M or S on center of web
C	American Standard Channel / Sloped inner flange	HP	Bearing Pile / Parallel flanges and equal flange and web thickness

Common drawing designations follow:

W Shape
W   18 x 35
— Weight in Pounds per Foot
— Nominal Depth in Inches (Actual 17-3/4")

American Standard Beam
S   12 x 31.8
— Weight in Pounds per Foot
— Depth in Inches

Miscellaneous Beam
M   8 x 6.5
— Weight in Pounds per Foot
— Depth in Inches

American Standard Channel
C   8 x 11.5
— Weight in Pounds per Foot
— Depth in Inches

Miscellaneous Channel
MC   8 x 22.8
— Weight in Pounds per Foot
— Depth in Inches

Angle
— Length of Long Leg in Inches
L   6 x 3-1/2 x 3/8 — Thickness of Each Leg in Inches
— Length of Other Leg in Inches

Tee Cut from W16 x 100
WT   8 x 50
— Weight in Pounds per Foot
— Nominal Depth in Inches (Actual 8-1/2")

Tee Cut from S12 x 35
ST   6 x 17.5
— Weight in Pounds per Foot
— Depth in Inches (Actual 6-1/4")

Tee Cut from M10 x 9
MT   5 x 4.5
— Weight in Pounds per Foot
— Depth in Inches

Bearing Pile
HP   12 x 84
— Weight in Pounds per Foot
— Nominal Depth in Inches (Actual 12-1/4")

Figure 10.1 Common Steel Sections

classification of the piece by shape. The first number refers to the nominal depth in inches of the section. The second number refers to the weight in pounds per linear foot of the section, which is critical in estimating the costs of the section—both in material and labor. For example:

> W12 × 22: The W designates that the member is a wide flange section, the 12 indicates its approximate depth in inches, and the 22 refers to its weight in pounds per linear foot (LF).

Angle shapes are an exception to this rule. The first two numbers in an angle designation stand for the lengths of the "legs" of the particular angle, in inches. The third number is the thickness of each leg, in inches. Some shapes, such as angles, do not include a weight in their designation. Determining the weight of angles, plate steel of various thicknesses, and tube steel requires the use of a table or online.

Structural steel is also available in different strengths, expressed as *yield stress*. Simply put, the yield stress, typically defined as *kips per square inch* (KSI), is the maximum allowable stress that can be exerted on a material before it fails. (A *kip* is a unit of measure equal to 1,000 pounds.) The different grades of steel are named according to the number of the test conducted by the American Society of Testing and Materials (ASTM) to determine the characteristics of the species. Not all the previously mentioned shapes are available in some of the more specialized grades of steel. Figure 10.2 lists common structural steel grades.

Drawings

Structural steel work is shown on structural drawings in plan view. Elevations, sections, and details are often added for clarity. Using schedules provided within the drawings to list columns, lintels, or repetitive steel features saves time during the takeoff. The details show the connections of individual pieces. Figure 10.3 is a simple structural steel drawing shown in plan view with the corresponding details to help illustrate some of the more common features.

Steel type	ASTM designation	Minimum yield stress in KSI	Shapes available
Carbon	A36 A529	36 50	All structural shape groups, and plate & bars through 8" thick
High-strength, low-alloy quenched and self-tempered	A913	50 60 65 70	All structural shape groups
High-strength, low-alloy Columbium-Vanadium	A572	42 50 55 60 65	All structural shape groups, and plates & bars through 6" thick All structural shape groups, and plates & bars through 4" thick Structural shape groups 1 & 2, and plates & bars through 2" thick Structural shape groups 1 & 2, and plates & bars through 1¼" thick Structural shape group 1, and plates & bars through 1¼" thick
High-strength, low-alloy Columbium-Vanadium	A992	50	All structural shape groups
Weathering high-strength, low-alloy	A242	42 46 50	Structural shape groups 4 & 5, and plates & bars 1½–4" thick Structural shape group 3, and plates & bars ¾–1½" thick Structural shape groups 1 & 2, and plates & bars ¾" thick
Weathering high-strength, low-alloy	A588	42 46 50	Plates & bars 5–8" thick Plates & bars 4–5" thick All structural shape groups, and plates & bars through 4" thick
Quenched and tempered low-alloy	A852	70	Plates & bars through 4" thick
Quenched and tempered alloy	A514	90 100	Plates & bars 2½–6" thick Plates & bars through 2½" thick

Figure 10.2 Common Structural Steel Specifications

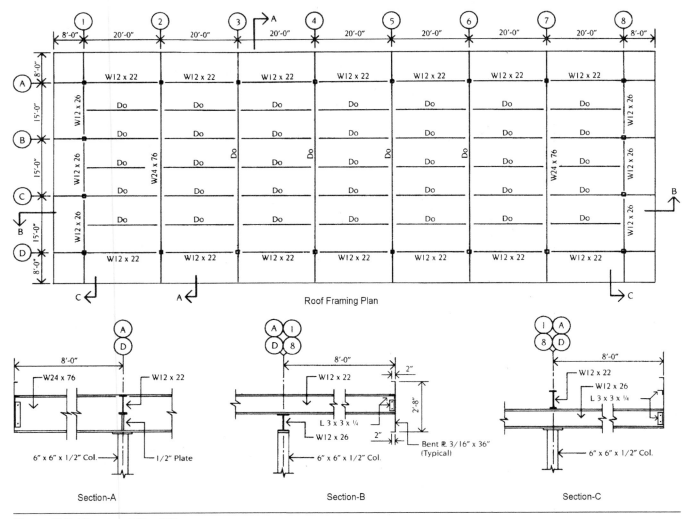

Figure 10.3 Structural Steel Plan

Structural steel drawings include nomenclature unique to the work. One of these is the term *Do,* which indicates duplication, similar to the expression *ditto.* It is used to show repetitive use of the same designation beam, girder, joist, and so forth. The symbol < is used on drawings to identify angles. The number "2" before the symbol refers to a pair of angles.

Taking-off and Pricing

Structural steel is taken off by the length of the piece, converted to weight in pounds and extended to tons (TNS) for pricing. Individual components should be listed according to shape and designation. Start with the largest member of the designation, and work down to the smallest. Pieces can be further classified by use (for example, beams, columns, and girders). Separating the pieces by application allows for pricing the erection and any special coatings that apply, rather than the materials. Referring to Figure 10.3, the weight of the steel can be calculated for each individual member. For example:

Calculate the weight of the W12 × 22 beams between column lines 1 and 2 shown in Figure 10.3. From the designation number, it can be determined that the W12 × 22 beam weighs 22 lb. per LF, and the plan shows 7 beams 20'-0" long. Therefore, the following calculation can be made:

$$7\,\text{beams} \times 20\text{'-0"}\ \text{each} \times 22\,\text{lb.}/\text{LF} = 3{,}080\,\text{lb.} = 1.54\,\text{TNS}$$

Many structural steel drawings provide column lengths. The highest elevation on a column or beam is called the *top of steel* and is shown on the drawings as TOS. The proposed elevation for the top of the leveling plate is also given. The difference between the two is the length of the column. All structural steel is priced by weight in tons. This includes costs for milled materials, fabrication labor, erection, and all other miscellaneous tasks, such as priming, coatings, and trucking to the site.

Carefully take off the lengths of the various members, from centerline of columns to centerline of columns. No deduction is required for connections in estimating. Material costs change frequently since steel is a commodity, so material prices should be contemporaneous whenever possible.

OPEN-WEB STEEL JOISTS

Open-web steel web joists, also known as *bar joists*, are structural members in a steel framing system that carry roof or floor gravity loads to other members in the system such as beams or columns. Steel joists are manufactured by welding hot-rolled or cold-formed sections to angle web or round bars to form a truss. Standard open-web and long-span steel joists were developed as a cost-effective alternative to wood frame construction. Steel joists' capacity to carry loads spanning greater distances has made them popular for all types of light-occupancy construction. They also can be used in fire-rated construction, and their open webbing allows the passage of mechanical piping and electrical conduits without drilling or coring holes.

Classification of Steel Joists

Steel joists can be classified according to one of the following three categories or series:

- **K-Series (including KCS):** Open-web, parallel-chord steel joists manufactured in standard depths of 8" 10", 12", 14", 16", 18", 20", 22", 24", 26", 28", and 30" with lengths up to 60'.
- **LH-Series**: Long-span, open-web steel joists manufactured in depths of 18", 20", 24", 28", 32", 36", 40", 44", and 48" with lengths up to 96'.
- **DLH-Series**: Deep, long-span, open-web steel joists manufactured in depths of 52", 56", 60", 64", 68", and 72" with lengths up to 144'.
- **CJ-Series**: Composite steel joists have a sheer connection between the top chord of the joist and the overlying concrete slab. CJ series have uniform depths of 10" to 96" with spans up to 120'.
- **Joist girders**: Primary structural members that typically bear on columns and carry the loads imposed by bar joists that run perpendicular to them.

Figure 10.4 shows the typical details for K, LH, and DLH series' open-web joists and the different types of designs available.

LH- and DLH-Series are also available with top chords that are pitched or parallel to the bottom chords. The ends of the joists can be square ends or *underslung*. Joist designations are defined as follows:

24K10

The first number, 24, is the depth in inches of the joist.

The letter K indicates that it is a K-Series joist.

The last number, 10, indicates the load capacity/size of the chords.

Taking-off and Pricing

In Figure 10.5, open-web steel joists are shown as they would appear in plan view on a typical structural drawing. They are quantified by the pound and then extended to tons (TNS) for pricing. They are typically taken off by the LF and quantity of each series and designation, and then converted to weight,

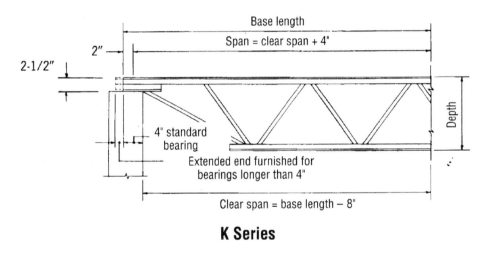

K Series

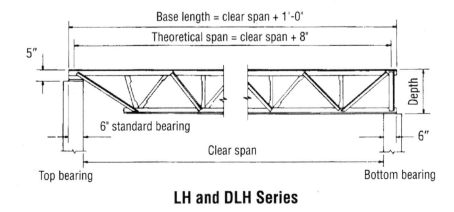

LH and DLH Series

Figure 10.4 Standard Joist Details

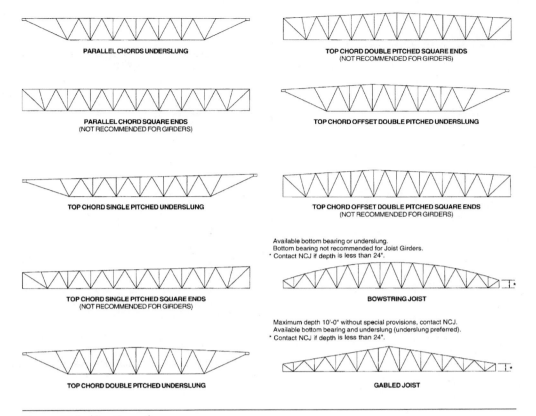

PARALLEL CHORDS UNDERSLUNG

TOP CHORD DOUBLE PITCHED SQUARE ENDS
(NOT RECOMMENDED FOR GIRDERS)

PARALLEL CHORD SQUARE ENDS
(NOT RECOMMENDED FOR GIRDERS)

TOP CHORD OFFSET DOUBLE PITCHED UNDERSLUNG

TOP CHORD SINGLE PITCHED UNDERSLUNG

TOP CHORD OFFSET DOUBLE PITCHED SQUARE ENDS
(NOT RECOMMENDED FOR GIRDERS)

Available bottom bearing or underslung.
Bottom bearing not recommended for Joist Girders.
* Contact NCJ if depth is less than 24".

TOP CHORD SINGLE PITCHED SQUARE ENDS
(NOT RECOMMENDED FOR GIRDERS)

BOWSTRING JOIST

Maximum depth 10'-0" without special provisions, contact NCJ.
Available bottom bearing and underslung (underslung preferred).
* Contact NCJ if depth is less than 24".

TOP CHORD DOUBLE PITCHED UNDERSLUNG

GABLED JOIST

Figure 10.4 *(Continued)*

in TNS. This requires referring to a table that lists the weights per foot of the different species of bar joist. Consulting the plan view of a typical open-web, K-Series steel joist (Figure 10.5 and Figure 10.6), it is possible to calculate the total weight of the K-Series steel joists for pricing. For example:

> *Calculate the weight of the K-Series steel joists in Figure 10.5.*
> *Joists shown are 16K4. A takeoff shows a quantity of 10 with a span of 30'-0".*
> *Locating a 16K4 joist in the table in Figure 10.5, we can see that it has a weight of 7.0 lb. per LF.*

> 10 EA × 30 LF = 300 LF × 7.0 lb./LF = 2,100 lb. = 1.05 TNS.

Once the total weight of the bar joists has been calculated, the estimator can multiply the weight by the cost of material, fabrication, and erection to determine the in-place cost.

Separate bar joists by designation number and series for accurate pricing. Steel joists that are welded in place should be listed separately in the estimate from those that are connected by nuts and bolts. Check the documents carefully and coordinate mechanical drawings for doubling joists at rooftop equipment. Once the bar joists have been set and fastened, a brace is added to protect against lateral movement called *bridging*. It provides

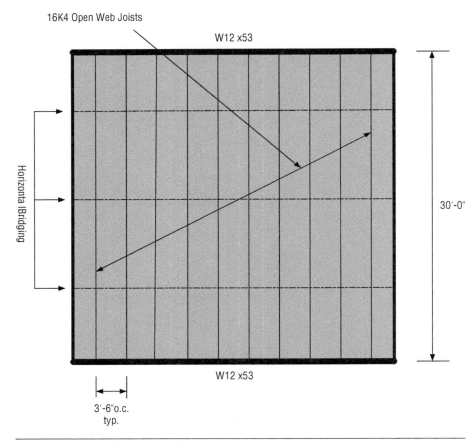

Figure 10.5 Open-Web Joist Plan

stability to prevent wracking of the joists. Bridging is most often accomplished with small, lightweight steel angles bolted or welded perpendicular to the span of the joist. The locations of the courses of bridging are shown on the structural plans and are determined in accordance with the joist manufacturer's recommendation or the structural design.

There are two main types of bridging:

- **Horizontal:** Two continuous steel members, one fastened to the top chord and one fastened to the bottom chord.
- **Diagonal:** Members running diagonally from the top of one chord to the bottom chord of the adjacent joist.

It is important that the terminations of bridging are securely fastened to the wall, column, or beam, regardless of the bridging installation method. This helps tie the bar joists to the structure. Both horizontal and diagonal bracing are taken off and priced by the LF. Bridging mechanically fastened with nuts and bolts should be listed separately in the estimate from welded bridging, as they have different installation costs.

Joists that have sheer pins at the top chord should be separated in the estimate as well, as there are different installation costs.

Joist Designation	8K1	10K1	12K1	12K3	12K5	14K1	14K3	14K4	14K6	16K2	16K3
Depth (in.)	8	10	12	12	12	14	14	14	14	16	16
Approx. Wt. (Lbs./Ft.)	5.1	5.0	5.0	5.7	7.1	5.2	6.0	6.7	7.7	5.5	6.3
Joist Designation	16K4	16K5	16K6	16K7	16K9	18K3	18K4	18K5	18K6	18K7	18K9
Depth (in.)	16	16	16	16	16	18	18	18	18	18	18
Approx. Wt. (Lbs./Ft.)	7.0	7.5	8.1	8.6	10.0	6.6	7.2	7.7	8.5	9.0	10.2
Joist Designation	18K10	20K3	20K4	20K5	20K6	20K7	20K9	20K10	22K4	22K5	22K6
Depth (in.)	18	20	20	20	20	20	20	20	22	22	22
Approx. Wt. (Lbs./Ft.)	11.7	6.7	7.6	8.2	8.9	9.3	10.8	12.2	8.0	8.8	9.2
Joist Designation	22K7	22K9	22K10	22K11	24K4	24K5	24K6	24K7	24K8	24K9	24K10
Depth (in.)	22	22	22	22	24	24	24	24	24	24	24
Approx. Wt. (Lbs./Ft.)	9.7	11.3	12.6	13.8	8.4	9.3	9.7	10.1	11.5	12.0	13.1
Joist Designation	24K12	26K5	26K6	26K7	26K8	26K9	26K10	26K12	28K6	28K7	28K8
Depth (in.)	24	26	26	26	26	26	26	26	28	28	28
Approx. Wt. (Lbs./Ft.)	16.0	9.8	10.6	10.9	12.1	12.2	13.8	16.6	11.4	11.8	12.7
Joist Designation	28K9	28K10	28K12	30K7	30K8	30K9	30K10	30K11	30K12		
Depth (in.)	28	28	28	30	30	30	30	30	30		
Approx. Wt. (Lbs./Ft.)	13.0	14.3	17.1	12.3	13.2	13.4	15.0	16.4	17.6		

Figure 10.6 Weights of Open-Web Steel Joists (K Series)

METAL DECKING

Metal decking consists of specially formed sheets of steel applied perpendicular to the span of the joists. These sheets serve as a substrate for the installation of roofing materials, such as rigid insulation and membrane, or as permanent forms for concrete floor slabs. Metal decking is fabricated from steel sheets in 18-, 20-, or 22-gauge thicknesses. The typical width is 30" with standard lengths ranging from 14' to 30', depending on specific project needs. The depth of the decking section can vary with type and application, but is usually 1½" to 2½". Special coatings and colors are available, but the most common finish is galvanized to reduce deterioration. Individual sections are fabricated with interlocking end and side laps for added rigidity.

Classification of Metal Decking

Metal decking can be classified in two main categories for estimating purposes:

- Corrugated, undulated, or "corruform" concrete-fill permanent forms, also used for roofing and siding of industrial-type buildings.

- Cellular-type with well-defined, bent contours in trapezoidal or rectangular pitches or depths, used as permanent concrete forms for larger-span spacings or heavier live load applications.

Other, more unique types of decking include long-span decking that ranges in depth from 4½" to 7½" fabricated from 14-, 16-, 18-, or 20-gauge sheet steel and acoustical decking which provides sound-deadening capabilities, while still maintaining a structural load capability.

Taking-off and Pricing

Metal decking is taken off by the square foot of area to be covered and extended to the square (SQ), where 1 SQ is equal to 100 SF. Allowances should be made for overlap at the sides and ends of the individual sheets, in accordance with the specification section of the design requirements noted on the drawings. Metal decking of different gauges, shapes, or finishes should be listed separately in the takeoff, as this will affect the material cost and potentially the installation cost as well.

Metal decking that serves as a permanent concrete form requires the use of sheet metal angles fastened to the perimeter of the decking and openings to act as an edge form for the placement of the concrete. Gauge of the angles and leg dimensions should also be indicated on the plans. These angles are taken off by the LF and should be listed by size and method of installation for proper pricing. The angles can be fastened by both welding and self-tapping screws. They are priced by the LF for both material and installation costs. (See "Edge Forms," Chapter 8.)

Another special task to check for is the installation of *shear studs*, which connect the composite floor systems to the structural steel through the decking. This is a separate task often performed last, just prior to the placement of the concrete. Shear studs are taken off and priced by the piece (EA), based on their spacing and locations. Material costs change frequently since steel is a commodity, so material prices should be contemporaneous whenever possible.

Labor to Install Metal Decking

Review the specifications carefully for installation requirements. For example, thinner-gauge decking that is welded may require the use of thickening washers placed at the point of fusion, to avoid burning through the decking. This practice is common for metal of 20-gauge or higher (thinner metal). Decking that is fastened to the structure via welding is often required to be "stitched" with self-tapping screws at the side and end laps at specific spacing. Also analyze the specifications for field touch-up requirements for the galvanized finish at the welds. This can be a time-consuming process and should be listed separately in the takeoff. It can be quantified by square foot area (SF) or labor-hours (LH) to perform the work. Installation is also priced by the square, (SQ).

Since decking is typically loaded to the roof or floor with a crane, there is an equipment component in the price. The loading of decking is coordinated with the erection of the steel so that a separate crane is not required. It is often placed on joists that have been positioned and fastened. The installation and welding tasks require a crew. The decking must be spread, with the required overlap and side lap, while welders follow behind and tack it to the deck. Review the specifications carefully for cutting the deck for openings for items such as ducts, curbs, hatches, and miscellaneous penetrations. This can often be time consuming and can require the opening to be reinforced from underneath with angles. Material prices change frequently since steel is a commodity, so material prices should be contemporaneous whenever possible. Also consider labor to perform punchlist tasks, which are almost inevitable with this type of work. Figure 10.7 illustrates the various types of steel decking.

LIGHT-GAUGE METAL FRAMING

Light-gauge metal framing, or LGMF refers to the method of construction that uses a high-tensile-strength, cold-rolled steel formed in the shape of joists, studs, track, and channel. It is listed in CSI 05 40 00 as Cold-Formed Metal Framing. All components are fabricated of structural grade steel in 12-, 14-, 16-, and 18-gauge thicknesses. The various sections are designed to provide the load-bearing characteristics of steel or wood framing at a reduced weight and cost. Light-gauge framing is also noncombustible and does not warp, shrink, or swell in contrast to wood. Sections are available with a galvanized coating or red zinc chromate paint that resists rusting. Slots or channels are factory-punched within the web of the section to allow the passage of wiring, piping, and horizontal bracing. LGMF can be shop-fabricated and delivered ready for assembly, thereby reducing on-site labor costs, and increasing quality control.

On-site cutting of the sections is done with a "chop" saw outfitted with a high-speed metal-cutting blade. Layout and erection are similar to the procedures used for wood framing components. Fastening the various components can be done by bolting, screwing with self-tapping screws, or welding. Again, be sure to review the documents for touching up welds with a zinc-based coating, a frequent requirement to maintain the integrity of the galvanizing. This can be a time-consuming process if done in the field.

Many LGMF projects require shop drawings that are designed, reviewed, and stamped by a registered structural engineer. This can be for both commercial and residential projects. The estimator is advised to consult the specification section governing LGMF work to determine the extent of the shop drawings as they can have a significant cost. They should be a separate line item in the estimate.

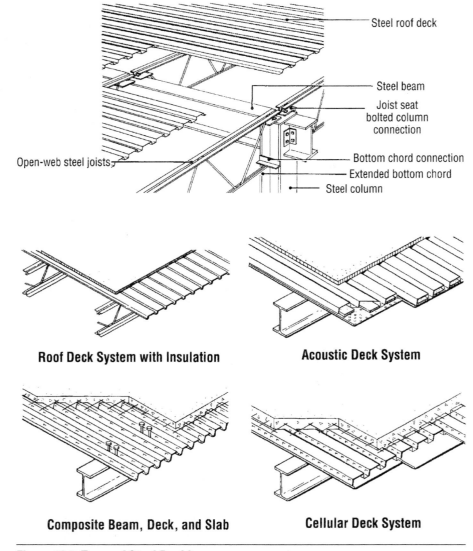

Figure 10.7 labels:
- Steel roof deck
- Steel beam
- Joist seat bolted column connection
- Bottom chord connection
- Extended bottom chord
- Steel column
- Open-web steel joists

Roof Deck System with Insulation

Acoustic Deck System

Composite Beam, Deck, and Slab

Cellular Deck System

Figure 10.7 Types of Steel Decking

Joists

Joist sections are available in 6", 8", 10", and 12" depths, and flanges in $1\frac{5}{8}$"
or $2\frac{1}{2}$" widths. The most common lengths are from 8' to 30' in 2' increments.
Joists are used in floor and roof construction with the typical 12", 16", and
24" on-center spacings familiar in wood frame construction. Joist and wall
bridging can be done with special C-shaped channels that are inserted into the
aligned punched slots, provided by the manufacturer. The spacing and
location of the bridging components are found in the specifications or on the
structural drawings.

Taking-off and Pricing

LGMF components are taken off by the LF. Special channels for use as box
joists (sill or head plates in wood construction) are also taken off by the LF, as

well as the C-shaped stiffener used within the walls or joist framing system. Sections of different shapes, sizes, gauges, and coatings (as well as the method of fastening) should be listed separately for accurate material and installation pricing. For projects that require fabricating panels off-site, the costs for materials and fabrication can be converted to the SF area of the panelized system. Labor for fabrication off-site must be added to the cost of erection on site, as well as transportation and handling at the site. The two costs should be figured separately, as the labor rates and productivities can be different. Again, the estimator should consider the specifications for the extent of the shop drawings and submittals.

STUDS AND TRACK

Studs are the vertical components of a wall system and are C-shaped with folded flanges. The specifications or plans determine the on-center spacing requirements. Steel track refers to the horizontal component to which the steel stud is fastened. It is typically located at the top and bottom perimeters of the wall. Figure 10.8 illustrates a load-bearing wall. Steel studs and track are available in 1⅝", 2½", 3⅝", 4", 6" and 8 depths. Standard gauges are 14, 16, and 18. They are also available in 20- and 25-gauge thicknesses, but these are not considered to be of load-bearing capacity and are not part of Division 5. Lighter gauge framing can be found in Division 9 (discussed in Chapter 14, "Finishes").

Taking-off and Pricing

There are four basic methods for estimating metal studs and track:

1. Take off the stud and track separately by the LF. Price studs and track totals separately for materials and labor costs.
2. Take off the wall by the LF, where 1 LF of wall is equal to 1 LF of length by the height. For example, 76 LF of wall that is 8'–0" high is quantified and listed as 76 LF. Convert material and labor costs to LF of wall based on a specific height. This requires wall of consistent height.

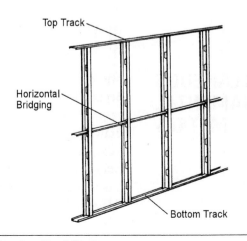

Figure 10.8 Load-Bearing Steel Studs

3. Calculate the SF area of the wall (length × height). For example, the wall cited in Method #2, 76 LF × 8' high, would be listed in the takeoff as 608 SF. Convert material and labor costs to SF.
4. Calculate the SF area as in Method #3. Estimate the cost of panel fabrication and the cost of erection separately. Include equipment costs required for erection.

Regardless of the method selected, be sure to separate walls of different stud widths, gauges, heights, methods of fastening, on-center spacings, and applications. Added materials for openings in the wall must also be calculated. Similar to wood construction, openings in light-gauge, load-bearing metal stud walls must be framed to support the load transmitted from above. This involves the use of headers, short studs called "cripple" studs, sills, and doubled jamb studs. Openings and their components may be taken off and priced by the opening (EA) based on size, or by the individual component (e.g., doubled studs per LF, headers per LF, or sills per LF). The choice of estimating method should be based on the type of openings. For example, openings for numerous doors and windows of the same size may best be quantified per opening (EA).

Material costs change frequently since steel is a commodity, so material prices should be contemporaneous whenever possible. Specialized equipment, such as staging for installation, may also be needed. Equipment should be listed separately, based on length of time required. Charges for erection, dismantling, or moving should also be included.

Many wall-framing applications require some level of scaffolding or lift to attach the top track and top of studs. The estimator should consider the equipment as part of the cost of the work.

A last thought on LGMF work in general: Most applications require some level of layout work. The LGMF estimator should clarify who will do the work – whether by a separate party such as a field engineering team or by the LGMF contractor. If layout is the responsibility of LGMF contractor, it often takes a crew of two and can be time consuming if there are lots of line changes in the structure. In this case, it is the responsibility of the GC or the field engineer to establish control lines for the LGMF contractor to start the layout from.

MISCELLANEOUS AND ORNAMENTAL METALS

Miscellaneous and ornamental metals are frequently specified under CSI Section 05 50 00—Metal Fabrications and 05 70 00—Decorative Metal, respectively. The work can be shown on both the architectural and structural drawings. Exterior railings are also shown on site improvement drawings. Miscellaneous and ornamental metals are typically separated into two categories for takeoff purposes. *Miscellaneous* refers to the various items of metal fabrications that do not fit into one specific classification. *Ornamental* refers to decorative railings and special metal fabrications for stairs as well as more unique metal fabrications.

It is not uncommon for projects with limited structural steel to still have a considerable number of miscellaneous and ornamental metal items. This category encompasses a diverse range of items that are fabricated, delivered, and installed or simply provided for other trades to install. The following sections describe the more common items and their respective takeoff and estimating procedures. It is by no means a comprehensive list.

Structural Steel Lintels

These are structural members, typically installed in masonry walls over windows or doors, to support the load of the masonry above the opening. The most common type of lintel is the steel angle. These can be cut from stock lengths of various-sized steel angle stock. They are longer than the width of the opening in order to span the opening and provide bearing on either side. The amount of bearing is dictated by the specifications and/or applicable building code, contingent on the size of the opening and the load above it. Angles can be supplied as either individual units or composite units of multiple angles, welded together for walls of multiple masonry widths.

Taking-off and Pricing

Like most steel items, lintels are taken off by the individual piece (EA) and converted to weight (TNS) for pricing. This is done by adding the bearing requirement for each side to the width of the opening in LF, multiplying the LF by the number of pieces, and then multiplying by the weight per LF to obtain a weight in pounds or tons.

For buildings with a substantial masonry scope of work, the structural documents may list the lintels in a lintel schedule, which refers to the lintel by name, such as L1, L2, L3, and the species of angle. It also specifies the size in length and whether the unit is an individual or a composite of angles welded together. Lintels should be listed separately from structural steel items on a takeoff, because the unit price per pound for smaller sections is considerably more than for larger structural sections. They also are frequently required to be galvanized as a finish coating to protect against the weather. This process has an added cost impact. Lintels are a furnished-and-delivered item only. Installation is typically included as part of the masonry scope of work because the lintels must be installed as the masonry courses progress. The estimator should confirm the party responsible for installation in the technical specifications.

Pipe Railings

Pipe railings are commonly found at interior and exterior stairs or ramps for pedestrian traffic. They are constructed of welded-steel pipe in configurations to match the profile of the stair or ramp. Diameters of the pipe can range from 1¼" to 1½". The welds are ground flush, and the railing is often primed, galvanized, or finish-painted ready for installation. They are installed by the use of sleeves, which are slightly bigger than the diameter of the pipe, embedded in the wet concrete. They can also be installed by coring a hole in the concrete and grouting the rail in place with the use of hydraulic cement. For applications in materials other than concrete, the railings may incorporate the use of a flange welded to the point of attachment that has holes punched for bolted fastening.

Taking-off and Pricing

The most common unit of takeoff and pricing for railings is by the LF of completed rail. Railings of various diameters and different methods of installation and finishes should be listed separately. Review the documents carefully for all types and locations of railings. It is not uncommon for railings to occur at grade or on the roof at the perimeter of an access hatch. The method of attachment and finishes are among the primary determinants in the cost. Stainless steel or shop finishing should also be noted to eliminate duplication under the Painting section in Division 9. Coring holes for the railing anchorage is estimated by the quantity and the diameter as previously noted in Chapter 8, "Concrete," as is the anchoring cement.

Ladders

Permanently installed ladders for access to high roofs from low roofs, from elevator pits, and from the interior to the roof are typically constructed of flat steel bar stock, with rungs of round steel bars. They are fabricated by welding the rungs perpendicular to the flat stock at the specified spacing. Angles or brackets are welded to the flat stock for fastening to the masonry or concrete wall. Standard finishes include primer, finished paint, and galvanizing ready for on-site finish paint.

Taking-off and Pricing

As with railings, the cost of ladders is impacted by the specific method of attachment and finish. Ladders are typically taken off by the vertical linear foot (VLF) or by the individual piece (EA). They are priced by the same units. Shop finishing should also be noted to eliminate duplication under the Painting section. Vertical ladders often have safety cages when the ladder exceeds a specific vertical height. These can be a standard prefabricated size or custom fabricated. The safety cage is often taken off and priced by the VLF for fabrication and painting as well. Frequently the cage is installed separately in the field. This allows ease of handling of the ladder as a result of reduced weight.

Grates

Grates are fabricated items used to allow the flow of air or water while supporting vehicular or pedestrian traffic. They are composed of parallel flat bar stock with intermediate round bars welded between. Grates typically include angles for support that are bolted to the adjacent masonry or concrete surfaces. Grating is also used for the intermediate landings and treads in industrial application stairs and for exterior fire escapes. In these applications, other parts need to be included in the estimate, such as toe plates and metal stringers to support the treads.

Taking-off and Pricing

The standard unit of takeoff and pricing for grating is SF. The LF measurement of support angles should be included as part of the takeoff and price. Figure 10.9 illustrates the use of grating as landings and treads on a stair. Review the specifications for coatings, such as galvanizing or special finishes that can have a significant impact on the cost. Railings that are part of a stair platform should be estimated separately.

Metal Pan Stairs

Steel stairs may be of the metal pan type, which requires placing concrete into the treads and landings as the wearing surface. They can also be constructed of diamond plate steel or metal grating (see previous section on grating). Metal pan stairs are typically fabricated off-site in multiple pieces, which usually include stringers with treads and risers attached to landing

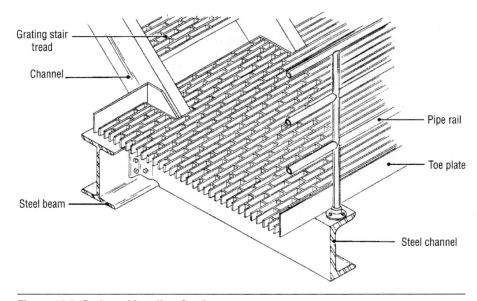

Figure 10.9 Stair and Landing Grating

components. They are then delivered and assembled or installed to complete the application. Many of the components to the assembly may be too heavy to handle without some level of equipment.

Taking-off and Pricing

The cost of fabrication should be kept separate from the cost of installing the stairs. Steel stair takeoff and its supporting structural members should be kept separate from the other miscellaneous metals categories. Stairs are frequently priced per riser (RSR), which would include the treads and supporting stringers. The quantity and size of the tread and riser combination will affect the price. Sizes of landings should be taken off and priced separately by the square foot of area (SF). Other parts, such as railings, toe rails, or miscellaneous filler pieces, are taken off and priced separately, by LF or EA depending on the specific application. Also remember to include the cost of the concrete to fill the pan and the landing deck. These costs can be included as part of the stair estimate, or they can be included in Division 3—Concrete.

The placement of concrete in the stairs can be labor intensive. It may also require the use of a concrete pump, if there is sufficient quantity. Since the process is fairly messy, be sure to include labor to clean the concrete from adjacent work. Figure 10.10 illustrates the basic components of a metal pan stair and the typical associated labor-hours.

Joist-Bearing Plates/Embedded Items

Joist-bearing plates are small pieces of steel of varying thicknesses, the most common being ½" or ¾", with approximate dimensions of 4" × 6". Welded to the bottom of the flat plate are hooked anchor bolts. The anchor bolts of the joist-bearing plate are embedded in the bond beam course of a masonry wall. Once the grout has cured, the joist-bearing plate is held secure for the attachment of bar joists or beams that bear on CMU walls. Bar joists are welded to the exposed portion of the joist-bearing plate. Similar versions of longer lengths, called *end wall-bearing plates,* which run parallel to bar joists, are fabricated and installed to support the metal deck at the masonry end walls of the structure. While simple in concept, placement and embedment must be coordinated with the fresh placement of grout in the bond beam or pocket. It also requires the layout in advance of the placement at a specific elevation. Installation is often the responsibility of the masonry subcontractor, but the estimator should check the specifications sections for both scopes carefully.

Seismic clips can also be part of the Miscellaneous Items section. *Seismic clips* are short pieces of steel angles 8" to 12" long with holes for threaded rods. They are installed on either side of an unreinforced CMU partition in renovation applications. They are installed at the top of the partition and the adjacent leg of the angle is welded or bolted to the structure above. They are bolted through with threaded rod. This reduces the collapse of the CMU in a seismic event.

Description	Labor-Hours	Unit
Concrete		
Stairs CIP	.600	LF tread
Landings CIP	.253	SF
Steel Custom Stair 3'-6" Wide	.914	riser
Steel Pan		
Stair Shop Fabricated 3'-6" Wide	.914	riser
Landing Shop Fabricated	.200	SF
Concrete Fill Pans	.070	SF
Spiral Stair		
Aluminum 5'-0" Diameter	.711	riser
Cast Iron 4'-0" Diameter	.711	riser
Steel Industrial 6'-0" Diameter	.800	riser
Included Ladder (Ships' Stair) 3'-0" Wide	1.067	VLF
Wood Box Stair Prefabricated 3'-6" Wide		
4' High	4.000	flight
8' High	5.333	flight
Open Stair Prefabricated 8" High	5.333	flight
Curved Stair 3'-3" Wide		
Open 1 Side 10' High	22.857	flight
Open 2 Sides 10' High	32.000	flight
Railings 1-1/2" Pipe, 2 Rail		
Aluminum	.200	LF
Steel	.200	LF
Wall-Mounted	.150	LF
Rails Ornamental		
Bronze or Stainless	.611	LF
Aluminum	.767	LF
Wrought Iron	.834	LF
Composite Metal, Wood, or Glass	1.467	LF

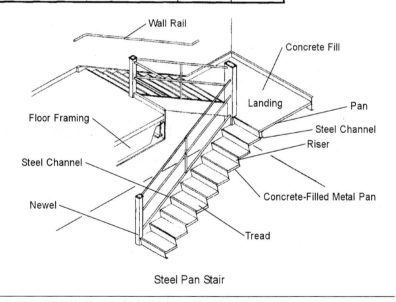

Steel Pan Stair

Figure 10.10 Installation Time in Labor-Hours for Stair Systems

Taking-off and Pricing

Joist-bearing plates are taken off and priced by the piece (EA). This is accomplished by counting the number of joist ends that bear on masonry walls. The end wall-bearing plates are taken off and priced by the LF. The major portion of the cost is the fabrication.

Seismic clips are taken off and estimated by the individual piece. Their installation costs can be labor intensive and may require scaffolding or work platforms to access the top of the partition. A fire watch may also be required if the deck attachment is welded.

Consult the documents for installation requirements. Most joist-bearing plates are furnished-and-delivered items only. Also consider costs of galvanized shop finishing or priming. Their installation is typically part of the masonry scope of work and proceeds as the top course is filled with grout. Specifications may require that the welded spots be touch up with galvanizing after installation.

Lally Columns

Lally columns for residential construction are used to transmit live and dead loads from the structural members above to the isolated footings. They are typically 3½"-diameter concrete-filled steel pipe columns, with separate cap and baseplates. Lally columns are available in a variety of stock lengths from 7' to 12' in 1' increments. Columns can be cut on-site using a heavy-duty pipe cutter. The concrete filling is part of the product and not added on site.

Taking-off and Pricing

Lally columns are taken off and priced by the individual piece based on the even foot length (EA). To obtain the length of the column, calculate the distance from the bottom of the load-bearing beam to the top of the footing, and round up to the nearest foot. The cost of cutting, usually minor, should be included as part of the installation.

Steel Window Guards

Steel window guards or grates are fabricated from a variety of direct-sized bar or round stock, most commonly ½" × ½" bar or ½"-diameter rod welded to flat stock at the perimeter. Other types of security grating include heavy-gauge, diamond-shaped wire mesh welded to a frame constructed of 1½" × 1½" angles welded as a picture frame. They are typically fabricated and painted off-site, then delivered and installed by bolting.

Taking-off and Pricing

Window guards are taken off and priced by the SF. Multiple guards of the same size can be priced by the piece (EA) especially for installation purposes. The cost of bolts, anchors, shields, and fasteners in general should be included as part of the installation cost. Review the specifications for shop finishes, as they will impact cost. Installation crews are typically two or more tradespersons and may require minor on-site welding and paint touchup.

Steel Bollards

Bollards are plain steel pipe filled with concrete. They come in various diameters starting at 4" and going to 12" in 2" increments. Bollards are used to protect site features such as transformers from damage caused by vehicular traffic. Bollards can be of an ornamental design as well. They are set with 36" to 48" below grade in concrete, and the remaining 42" to 48" above grade. Finishes can be applied in the field or in the shop. It is not uncommon to have a shop-applied galvanizing with field-applied finish paint or a high-density polyethylene cover. Bollards are shown on site improvement drawings but are typically specified in Division 5. Bollards can come prefilled with concrete or for larger diameter bollards, filled on site while being set.

Taking-off and Pricing

Quantities are taken off and priced by the individual piece (EA) based on diameter and length. Excavation for bollards is typically specified under the excavation scope in Division 31—Earthwork, but can be required as part of this work. Equipment for the excavation, handling, and installation is required for all but the smallest-diameter bollards. Traditionally, all bollards are set at the same time, thereby avoiding short load charges for the concrete. Site finishing is most often specified under the Painting section of Division 9—Finishes.

MISCELLANEOUS COSTS

Estimating structural steel and miscellaneous metals often requires items that are not easily categorized. Among them are shop drawings, shop painting, fire watch, and field erection.

Shop Drawings

Prior to the actual fabrication of any structural steel components, the steel fabricator is required to produce a set of working drawings, or shop drawings, that show the actual connection details, heights, and lengths of the various structural members, such as beams, columns, and joists. This is part of a checks-and-balances system between engineers and contractors to avoid costly errors. In addition, shop drawings provide lintel schedules and show base and cap plate details. Shop drawings are reviewed by the architect or structural engineer for conformance and are approved for fabrication. The fabricator then uses them as a guideline for fabrication and, ultimately, erection in the field.

Taking-off and Pricing

Pricing the production of shop drawings is a difficult task with many variables and is a direct function of the complexity of the steel design. The more complicated the steel fabrication, the more details, in the form of shop drawings, are required. Shop drawings can be priced by the individual sheet required; 24" × 36" sheets are considered standard. The alternate method involves pricing the shop drawings per ton of steel. Shop drawing detailing can range from a low of 2 hours per ton to upward of 10 hours per ton for very complex projects.

Shop Priming of Structural Steel

After the various steel components have been fabricated, they are generally primed prior to leaving the shop, unless the steel will be spray-fireproofed or encased in concrete. The type of primer paint is typically identified in the specifications. Primers can vary in cost dramatically depending on the coating type, so research the primer product carefully. Consider any touch-up that may be required in the field if primer has been damaged by welding or handling. The cost of shop priming is based on the amount of steel to be primed in tons, TNS. As a guide, ordinary structural steel members contain approximately 175–250 SF of surface area per ton. Based on the cost of the specified paint and its recommended coverage per gallon, the estimator can calculate the cost of materials per ton.

Labor costs depend on the application method used, most commonly spraying. Review the specifications for the number of coats required and any specified coating film thicknesses. Most structural steel members require some level of cleaning off the rust or shop dust prior to priming. Field touch-up is priced as a separate item and is quantified by labor-hours. Figure 10.11 provides guidance on the coverage and labor-hours for priming various types for structural steel.

Field Erection of Structural Steel

All of the components of a structural steel frame are erected by a crew rather than a single individual. The size of the crew is determined by the complexity

Coating Structural Steel

On field-welded jobs, the shop-applied primer coat is necessarily omitted. All painting must be done in the field and usually consists of red oxide rust inhibitive paint or an aluminum paint. The table below shows paint coverage and daily production for field painting.

Type Construction	Surface Area per Ton	Coat	One Gallon Covers		In 8 Hrs. Person Covers		Average per Ton Spray	
			Brush	Spray	Brush	Spray	Gallons	Labor-hours
Light Structural	300 SF to 500 SF	1st	500 SF	455 SF	640 SF	2000 SF	0.9 GAL	1.6 LH
		2nd	450	410	800	2400	1.0	1.3
		3rd	450	410	960	3200	1.0	1.0
Medium	150 SF to 300 SF	All	400	365	1600	3200	0.6	0.6
Heavy Structural	50 SF to 150 SF	1st	400	365	1920	4000	0.2	0.2
		2nd	400	365	2000	4000	0.2	0.2
		3rd	400	365	2000	4000	0.2	0.2
Weighted Average	225 SF	All	400	365	1350	3000	0.6	0.6

Figure 10.11 Coating Structural Steel

Source RSMeans Building Construction Cost Data

and scope of the project. For larger, more complex projects the erection crew follows an erection plan. The crew is most often composed of a foreman, multiple ironworkers, operator (crane), and a welder(s). A crane and welding machine are also part of the crew costs. The erection of structural steel must include the handling and distribution on-site, crane setup and moving, crane rental, and the materials and labor for on-site welding. Most erection crews provide a foreman or other form of supervision during the initial phase of the process. Crews can start off small to erect and plumb columns or set a few beams, but typically, steel erection crews come en masse to erect the frame as quickly and efficiently as possible to minimize crane time on site. Once the frame has been assembled and the joist and deck set on the frames the detail work can begin. The "buttoning-up" work takes more time with a smaller crew. There are a variety of items that affect the productivity of an erection crew. The most common ones are weather and temperature, on-site accessibility, experience of the crane operator, and number of connections. Always review historical data as a basis for pricing erection. Steel erection is a special task that should be left to those who are familiar with the work and have the appropriate tools and safety equipment.

For projects with a steel frame, before erection work can begin there is a process called *shakeout* beginning just after delivery. The staging area for the steel is prepared with wood dunnage to set the steel on and a general layout of which pieces will go where on the ground. This expedites the erection process by identifying where specific pieces are on the ground when they are needed. The delivery of steel is checked for correctness and unloaded with equipment. The labor for the shakeout process depends on the size of the steel package. It is, however, a critical part of the process that requires time and should be included in the estimate.

For smaller residential projects the entire steel scope of work may be a single wide flange member for use as a main girder. In this application, it is best to try to coordinate the delivery with the installation so that it goes from the truck to its installation location. Many smaller fabricators will ship with a truck that has a boom mounted on it for unloading.

Taking-off and Pricing

Structural steel erection is taken off and priced by the ton, TNS. The quantity of steel in tons can be obtained from the material takeoff. Structural steel members should be separated from joist and decking for accurate pricing. The major portion of the steel is first erected and bolted or welded. The joists are then set and the deck landed on the various floors or roof as quickly as possible so that on-site crane time can be minimized. As the frame is squared and plumbed, it is braced for safety purposes. The crew installs fall protection at the outside perimeter and interior openings in accordance with OSHA guidelines. The remaining tasks of installing and tightening bolts to the required torque, welding bar joists, and fastening decking take somewhat longer and are usually done by a detail crew after the main erection crew is finished. Erection is based on the cost of a crew day and the anticipated daily output. This is subject to change based on the individual project.

Equipment costs include the daily costs of the crane for erection. Crane costs can be estimated by the hour, half day, or full day, depending on the capacity (size) of the crane and the expected output (in TNS) per day. Cranes have mobilization and demobilization costs – that is, the setup prior to the erection and the breakdown of the crane after the erection has been completed. These costs must be considered in the estimate as well as travel time to and from the crane yard. This is referred to as a "portal-to-portal" cost. The capacity of the crane needed is not solely about weight being lifted, but may be dictated by the reach required. The estimator should consider reach when calculating the cost of the crane needed.

Fire Watch

When welding occurs around combustible materials, there is a clear and present danger of fire. To reduce the chance for sparks to develop into fire, an individual is assigned to stand by and watch for fire caused during the welding process. This most often occurs during renovation work in an existing structure. This preventative measure is called a *fire watch* and is a requirement under the law in many places. Fire extinguishers are also needed during the fire watch. This is typically accomplished with an additional person being attached to the welding crew.

Taking-off and Pricing

Take-off and pricing units of the fire watch are by the labor-hour. They are typically in day or half-day increments. Equipment costs also include extinguishers.

SUMMARY

As a common component in the commercial construction marketplace – and to a somewhat limited degree, residential – steel often accounts for a large portion of the cost of a project. For that reason, a detailed and comprehensive takeoff and estimate may require additional time to check calculations and verify quantities.

Steel frame projects have special conditions in the field, such as access and overhead clearance, that can have substantial impact on costs of erection. Shop coatings can have a wide range in price and should be checked carefully. It is also important to remember the costs of shop drawing production and revision and some ancillary costs such as traffic details during erection or delivery on an urban site, the shakeout process, staging areas, fall protection, mobilization and demobilization of the crane, and coating touch-up.

Steel erection is a crew task, and its cost is based on the size of the crew and its daily output in tons of steel: structural, bar joist, and decking. Crew sizes and outputs can change over the course of the erection and are typically downsized after the main frame has been erected and braced. The selection of the crane can be dictated not only by weight but also reach of the crane.

Light gauge metal framing (LGMF) is the use of metal shapes such as stud, track, and joists that take the place of wood-framing members. It is popular for its load-bearing capacity and fire resistivity. Takeoff and pricing can be by the individual piece or by completed assembly.

Division 5—Metals has many components fabricated from steel that are not structural. These are classified as miscellaneous and ornamental metals. They include railings, steel pan stairs, grating, ladders, and lintels, to name a few. Some are furnished and installed, and some are furnished only to be installed by another trade. The estimator is cautioned to read the specifications carefully to determine the scope. Miscellaneous metals can also be included on drawings such as site plans.

The steel scope can also include miscellaneous costs such as shop priming and field touch-up of coatings, a fire watch for renovation work, as well as shop drawings and details.

QUESTIONS/ PROBLEMS

1. A wide flange member with the designation W27 × 161 is 54' long. If the cost for fabrication, priming, and delivery is $1,230 per ton, what is the cost of this member?
2. An erection crew has a daily cost of $5,678 and will erect 28 tons of steel in a day. What is the cost to erect 467 tons of steel, and how many days should it take?
3. Using the structural steel plan in Figure 10.3, how many squares of decking are required? Round up the answer to the nearest whole number of squares.
4. Using Figure 10.3 Structural Steel Plan, calculate the weight in tons of all the wide flange members along the numbered column lines only.
5. If the span in Figure 10.5 Open-Web Joist Plan is increased from 30' to 38' and the species of K-series joist is changed from 16K4 to 28K6, what is the weight of the joist in tons?
6. In question #5, if the erection crew can set 1,700 lb. of bar joist per hour, how long will this installation take? Round up to the nearest whole hour.
7. Using Figure 10.11, Coating Structural Steel, and Figure 10.3, Structural Steel Plan, calculate the cost of the material (gallons of paint) to prime by spraying two coats on all of the beams shown in plan view on Figure 10.3. A gallon is $86 and the steel is considered light structural steel.

8. Considering the quantity of steel in tons in question #7 how many labor-hours will it take to spray the two coats?

9. Calculate the vertical length of the 6" × 6" × ½" columns in section A of Figure 10.3 Structural Steel Plan if the TOS for the W24 × 76 beam is elevation 174.00 and the elevation of the leveling plate for the column is 152.40 (depth of W24 can be found on line).

10. Using the appropriate figure from the chapter, how many labor-hours should it take to install 80 LF of 1½" diameter two-rail steel railing? If the labor cost per hour is $99.50, what is the installed cost of 80 LF?

11 | Wood, Plastics, and Composites

W ood and plastics are covered in CSI MasterFormat®, Division 6. Although there are numerous sections within this division, for estimating purposes all tasks can be classified into one of four major categories:

- Rough carpentry, framing, and blocking
- Finish carpentry
- Architectural millwork
- Casework and cabinetry

Within these four major groups are most of the tasks an estimator would encounter in a residential or light-commercial construction project. While not essential, it is beneficial to have a working knowledge of the carpentry process, especially framing.

For the purpose of this chapter, we will use the terms *rough carpentry* and *framing* interchangeably to refer to building structures with wood structural components, including framing-grade lumber, composite materials, and sheathings. Wood trusses are also included in framing. Pre-engineered and prefabricated structural components, such as composite joists, laminated veneer lumber (LVLs) and glue-laminated members are often used in place of conventional framing lumber. Since their inception, these products have gained in popularity and are the rule rather than the exception. Their greatest popularity is for framing floor and roof systems.

Finish carpentry refers to materials and techniques for finish-grade wood trims for windows, doors, baseboards, and moldings on the building's interior. It also includes installing details, both interior and exterior, from standard lumberyard stock, as well as paint- and stain-grade wood species. It refers to standard profile trims: both standing and running.

Estimating Building Costs for the Residential and Light Commercial Construction Professional,
Third Edition. Wayne J. Del Pico.
© 2023 John Wiley & Sons, Inc. Published 2023 by John Wiley & Sons, Inc.

Architectural millwork is classified as custom-fabricated moldings and trims, most often used in the structure's interior. This work requires first-class workmanship. These profiles are milled from rough stock, typically less common hardwoods, and are often prefinished off site.

Casework and cabinetry refers to the installation of prefabricated (mass-produced) or custom-fabricated casework, such as bookshelves, countertops, and cabinets that can be store bought and require some assembly. Please note that manufactured cabinetry can also be specified in section 12 30 00—Casework.

It should be noted that the dimensions used in this text are based on the English or *Imperial* system of feet and inches in contrast to the SI Metric System. For most US domestic projects, the Imperial system now referred to as the US Standard System of units is still the predominant form of measurement.

ROUGH CARPENTRY AND FRAMING

In light-commercial construction, most rough carpentry and framing details are shown on a portion of the structural drawing set called *framing plans*. As the name implies, the majority of the information is provided in plan view, supported by details and sections. The framing plan, a modified version of structural drawings showing the "skeleton" of the structure, illustrates the structural wood members used to construct the frame of the building. The building cross section and wall section provide essential information for taking off and accurately pricing the wall framing and sheathing, the roof framing and sheathing, and the roof/ceiling assembly. Note that in residential construction, specifically renovation work, there may be no framing plan(s). In this circumstance, the estimator must rely on the building code and experience as a guideline to frame within the parameters of the architectural drawings. This can pose a challenge for the estimator in light of ever-evolving seismic and wind loading code criteria.

Taking-off and Pricing

Prior to the start of the lumber takeoff, begin with a thorough review of the plans. This includes both the architectural and the structural (framing) plans. The purpose is to become familiar with the structure and coordinate the work between both design disciplines. Next, the estimator should read the specifications for the grade and species of lumber to be used. *Grade* can be defined as quality, both visual and, more important, structural. Different species have different levels of structural integrity (strength) and, as a result, different prices. This is also true of sheathings. The species of wood that makes up the sheathing is a major determinant of the price. Define all materials in the lumber list by species, grade, and any other special requirements, such as pressure-treated or fire-retardant woods, as these affect the price.

The most accurate procedure for estimating lumber quantities is to perform a detailed and comprehensive takeoff of each piece of lumber, sheathing, and rough hardware included in the work. Shortcuts to this procedure can often result in costly errors or omissions. The completed *lumber list,* as it is called, is then sent to various lumber suppliers for price quotes. When this is done correctly, it provides current pricing of wood materials. Performing the takeoff in this manner allows the estimator to gain a full understanding of the project and is essential when determining the cost of associated labor.

In an economic climate that is constantly shifting, lumber and sheathing prices have a limited shelf life. As a commodity, the prices change weekly or in some locations daily. Before including any lumber list in the estimate, the estimator would be wise to check with the vendor for the latest pricing the day of the bid. Escalation costs can be added if the bid is expected to be held firm for a period of time. Do not forget that with high fuel costs, delivery can include premiums for fuel.

Platform Framing

All of the estimating techniques described in this chapter are based on *platform* or *western framing.* In platform framing, the load-bearing walls start at the top of the subfloor sheathing and continue to the underside of the platform or floor above. An alternate method, called *balloon framing,* starts the exterior and load-bearing walls at the top of the pressure-treated sills and continues to the underside of the ceiling joists. Platform framing is the more common method employed for residential/light-commercial wood-framing projects.

Dressed Lumber

Lumber prepared for frame construction is called *dressed* lumber. The term *dressed* refers to the surfacing of the rough lumber, accomplished by a thickness planer. This process provides a uniform size and shape with a smooth surface. Lumber sold at lumberyards that has been surfaced on all four sides is called *S4S.* Once the lumber has been surfaced, it is smaller than its original dimensions. For example, a 2" × 4" piece of rough lumber measures 1½" × 3½" after being dressed. The 2" × 4" dimension is called the *nominal* dimension, and 1½" × 3½" is the *actual size.* Most rough carpentry work can be classified as frame construction, which includes the use of sheathings and structural-grade lumber, such as 2" × 4", 2" × 6", and 2" × 8". Framing lumber has nominal dimensions from 2" to 6" in thickness and 4" to 14" in width. The length of the lumber differs by the species and availability, but most commonly ranges from 8' to 20' in 2' increments. Larger-dimensioned lumber, such as 2" × 10", 2" × 12", 4" × 6", and 6" × 8", is available in lengths up to 24', but may be a special order.

Dressed lumber is milled from rough lumber and not manufactured or engineered. Products like laminated veneer lumber or engineered lumber will be discussed later in this chapter. Dressed lumber is stable from a moisture content as well, which reduces dimensional changes once incorporated in the work.

Framing Assemblies

The framing of a structure can be broken down into four main systems, or assemblies. An *assembly* is a series of components that together form a larger system or portion of work. The four main assemblies in a frame are:

- Floor framing
- Wall/partition framing
- Roof/ceiling framing
- Exterior trims and miscellaneous items

As with most estimating procedures, it is best to start at the beginning of a task and proceed as the project would be built. It is also important to include all incidental materials that are part of the assembly, so that nothing is omitted. Seasoned estimators develop a checklist – mental or written – to ensure that all components are included in the takeoff.

For the sake of clarity, the following discussion covers the procedure for the material takeoff portion of the work only. Estimating labor follows after the takeoff of all four assemblies has been discussed.

Floor Framing Assembly

The floor framing assembly consists of all framing materials from the top of the subfloor down. This includes the subfloor sheathing, joists, box or band joists, cross bridging, girders, sills, and sill sealers. In addition, ancillary items, such as fasteners, adhesives, and joist hangers, should be included in the assembly in which they are required, either by design or code.

Figure 11.1a and b illustrates the floor framing assembly with the floor framing plan and a cross section through the floor frame.

GIRDERS AND SILLS

When the width of the structure is greater than the floor joist can span, a *built-up* or single-piece member called a *girder* is used. The term *built-up* refers to multiple pieces of the same size dressed lumber nailed together to create a monolithic member. The girder is located at a specific point, frequently the midpoint in the width, to reduce the overall span of the joists. The girder is supported at specified intervals by lally columns that transmit the load to the footings. (Both lally columns and footings are defined in prior chapters in this text.) Sills are anchored to the top of the foundation wall by anchor bolts to provide a nailable bearing surface for the joist and box joist.

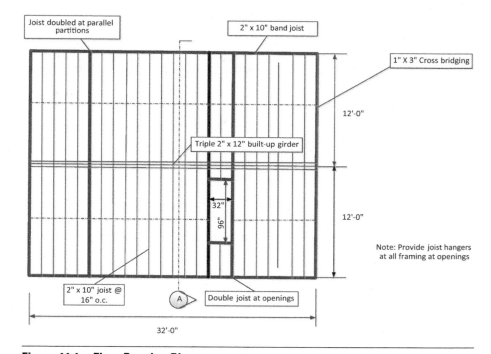

Figure 11.1a Floor Framing Plan

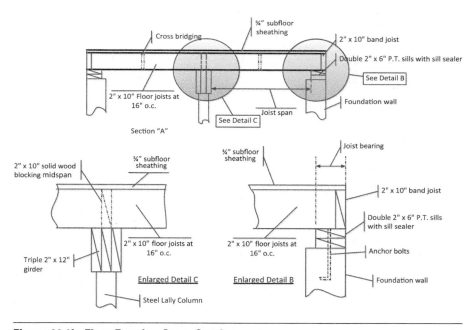

Figure 11.1b Floor Framing Cross Section

Sills, or sill plates, are typically 2" × 6" or 2" × 8" and are doubled up on some plans. Because they are bolted to the concrete or masonry foundation, sills are *pressure-treated*. Pressure-treated refers to the process of treating wood with chemical preservatives under high pressure to protect against decay. The thin, compressible material installed between the sill and the top of the foundation wall is called the *sill sealer,* which acts as a barrier against insects and air infiltration and is made of synthetic materials resistant to decay.

JOISTS

The structural members that provide support for the floor are called *joists*. Joists are spaced at regular intervals, most commonly 12", 16", or 24" on center. They span the space between the girder and the sill on the foundation wall. A *box* or *band joist* (also known as a *rim joist*) runs perpendicular to the joists at their ends. It completes the platform or box that the floor framing resembles. To maintain the spacing at the midpoints and to prevent the joists from rolling or buckling, *cross bridging* is installed, located at approximately 8' on center for joists that span more than 16'. Cross-bridging pieces are typically 1" × 3" or 1" × 4" cut to fit diagonally between the top of one joist and the bottom of the next. They can be metal pieces as well. In addition to cross bridging, solid blocking can be installed at right angles to the joists between them. Solid blocking is used as a firestop between joists at the center of a bearing partition or girder. The flooring laid perpendicular to the joists is called the *subfloor* and is typically plywood or a similar structurally rated sheathing with a tongue and groove edge. The specifications frequently require that the sheathing be glued and nailed to the joists.

TAKING-OFF AND PRICING

Girders are taken off by the linear foot (LF). Built-up girders with multiple members of the same size should be counted, then multiplied by the length to arrive at the total LF. The length should include the clear span plus the required bearing at each end of the girder. The girder's lumber should be converted to stock sales lengths, which will be discussed in more detail in the following sections.

Sill sealer is taken off by the LF of the foundation wall requiring sills. The total quantity should be rounded to the nearest sales unit. For example, if the nearest sales unit is one roll (120 LF per roll) and the total linear footage required is 110 LF, then the lumber list should reflect one roll. Sill pieces are taken off by the LF and multiplied by 2 for sills that are doubled. Then convert them to stock sales length. If 240 LF are required, this could be listed in the takeoff in several ways, such as 20 pieces at 12' or 15 pieces at 16'. A combination of different lengths could also be used.

To determine the quantity of floor joists, divide the length of the floor frame (in feet) by the spacing (in feet), plus one joist for the end. The takeoff should list the quantity of each length of joist separately and be reported in the lumber list as the total quantity of pieces of a specific length. The length refers to the span of the joist plus the required bearing at each end. For example, referring to Figure 11.13:

> The length of the floor frame is 32' divided by 1.33' (16"o.c) = 24 joists + 1 joist at the end = 25 joists.

> The overall length of the span from the outside foundation wall to the center of the girder is 12'. The result is 25 – 12' pieces of joists on each side of the girder, or a total of 50 – 12' pieces of joists.

To complete the frame, the band joist must be calculated. Band joists run perpendicular to the joists and can be taken off and extended to the total quantity of pieces of a specific length. Referring to Figure 11.1a, the joists are perpendicular to the 32' length of the frame.

Therefore, 32' × 2 sides = 64 LF band joist. This can be extended to 8 pieces at 8' long or 4 pieces at 16' long.

The quantity of floor joists and band joists to be reported in the lumber list would be:
50 pieces at 12'
4 pieces at 16'

Note that, in order for the floor framing lumber to be complete, the estimator must be sure to include the joists that are doubled at the stair openings and under the partitions, as shown in Figure 11.1a. The actual size of the joists required should be extended to the stock sales length. For example, if the actual length of the joists needed is 11'-6", report it in the lumber list as a 12' length.

Cross bridging is shown at midspan between the foundation walls and the girder. It is taken off by the LF and again extended to typical practical stock lengths. To determine the actual LF needed, the length of a typical piece is calculated by using the Pythagorean theorem. Once calculated, most estimators will remember the length of the diagonal based on spacing or it can be part of a checklist noted previously. Referring to Figure 11.2, the following calculation can be made:

The length of a piece of cross bridging *C* is *C* = 17.2" or 17¼", since there are two pieces per bay: 2 × 17 ¼" = 34½". The 34½" could be rounded to 36" or 3'.

There are 32' of joists on each side of the girder, totaling 64 LF. To determine the number of bays, divide 64 LF by the 1.33" o.c. spacing, which equals 48 bays.

48 bays × 3' per bay equals 144 LF of bridging material. The 144 LF could then be converted to 12 pieces at 12', (or 144' ÷ 64' = 2.25' per linear foot of 64' of frame).

The subfloor sheathing is taken off by the SF. Subflooring is no longer planks but a form of sheathing. The estimator can determine the quantity of sheets required by dividing the SF area by the area of an individual sheet. Sheathing is typically manufactured in 4' × 8' sheets, with an area of 32 SF. The subfloor sheathing is laid with the 8' dimension of the sheet perpendicular to the joists. For the example shown in Figure 11.1a, the subfloor requirements would be determined by a simple calculation of the total area divided by 32 SF per sheet.

The floor area = 32' × 24' = 768 SF, divided by 32 SF per sheet = 24 sheets.

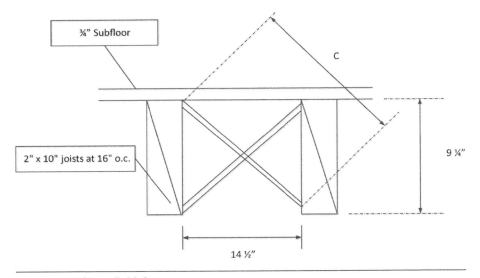

Figure 11.2 Cross Bridging

Since the plywood for all floor frames does not work out to the exact sheet (as in our example), the estimator may need to make adjustments for odd-dimensioned frames. This may require counting actual sheets placed over the frame. When estimating composite lumber, the same process should be followed; however, most composites are available by the foot versus two-foot increments.

To reduce squeaking caused by movement in the subfloor, and to provide added rigidity, the sheathing is typically glued to the deck with subfloor adhesive. Calculating the amount of adhesive needed often depends on experience. A general rule is that a one-quart tube of adhesive should cover approximately five sheets of sheathing. This rule is subject to change with the weather, handling, and care taken in the application. Colder weather or poor handling will reduce the number of sheets covered by as much as one-third.

Study the drawings for the ends of joists that "hang" or do not rest on anything. Building codes require joist hangers to support these members. Fasteners, including nails and screws to complete the hanger system, are most often sold by the pound or by a specific size box. Check the specifications and/or manufacturer's recommendations for nailing and fastening schedules that indicate the spacing on center for fasteners. Most estimators quantify nails as a lump sum (LS) item but can be calculated as a dollar value per SF of frame. As a reminder, *always* use the dimensions shown on the drawings for calculating quantities. Scaling should be a last resort when no other options are available. Some anchors or fasteners, especially those for seismic or wind resistance, called *holddowns* or *tiedowns*, can be expensive. The estimator is advised to make an accurate count of the rough hardware and fasteners.

Pricing of the floor frame materials will be part of the overall pricing of the lumber list.

Wall/Partition Framing Assembly

The wall and partition framing consists of the exterior walls and sheathing, the interior load-bearing partitions, and the interior non-load-bearing partitions. Load-bearing walls carry live and dead loads from a part of the structure above, such as a floor, roof, or ceiling. A major component of the load-bearing wall or partition is the *header,* which spans the openings in load-bearing walls above windows and doors. Headers are structural members that transmit the load from above the opening to the framing on either side. Typical header construction consists of 2" × 6", 8", 10", or 12" (nominal) framing lumber nailed together with ½" plywood spacers to equal the thickness of the wall. Figure 11.3 illustrates a typical wood header.

The vertical members that support the header are called *trimmers* or *jack studs.* Jack studs are nailed to full studs at each side of the header, sometimes referred to as *king studs.* They may be known by other names based on local nomenclature.

All partitions and walls have horizontal members that hold the studs, or vertical members, at the desired spacing, called *plates.* The plate at the top of the wall is referred to as the *top plate,* and the one at the bottom is called the *sill plate* or *sole plate.* Most load-bearing wall construction requires that the top plate be doubled. The horizontal member that runs parallel to the header at the windowsill height is called the sill. The short studs that fill in under the windowsill or above the header are called *cripples.* Figure 11.4 illustrates typical load-bearing wall framing.

To complete the exterior wall system, plywood or similarly rated sheathing is installed over the framed wall. This helps give the wall rigidity and braces it

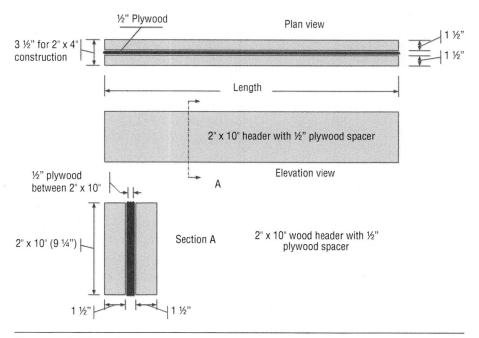

Figure 11.3 Wood Header

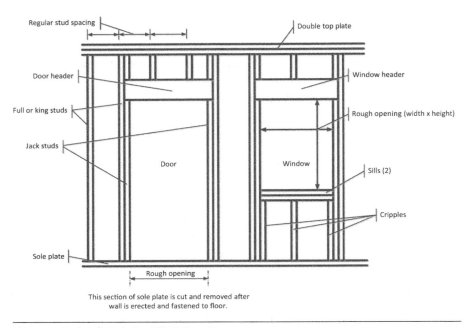

Figure 11.4 Load-Bearing Wall Framing

against the wind. Just as in floor framing, wall sheathing is nailed with the 8' length perpendicular to the studs.

TAKING-OFF AND PRICING

Wall and partition takeoff starts with the exterior walls. The soleplate and double top plate are taken off by the LF by first determining the length of the wall and multiplying it by 3 (if there is a double top plate, 2 if not). Determining the quantity of studs required is more a matter of experience than a mathematical formula. When determining the number of studs required, provide additional studs for outside corners, backer studs that provide nailing for intersecting partitions, and jack/king studs for openings. Since counting individual studs can be time consuming, it is common practice to use a multiplying factor to account for these additional studs, based on the number of intersecting partitions, openings, and corners. Factors can range from 1.10 to 1.35 studs per LF of wall length based on a 16" o.c. spacing. A reasonable average would be 1.25 studs per LF of wall. For example, if the perimeter of the exterior wall were 100 LF, then the number of studs required would be:

100 LF × 1.25 studs per LF = 125 studs

The height (length) of the stud is calculated by referring to the wall section on the architectural drawings for the soleplate to top plate dimension and subtracting the width of the three plates. Figure 11.5 shows a partial wall section for determining the stud height.

The example in Figure 11.5 shows that the height of the stud is the plate-to-plate dimension of 8', less the thickness of three 2" × 4" studs: 3 times 1½"

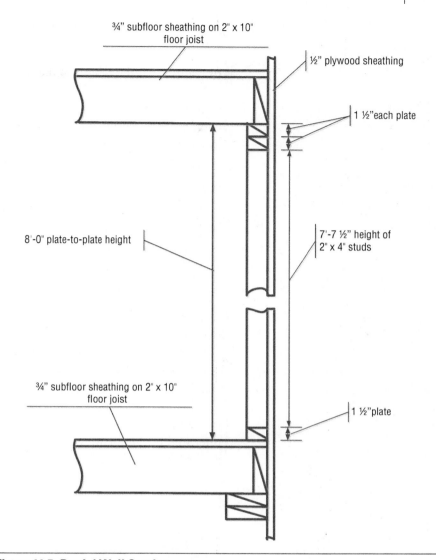

¾" subfloor sheathing on 2" x 10" floor joist

½" plywood sheathing

1 ½"each plate

8'-0" plate-to-plate height

7'-7 ½" height of 2" x 4" studs

¾" subfloor sheathing on 2" x 10" floor joist

1 ½"plate

Figure 11.5 Partial Wall Section

or 4½". The stud height is 96" (8') minus 4½" or 91½" (7'–7½"). The actual stud length would have to be extended from 7'–7½" to 8' stock lengths, unless *precuts* were available. Precuts are studs that have been precut at the lumberyard to standard lengths for common wall types.

Window and door header lengths are determined by adding the rough opening width to the required bearing on each side. This is typically the thickness of a 2" × 4" or 2" × 6" stud, or 1½". The *rough opening* is the clear width and height between the framing into which the window or door will fit. It is expressed as *width by height* and can be found on the window or door schedules. The size of the stock (2" × 6", 2" × 8", 2" × 10", etc.) for the header should be shown on the architectural wall section, the structural framing plan, or as a note on the structural detail drawings. In the absence of a specified stock size, consult the engineer or, as a minimum, refer to the applicable building code. To convert the length of the header to the correct stock length, verify the quantity of pieces needed based on the wall thickness. Figure 11.6 illustrates the rough opening and header size.

In Figure 11.6, the length of the header is the width of the rough opening, 36" plus the bearing dimension of the jack studs on either side of the opening, or 2" × 1½" = 3", for a total of 39". However, since the header is composed of two pieces that are each 39" long, the length could be extended to 78", or 6'-6". The stock length for a single header, in this example, would be one 8' piece, or 14' if there were multiples of this opening.

Exterior wall sheathing is taken off by the SF and can be extended to the number of sheets required, in the same manner as subfloor sheathing. Note that the vertical height of exterior wall sheathing does not start at the soleplate, but at the bottom of the pressure-treated sills. To determine the vertical height of the wall sheathing, again refer to the wall section on the architectural drawings and calculate the vertical dimension from the pressure-treated sills to the top plate of the uppermost story. This dimension is, in turn, multiplied by the perimeter of the exterior wall. This dimension may not be consistent for the entire perimeter because of changes in story heights and rooflines. Deductions for windows, doors, and other exterior openings should be based on the size of the opening and the method of framing. Common field practice calls for sheathing over openings and cutting them out afterward. In this case, deductions do not apply, and the cut-out pieces are considered waste. To check the quantity of sheathing, refer to the exterior wall elevations on the architectural drawings. Additional sheathing for gable ends that extend beyond the top plate must be taken off

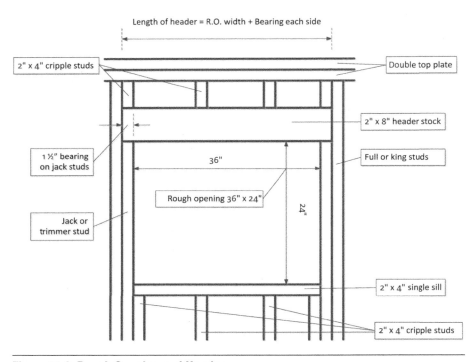

Figure 11.6 Rough Opening and Header

with consideration for waste resulting from cutting at angles to match the roof pitch. There is no standard waste factor that can be applied; it must be determined on an individual basis. Ranges for waste can vary from 3% to 10%, based on the complexity of the frame.

The procedure for taking off interior load-bearing partitions is the same for exterior walls, except that there is typically no wall sheathing. (Special partitions for seismic bracing may have sheathing on the interior.) Non-load-bearing partitions are similar, but do not require double top plates or structural headers. Increase the length of the stud by 1½" for interior partitions with the same plate-to-plate height. It should also be noted that the multiplier for interior partitions is usually higher than the recommended 1.25 studs per LF used on exterior walls. Due to the increased number of intersecting partitions at the interior of the space, more corner and backer studs are typically used. A multiplier of 1.4 to 1.6 is recommended.

Additional stock for items such as braces or springboards for "plumbing up" exterior walls should be acknowledged. This additional stock should be taken off and included as part of the lumber list. Pricing of the wall and partition materials will be part of the overall pricing of the lumber list.

Roof/Ceiling Framing Assembly

Conventional, or *"stick"* framing uses individual members, rather than trusses, to construct the roof. Before starting the takeoff, the estimator should become familiar with the roof and ceiling framing system by studying the architectural and structural drawings, specifically the building cross sections, wall sections, elevations, roof framing plans, and corresponding details. The takeoff of the roof and ceiling framing assembly can often present the greatest challenge to the estimator due to the added level of calculations. This challenge can be reduced by a detailed step-by-step process.

ROOF PITCH

Rafters are structural members that follow the slope, or pitch, of a conventionally framed roof. *Pitch* is the angle or inclination of the roof, expressed as a ratio between the horizontal *run* of the roof and its vertical *rise*. It is typically noted on the architectural drawings as an incremental ratio per 12" of run.

Figure 11.7 illustrates a typical symbol used to designate pitch, where the pitch of the roof is 8" of rise for every 12" of run. The rafters are supported at the base by the top plate at the top of the exterior wall. In the same way that floor joists are spaced along the pressure-treated sills, rafters are spaced along the top plate. Common spacing is 12", 16", or 24" on center similar to floor and wall/partition assemblies. The part of the rafter that extends beyond the face of the exterior wall is called the *rafter tail* or *tail*. It provides the nailing

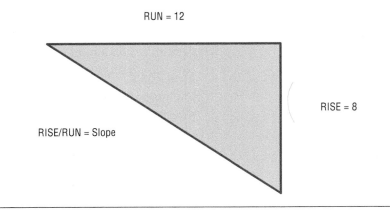

Figure 11.7 Roof Pitch Symbol

for the fascia and soffit (discussed later in this chapter) and constitutes the roof's overhang. The highest point of the rafter terminates at a perpendicular member in the horizontal plane called the *ridge* or *ridge board*.

CEILING JOISTS AND STRAPPING

To complete the triangular shape of the roof frame, horizontal members called ceiling joists provide the floor of the attic space or ceiling of the floor below. Ceiling joists extend from the top plates of bearing walls and span the space, much like floor joists that span from sill to girder or sill. Ceiling joists for gable-end roofs run parallel to rafters. *Strapping* typically comprises 1" × 3" (nominal) boards nailed to the interior side of the ceiling joists. It is used to maintain the spacing of ceiling joists between bearing points and to provide furring for the ceiling finish. Strapping runs at right angles to the ceiling joists in the same horizontal plane. Strapping is commonly spaced at 12" or 16" on center.

COLLAR TIES AND ROOF SHEATHING

To increase the rigidity of the roof frame, horizontal members called *collar ties* are installed from rafter to rafter on opposite sides of the ridge. Collar ties are typically located in the top third of the imaginary triangle created by the roof frame. The *roof sheathing* extends from the rafter tail to the ridge board along the top surface of the rafter and provides a substrate for the application of the roofing. Figure 11.8 shows a typical roof frame as viewed in a building cross section.

HIP ROOFS

Another type of roof with intersecting roof planes is a *hip roof*. The member that forms the outside intersection of the hip roof is a *hip rafter*. The member at the inside intersection is referred to as the *valley rafter*. In addition to the common rafters, smaller rafters called *jack rafters* rise from the top plate and terminate at the hip or valley rafters to maintain the spacing. Hip roofs can be more time consuming to take off and calculate materials as a result of the added components and measurements to consider. Figure 11.9 illustrates a hip roof frame.

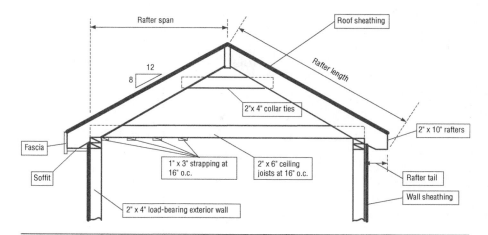

Figure 11.8 Section at Roof Frame

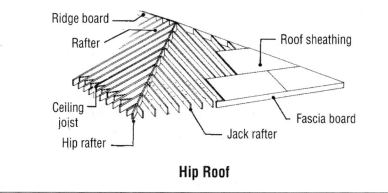

Hip Roof

Figure 11.9 Hip Roof Frame

TAKING-OFF AND PRICING

The quantity of common rafters required is determined by the same method used to calculate the number of floor joists. (Refer to the floor frame assembly discussed previously in this chapter.) Divide the overall length of the structure by the spacing (both in feet) and include one additional common rafter for the end. Also be sure to include double rafters at openings in the roof for dormers, skylights, or other requirements noted on the drawings. It is important to include the rafters from both sides of the ridge. Once the quantity has been determined, calculate the length of the rafters and stock to be included in the lumber list.

CALCULATING COMMON RAFTER LENGTHS

If a vertical line passing through the center of the ridge were intersected by a horizontal line from the rafter tail to the center of the ridge, the intersection would form a right angle (see the dashed lines in Figure 11.10). By adding the rafter to these two imaginary lines, the shape would be a right triangle, subject to the rules of the Pythagorean theorem. (See Chapter 3.) Figure 11.10 illustrates the imaginary triangle from the roof frame in Figure 11.8.

To determine the length of the rafter, first calculate the total run *B* by adding the rafter span to the overhang. The *rafter span* is the horizontal distance

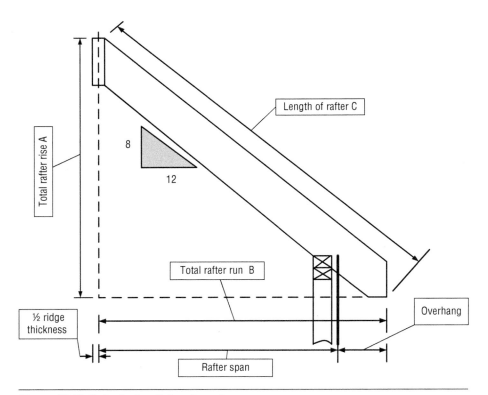

Figure 11.10 Calculating Rafter Length

from the face of the ridge to the outside face of the exterior wall sheathing. Determine the distance from the outside of the wall sheathing to the center of the ridge using the dimensions on the architectural drawings. For example, in Figure 11.10, if the dimension from the center of the ridge to the outside of the sheathing is 14', the estimator can calculate the total run of the rafter by the following calculations:

$$B = + 14' \ (\textit{center of the ridge to outside of the sheathing})$$
$$+ 1' \ (\textit{overhang})$$
$$- \tfrac{3}{4}" \ (\textit{half the thickness of the ridge})$$
$$B = 14'\text{-}11\text{-}\tfrac{1}{4}" = 14.94' = \textit{total run of the rafter}$$

Once the total run of the rafter *B* is determined, calculate the total rise of the rafter *A*. Since the pitch, or rise-to-run ratio, remains constant over the length of the rafter, the pitch can be used to determine *A*.

The 8:12 pitch used in the example can be expressed as a decimal (8 divided by 12 = 0.67). This decimal equivalent of the pitch is a *unitless* value, with no units such as feet or inches. It is a constant used as a multiplier. The calculation for determining *A* is as follows:

$$A = 0.67 \times B = .67 \times 14.94' = 10.0'$$

With values for *A* and *B*, use the Pythagorean theorem: $A^2 + B^2 = C^2$.

$$\textit{Where } C = \sqrt{A^2 + B^2} = \sqrt{(10)^2 + (14.94)^2} = 17.97 \text{ or } 17'\text{-}11\text{-}\tfrac{11}{16}"$$

After determining the length of the rafter in the previous example, we can now conclude that a stock length of 18' is required. Note that not all roofs have equal pitches for both sides of the ridge. Roofs with varying pitches require separate calculations for each set of rafters. For most gable-end roofs, the measurement used to determine the quantity of rafters and sheathing is the length of the building. Some designs have end rafters that project beyond the plane of the gable end with an overhang. This is called a *lookout frame*. Refer to the elevations, sections, and details to determine whether additional rafters, sheathing, or other lumber are needed.

The quantity of ceiling joists is determined by the same method used to determine floor joists. Divide the length of the building (in feet) by the spacing of the ceiling joist (in feet). Add one joist for the end. The actual length of the joist is determined by the dimensions provided on the plans. Since ceiling joists typically bear on exterior walls and interior partitions, the length is the dimensions between various walls of rooms, plus sufficient bearing. Again, the actual length is extended to the stock length for inclusion in the lumber list.

Collar ties are taken off using the same method as ceiling joists. The length is determined by their placement on the drawings. The Pythagorean theorem can be used if necessary for determining the exact length. It can then be extended to the stock length of the material required.

Strapping is taken off and listed by the LF. First, calculate the length and width of the ceiling joists to be strapped. Since strapping is nailed perpendicular to the ceiling joists, determine the quantity of pieces needed by dividing the length (in feet) by the spacing (in feet). Multiply the quantity by the width of the building (length of the strapping). The total linear footage can then be added to the lumber list. Strapping is typically priced by the LF, so it is not necessary to extend it to stock lengths. Since strapping is installed end to end, the materials can be sold as random length stock. *Random length* refers to various pieces of different standard stock length adding up to the total required footage.

To take off roof sheathing, multiply the length of the building by the length of the rafter. (Both must be in the same units.) For practical purposes, the length of the rafter can be rounded up to the nearest foot. In the previous example, the 17'-11$^{11}/_{16}$" would naturally be rounded to 18' stock. Just like subfloor and sidewall sheathing materials, extend the SF area to sheets by dividing the total area by 32 (SF per sheet). Round up to the nearest full sheet. Do not deduct for openings such as skylights since these are often cut out in place, and the sheathing becomes waste.

DETERMINING HIP/VALLEY RAFTER LENGTH

Determining the length of the hip and valley is similar to the procedure for common rafters, but with one additional calculation. Since hip and valley rafters are most often at 45° angles to the common rafter, the dimension of *B* must be adjusted to reflect the additional length, using the Pythagorean theorem. If the *B* dimension of 14.94' for the common rafter in the previous example is used, determine the total run, or *B* dimension, for the hip rafter in question:

Since the run of the hip rafter is 45° to the run of the common rafter (in the horizontal plane as measured at the top of the wall), the length of each leg of the triangle would be the same. Using the Pythagorean theorem, the hypotenuse is the total run, or B_h dimension, for the hip rafter.

For example:

$$B_h = \sqrt{(14.94)^2 + (14.94)^2} = 21.12$$

This represents the total run of the hip rafter. The total rise of the hip rafter A_h must equal the total rise of the common rafter since they both meet at the ridge. This means that total rise *A* of the common rafter and total rise A_h of the hip rafter are the same.

> *If the previous A value of 10' is used with the Bh value for the hip, 21.12', then:*

> *Length of the hip rafter$_{Ch}$ = $\sqrt{(10)^2 + (21.12)^2}$ = 23.37 = 23' – 4 – 5/16"*

This would require 24' stock for the hip rafter. This same calculation applies in determining the length of the valley rafter and is applied for calculations for composite beams for hip and valley rafters.

Most building codes require rafter clips to anchor the rafter to the top plate, including common, hip, valley, and jack rafters. The quantity can be determined by counting the number of rafters and multiplying by two for both sides of each rafter. This requirement would be included in the specifications governing the work of the framing.

Occasionally, hip rafters intersect with common rafters at angles other than 45°; these are irregular and are sometimes called *bastard hip rafters*. The length of irregular hip rafters can be calculated in the same way as conventional hip rafters, except that the length of the triangle leg perpendicular to the rafter (in the horizontal plane) is not the same as the run of the common rafter and must be determined separately. This is a mathematical calculation based on the angle of the irregular hip rafter.

The length of jack rafters is calculated by the same method as common rafters, but this can be time consuming. Consult framing texts with tables for jack rafter lengths based on the pitch and run of the common rafter. Handheld calculators designed for contractors are also helpful in calculating the length of common, hip, valley, and jack rafters at the push of a button.

The ridge board is taken off by the LF. The length of the ridge in a gable-end framed structure is determined by the overall length of the building, plus the length of the lookout frame, if any. The *lookout frame* is the overhang at the gable ends of the building. For a hip roof, the ridge is the length between the intersections of the hip or valley rafters at either end of the roof. Ridge boards, like other framing members, should be extended to the stock lengths and quantity required and included in the lumber list for pricing.

BLOCKING

Blocking consists of small pieces of dressed framing lumber installed between studs or other structural members for reinforcing or installing other work, such as cabinetry, trim, windows, doors, or mechanical equipment. Blocking is not always shown on drawings, but the experienced estimator knows that it is required for certain items of work. Review the architectural drawings for such items as cabinetry, finish trim, millwork, toilet accessories, toilet partitions, handrails at stairs and corridors, windows or doors installed in masonry or metal stud-framed openings, or similar installations that would require special wood blocking for anchoring the work. Review the electrical and mechanical drawings for blocking that may be required for the installation of this work. Blocking is also required to attach water piping as it exits the wall for plumbing fixtures, such as sinks, toilets, or tubs. It is also needed to support piping within walls or joists. Electrical devices in walls or floors between existing framing may also require blocking, along with backer boards for electrical panels. Study the specifications for specific items that may require blocking. Frequently, blocking is required for the installation of work excluded from the contractor's agreement, such as equipment installed by the owner or under separate agreement. There may also be code requirements for blocking at seams of plywood sheathing or as truss bracing. This is not always illustrated on the plans or written in the specifications other than a general note requiring compliance with all code or ordinances having jurisdiction.

Dimensional lumber, such as 2" × 6", 2" × 8", and 2" × 10", used for blocking is typically taken off by the LF and can be converted to random-length materials or specific lengths if required. Sheathing used for blocking or backer boards, or as spacers within headers, is taken off by the square foot area (SF) and, again, converted to full sheets.

Certain classifications of construction, such as fire-resistant work, may require fire-treated wood. Quantities of fire-treated wood should be listed separately, as the cost for such materials is considerably more. In addition to regular blocking, *fire blocking* is required by building codes to slow the transmission of fire through a wall or stair construction and interior cavity. Again, in the interest of clarity, fire blocking is not always shown on the drawings. In structures with balloon framing, it is required at the floor level of all walls passing through the floor. Determining the need for blocking is often a matter of experience or best practice. The estimator should be familiar with the local codes having jurisdiction over the project.

As with all other framing components, the blocking materials should be converted to the optimal usable lengths and added to the lumber list for pricing.

STAIR FRAMING

Stairs are a combination of incremental vertical members, called *risers,* and incremental horizontal runs, called *treads*. The structural members that support the individual treads and risers are called *stringers*. At the framing stage of the work, stringers are installed with a temporary tread as a means of access for the workers. Finish treads and risers are discussed later in this chapter in the finish carpentry section. Stairs are shown on the architectural drawings. Refer to the various floor plans to determine the location, the size of the stair, the number of treads and risers, and the incremental rise and run of the stair. Riser and tread sizes are designated on the plan of the stair itself. Figure 11.11 shows a stair in plan view with the designated riser height. If the same stair is sectioned, it will be evident that the sum of the risers equals the floor-to-floor dimension of the stairs. (See Figure 11.12.)

To calculate the length of the stair stringers, determine the diagonal measurement between the total rise and total run of the stairs. This is the hypotenuse of the right triangle created by the rise and run of the stair. For stairs that have been designed to a specific riser and tread size, multiply the number of risers by the incremental riser height. The same calculation can be used for the treads. For example, in Figure 11.11, the following calculations would result:

13 risers × 7" = 97" or 8'-1" total rise

12 treads × 10" = 123" or 10'-3" total run

Once the total rise and run of the stairs have been determined, calculate the length of the stair stringers using the Pythagorean theorem. The vertical leg A of the triangle is the total rise of the stair, and the horizontal leg B of the triangle is the total run of the stair. Substituting the numbers from the previous example, the following calculations result:

Length of stringer

$$C = \sqrt{(8.125)^2 + (10.25)^2} = 13.07 \text{ or approx. } 13'-1"$$

Therefore, the length of stock needed is 14'.

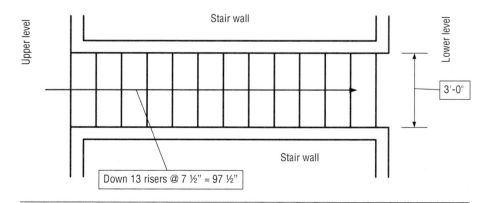

Down 13 risers @ 7 ½" = 97 ½"

Figure 11.11 Stair in Plan View

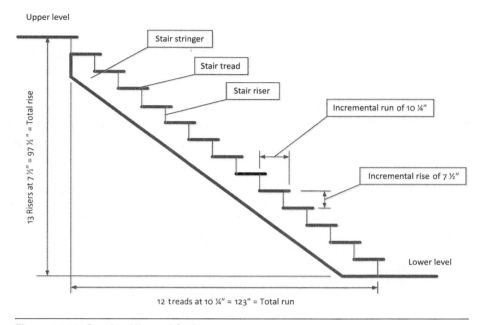

Figure 11.12 Section View of Stairs

The number of stringers required for a stair is based on the width of the stair. In the absence of a specific number of stringers on the drawings, refer to local building codes. Typically, there is a minimum of three stringers for a 3' wide stair.

Stair stringer stock is taken off and listed by the LF, converted to a multiple of certain length members, such as 3 pieces at 14'. Takeoff units for temporary stair treads are by the piece (EA) or by the LF. For example, from Figure 11.12:

The stair width is 3'. Therefore, 12 treads are required at 36" or 3' EA, or a total of 36 LF.

COMPOSITE MEMBERS FOR FRAMING

As an alternative to conventional structural-grade lumber for framing floor and roof systems, special members have been fabricated from structural woods and plywood, under strict engineering guidelines. These are generally referred to as *composite members* and are used in place of floor and ceiling joists, roof rafters, headers, and other load-bearing members. Composite members used in floor, ceiling, and roof framing look similar to structural steel I beams, with flanges constructed of structural-grade lumber and webs of plywood or *oriented strand board* (OSB). Headers and similar load-bearing members that act as beams are generically referred to as *laminated veneer lumber* (LVL), composed of veneers of structural plywood laminated together under pressure to form a structural member. They resemble dimensional lumber and area easily distinguished from the load bearing members that resemble I-joists.

Composite members offer several advantages over conventional framing lumber. Because of their engineered design, strict manufacturing standards, and quality control, they can span greater unsupported lengths with more spacing in between. They are not subject to the shrinking and movement normally associated with conventional lumber, thereby reducing "squeaking" and deflection. They are not subject to the flaws and imperfections found in dimensional framing lumber. Composite members are available in a variety of cross-sectional sizes and lengths up to 60'. They are handled, cut, and installed using methods similar to conventional framing. (See the sections that discuss floor and roof framing systems earlier in this chapter.) Review the structural drawings, paying special attention to floor and roof framing plans, sections, and details showing load-bearing members and exterior wall openings. Specifications should be reviewed for the required product or manufacturer.

Since composite members are a manufactured product, special accessories or components are required. The estimator must become familiar with the particular product specified, so that no components are left out. As a result of the increased spans allowed by composite members, some form of equipment may be required to handle or set in place individual joists, beams, or rafters. Carefully review the lengths of larger composite members, such as girders, hip and valley rafters, and ridge beams. The equipment costs must be included in the estimate. The estimator would be prudent to consider the blocking of the webs of I-beams as this can be time consuming.

Composite members are taken off by the LF and listed for pricing according to depth, width, and length of the member. List the various members according to the manufacturer's model or designation number, if available, for accurate pricing. Many manufacturers have model numbers that indicate load-carrying capabilities. Composite members have increased dramatically in popularity and availability in the last two decades and are now regularly stocked by most lumber companies. Since composite members are used in place of conventional framing members, the methods, and procedures for determining the lengths of common rafters, hip rafters, floor joists, headers, cross bridging, and beams will apply. The estimator should bear in mind that many forms of engineered lumber need only be rounded to the nearest foot versus the 2' increment in dimensional lumber.

TRUSSES

Trusses are prefabricated structural components composed of a combination of wood members, usually in a triangular arrangement, to form a rigid framework. They are used as an alternative to the roof and ceiling framing in conventional "stick" framing. The architectural roof framing plans or the structural framing plans show the roof truss system. The plan view should show the span, on-center spacing, and the number of trusses required to complete the system. Details are often included to show the shape of the truss in elevation view. This view should provide the configuration of the truss members, the span and overall length of the trusses, the pitch of the top chord of the trusses, and the bracing requirements.

Installed trusses require lateral bracing, or bracing parallel to the length of the building, applied to the underside of the top chords of the trusses. Additional bracing applied to the vertical members closest to the center is called *cross bracing*, which prevents lateral movement in the same way that cross bridging does for floor frame systems. *Diagonal bracing* or bracing of the bottom and top chords against buckling, may also be required. A combination of all three types of bracing may be needed, depending on the design. The bracing requirements and associated materials are not always shown on the contract documents, and they are sometimes referred to only as "in accordance with manufacturer's specifications."

TAKING-OFF AND PRICING

Trusses are taken off by the individual piece, EA, and must be detailed in the lumber list separately according to length, span, pitch, and type. To determine the number of trusses, divide the length of the trussed area by the spacing. Special trusses for end conditions, such as gable-end trusses, should be identified separately. Trusses that are specifically designed for a particular project often necessitate pricing by the manufacturer or fabricator of the trusses. This can take time since engineering design is required first. The estimator is urged to submit trusses for pricing well in advance of the date that pricing is needed.

Bracing for trusses is taken off and listed by the LF. Typical sizes are 1" × 4" or 2" × 4", but review the design carefully for location, size, and type. Quantities can be reported as random length or converted to stock lengths. The estimator is directed to consult the truss fabricator's approved shop drawings for the location and size of bracing if available at bid time. Trusses often require shop drawings with an engineer's stamp as part of the scope of work. In reviewing any proposal from a truss fabricator, the estimator should qualify that the engineering requirements are part of the price.

Trusses are large and awkward to handle, and all but the smallest require some type of hoisting equipment to set correctly and safely. The type of equipment is based on the size of the trusses, access to the structure, and number of stories in the building, as well as project coordination. Much like the erection of steel (discussed in Chapter 10), truss erectors prepare a plan based on the crew. Other auxiliary types of equipment, such as spreader bars and multipoint lifting rigs, may be required. The procedure detailed in Chapter 10 for estimating the costs of a crane should be consulted.

Exterior Trims

To enclose the framing system and make the work more visually appealing, exterior wood or synthetic trim is used. Exterior trims are typically 1" × 3", 1" × 4", and so on, up to 1" × 12" in size. The species of the trim depends on the specification or may be noted on the drawings. Refer to architectural drawings, specifically the elevations, wall, and roof sections, which should show the configuration and sizes of the exterior trims. Exterior trims run the full range from the top of the foundation to the roof. Vertical trim pieces at the corners of the structure are called *corner boards*. Horizontal trim at the foundation line is called the *water table,* and trim at the top of the exterior wall running around the perimeter of the building is called the *frieze board*. Horizontal trim attached to the vertical portion of the rafter tail is called the *fascia board,* and the trim at the underside of the rafter tail is called the *soffit*. Exterior trim that follows the rafter line on gable-end roofs is called the *rake board,* while smaller trim applied on top of the rake board is called a *rake molding*. This is by no means the full extent of the exterior trim, but these are some of the more common pieces.

TAKING-OFF AND PRICING

Exterior trim and moldings are taken off and priced by the LF. Exterior trims should be listed according to size in width and thickness, species of wood, manufacturer of synthetics, and type or function of the trim piece. Most trims are listed in the lumber list as the stock lengths. To minimize visible joints, the longest practical lengths are typically used. Review the details of the trim for any special blocking required for the installation of the trim boards. Frequently, special moldings are applied to exterior trims to enhance the design. These are taken off and separated on the lumber list by manufacturer, wood or synthetic species, and model number. Manufactured moldings are priced by the LF and are, again, ordered in the longest manageable lengths. Special pieces fabricated from synthetics such as PVC are also extremely popular in response to their longevity, maintenance-free characteristics, and availability. Due to the wide range of items, it is essential to identify products by manufacturer, model, dimensions, and quantity so that they may be accurately priced for both material and labor.

Estimating Lumber Waste

Estimating exactly what pieces and species of material will comprise the waste when bidding a project is a fool's errand. As professionals we know that certain waste will occur with wood framing and trim as a matter of experience, just not which pieces. A common practice therefore is to include in the estimate a dollar amount that represents the percentage of the waste based on the lumber list. For example, if a lumber list totals $50,000 and the estimator determined that 5% was an appropriate waste factor, it would be prudent to include in the estimate 5% of the $50,000. or $2,500. to accommodate the waste. This can be added as a separate line in the estimate. A less favored method is to inflate the selected quote.

Estimating Framing Labor

In general, framing, and rough carpentry is a crew task. Most framing crews consist of multiple personnel who perform a specific task within the crew, such as supervision, layout, cutting, and prefabricating items like headers, gable-end trusses, stair stringers, and so forth. The most efficient framing crews are those who have worked together and are familiar with one another's work habits and skill sets.

At the risk of using an overworked cliché, an experienced framing crew should work "like a finely-tuned Swiss watch." In order to be efficient, each member of the crew must meet the expected production set by the schedule. An average seven-member framing crew would likely have one working supervisor/foreman who, in addition to coordinating the crew and doing layout, also produces work. Three (or more) experienced carpenters typically lead the work, supported by two less experienced carpenters or apprentices. The apprentices assemble and stand walls, carry stock, and provide much of the labor to perform the more mundane or repetitive tasks, such as nailing plywood to framing. Last but not least, one mill carpenter would cut stock to specific lengths for headers, joists, rafters, and various other stock. This is typically a seasoned individual capable of multitasking and "feeding" pieces to other members of the crew, so as not to delay the work.

With a crew of this makeup, it is clear that not all crew members are on the same pay scale. Foremen or supervisors typically earn 10% to 20% more than lead or production carpenters. Organizations with collective bargaining agreements also acknowledge supervisory skills with additional compensation. Lead or experienced carpenters in the same crew receive roughly the same compensation. Apprentices' pay scales run the gamut from 50% to 80% of the lead carpenter's pay, depending on experience and reliability. Apprentice programs for union employees are rigidly structured. A mill carpenter's pay may vary based on performance and reliability, with earnings at least matching the lead carpenter's rate, and frequently somewhere between that of a lead carpenter and a foreman. While this all may seem rather vague, it is meant to serve as a guideline to estimate framing labor. Larger projects with repetitive work may require larger crews with still only one supervisor. Historical data is always the best guideline for determining both correct crew size and productivity.

The labor costs for framing a structure are typically broken down into crew-days based on each level of assembly, just as with the takeoff portion of the work. Price the labor portion of the four major assemblies *after* the lumber list has been done. Once the lumber list has been completed, the estimator should be thoroughly familiar with the intricacies of the structure and ready to start the labor portion of the estimate. Recognizing that crew size is dependent on the individual project. The estimator is free to make adjustments as needed. Too large a crew becomes counterproductive, with not all hands producing. Too small a crew causes the project to drag on with measurably slow progress. Study the details of the project, and fit crew members into various positions as if the project were actually being built, rather than estimated. It should also be noted that with production framing crews that move from one building or house to the next, there may be different crews for different parts of the process. For larger framing companies with more than one crew working at a time, it is not uncommon for the crew size to be reduced once the first three assemblies have been completed. A reduced crew size allows focus on miscellaneous or "pickup" items, such as strapping, stair stringers, fire blocking, blocking details at eaves, and other odds and ends.

It is important to construct the ideal crew for each assembly. For example, the cost for an optimal crew of seven could be assembled as follows (assuming all labor burdens have been including in the sample wages):

1 Supervisor/carpenter at $55.00 per hour	= $ 55
3 Lead carpenters at $50.00 per hour	= $150
2 Apprentice carpenters at $28.00 per hour	= $ 56
1 Mill carpenter at $52.00 per hour	= $ 52
Total crew cost per hour	= $313
Total crew cost per day = $313.00/hr. × 8 hr. = $2,504 per day.	

Now analyze the time in crew-days that it would take to frame each of the four assemblies. Again, if we create a hypothetical example:

Floor frame assembly (1st floor)	2 crew days
Wall and partition assembly (1st floor)	2 crew days
Floor frame assembly (2nd floor)	1 crew day
Wall and partition assembly (2nd floor)	2 crew days
Roof and ceiling assembly	4 crew days
Stair frame and misc. blocking	1 crew day
Total crew-days to frame	12 crew days
Total crew-cost to frame	12 crew days × $2,504 = $30,048

If the crew was reduced to two lead carpenters and one apprentice to install exterior trim, recalculate the crew cost to $128 per hour, or $1,024 per day. If it was determined that it would take four crew-days to trim the exterior, the process could be repeated:

$$4\ crew\text{-}days \times \$1,024 = \$4,096$$

For large or small wood frame projects the estimator is encouraged to prepare a plan on how the work will proceed, much like the erection of steel. The plan should consist of a day-to-day summary of what is anticipated for production and the labor commitment for the day. This can often force detailed thinking about the labor requirement, thereby avoiding "robust optimism."

Fasteners and Seismic/Wind Requirements

As with each successive change to the building codes, wood framing within the United States has seen some substantial changes, many of which have been in reaction to some of the devastating natural disasters that have occurred, specifically those attributable to seismic and wind-borne calamities.

As a result, there are constantly evolving new requirements for bracing and fasteners to counteract the damage caused by these events. Some plans, predominantly commercial, will detail the connectors in both type and location. This may be a requirement for the structural engineer to stamp the plans. This is not always true regarding residential. The estimator is advised to investigate the requirements for tiedowns and holddowns for the project location.

In addition, many jurisdictions now require shop drawings stamped by a registered structural engineer illustrating the details for wood framing systems in much the same manner as structural steel shop drawings. This is even required on residential projects in many jurisdictions. Estimators are *strongly* encouraged to research and become familiar with the codes governing their particular project prior to estimating the cost of the work.

For many projects, the hold-downs and tiedowns are installed as the work proceeds. For other projects, these can be the work of the trailer crew. Again, the estimator is encouraged to decide who will do what via a plan.

FINISH CARPENTRY

Finish carpentry can be defined as the use of finish woods or synthetic materials (in contrast to dimensional framing lumber) in a variety of shapes, sizes, profiles, and species, applied to provide a finished appearance to windows, doors, stairs, and other features on the interior and exterior of the building. In contrast to architectural millwork (to be discussed later in this chapter), finish carpentry is the application of stock trim pieces versus custom-fabricated woodwork. For the purposes of this text, it is classified as the materials that can be purchased at the local lumberyard or home improvement store. Finish carpentry also includes the use of square edge stock commonly referred to as *dressed wood* and synthetic moldings and trims.

Taking-off and Pricing

Study the architectural drawings for the location, size, and configuration of the interior finish carpentry. The estimator should pay special attention to the interior elevations and the wall sections that may illustrate finish carpentry work. Finish carpentry details are often used to clarify or enlarge a particular section or elevation. Interior trim materials, such as casings, baseboard, cornice, chair rails, railings, skirt, and cheek boards, are taken off and priced by the LF. Quantities can be determined by measuring the perimeter of the room, door, window, or wall. Other interior finish woodwork, such as wainscoting, is taken off and priced by the SF. Stair parts, such as newel posts, balusters, treads, and risers, are taken off by the individual piece (EA), according to size and spacing.

As in all types of carpentry work, the takeoff for finish carpentry items should be separated according to the size, species, grade, and intended

Terminology for Finish Carpentry

The trim around window perimeters and the sides and head of doors is called casing. *The wood trim at the base of the wall that runs along the perimeter of the room is called the* baseboard. *Horizontal trim that runs parallel to the baseboard approximately 3' above it is called the* chair rail. *A wood surface may be applied between the chair rail and the baseboard, in the form of paneling, raised panels, or vertical or horizontal strips, referred to as* wainscoting. *Horizontal trim at the intersection of the walls and ceilings is called the* cornice.

To complete stairs, finish treads and risers are necessary. Additional stair parts, such as newel posts *to support the railings, are also considered finish carpentry, as are the* railings *that run between the newel posts and provide enclosure to the stair.* Balusters, *or individual vertical members, are attached to the railing.* Skirt boards *are trim pieces at the exterior of an open stair running parallel to the stringer. A similar trim at the interior of a stair is called a* cheek board.

application of the piece. Waste factors for materials should account for both the quality of the product and the difficulty of the application. Waste factors for finish carpentry materials range from a conservative 4% to a liberal 15% depending on the task. Materials pricing should follow that same procedure as the framing lumber by developing a stock list of finish materials and then soliciting current pricing from suppliers.

Labor for Finish Carpentry

Estimating labor costs for installing trim work is calculated by the LF, based on two factors:

- Production rate of each individual carpenter
- Difficulty of the application

In contrast to framing, finish carpentry tasks are measured by the production rate of the individual carpenter versus the production rate of the entire crew. Even if multiple carpenters are performing the same task, the production rate is measured as an average of the total output. This allows the estimator to price the labor based on reasonable expectations supported by past performance. This is where historical data is extremely valuable.

Assessing the level of skill required by the craftsperson to perform the task, however, can be difficult. For example, casing a window without a sill or apron is easier than installing cornice moldings at the ceiling line. The pay scale would reflect the skill level of the carpenter, and a daily rate would have to be calculated. If the cost for the finish carpenter is $65 per hour, the daily cost would be $520. If the same carpenter installed an average of 200 LF of wood base in an 8-hour day, the labor cost per LF would be:

$$\$520 \text{ per day} \div 200 \text{ LF per day} = \$2.60 \text{ per LF}$$

Most finish carpentry crews can be priced similarly to that of the framing crew. The main difference is that there is traditionally no trailer crew that comes behind. The estimator is encouraged to analyze the finish carpentry scope of work in detail and assign labor hours to each task or group of tasks. This allows the estimator to summarize the hours required and determine a crew size. Bear in mind that some tasks such as installing baseboard or stair parts are an individual task in comparison to cornice moldings, which frequently require a second set of hands.

ARCHITECTURAL MILLWORK

Custom-fabricated trims, molding, casings, cornices, and any other woodwork components from rough stock is referred to as *architectural millwork*. Work is typically done in a millwork shop according to quality control guidelines using machinery and equipment that are not portable. Unique or one-of-a-kind designs can be prepared specifically for the project. The woodwork is fabricated from a higher grade or species of hardwood, as compared to the softer woods that constitute the major portion of finish carpentry products. The mill shop selects the rough stock and planes, fits, shapes, sands, and even finishes the trims in the shop to fulfill the design. Quality control measures, such as maintaining temperature and humidity requirements to stabilize the work, matching grain to provide continuity, and pretreating or prefinishing the wood for the uniformity of color, are all considered within the realm of architectural millwork. Architectural millwork may even require the cutting of special knives for shapers to match a profile of trim.

Taking-off and Pricing

Takeoff procedures parallel those of finish carpentry work. However, translating the LF requirements of the finish materials into the quantity of rough stock needed is somewhat more difficult. The difficulty lies in that the estimator must determine how much rough material should be purchased in order to mill the required quantity of dressed wood, while accounting for natural imperfections, waste, and losses due to fabrication. Hardwood, or finish woods "in the rough," in general have their own system of measurement called *board foot measure* (BFM), that is the basis for pricing. One board foot is equal to the volume of a piece of wood 1" thick by 1' square. When calculating in BFM, nominal dimensions are used.

To calculate the board footage of any piece of wood, the following formula is applied:

$$BFM = \frac{t \times w \times l \times n}{12} \; where:$$

BFM = *board foot measure in feet*

t = *nominal thickness in inches*

w = *nominal width in inches*

l = *length in feet of the individual piece*

n = *number of pieces*

For example: How many board feet are in a 2" × 6" board that is 16' long?

$$BFM = \frac{2" \times 6" \times 16' \times 1 \; piece}{12} = 16 \; board \; feet \; (BFM)$$

Estimating architectural millwork can be broken down into three main phases of the process:

- Milling (material and shop labor)
- Shaping and finishing (material and shop labor)
- Installation (minor material and field labor)

Milling is the process of converting wood from a rough state to a dressed product. It includes *planing*, or dressing, the horizontal or widest portion of the stock. Additionally, the stock is passed through the joiner, which squares the edges with reference to the widest portion of the material. The material can be cut to a specific size and length of square-edged stock, then run through various sanding machines to prepare it for finishing, or it can proceed to additional milling steps. These additional steps may include cutting a profile on the face of the wood, as in the case of a cornice or sculptured molding. This process is called *shaping*, and, depending on the elaborateness of the profile, can require multiple passes through the shaper to achieve the finished product. Once the profile has been completed, it is then sanded and prepared for finishing.

Finishing is the process by which the square-edged stock or profiled trims are stained and/or sealed. Frequently, the work is conditioned to accept the stain more uniformly. The stain is applied and allowed to dry, then finished with a variety of sealers, such as urethanes or lacquers. Once the material has been finished, it can be shipped to the project for installation.

Waste factors for the finished product can be higher than its finished carpentry counterpart ranging from 15% to 40%.

Labor for Architectural Millwork

All of the shop processes – milling, shaping, and finishing – are labor intensive, as a result of the care and attention required to produce the finished product. Be sure to analyze each step in the process to assign labor-hours and, ultimately, costs for the work. Installation labor costs follow the same guidelines as those for finish carpentry. They are based solely on the production of the individual carpenter. Note that installation production rates on finished architectural millwork can be significantly less than for its finish carpentry counterpart. Often, this is due to the level of workmanship that is expected of a custom millwork project and the added care in the joinery. Also consider on-site staging or platforms for installing the work, as well as touch-up of the finish. Billing rates for millwork carpenters are at the high end of the industry.

The estimator is urged to solicit and secure prices for architectural millwork; materials, fabrication and finishing labor and installation whenever possible. This is considered a specialty trade and requires an in-depth knowledge of the process to price accurately.

STRUCTURAL PLASTICS AND PLASTIC FABRICATIONS

With the advent of mainstream recycling, a tremendous variety of items are now being fabricated from new and recycled plastics as well as fiberglass. They are readily available at big box home improvement stores as well as lumberyards. These products run the gamut from trim to panels to lattice to molded items. They are most often easy to work with and are readily available. The estimator is directed to research the individual product as thoroughly as possible – the internet is a good source. These products are suggested alternatives to natural wood products and often have installation videos on the internet to help the estimator understand the installation.

Taking-off and Pricing

Takeoff procedures follow those of the product's natural counterpart. In addition, glues, adhesives, brackets, hangers, and finish trims are also part of the quantities to be estimated. Given their proprietary nature, some of these products can be costly. Whenever possible, the estimator is directed to obtain contemporaneous pricing from a manufacturer or distributor of the product. If the product is a special order, do not forget the costs of shipping.

Labor to Install Structural Plastics and Fabrications

The labor required to install the product most often follows that of its natural counterpart – for example, wood decking and synthetic decking. For first-time installations, the estimator would be wise to allow extra time to accommodate a learning curve. Synthetic wood components often behave differently than their natural counterpart. The estimator should inquire as to any unique characteristics or considerations when estimating a specific product. Many plastic fabrications such as exterior railings are systems that need to be cut and field assembled. Many of these systems require special hardware or fasteners for the system. Because of the proprietary nature of the fasteners or adhesives, they are often expensive. As with any proprietary system, the estimator is urged to research the product carefully and solicit installation advice from the manufacturer, beyond the internet whenever necessary.

Many of these tasks are completed by carpenters and as a result fall into the category of finished carpentry. Estimating the actual labor cost follows a process similar to finish carpentry tasks with a focus more toward assembly than installation.

CASEWORK AND CABINETRY

Casework and cabinetry, for the sake of this text, can be defined as the purchase and installation of production-line wood or laminate cabinets, bookcases, vanities, countertops, and the like. These products are mass-produced and sold as individual units in an enormous variety of shapes, sizes, compositions, and price ranges. Estimating fabrication and finish costs for custom cabinetry requires special estimating experience and follows a procedure similar to that outlined in the previous "Architectural Millwork" section.

Most cabinetry, including that for kitchens and bathrooms, can be classified as one of three main types for estimating purposes:

- **Base units:** Installed on the floor and terminate at the underside of the countertop.
- **Wall units:** Fastened on the wall above the countertop, terminating below the ceiling line.
- **Full-height units:** Continuous cabinets that start at the floor and terminate at the top of the wall cabinets.

Casework consists of modular or prefabricated units, such as bookcases, retail display cases, vanities, storage cabinetry, and shelving. All production-line cabinetry and casework are prefinished at the factory. Cabinetry consists of the box or shell of the cabinet, called the *carcass,* and the door or drawers that are applied to the carcass, depending on its function. Wood cabinetry and casework are prefinished in much the same manner as architectural woodwork. The work is stained or painted and then sealed with a clear urethane or lacquer coating. An alternative method of construction and prefinishing of casework and cabinetry is to cover the exposed surfaces with a thin sheet of resin-impregnated paper, called *plastic laminate.* Plastic laminate, available in a variety of colors and textures, is also used to cover the surfaces of kitchen countertops and bathroom vanity tops.

Casework can also be fabricated and finished onsite when custom sizes and applications are required. Material requirements are summarized and shipped to the jobsite similar to finish carpentry materials.

It should be noted that manufactured casework and cabinetry along with countertops can also be specified under section 12 30 00—Casework. The estimator is advised to check the specifications carefully for proper location in the estimate.

Taking-off and Pricing

To begin the takeoff process, review the architectural drawings with specific attention to interior elevations, floor plans, sections, and details of the cabinetry. Many architectural drawing sets include separate plans or elevations for cabinetry and casework. These plans provide detailed dimensions that allow you to determine stock sizes offered by manufacturers.

Review the specifications for the manufacturer, model, and series of the cabinetry, casework, or plastic laminate shown, as this is a major determinant in the materials' cost. Cabinetry and casework are taken off and listed by the quantity of each piece and priced by the individual piece (EA). Individual pieces should be separated according to size, function, and type. A sample cabinet takeoff might look as follows:

> *ABC Manufacturing Co.; Premium line, "Colonial Series" in Medium Oak finish*
>
> *36", 2-door sink base unit—1 EA*
>
> *42", corner base unit—1 EA*
>
> *24", 2-door base unit with 1 drawer—3 EA*
>
> *30" wide × 36" high wall cabinet with 2 door—3 EA*
>
> *36" wide × 84" high full height pantry with 2 doors—1 EA*
>
> *3" × 36" filler pieces—3 EA*

Special accessories, such as sliding trays, baskets, sliding cutting boards, or adjustable shelving within the unit, should also be noted. This list is then submitted to the sales representative or retailer for pricing. Review the quote carefully for completeness, accuracy, and inclusions, such as freight or handling costs.

Cabinet hardware, such as knobs or pulls, should be counted, and listed separately from the cabinets, as these are not always provided with the cabinetry and may constitute an additional cost for both materials and installation. For onsite fabrication draw slides and similar hardware may need to be taken off and priced by the piece, EA.

Plastic laminate countertops with a standard 25" depth are taken off and priced by the LF for fabricated cost. This is typically done by measuring the countertop from the abutting wall. Alternate units of the takeoff and pricing include the SF of countertop surface. Plastic laminate sheets for field installation as backsplashes are taken off and priced by the SF. Plastic laminate for field applications should be converted to the manufacturer's available sheet size. Not all textures and colors are available in every size. Waste must also be considered and will vary in accordance with the availability of product, size of the sheet, and application.

Also include material costs for contact cement and any special tools or equipment required for installation. Countertops can be manufactured with a backsplash, an integral vertical return that abuts the wall. Separate unattached backsplashes may also be used and may require a separate listing in the takeoff. They are taken off and priced by the LF based on the height of the backsplash. The standard height is 4", although custom heights can vary. Other types of backsplashes can include sheet plastic laminate adhered to the wall surface. The material costs are priced by the SF. Convert the required SF to the available sheet size. Sheets can vary in size from 30" × 72" to 60" × 144". Not all colors and patterns are available in every size. Check manufacturers' stock when pricing the needed materials. This same process is applicable to custom-fabricated tops that exceed standard widths and shapes. Field cutouts for sinks and cooktops and holes for faucets are frequently additional costs to the installation.

In addition to plastic laminates that are numerous other countertop materials available such as concrete, natural stones: such as granite and soapstone, and synthetics that resemble stone. The estimator is urged to solicit and secure a contemporaneous price to template, fabricate, and install, complete with cutouts, any specialty countertops or surfaces. This can include windowsills and other unique applications on commercial projects. While it is adequate to approximate the cost of this work, it is better to have a contemporaneous price to include within a bid.

Labor to Install

Labor costs are calculated by the labor-hours expended to install the casework and countertops. Due to the limited amount of space in most applications, the installation crew typically consists of two individuals. Include unloading, assembling (if required), and installing the cabinets, casework, and countertops, as well as the installation of the hardware, such as pulls and knobs. While most casework comes completely assembled, this is not always the case. It is advisable to ask about the level of field assembly required. Installing casework requires first-class workmanship, yet a skilled crew can be productive. Installation in existing spaces can be more time consuming due to the shimming of cabinets and scribing of fillers needed to meet existing walls and floors that are out of level and plumb. Also note the substrate to which the cabinetry will be attached. Masonry substrates require more time than wood or drywall substrates. Detail work, such as scribing fillers that are adjacent to walls and ceilings, can also be time consuming and should be reviewed carefully.

Labor to install trims, such as valances, skirt boards, kicks, and cornices, should be estimated separately as an addition to the cost of installing the carcasses. Also include labor for adjusting drawers, doors, latches, and the like. Labor for countertop installation can be expressed by the LF of countertop for standard 25" deep tops or by the SF for irregularly sized surfaces. Special profiles on the leading edge of the countertop can impact costs by adding an LF cost to the fabrication.

Simulated stone or synthetic countertops may be field measured by a laser and a computer or may require a template cut from ¼" plywood in order to fabricate the finished product. Heavy countertops fabricated from natural stone or architectural concrete may also require that the cabinetry be reinforced with wood or steel or may require a large crew to handle. Similarly, leading edges of synthetic or natural stone tops with special profiles add to the cost of the finished product. Lastly, some stone surfaces require a liquid sealer that is applied after installation. The estimator should review the specifications to determine if this is required.

SHOP-FABRICATED STRUCTURAL WOOD

Shop-fabricated structural wood panels are components that are preassembled in a shop environment that are later field assembled into the walls, floors, and ceilings/roof of a structure. Rigid frames are fabricated with studs and plates along with rigid or batt insulation and plywood or similarly rated sheathing on one or both sides. They are fabricated complete with structural components such as headers built into the panel. Once on site, they are unloaded and erected to form the complete frame. This process is often selected as it has better quality control and can save the time associated with field framing.

Shop-fabricated structural wood panel systems require shop drawings that can add to the cost of the project. They also require a level of coordination not required with onsite stick framing methods.

Taking-off and Pricing

Prefabricated panels are taken off and priced by the SF area of the panel. Each opening for a window or door adds to the cost. R-values for the insulation within the panelized system impact cost. Costs per SF for floor and roof systems increase with the loading capacity of the assembly. This type of structure adds a shipping or delivery cost as well as the cost of field assembly. Panels can include embedded electrical boxes and conduits for wiring, or a variety of other items.

The estimator is encouraged to solicit and secure contemporaneous pricing for a complete system; fabricated, delivered, and installed whenever possible. This type of frame system can have a high fabrication cost due to the expensive and specialized equipment needed to fabricate the panels. It may also require a structural engineer's services in the design and stamping of the shop drawings, plus the cost of delivery. Field assembly always requires some level of crane or hoisting equipment. Field crews and fabrication crews are typically different and have different costs.

OTHER FRAMING SYSTEMS

Other wood framing systems include post and beam, plank and beam, heavy timber, and laminated timber construction. They are all considered specialty types of framing and not typically within the realm of the general contractor estimator. That having been said, they should be introduced and discussed briefly. Post and beam use larger cross section timbers for structural support. Sizes start at 4" × 4" and increase from there. Spacing is also increased between structural columns and infilled with regular 2" × 4" wall framing, sheathing and finishes. Post and beam systems are often integrated with plank and beam roof and ceiling assembly systems. Plank and beam systems use 2" nominal or thicker floor or roof decking with greater spacing between structural beams. Planking is typically tongue and groove for added rigidity and to facilitate the transfer of loads from one plank to another. Plank and beam floor systems typically use 2" × 6" or 2" × 8" tongue and groove decking.

Heavy timber was the normal method used to build most wood framed buildings prior to dimensional lumber. Much of the heavy timber used is considered old-growth timber, which is in short supply and may even be prohibited from harvesting for construction. Typical heavy timber columns were 8" × 8", 12" × 12", or even 16" × 16" with heavy timber roof beams of 8" × 10", 8" × 12" with 3" thick planks. Frequently, this was used in commercial applications such as factories and warehouses.

Laminated timber construction relies on heavy timber principles with laminated (glued) structural members to make the large sizes post and beams needed. Laminated members are custom fabricated from pieces of structural grade lumber bonded together with exceptionally strong glues. They provide large clear spans and carry heavy floor and roof loads.

Taking-off and Pricing

For all of the components in this category, takeoff and pricing is by the LF of the various cross sections of timber. The individual pieces are separated in the takeoff and estimated by size, length, cross section, application (column or beam), and the species of lumber. Planking should be handled in a similar manner for size and application. Connectors for individual beams and column connections are heavy steel and most often expensive. Each piece of varying size and shape should be counted and listed as EA. This is also true for the nuts, washers, and bolts for the connection plates.

The estimator is encouraged to solicit and secure contemporaneous pricing for a complete system of both varieties of frame. The quote should include fabrication, delivery, hardware, and installation whenever possible. This type of frame system can have a high fabrication cost due to the expensive lumber cost, especially with species of high quality. It may also require a structural engineer's services in the design and stamping of the shop drawings, plus the cost of delivery. Field erection always requires a crane. Field crews develop and follow an erection plan in a similar manner to steel erection or can require assembly on the ground and lifting or tilting into place.

SUMMARY

In order to accurately estimate carpentry tasks, a thorough review of the documents is required. In addition, a clear understanding of the different types of tasks and the level of workmanship required is helpful in correctly determining labor costs. Wood structures can have a variety of framing types based on the use of the structure. Some can use larger dimensional lumber with connecting hardware, or more modest holddowns or tiedowns to resist wind loading and seismic events. The estimator should also be familiar with the building code governing the project for locale specific framing details or requirements.

For many of the framing techniques a framing erection plan and a crane are required. In addition, the project may require shop drawings with a structural engineer's stamp.

Finish carpentry is often carried out by smaller crews with productivity based on the individual carpenter's output. Special configurations of trims from hardwoods can be created specifically for a project. Architectural millwork can often have fabrication and finishing cost in addition to the material and installation costs.

Cabinetry and casework can be manufactured off site and installed on site, or it can be custom fabricated on site to meet the specific needs of the project. The estimator is urged to check the specifications for determining if the product is manufactured or custom built.

Shop-fabricated structural wood and timber frame systems are also found on projects. The estimator is encouraged to seek comprehensive subcontractor pricing for any application that requires special shop fabrication or unique erection techniques.

QUESTIONS/ PROBLEMS

1. What are the four (4) major categories of Division 6 for estimating purposes?

2. What is the difference between a nominal and actual dimension?

3. What are the four (4) main assemblies in a frame? Explain each briefly.

4. What is the subfloor? How is it taken off and priced? Explain in detail.

5. A foundation is 64' long × 28' wide with a girder running down the center (parallel to the 64' dimension). Calculate the number of 2" × 10" floor joists needed if they are spaced at 16" o.c. Calculate the band joist and the solid blocking over the girder. Note the length and quantity of each member. (Assume no openings in the deck.)

6. How many sheets of ¾" tongue-and-groove (T&G) subfloor sheathing are required to cover the floor frame in question #5? If a sheet of ¾" T&G subfloor is $80.50 what is the total material cost for the subfloor sheathing?

7. For the building in question #5, calculate the length of the rafter and the quantity of the rafters if the roof pitch is 6:12 (each side of the ridge), with ½" sidewall sheathing and a 10" overhang (measured from the face of the sheathing). It has a 2" × 12" ridge. What is the stock length needed?

8. Calculate the amount of board feet (BFM) in 236' of 2" × 8" (actual dimension) of poplar.

9. The average daily output for a crew of three carpenters to picture frame trim (no sill or apron) windows is 422 LF. If the billing rate for a carpenter is $65.50 per hour, what is the cost to trim out a window that is 4' × 6'?

10. Calculate the header stock to be purchased if the rough opening for a window that is 6'-2" × 4'-6"? The wall is 2" × 4" construction and requires a minimum of 3" of bearing.

12 | Thermal and Moisture Protection

CSI MasterFormat® Division 7—Thermal and Moisture Protection covers work that protects the structure from the elements and outside temperatures. *Waterproofing* and *dampproofing* include coatings below and above grade to prevent water and moisture migration, respectively. *Insulation* materials reduce the transmission of heating or cooling through the exterior envelope. *Roofing* provides a watertight surface for the uppermost surface of the structure. *Siding* provides protection from the elements on the vertical surfaces of the structure. To provide a water- and moisture-tight envelope, *caulking* and *sealants* are also used to fill in spaces and seal surface areas. Miscellaneous items such as skylights, roof accessories, gutters, and downspouts are also part of Division 7.

Most of the work in this division is detailed on the architectural drawings. The estimator is urged to check the roof plan for the location and extent of roofing work. Also consult:

- *Details and sections of the roof system* for flashing and sheet metal details, insulation at the roof envelope, and roof accessories.
- *Wall sections and details* for information about thermal insulation, air barriers, and vapor barriers at the exterior walls.
- *Building sections* for insulation materials at the floor, wall, and ceiling levels.
- *Wall and foundation sections and details* for surfaces to be waterproofed, dampproofed, or insulated below grade.
- *Division 7 of the technical specifications* to help determine appropriate products and installation methods. Items such as caulking and sealants are not always detailed at every required location on the drawings. Consult the specifications to verify the location and extent of caulking and sealants.

Estimating Building Costs for the Residential and Light Commercial Construction Professional, Third Edition. Wayne J. Del Pico.
© 2023 John Wiley & Sons, Inc. Published 2023 by John Wiley & Sons, Inc.

WATERPROOFING

The purpose of waterproofing is to prevent water from penetrating through structures. Concrete and masonry are highly susceptible to water penetration due to their porosity. Waterproofing is most often applied to concrete or masonry construction below grade when standing water is expected or above grade in areas of high precipitation. Waterproofing is not to be confused with dampproofing, which are very different products. Check the specifications for particular waterproofing products and application methods.

Types of Waterproofing

Waterproofing is a membrane or coating used in specific locations to prevent water from entering the structure. The are several effective methods of waterproofing. The following are the most common. Below grade waterproofing is typically used in conjunction with a foundation drainage system. *Membrane waterproofing* is applied to the surface of the protected area. Membranes can be sheet, liquid, or bituminous and are usually applied to the vertical surface being protected. The surface is prepared, primed, and then either sprayed or mechanically applied. Bituminous coatings can be troweled or sprayed on to the surface depending on the particular product.

Clay waterproofing is a layer of expanding material called bentonite clay. When mixed with water bentonite expands to more than 10 times its dry volume. It can be sprayed or applied in sheets. In some cases, it requires a cementitious parging course before it can be applied to masonry. Bentonite sheets can be mechanically fastened to the surface or can be adhered using bitumen.

Cementitious waterproofing can be applied directly to the surface of the concrete. The compound is typically applied with trowels or is sprayed and fills the pores of the concrete or masonry. It is a mixture of Portland cement, fine aggregates, and an acrylic or plastic admixture. Many types also contain minute iron fillers. This is called *iron oxide* waterproofing. The iron-oxidizing agent in the mix causes the iron fillers to rust and expand to fill any pores in the coating. Another form of cementitious waterproofing uses hydraulic cement, which is a mixture of rapid setting cement and nonshrinking hydraulic compounds.

It should also be noted that many waterproofing systems are used in conjunction with rigid insulations and hardboard protective layers to create a proprietary system. The estimator is urged to check the internet for videos illustrating the particular system's installation.

Sheet Membrane Waterproofing

Membrane waterproofing consists of one or more layers of asphalt-saturated fabric, butyl, EPDM, or rubber membranes applied with an adhesive after the surface has been cleaned and primed and all holes have been filled. The adhesive can be synthetic- or asphalt-based, compatible with the primer and fabric. Lapping at sides and ends is as dictated by the specifications of the manufacturer. Additional materials (required for lap) will vary from 6% to

10% of the area to be waterproofed. The estimator should become familiar with the particular product and its application for accurate pricing. Some products are applied cold, and others at specific temperatures, and some products have drying times between coats that reduce productivity.

Liquid Membrane Waterproofing

Liquid membrane waterproofing requires cleaning the substrate and filling cracks and holes prior to application. The preparation time involved can often equal or exceed that of the product application. Liquid membrane application procedures vary significantly by product. The estimator is urged to investigate coverages, in SF per gallon of the specified product. Manufacturer's recommendations for individual products must be considered when pricing a specific product. Verify the number of coats specified and convert this information to gallons of material. The number of coats may be determined by final thicknesses of the surface in mils or one-thousandths of an inch.

Clay or Bentonite Waterproofing

Clay or bentonite waterproofing requires cleaning the substrate and filling cracks and holes prior to application. It may also require the removal of any oils or release agents from the surfaces to be coated. Masonry substrates often require a cementitious parge coat prior to the application of the bentonite. Bentonite is especially effective in tight locations where access is not available for membrane applications. As noted previously bentonite can be applied mechanically in sheets as well as sprayed.

Cementitious Waterproofing

Cementitious waterproofing requires specific preparation of the concrete or masonry surface prior to application. The surface should be thoroughly washed to remove residual materials or coatings. Holes, cracks, and penetrations should be filled and patched. This can be labor intensive and costly. Many trowel-coat or brushed products require multiple coats. The minimum is typically two coats, although some products require as many as five coats. Manufacturer's recommendations for individual products must be considered when pricing a specific product. Cementitious waterproofing is mixed on site with limited workability times for hand (trowel or brushed) applied installations. Sprayed applications are frequently less labor intensive but involve equipment costs.

Taking-off and Pricing

While all waterproofing products perform the same basic function, labor and material costs vary. Quantities for each method should be taken off, listed, and priced separately. All types of waterproofing are taken off by the square foot (SF) of area to be protected. Quantities and pricing of preparation should be kept separate from actual application of the waterproofing. Use conservative production rates for preparation of older surfaces. Waterproofing is frequently a crew task requiring a minimum of two workers, although smaller surface areas such as elevator pits may prevent more than one individual working at a time.

The standard units for pricing for the membrane, cementitious, clay, and liquid methods are the square foot (SF), which can be converted to squares (SQ), where 1 SQ = 100 SF. For below-grade applications, the SF area to be waterproofed is determined by the amount of surface area that is in contact with the backfill. Consult the grading and drainage drawing (part of the site plans) to determine accurate locations and heights of foundation walls. For foundation applications, multiply the foundation perimeter by the distance from the finish grade at the exterior to the bottom of the footing.

Keep in mind that many of the products used in waterproofing require personal protective equipment (PPE) for the applicator. These costs should be included when pricing the installation. For products that require spray and/or mixing equipment, the estimator should consider added time to set up, breakdown, and clean the equipment for each application. Limited access or tight spaces for hand-applied products may reduce daily outputs or even require additional labor as support staff.

Many waterproofing products, especially proprietary systems, carry significant liability. As a result, some systems require that a manufacturer's trained representative be present during the application process to ensure the product standards are met. This cost should be included in the price for the work.

DAMPPROOFING

Dampproofing is most often a bituminous coating applied to the foundation area below grade to prevent moisture penetration, is not intended to resist water pressure and should not be confused with waterproofing. The most common methods are spraying, brushing, or troweling a bituminous-based, tarlike coating on the areas to be protected, uniformly covering the areas below grade. Bituminous dampproofing is either applied hot or cold and can include a membrane. Another method, called *parging*, involves plastering a cement-based mixture over the area to be protected. Liquid-applied dampproofing is also common and follows the same procedure for estimating liquid-applied waterproofing.

Taking-off and Pricing

Dampproofing can be performed by a crew of multiple individuals or by a single worker, depending on the particular project size and product. Standard units for takeoff for all types of dampproofing are SF and can be extended to squares (SQ). Determining areas for dampproofing follows the same procedure as waterproofing. Quantities for bituminous applications should be separated according to method of application, as labor costs will differ. Parging should be listed by the thickness and number of coats. The standard thickness is ½", applied in two coats. Like waterproofing, dampproofing requires surface preparation, also taken off and priced by the SF of surface area to be prepped.

Special preparation tasks, such as cleaning or roughing the surface area, as noted in the technical specifications or required by the manufacturer, should be included separately. This work should be taken off and estimated separately, also by the SF. Some dampproofing products require a primer coat prior to dampproofing, also estimated in SF of surface area. Hot applied dampproofing requires equipment for both heating and spraying, and potentially additional personnel to staff the equipment.

The estimator is strongly recommended to research the product specified, including productivities and material costs as there is a significant difference in price, especially for sustainable products.

THERMAL INSULATION

Thermal insulation is used to reduce heating or cooling loss through the exterior of the structure or to unconditioned areas within the structure. Insulation materials are numerous and varied depending on the application, but all share a common theme: they are rated for *thermal resistance*. Thermal resistivity is expressed as the material's *R-value*. The higher the R-value, the more effective the insulation. Higher R-values usually translate to higher costs for materials. The estimator should consult the architectural plan views, elevations, sections, and details that provide the location of the materials. In addition to the building envelope, insulation information can be found on building sections, wall sections, and details, particularly on the sections through the exterior envelope of the structure. The specification should also be reviewed for product information that may provide pricing criteria. Figure 12.1 lists the location of the various types of thermal insulation.

Insulation is available in various forms, sizes, thicknesses, and R-values. In wood and light-gauge metal frame construction, blankets, or rolls of insulation, called *batts*, are installed between the studs or joists at the exterior walls, roof, or floor of the structure. Insulation is available with an attached paper or foil facing that act as a vapor barrier. (Vapor barriers are discussed in detail later in this chapter.) Rigid insulation is used where a particular shape must be maintained, such as under concrete slabs, at the exterior of foundation walls, for roof decks, and over exterior sheathings under siding materials. Some common types include expanded polystyrene, often called *beadboard,* and extruded polystyrenes, such as Styrofoam™. More complex rigid insulations such as *polyurethanes* and *polyisocyanurates* can be used alone or as part of other systems such as roofing. Polyisocyanurates are faced with asphalt-impregnated glass fiber felts and generally have a high R-value per inch of thickness; they are a common component in flat roof systems. These are typically estimated as part of the roof system components. Other rigid insulation products are available with foil facing for reflecting heat.

Location within Structure	Type of Insulation	Attachment
Basement foundation walls	Rigid insulation on interior or exterior of foundation walls	Typically mechanically attached or with adhesive
Basement foundation walls	Batt insulation with or without vapor barrier	Within framing or furring on exterior wall
Under slab basement	Rigid insulation under slab over full area	Placed on sand bed, seams taped, with vapor barrier
Stem or frost walls at crawl space	Rigid insulation from underside of sheathing down to top of footing	Typically mechanically attached
Floor cap at basement or crawlspace	Batt insulation without vapor barrier	Set between joists
Floor cap at basement or crawlspace	Open or closed cell spray foam insulation, based on application	Sprayed between joists
Exterior Walls	Batt insulation with vapor barrier and possibly rigid insulation board	Within framing or furring on exterior wall
Exterior Walls	Open or closed cell spray foam insulation, based on application	Sprayed between studs
Ceiling	Batt insulation with or without vapor barrier	Set between joists
Ceiling	Open or closed cell spray foam insulation, based on application	Sprayed between joists
Roof Rafters	Batt insulation with or without vapor barrier	Set between rafters with air space
Roof Rafters	Open or closed cell spray foam insulation, based on application	Set between rafters
Trusses	Batt insulation with or without vapor barrier	Set between bottom chord of trusses
Trusses	Open or closed cell spray foam insulation, based on application	Spray between bottom chord of trusses
Misc. Areas	Batt or foam insulation in cracks along window, door, or any penetration perimeter	Manually stuff or spray cracks full of insulation

Figure 12.1 Location of Various Types of Insulation

Granular insulation such as perlite and vermiculite are used in areas that have access for pouring. Loose granular insulation requires a form to retain the insulation in place. This makes these products ideal for filling the cores of CMU in exterior walls. While still insulation, they are part of the CMU wall system, and their costs are often captured in Division 4—Masonry portion of the estimate.

Loose-fill insulation is composed of fibers or chips that are either blown in or poured in place. Most loose-fill insulation are byproducts of mineral- or glass wool or cellulose. Their most common application is in open areas of attics covering the ceiling below.

A wide variety of foamed-in-place insulating products are also being commonly used. They are referred to as *spray-foam insulation*. The most popular types are spray polyurethane, polyisocyanurate, and phenolic-based. The insulation is created by a chemical reaction that expands the mixture as much as 30 times to fill the cavity. When cured, the products solidify into a cellular plastic insulation of high R-value. Products can be *open-* or *closed-cell*, which define the composition of the final installation. Many of the common products in today's market are proprietary and require specialized training, licensing to purchase, and equipment to install. There is also a clean up component to the cost of this work that is not required in other insulation installations.

Taking-off and Pricing

Batt or Blanket Insulation

Batt or roll insulation is taken off and estimated by the SF and should be listed separately according to width, thickness, R-value, and its location within the building (roof, walls, floors, etc.). For wood or metal framing, the width refers to the space between each stud or joist. For example, wood studs spaced at 16" o.c. require 15"-wide material. Materials should also be listed separately by vapor barrier facings, such as *kraft* paper and foil facing. Faced and unfaced materials differ in price. Consult the specifications carefully for the required R-values and specific product information.

Labor costs reflect the amount of material that can be installed per day. Labor to install insulation is based on individual (versus crew) productivity. Consider costs for scaffolding to access high ceilings or roof caps. Large, open areas should be separated from smaller "cut-up" areas in the takeoff, as this affects productivity. A single skilled worker should install 900–1,600 SF per day of batt insulation in walls. Productivity on ceiling or floor installations ranges is typically 400–700 SF per day per worker. The estimator should note that these are ranges and specific project conditions will dictate actual productivity.

In addition to production areas, there are often small spaces or crevices around windows and doors that require insulation to complete the work.

Filling these areas by hand can also include covering them with a tape to prevent the transmission of air or vapor. For walls with a significant number of openings, in excess of 15% of the gross area, the estimator may want to add labor to the installation on a per-opening basis.

Rigid Insulation

Rigid insulation is taken off and priced by the SF, according to the type of material, application, method of installation, R-value, and size (length and width of sheet). The estimator often converts the square feet to sheets of the individual product based on the size needed plus a waste factor. The most common size sheets are 24" × 96" and 48" × 96", although other sizes are available.

Labor for installing rigid insulation is based on individual productivity. The most common installation methods are mechanical fastening with screws and washers, using adhesives, and laying the material in place for under slab applications. Check the specifications for the method of attachment, and ensure all fasteners and adhesives have been included, which may include tape to cover the seams between sheets. Installation costs vary significantly with the method of installation. Consider costs associated with scaffolding or hoisting of materials. These costs should be listed separately in the estimate.

Productivity for installing rigid installation varies significantly by product and application method. For vertical applications, a single skilled worker should install between 600 and 1,000 SF per day. For horizontal applications, such as under concrete slabs where the method of installation consists of placing the insulation on the subgrade, a single skilled worker should install between 1,000 and 1,600 SF per day, depending on the amount of penetrations through the slab.

Many roofing- and masonry wall- systems include insulation as part of the system specifications. The cost of this insulation is part of the roof or masonry estimate and care should be taken not to duplicate quantities in the takeoff. The same is true for underslab or foundation wall applications, while the material and execution are specified in Division 7, the work is performed while the foundation work is in process and may be best suited to include in Division 3 work.

Loose-Fill Insulation

Loose-fill insulation in the form of mineral wool (a form of molten rock) and expanded volcanic rock materials, such as perlite and vermiculite insulation, is often used to insulate cavity spaces, cells of concrete blocks, and attic spaces. It can be installed by spraying, pouring by hand, or machine-blowing. Loose-fill insulation is sold in bags of varying volume, which can be converted to the number of bags required for accurate material pricing. Quantities can be listed by SF or cubic foot (CF), depending on the application and the depth of material, or in some cases the settled R-value. Frequently, the task may include ancillary components such as baffles to hold the product in place.

Labor costs are based on the amount placed per day. Hand-placement is considerably slower than machine-blowing, which is more productive using a two-person crew than an individual. Be sure to include the cost of equipment for machine-blowing. Regardless of the placement method, loose-fill insulation can be a messy operation. Waste factors can be in excess of 20% depending on the method of installation. Include time for cleanup after the work has been completed as well as PPE for the installers.

For loose-fill insulation consisting of perlite, wood fiber, cellulose, mineral wool, or similar material poured by hand, a single skilled worker should place approximately 200–400 CF per day, including prep and clean up.

For applications of loose insulation installed within the cores of CMU, a crew of two skilled workers should install between 1,200 and 4,800 SF per day based on wall thickness, height, and access. Completed applications may require some level of protection from precipitation if the wall is not protected from the weather at completion of the insulation.

Applications consisting of loose insulation blown in ceilings with open access vary by thickness (depth) and material used. A crew of two skilled workers can install between 900 and 5,000 SF per day.

Re-insulating existing structures, such as attics can reduce productivity significantly due to removal of floor boards or working around equipment. The same is true for exterior walls. Both can require significant labor to prep before actual insulation. In the case of walls where the insulation is blown in through the exterior, layout, and drilling of holes as well as patching of siding can involve significant labor. The estimator is advised to review existing conditions carefully.

These productivities are offered for guidance only and should be evaluated under the specific conditions of the project being estimated.

Foamed in Place Insulation

Sprayed on foam insulation is typically taken off by the square foot area (SF) and categorized by depth of thickness, R-value, and application. Horizontal and vertical applications will differ significantly. A two-part system of chemicals is mixed at the nozzle of the sprayer and cures into a cellular plastic that fills the cavity being insulated. A crew of three trained workers can install between 1,000 and 6,000 SF per day with the appropriate equipment, depending on thickness, application, and the material being used. Access to the space is a key factor in productivity. Crawl spaces with limited access can reduce productivity dramatically. This would include screeding the over sprayed material to the wall thickness. Open-cell and closed-cell products may be used on the same project, so care should be taken to separate them in the estimate for more accurate pricing. Labor to set up, breakdown, and clean the equipment should also be included. Much of the cost of this work is dependent on the product and manufacturer requirements. The estimator should solicit subcontractor pricing whenever possible due to the proprietary nature of the work.

VAPOR BARRIERS

Vapor barriers, or retarders, prevent the transmission of moisture (caused by temperature and humidity changes) between the building interior and its exterior. They are typically applied to the warm side of walls, floors, roofs, and ceiling construction. The most common vapor barrier is *polyethylene sheeting,* a plastic film available in thicknesses designated by *mils* (1/1000"), the most common of which are 4- and 6-mil thicknesses.

Vapor retarders can also be attached to the insulation as either a kraft- or foil-faced paper. Vapor control can also be achieved by foil-backed gypsum sheets. Both options eliminate the need for a separate material or installation labor cost in the estimate.

The estimator is directed to research the specifications carefully, as there are a wide variety of products and costs associated with vapor control.

Taking-off and Pricing

Vapor barriers are taken off by the SF and separated by thickness. The SF area can be extended to number of rolls and the required roll size. Be sure to consider side and end laps, which will vary from 5% to 10% of the overall area. If the vapor control is part of the insulation or gypsum facing, it should be noted in the estimate.

Other material costs include attachment, such as staples and tape to seal the seams at the lap. The specifications should stipulate lap dimensions, method of attachment, and requirements for sealing the seams. Do not deduct for openings, which are usually cut out afterward, with the residual pieces treated as waste. Review the specifications carefully for other types of vapor barriers in addition to polyethylene. Occasionally, the under-slab vapor barrier material and installation may be specified in Division 3—Concrete. These products and installations tend to be of a more sophisticated nature, for both product and installation.

Labor to Install Vapor Barriers

Labor to install vapor barriers is priced by the SF of area to be covered. Labor costs for under-slab applications are typically less than wall or ceiling applications. Installation of polyethylene sheeting is most efficient as a two-person task. Staging or lifts for access to high areas should be figured separately. Labor costs should reflect reduced productivity when staging is used.

Note also that many of the products being used require that penetrations through the vapor barrier be sealed. This is to maintain the integrity of the vapor barrier. This can constitute significant labor costs if each penetration is taped or sealed.

For a single-ply polyethylene vapor barrier in an above-grade application, a single skilled worker should place approximately 2,000 to 3,700 SF per day, again based on access and application. These productivities are offered for guidance only and should be evaluated under the specific conditions of the project being estimated.

AIR INFILTRATION BARRIERS

Air infiltration barriers, sometimes called weather barriers, protect against drafts caused by wind and water penetration. They reduce the number of air changes within a building to create a more energy efficient interior environment. Typically, air infiltration barriers are installed at the exterior of the sidewall sheathing under the siding. Since the efficiency of the building insulation is influenced by humidity and the exterior temperature, the best scenario is a drywall cavity free of moisture, which allows air to migrate through the wall.

With constant changes to energy codes in many states, air infiltration barriers have become a focal point of design in light-commercial projects. Designs often include expensive membrane products that require specific application temperatures, priming of the substrate, and integral terminations at exterior wall openings. These products and their related details are expensive and labor intensive to install. Review the specified product carefully and search for any installation details to help determine pricing. Additional sources of product information can be obtained from manufacturers' literature or websites. Less sophisticated products, such as 15 lb. asphalt felt paper or Tyvek® HomeWrap® are also commonly used.

Refer to the architectural wall sections for the location and the exterior elevations for the limits of air infiltration barriers. Check the specifications for the particular product and installation criteria. Deductions for window or door openings are not usually considered.

In addition to separate products, air infiltration barriers can be a coating or surface on exterior sheathings. The seams between sheets are taped with a proprietary product to maintain the integrity of the barrier.

Taking-off and Pricing

Air infiltration barriers are taken off and priced by the SF, measuring the area of the exterior wall sheathing to be covered. Convert the area to roll size, if appropriate, according to the manufacturer or specifications. Include additional materials for side and end laps. Installation is most productive with two people. For framed-in-place walls, consider supplemental staging to access the building exterior. For wood-framed walls that are tilted up after fabrication, labor costs will be reduced if the air barrier is installed prior to standup. Installation labor varies dramatically with different products. Obtain manufacturers' recommendations for installation times or productivity rates, when available.

A crew of two skilled workers can install between 3,000 and 7,000 SF per day of a spun, bonded polypropylene exterior air barrier, excluding costs for staging or lifts.

Some commercial air barrier products that require special prep and primers applied to the surface should be researched carefully. Many of these systems are highly proprietary and may require the presence of a full-time inspector during the installation or a specially trained installer. The estimator should solicit subcontractor pricing whenever possible due to the proprietary nature of the work.

EXTERIOR SIDING

Exterior siding is available in a variety of materials, the most common being wood, metal, cement fiber, and vinyl. Wood siding is manufactured in bevel, shiplap, and tongue-and-groove patterns to be installed horizontally as exterior cladding. Bevel siding, commonly referred to as *clapboard,* is installed with exposures ranging from 3" to 6". (*Exposure* refers to the portion of the clapboard that is exposed to the weather.) Clapboard is available in lengths from 3' to 20' in 1'-0" increments. Cedar is the most common wood for clapboard siding, but redwood, pine, fir, and spruce are also available. (See Figure 12.2.)

Shakes and shingles are other types of wood siding that can be installed at the exterior sidewall with typical exposures from 4–9". Shakes and shingles range in length from 12" to 24" and can be from 2½" to 14" in random

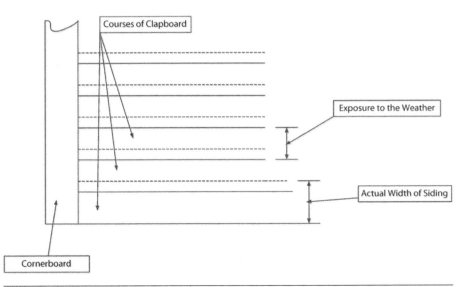

Figure 12.2 Clapboard Siding

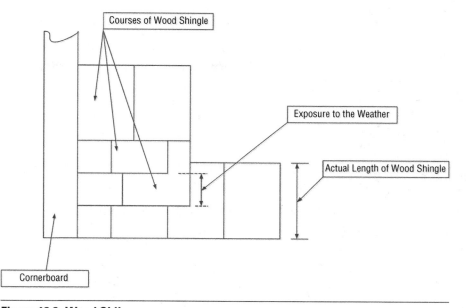

Figure 12.3 Wood Siding

widths or sawn to widths of 4", 5", or 6". Shakes and shingles are most often red or white cedar, and they vary in grade and price. (See Figure 12.3.)

Cement fiber siding is a composite material. It is strong, durable, and reasonably easy to work with. It is available in a variety of sizes, textures, and colors. Siding can be lap, vertical, or shingle type with trims and fasteners to match. Lap siding can be smooth or textured with widths ranging from 5¼" to 7¼" in 12' lengths. Vertical siding is available in strips, planks, or sheets, in a variety of sizes, colors, and textures. Shingle-type siding is available in individual shingles or panelized.

Simulated wood clapboard is also available, manufactured from metals such as thin-gauge steel and aluminum, and vinyl siding (extruded polyvinyl chloride, or PVC). This type of siding is installed horizontally with preset exposures from 4" to 8". Most panels are approximately 8" to 9" in height and 12' to 12'-6" in length. Finishes range from smooth to wood-grain textures in a variety of colors. Preformed siding requires accessories for proper installation that must be included as part of the takeoff. These include inside and outside corners, "J" channels for terminations of siding, horizontal starter strips, and finish trims, as well as soffit and fascia pieces and trims.

Other more commercial siding materials include:

- **Corrugated metal panels:** Fastened to a girt system with a screw and washer assembly.
- **Insulated metal panels:** Smooth in appearance, with a concealed fastening system.

Custom shapes and architectural details, such as interior and exterior corners, window and door head and jamb trim, and special termination trim pieces, are also available.

Taking-off and Pricing

Consult the architectural drawings for wall sections showing details of the siding. Sections will provide dimensions for calculating heights. Exterior elevations should confirm the dimensions for the takeoff. Floor plans can be used to check locations and running dimensions obtained from the elevations. Familiarity with the specified products helps to accurately price material and labor costs.

Individual siding components are taken off separately, using the following guidelines:

- **Wood siding:** By the SF or square (1 SQ = 100 SF) for pricing. Note the grade and species of the siding and the exposure. Changing the exposure from 5" to 4" for white cedar shingles, for example, requires a 20% increase in material.
- **Preformed metal and vinyl siding:** By the SF, converted to the square for pricing.
- **Cement fiber:** By the SF or square (1 SQ = 100 SF) for pricing. Note the color, style, and exposure of the siding. Review the manufacturer's specifications and installation instructions to become familiar with the individual product. Include components such as trims, inside and outside corners, and starter components.
- **Trim pieces:** By the linear foot (LF). Trims for vinyl siding, such as starter strips, inside and outside corners, and "J" channels, are also taken off and priced by the LF but may be converted to the manufacturer's individual sales length.
- **Accessories:** By the SF for soffit and fascia materials. Other accessories and trim pieces, such as "F" channels, or metal channel moldings, are taken off and priced by the LF.
- **Decorative pieces:** Vinyl shutters by the individual piece or pair (PR) according to size, style, and color. Number blocks, hose bibb ports, or light blocks are by the individual piece, EA.
- **Caulking and sealants:** Many siding products have caulking and/ or sealants that complete the assembly. Caulking is completed along with the installation and can be time consuming. The estimator is advised to become familiar with the individual system specified.

Special Considerations

Siding is most often installed over the exterior sheathing and air infiltration barrier. Shingles without corner boards require mitering. Take off the vertical length of exterior corners to be mitered separately, and list them by the LF. This work is priced by the LF. This type of application is considerably more labor intensive, which must be considered for accurate pricing. When an exterior wall intersects a roof, there is generally flashing, and custom-cutting of siding materials is required. This fitting, or *scribing*, should also be calculated in LF of the abutting surfaces. The estimator should consider double courses for starter rows and any underlayment, such as fiberboard strips, for undercoursing. As noted above, sealants are often required at openings or the intersection of dissimilar materials and must be done on a course-by-course basis, this can reduce productivity.

Waste must be included for all types of siding and associated trims. Waste varies with the quality of the product. Natural products, such as clapboards and wood shingles, have higher waste factors because of natural imperfections. Lower grades of natural wood and manufactured products also have higher waste factors. Generally speaking, as the quality of the product decreases, the amount of waste increases. The final or net quantity of siding materials should include deductions for openings, such as doors or windows. Openings smaller than 4 SF are usually not deducted.

Both hand- and gun nails should be included as part of the material cost. Most nails are calculated as an added cost per SQ of siding material.

Labor to Install Siding

Labor is priced by the square (SQ), or 100 SF. Production rates vary dramatically depending on the product. Wood siding is typically installed by one person. Productivity is measured by the individual output. Vinyl siding can be a crew task, depending on the size of the façade. Vertical metal siding is installed by a multiple-person crew. Production in squares per day is based on the combined effort of the crew.

Most siding applications require staging or a lift to access areas above ground. This must be included in the cost. Note that façades with many interruptions, referred to as *cut-up*, dramatically reduce productivity. Include trims at windows, doors, and exterior openings. Cut-up areas with reduced production rates should be separated from wide-open areas in the takeoff, as each will have different productivity rates.

Building façades with decorative patterns from wood shingles or wood siding can be labor intensive, especially if wood shingles are scalloped or cut into patterns by hand. Some patterns require time to layout for accurate placement. This should be added to the estimated labor cost.

ROOFING

A roofing system consists of many different parts that work together to form a watertight envelope at the top of the building. The major components are:

- Insulation (applied to the roof deck if required by system)
- Waterproofing membrane or layer
- Protective surfacing
- Flashings and counterflashings
- Metal perimeter termination pieces

See Figure 12.4 for the standard terminology associated with flat roofing systems.

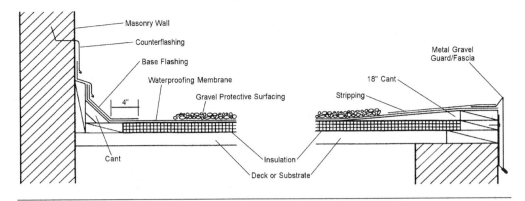

Figure 12.4 Roofing Components

For all roof systems, a suitable substrate is required to fasten the roof system to. This will be assumed for all applications. The estimator should check the architectural drawings – specifically, building sections, roof plans, and larger-scale plans showing the details at the roofline. Specifications also identify the type of roofing system and information on individual components, such as the membrane, insulation, flashings, and the accessories needed to complete the system.

Figure 12.5 shows a typical roofing plan and details for a shingle roof. The importance of understanding the manufacturer's particular requirements cannot be stressed enough. Many specifications require a manufacturer's representative to be on-site during installation in order to warranty the installation. This cost, in most cases, is borne by the contractor and should be carried in the estimate.

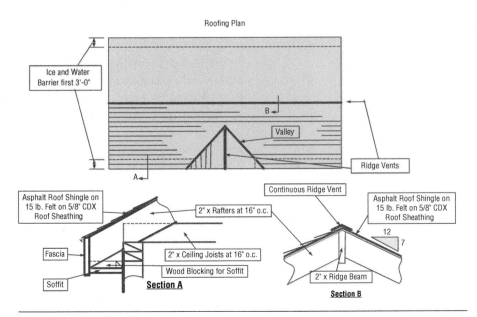

Figure 12.5 Roofing Plan

For basic estimating purposes, roof systems can be categorized into four basic types:

- Built-up asphaltic
- Single-ply membrane
- Metal panel
- Shingles/tiles

Each is discussed in detail in the following sections. A combination of one or more roof systems in one project is not uncommon.

Built-up Roof Systems

A built-up roof consists of felt, bitumen, and surfacing. (See Figure 12.6.) The felts, which are made of glass, organic, or polyester fibers, serve much the same purpose as reinforcing steel in concrete. They provide tensile reinforcement to resist pulling forces in the roofing material. Felts installed in layers allow more bitumen to be applied to the whole system. Bitumen (coal-tar pitch or asphalt) is the "glue" that holds the felts together and serves as waterproofing.

Surfacing materials are smooth gravel or slag, mineral granules, or a mineral-coated cap sheet that protects the membrane from mechanical or pedestrian damage. Gravel, slag, and mineral granules may be embedded into the still-fluid flood coat. Built-up roof systems can contain two to four plies, or layers of felt with asphalt bitumen or coal-tar pitch. All systems may be applied to rigid deck insulation or directly to the structural roof deck. Flashings, counterflashings, metal gravel stops, and treated wood cants are also needed to complete the system.

It should be noted that built-up roof systems occur more often in renovation work than in new structures. It is quite common to see existing built-up systems cut and patched when adding equipment to a roof and new membrane systems for additions and new structures.

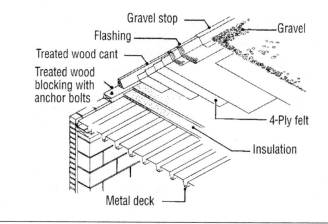

Figure 12.6 Components of a Built-up Roof

Taking-off and Pricing

Take off individual components in the built-up roof system separately, using the following guidelines:

- **Roof insulation and felts:** By the SF and extended to the SQ (100 SF), often providing addition materials for vertical transitions at rising walls and equipment curbs.
- **Bitumen:** Based on the coverage required in the specifications. (Can be listed by weight based on the weight per SF specified by the manufacturer.) Bitumen should be rounded to the nearest sales unit.
- **Mineral surfacing:** By the weight specified per SF and converted to total weight in pounds or tons.
- **Flashings, counterflashings, gravel stops, and treated wood or fiber cants:** Taken off and listed by the LF or piece EA on a per application basis.

Separate all items according to type of metal, gauge of metal, and overall size of the flashings. Materials should be priced separately from labor. Individually priced materials can be combined to develop a price for materials per square (SQ) for the overall roof system.

Include additional items, such as collars, sleeves, or penetrations through the roof, in the takeoff. Count and list them individually as EA, according to the size of each item and its use. Items supplied and installed by other trades, such as rooftop curbs for HVAC units or smoke and access hatches flashed into the roof system, may also need to be included in the estimate. Roof accessories have added material and labor costs to install and should be reviewed carefully. Some flashings and counterflashings at rising masonry walls can be labor intensive not to mention multiple components to complete the system. Plans should be adequately detailed to illustrate how the system is configured.

Labor for Built-Up Systems

With the exception of very minor repairs, built-up roofing is always a crew task. Productivity is based on the work of all members of the crew, typically calculated as squares per day. The roof assembly can be broken into individual components such as insulation, fire barrier layer, vapor barrier, and various courses of felts and bitumen. Roof demolition should be taken off and priced separately, as it is labor intensive. Costs for disposing of old roofing or debris from new work should be calculated separately, including the dumpsters or containers to transport the debris.

Built-up roof work is most often performed by large crews. Detail work, such as flashing in vents through the roof and curbs, is frequently only "temped-in" until a pickup crew can complete some of the more time-consuming details. Roof metal installation rates are based on linear foot of production per day. (This is discussed in greater detail later in this chapter.)

Weather conditions are important, as extremes can reduce productivity and quality. Anticipated production rates should be scaled back during seasons when these conditions are likely. As with most flat roof systems, some type of equipment, typically a crane is required for removal of roofing debris on re-roof projects, and the stocking of materials for the new roof. See Figure 12.7 for labor productivity guidelines.

Description	Labor-Hours	Unit
Built-up Roofing		
Asphalt Flood Coat with Gravel or Slag		
Fiberglass Base Sheet		
3 Plies Felt Mopped	2.545	SQ
On Nailable Decks	2.667	SQ
4 Plies Felt Mopped	2.800	SQ
On Nailable Decks	2.947	SQ
Coated Glass Fiber Base Sheet		
3 Plies Felt Mopped	2.800	SQ
On Nailable Decks	2.947	SQ
Organic Base Sheet and 3 Plies Felt	2.545	SQ
On Nailable Decks	2.667	SQ
Coal-Tar Pitch with Gravel or Slag		
Coated Glass Fiber Base Sheet and		
2 Plies Glass Fiber Felt	2.947	SQ
On Nailable Decks	3.111	SQ
Asphalt Mineral Surface Roll Roofing	2.074	SQ
Walkway		
Aslphalt Impregnated	.020	SF
Patio Blocks 2" Thick	.070	SF
Expansion Joints Covers	.048	LF

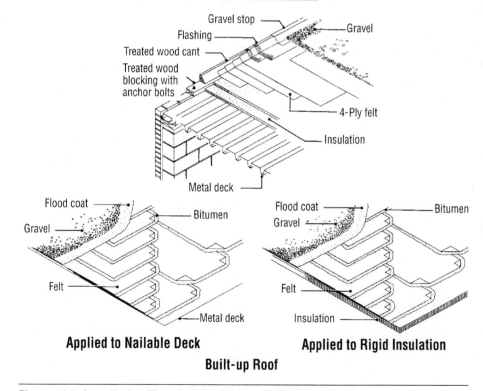

Applied to Nailable Deck | **Applied to Rigid Insulation**

Built-up Roof

Figure 12.7 Installation Time in Labor-Hours for Built-up Roofing

Single-Ply Roofing Systems

Single-ply, or elastomeric, roofing can be broken down into three categories:

- Thermosetting
- Thermoplastic
- Composites

All three use a PVC, synthetic rubber, rubber sheet material, or *membrane,* applied over the top of the insulation or approved substrate to serve as a waterproofing surface. The latest technology includes a fire-resistive membrane, reinforced or unreinforced, and factory applied self-sealing seams.

Single-ply roofing can be applied loose-laid and ballasted, partially adhered, and fully adhered. Loose-laid systems involve fusing, welding, or gluing the side and end laps of the membrane to form a continuous non-adhered sheet, held in place by a layer of gravel (ballast). Partially adhered single-ply membrane is attached (with strips or plate fasteners) to the roof substrate, which has the benefit of reducing the dead load. Because the system allows movement, ballast is not required. Fully adhered systems are uniform and continuously adhered to the manufacturer-approved base. Figure 12.8 shows a single-ply roofing system and estimating guidelines for labor.

Taking-off and Pricing

The takeoff procedure and units are the same for single-ply roofing as for built-up systems. The roof is calculated in square feet and then extended to the square, 1 SQ (100 SF). The components to be priced are:

- Insulation (species) and method of attachment
- Tapered insulation components (if applicable)
- Membrane and the method of attachment to the insulation
- Hardboard base (if applicable) over the insulation
- Metal fascia, or *gravel stop*
- Accessories: cants, flashing and counterflashing, flashings for penetrations, expansion joints, etc.

Take off and price metal fascias, flashings, counterflashings, and termination strips by the LF. Include pressure treated wood nailers (for both built-up and single-ply systems) for support at the edge of the rigid insulation and for nailing the fascia or gravel stop. This is not always the work of the roofing trades. The roofing scope of work may include furnishing and installing wood blocking at the roof perimeter or under curbs, or it can be included under Section 06 10 00—Rough Carpentry. The estimator is urged to review the specifications to determine the section the wood blocking work is included in and its method of attachment.

Other components such as "witch hats" for pipe or conduit penetrations, "pitch pockets" also for conduit penetrations, corner reinforcing at curbs or similar cornered penetrations are taken off and priced by the individual

Description	Labor-Hours	Unit
Single-Ply Membrane, General		
Loose-Laid and Ballasted	.784	SQ
Mechanically Fastened	1.143	SQ
Fully Adhered, All Types	1.538	SQ
Modified Bitumen, Cap Sheet		
Fully Adhered Torch Welding	.019	SF
Asphalt Mopped	.028	SF
Cured Neoprene for Flashing	.028	SF

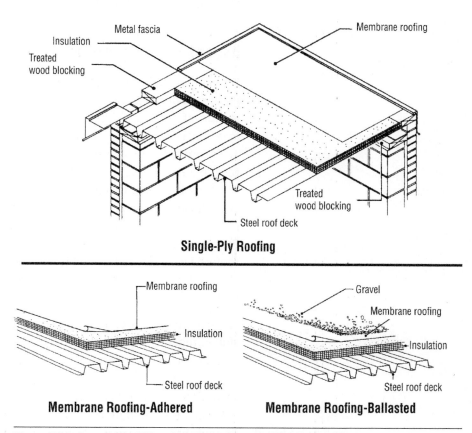

Single-Ply Roofing

Membrane Roofing-Adhered

Membrane Roofing-Ballasted

Figure 12.8 Installation Time in Labor-Hours for Single-Ply Roofing

piece, EA. Expansion joints at the roof are often included as part of the roof system to maintain the integrity of the system. These are taken off and estimated by the linear foot, LF. The estimator should review the roof details carefully to ensure all components are included.

As with all roofing work, it is important to consider how the materials will be hoisted to the roof. This can be done by a roof crane, a hired crane, or a variety of other methods. Hoisting is not always identified in the specifications as a direct responsibility of the roofer. Often, the responsibility for hoisting on a commercial project is included in Division 1—General Requirements, under Section 01 50 00—Temporary Facilities. Address the issue with the architect or engineer if the documents do not specify hoisting.

Labor-hours can be calculated based on each step of component in the assembly. As with most roofing installations the productivity is based upon the output of the full crew.

Many membrane roof systems are highly proprietary and require the installation be continually supervised by a factory-trained and authorized representative each step of the way. This cost should be included. In addition, many projects require that the completed system include a roof bond. Roof bonds can represent a significant cost, especially if the term of the bond is 20 or 25 years.

Metal Roof Systems

For estimating purposes, metal roofing systems can be divided into two groups: preformed and formed metal. *Preformed metal roofs*, available in long lengths of varying widths and shapes, are constructed from aluminum, steel, or composite materials, such as fiberglass. Aluminum roofs can be left natural or factory prepainted. Steel roofs are usually galvanized or factory painted. Most manufacturers require the product to be factory-painted in order to warranty its finish. Preformed metal roofing is installed on sloped roofs according to the manufacturer's recommendation for minimum pitch. Lapped ends may be sealed with a preformed sealant to match the deck configuration. Preformed roofing is a cost-effective, durable alternative to other types of roofing for large buildings with shallow pitch or flat roof surfaces. The metal sheet is typically attached to the substrate framing using self-tapping screws with a rubber grommet.

Formed metal roofing, typically selected for aesthetic reasons, is installed on sloped roofs that have been covered with a base material, such as plywood or concrete. Typical sheet materials include copper, lead-coated copper, galvanized steel, and zinc alloy. Flat sheets are joined by tool-formed batten-seam, flat-seam, and standing-seam joints. Formed metal roofing – both material and installation – is typically more expensive than preformed roofing. In most seamed applications the roof requires shop drawings that are custom for the roof. The approved shop drawings are then used to fabricate the individual pieces in accordance with the design.

Taking-off and Pricing

Takeoff is done by the SF and can be listed by the individual piece. Specific components are as follows:

- **Preformed sealant material:** By the LF. Determine quantities by calculating the SF of roof surface to be covered with the manufacturer's recommended allowances for side and end laps.
- **Formed metal roofing:** By the SF and converted to the SQ (each 100 SF).

- **Trim pieces (battens for the seams and finish end pieces):** By the LF or individual piece. Lining for valleys, hips, and ridges are also taken off and priced by the LF.

Formed metal roofing is most often used in custom applications and one-of-a-kind designs. It requires field measurement and shop drawings for fabrication. Be sure to include these costs in the estimate.

Takeoff for both types of metal roofing should be separated according to type of metal, manufacturer, finish, and particular application. Check the specifications for sheet material species, weight, and other characteristics, including color or special coatings on the exposed face of the metal, to help ensure accurate pricing. Custom colors can add significantly to both the cost of the material and the lead time for fabrication. Trim pieces can often increase time for installation due to field cutting and fitting. The estimator should include any hoisting and personnel lift requirements for detail work at the facia or rakes.

Unlike most roofing work, formed metal roofing has both fabrication labor and field labor. Be sure to include shop time to cut, bend, and fabricate the pieces – work typically performed by sheet metal workers. The estimator should include the cost of shipping to the site, if required. Separate field installation and shop fabrication costs for accurate pricing.

Roof Shingles and Tiles

Shingles and tiles are popular for sloped roofs with a pitch of more than 4" per foot. Both are *watershed* materials, designed to direct water away from the building by means of the slope, or *pitch,* of the roof. Shingles are installed in layers with staggered joints over roofing felt underlayment. Nails or fasteners are concealed by the course above. Shingle materials include wood, asphalt, fiberglass, metal, and masonry tiles. Asphalt and fiberglass shingles are available in a variety of weights and styles; three-tab is the most common. The types and styles of roof shingles for residential light-commercial applications are almost endless. The components and quantities of the roof system can be derived from the architectural roof plans, building sections, and roof details. The estimator is urged to review the details carefully for a full understanding of the system, especially when it comes to tiled roofs. (See Figure 12.9.)

Wood shingles may be either shingle or shake grade and are most commonly cedar. Metal shingles are either aluminum or steel and are generally prefinished. Natural and synthetic slate and clay tiles are available in a variety of shapes, sizes, colors, weights, and textures. Since these are heavier materials, they require specialized installation techniques and a stronger structural roof system. Product costs vary widely by grade of material and manufacturer. Natural slate products tend to be regional and

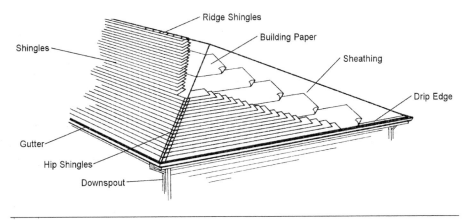

Figure 12.9 Shingled Roof

may have significant shipping costs if not local. Shipping costs should be separate but included in the estimate.

Shingle and tile roof systems require additional components to complete the assembly. Special metal trim pieces called *drip edge* protect the edge of the roof deck and allow water to drip off the roof edge. These are made of a corrosion-resistant metal, typically aluminum or galvanized steel, and may be perforated to allow air flow along the underside of the roofing substrate. Drip edge may be omitted with wood shingles or slates, so the estimator must check the details carefully. Preformed metal flashings may be specified for the valleys of shingled and tiled roofs. Valley flashing can be lead-coated copper, copper, or zinc alloy and may require fabrication off site. Ridge vents ventilate the attic space, allowing air to transfer from the attic or rafter space to the outside, and thereby prevent moisture build-up along the underside of the roof sheathing. Ridge-venting materials are available in a variety of styles and compositions.

Shingle and tile roof systems require special shingles, called *cap shingles,* at the ridge or hip of the roof. A cap shingle may be a regular three-tab shingle modified for use as a cap, as in the case of asphalt or fiberglass shingles, or a special prefabricated cap, as with some clay or metal tile designs.

Special membrane material installed under shingles at the eaves, rakes, and hips and valleys of the shingle or tile roof is called an *ice/water barrier.* Most products are a bitumen-based, self-adhering membrane for use in cold climates where ice and water may dam along the eaves and valleys and cause water to back up under the roof shingles.

Taking-off and Pricing

Shingle and tile roofing are taken off by the SF and extended to the square, SQ (100 SF). In the case of asphalt or fiberglass shingles, squares can be converted to individual bundles for pricing of materials. Additional takeoff guidelines are as follows:

- **Underlayment felts:** By the SF, converted to rolls, allowing for specified laps at sides and ends.
- **Wood shingles, shakes, and slate tiles:** By the SF, converted to cartons or bundles based on the individual product.
- **Special hip and ridge tiles or cap shingles:** By the LF, but can be converted to the individual piece, EA (as in the case of metal or clay tiles).
- **Drip edge (vented or nonvented):** By the LF but can be converted to the individual piece (most commonly 10'-0" long) and listed as EA. Separate drip edge by vented and nonvented.
- **Ridge vent:** By the LF but can be converted to the individual piece (EA) or roll.
- **Step and roll flashing:** 5" × 7" step flashing can be taken off by the piece, EA allowing 1 per shingle course against a rising wall. Roll flashings are taken off and priced by the LF in the required width and gauge.
- **Ice/water barriers:** By the SF, but can be extended to the manufacturer's size roll, allowing for specified side and end laps.

Be sure to include hand and pneumatic nails in the estimate. Hand nails are priced by the pound (lb.), and pneumatic nails by the box. Review the materials specified for manufacturer, weights per square, model numbers, end lap for continuous products, and any other information that will help to accurately price them.

Labor to Install Shingle Roofs

Labor cost to install shingle and tile roofs are based on crew productivity. Crews on shingle roofing projects can be smaller than on built-up and single-ply roofs, as the shingle installation is repetitive, and most roofers on the crew perform the same tasks. However, productivity should be adjusted based on crew size. For projects that are re-roofs, the cost of stripping shingles and disposal should be kept separate from the new work. For many crews the cost of hoisting shingles to the roof is done manually, especially in residential applications. Bundles are placed along the ridge as storage and used as needed. Figure 12.10 shows labor-hour guidelines for shingle or tile roofs.

Adjust production for steep-pitched roofs, weather extremes, or cut-up roofs (roofs with many penetrations, such as dormers, skylights, or pipes). The estimator should include the cost of fall protection and personnel protective equipment for workers, as well as labor to clean up debris generated by the process.

The estimator should also consider the quantity of squares (SQs) that can be accomplished per day and provide temporary protection for re-roof projects

Description	Labor-Hours	Unit
Shingles, Manual Nailed		
Aluminium	1.600	SQ
Ridge Cap or Valley, Manual Nailed	.047	LF
Fiber Cement		
500 lb. per SQ	3.636	SQ
Starters	2.667	CLF
Hip and Ridge	8.000	CLF
Asphalt Standard Strip		
Class A 210 to 235 lb. per SQ	1.455	SQ
Class C 235 to 240 lb. per SQ	1.600	SQ
Standard Laminated		
Class A 240 to 260 lb. per SQ	1.778	SQ
Class C 260 to 300 lb. per SQ	2.000	SQ
Premium Laminated		
Class A 260 to 300 lb. per SQ	2.286	SQ
Class C 300 to 385 lb. per SQ	2.667	SQ
Hip and Ridge Shingles	.024	LF
Slate Including Felt Underlay	4.571	SQ
Steel	3.636	SQ
Wood		
5" Exposure, 16" long	3.200	SQ
5-1/2" Exposure, 18" long	2.090	SQ
Panelized 8" Strips 7" Exposure	2.667	SQ
Ridge	.023	LF
Shingles, Pneumatic Nailed		
Asphalt Standard Strip		
Class A 210 to 235 lb. per SQ	1.143	SQ
Class C 235 to 240 lb. per SQ	1.280	SQ
Standard Laminated		
Class A 240 to 260 lb. per SQ	1.422	SQ
Class C 260 to 300 lb. per SQ	1.600	SQ
Premium Laminated		
Class A 260 to 300 lb. per SQ	1.831	SQ
Class C 300 to 385 lb. per SQ	2.133	SQ
Hip and Ridge Shingles	.019	LF
Wood		
5" Exposure	2.462	SQ
5-1/2" Exposure	2.241	SQ
Panelized 8' Strips 7" Exposure	2.000	SQ
Tiles		
Aluminum		
Mission	3.200	SQ
Spanish	2.667	SQ
Clay,		
Americana, 158 Pc/SQ	4.848	SQ
Spanish, 171 Pc/SQ	4.444	SQ
Mission, 192 Pc/SQ	6.957	SQ
French, 133 Pc/SQ	5.926	SQ
Norman, 317 Pc/SQ	8.000	SQ
Williamsburg, 158 Pc/SQ	5.926	SQ
Concrete	5.926	SQ

Figure 12.10 Installation Time in Labor-Hours for Shingle and Tile Roofing

where overnight precipitation could damage the interiors. The above table is to be used as a guideline in calculating productivity.

Natural Roof Coverings

Many projects being constructed are incorporating some type of green or sustainable roof to comply with the growing trend of environmentalism and code changes. Green roofs are sometimes referred to as living roofs, eco roofs, or vegetated roofs. While there is vigorous debate as to exactly what constitutes a green roof, they generally consist of vegetation, a growing medium, and a waterproof membrane. Green roofs are not a new concept but have been employed in roof construction for hundreds, or possibly thousands, of years. They provide several benefits to the building, including improved insulating qualities from exterior sound and helping to lower urban air temperatures, create a habitat for wildlife, reduce heating/cooling costs due to increased mass and thermal resistivity, and reduce stormwater runoff to name but a few.

Financially, there are increased product and installation costs as well as maintenance costs over the long term. Also, there are added costs to the structural components of the roof and related components to carry the added load from the medium and vegetation. Green roofs can be categorized as intensive, semi-intensive, or extensive, depending on the depth of the planting medium and the amount of maintenance they require. Green roofs can be flat or pitched (Figure 12.11).

Taking-off and Pricing

The estimator is advised to study the plans and specifications carefully to become familiar with the individual components of the system. All roofing components that constitute the waterproof membrane system should be kept separate from the medium and vegetation. Any plumbing components for irrigation or drainage should be separated, as these are typically part of the plumbing scope of the project. Testing of the waterproof membrane system prior to the installation of the medium and vegetation is also a requirement. They may be required by an independent testing agency.

Takeoff and pricing of the roofing membrane, flashings, and insulation components should follow the same procedure as described for membrane roofing systems. The medium is taken off and priced according to the type and thickness as well as the area being covered. The same is true for vegetative covers or sod. Small areas are reported by the SF. Areas for larger roofs can be reported by the SQ. Green roofs are often proprietary systems requiring strict adherence to the manufacturer's requirements. Systems can have drainage, filter, and protection layers as well as a nonsoil growth

Description	Labor-Hours	Unit
Vegetated Roofing		
Soil mixture for green roof 30% sand, 55% gravel, 15% soil		
Hoist and spread soil mixture to 4" depth, up to 5-stories-tall roof	.014	SF
6" depth	.021	SF
8" depth	.028	SF
10" depth	.035	SF
12" depth	.042	SF
Mobilization 55-ton crane to site	2.222	EA
Hoisting cost to five stories per day (avg. 28 picks per day)	56.000	Day
Mobilization or demobilization, 100-ton crane to site driver and escort	6.400	EA
Hoisting cost 6 to 10 stories per day (avg. 21 picks per day)	56.000	Day
Hoist and spread soil mixture to 4" depth 6- to 10-stories-tall roof	.014	SF
6" depth	.021	SF
8" depth	.028	SF
10" depth	.035	SF
12" depth	.042	SF
Green roof edging treated lumber 4" × 4", no hoisting included	.040	LF
4" × 6"	.040	LF
4" × 8"	.044	LF
4" × 6" double stacked	.053	LF
Green roof edging redwood lumber 4" × 4", no hoisting included	.040	LF
4" × 6"	.040	LF
4" × 8"	.044	LF
4" × 6" double stacked	.053	LF
Planting sedum, light soil, potted, 2-1/4" diameter, two per SF	.019	SF
One per SF	.010	SF
Planting sedum mat per SF including shipping (4,000 SF minimum)	.008	SF
Installation sedum mat system (no soil required) per SF (4,000 SF minimum)	.008	SF

Figure 12.11 Installation Time in Labor-Hours for Green Roof Systems

Source RSMeans Estimating Handbook, Third Edition. © 2009 John Wiley & Sons, Inc.

medium. The estimator is urged to study and understand the individual system not only on the plans, but outside sources such as the internet.

Costs for maintenance until the roof has stabilized are often part of the technical specification, as well as any ancillary costs associated with tasks such as scaffolding, the rigging and hoisting of materials to the roof, fall protection, distribution, planting, and cleanup. Bear in mind that vegetative roofs will frequently require more than one hoisting of materials.

Labor

Several different trades and expertise are required in the construction of a green roof. In addition, specifications may require that the progress be inspected at multiple stages in the process. The construction of living roofs is a crew task and can include multiple crews and coordination among crews. Again, the estimator is strongly urged to thoroughly understand the specified system and solicit contemporaneous subcontractor pricing when available.

Flashing and Sheet Metal Work

Flashing refers to pieces of sheet metal or impervious flexible membrane material used to seal and protect joints in a building and prevent leaks. It can be a separate system or part of another system such as roofing. It can be custom-fabricated, as in the case of rigid through-wall flashings, or purchased off the shelf, as with roll aluminum and step flashing materials. Sheet metal work involves pre- or custom-formed metal fabrications, such as parapet caps, gravel stops, and cleat strips, which make the roofing system weathertight. Concealed flashing is made of sheet metal or membrane material. Sheet metal fabrications and exposed flashings can be fabricated from a variety of materials, such as copper, lead-coated copper, aluminum, galvanized steel, lead, or zinc alloy. Flashing can be formed on-site or preformed in the shop prior to installation.

Flashing is also used for window/door heads and masonry applications. Flashing details are often shown in the plan, sections, and roof details. Plan view and elevation drawings show their locations. Consult the specifications for material characteristics such as metal gauge, color, or special coatings that will affect the price. Custom sheet metal fabrications may require shop drawings for review and approval, which can also add to the cost.

Taking-off and Pricing

Stock (off-the-shelf) flashing is taken off by the LF for rigid or roll form materials. Convert the LF into pieces or rolls for pricing. Premanufactured flashing for pipe and conduit penetrations is taken off and priced by the piece. For accurate quantities, cross-reference the mechanical and electrical drawings for penetrations for piping and conduits. Note that for many rooftop units the electrical conduits pass within the unit itself and do not require a separate penetration. Custom-fabricated sheet metal work often requires shop drawings based on dimensions verified in the field. To determine the quantity of sheet metal required to fabricate the various pieces, calculate the number of sheets needed. Be sure to include waste or lap for continuous applications. Sheet metal is available in a variety of different-sized sheets, depending on the manufacturer and product.

Labor to Install Flashings

Labor for flashing and sheet metal work must be evaluated separately. Off-the-shelf materials have an installation labor component only. The materials are purchased and installed as part of the normal roofing process, with only minor adjustments to their stock shapes. Most of this work is priced by the LF and may be included in the roofing scope of work. Custom-fabricated sheet metal work, such as parapet caps, gravel stops, and edge cleats, require both shop-fabrication and field-installation time. As discussed previously with formed metal roofing, separate these different labor costs for accurate pricing. Figure 12.12 provides guidelines for installation labor.

Description	Labor-Hours	Unit
Gutters		
Aluminum	.067	LF
Copper Stock Units		
4" Wide	.067	LF
6" Wide	.070	LF
Steel Galvanized or Stainless	.067	LF
Vinyl	.073	LF
Wood	.080	LF
Downspouts		
Aluminum		
2" x 3"	.042	LF
3" Diameter	.042	LF
4" Diameter	.057	LF
Copper		
2" or 3" Diameter	.042	LF
4" Diameter	.055	LF
5" Diameter	.062	LF
2" x 3"	.042	LF
3" x 4"	.055	LF
Steel Galvanized		
2" or 3" Diameter	.042	LF
4" Diameter	.055	LF
5" Diameter	.062	LF
6" Diameter	.076	LF
2" x 3"	.042	LF
3" x 4"	.055	LF
Epoxy Painted		
2" x 3"	.042	LF
3" x 4"	.055	LF
Steel Pipe, Black, Extra Heavy		
4" Diameter	.400	LF
6" Diameter	.444	LF
Stainless Steel		
2" x 3" or 3" Diameter	.042	LF
3" x 4" or 4" Diameter	.055	LF
4" x 5" or 5" Diameter	.059	LF
Vinyl		
2" x 3"	.038	LF
2-1/2" Diameter	.036	LF

Wood Gutter

Metal or Vinyl Gutter

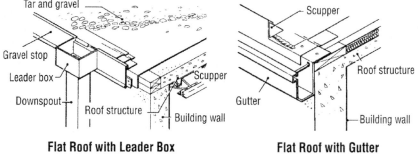

Flat Roof with Leader Box

Flat Roof with Gutter

Figure 12.12 Installation Time in Labor-Hours for Flashing, Expansion Joints, and Gravel Stops

Gutters and Downspouts

Gutters and downspouts are metal or wood fabrications used to channel precipitation from the roof surface to the ground. Most metal gutters and downspouts are prefabricated and available in a variety of colors and sizes. Custom-fabricated metal gutters and downspouts follow the installation procedure identified previously for sheet metal work. Wood gutters are shaped from a variety of wood species that are decay resistant and can be purchased through many lumberyards. Wood gutters are expensive and can often require lead times to obtain.

More unique pieces for roof drainage such as leader boxes, *scuppers*, or outlet boxes that drain water through a parapet may require shop drawings for review and approval, which can add to the cost.

Taking-off and Pricing

Gutters and downspout materials are taken off and priced by the LF, separated according to composition, size, gauge of metal, and application, as well as special colors or finishes. The method of gutter installation should be noted on the takeoff for accurate pricing. Simple installations may include fastening directly to the fascia with nails or screws, as in the case of wood gutters. Other installations require more complicated hanging/fastening systems. Scuppers are taken off and priced by the individual piece, EA. They also have a fabrication component. Check the specifications for materials, accessories, and method of installation required. Exterior elevation drawings and sections at the eave of the roof provide essential information for accurate takeoff and pricing. Figure 12.13 illustrates gutters and downspouts with normal labor-hours.

Roof Accessories

Roof accessories include skylights, roof hatches, HVAC curbs, and smoke vents. Most are taken off and priced by the individual piece (EA) according to size, function, and any special characteristics. Manufacturer, make, model, type of glazing (for skylights), and size are critical characteristics for accurate pricing.

Labor to install roof accessories is priced by the individual piece or per SF of surface area the feature occupies, as in the case of large skylights. Other cost considerations include assembly labor, if required. Many skyroofs (large, linear, or vaulted series of connected skylights) require some assembly prior to installation. Consideration should also be given to hoisting the materials to the roof. Most roof accessories are bulky and heavy, precluding hand-loading. Consider costs for "temping-in" roof accessories whose installation may extend beyond a single day. Be sure to examine the specification for hoisting responsibility. Figure 12.14 shows roof accessories and labor-hour guidelines.

Labor to install roof accessories is only part of the cost. The estimator must include the cost to tie the accessory into the roof system. This is most often part of the roofing costs, but the estimator should check to ensure that all accessories have a commensurate component in the roofing estimate and are being done in sequence.

Description	Labor-Hours	Unit
Flashing Aluminum Mill Finish	.055	SF
Fabric-Backed or Mastic-Coated, 2 Sides	.024	SF
Copper Sheets		
16 oz.	.070	SF
20 oz.	.073	SF
24 oz.	.076	SF
32 oz.	.080	SF
Paperbacked, Fabric-Backed, or		
Mastic-Backed, 2 Sides	.024	SF
Lead-Coated Copper, Paperbacked,		
Fabric-Backed, or Mastic-Backed	.024	SF
Lead 2.5 lbs. per S.F.	.059	SF
Polyvinyl Chloride or Butyl Rubber	.028	SF
Copper-Clad Stainless Steel Sheets Under 500 lbs.		
.015" Thick	.070	SF
.018" Thick	.080	SF
Stainless Steel Sheets or		
Terne Coated Stainless Steel	.052	SF
Paberbacked 2 Sides	.024	SF
Zinc and Copper Alloy, .020" Thick	.052	SF
Expansion Joint, Butyl or Neoprene Center with		
Metal Flanges	.048	LF
Neoprene, Double-Seal Type with		
Thick Center, 4-1/2" Wide	.064	LF
Polyethylene Bellows with		
Galvanized Flanges	.080	LF
Roof Expansion Joint with		
Extruded Aluminum Cover, 2"	.070	LF
Roof Expansion Joint, Plastic Curbs,		
Foam Center	.080	LF
Transitions, Regular, Minimum	.800	EA
Maximum	2.000	EA
Large, Minimum	.889	EA
Maximum	2.667	EA
Roof to Wall Expansion Joint with Extruded		
Aluminum Cover	.070	LF
Wall Expansion Joint, Closed Cell Foam on		
PVC Cover, 9" Wide	.064	LF
12" Wide	.070	LF
Gravel Stops		
4" Face Height	.055	LF
6" Face Height	.059	LF
8" Face Height	.064	LF
12" Face Height	.080	LF

Figure 12.13 Installation Time in Labor-Hours for Gutters and Downspouts

Joint Protection

Caulking and sealants provide a water, vapor, and air barrier between joints or gaps of adjacent, but dissimilar, materials and come under the heading of *joint protection*. A classic example is the joint between a steel door frame and a masonry wall opening. Caulking and sealants are manufactured for a full range of applications, such as interior or exterior use, expansion and contraction, service temperature range, and compatibility with the material to be sealed. Sealants can be used straight from the tube, or they may be multicomponents that require mixing before application.

Description	Labor-Hours	Unit
Roof Hatches with Curb		
2'-6" x 3'-0"	3.200	EA
2'-6" x 4'-6"	3.556	EA
2'-6" x 8'-0"	4.848	EA
Smoke Vents		
4'-0" x 4'-0"	2.462	EA
4'-0" x 8'-0"	4.000	EA
Plastic Roof Domes Flush or Curb Mounted		
Under 10 SF		
Single	.200	SF
Double	.246	SF
10 SF to 20 SF		
Single	.081	SF
Double	.102	SF
20 SF to 30 SF		
Single	.069	SF
Double	.081	SF
30 SF to 65 SF		
Single	.052	SF
Double	.069	SF
Ventiliation, Insulated Plexiglass Dome		
Curb Mounted		
30" x 32"	2.667	EA
36" x 52"	3.200	EA
Skyroofs		
Translucent Panels 2-3/4" Thick	.081	SF Horiz.
Continuous Vaulted to 8' Wide		
Single Glazed	.200	SF Horiz.
Double Glazed	.221	SF Horiz.
To 20' Wide Single Glazed	.183	SF Horiz.
Over 20' Wide Single Glazed	.160	SF Horiz.
Pyramid Type to 30' Clear Opening	.194	SF Horiz.
Grid Type 4' x 10' Modules	.200	SF Horiz.
Ridge Units Continuous to 8' Wide		
Double	.246	SF Horiz.
Single	.160	SF Horiz.

Roof Hatch

Smoke Vent

Circular Dome Skylight

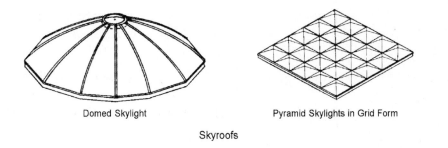

Domed Skylight   Pyramid Skylights in Grid Form

Skyroofs

Figure 12.14 Installation Time in Labor-Hours for Roof Hatches, Smoke Hatches, and Skylights

Joint sealants at exterior applications are normally applied over a backup material that controls the depth of the joint. They serve as bond breaks to allow free movement of the joint and prevent water penetration. The backup material, called *backer rod,* is available in a variety of compositions, such as butyl, neoprene, polyethylene, and rubber as well as diameters to fit the opening.

Check the architectural drawings (elevations, wall sections, and details) that show window and exterior door installations. Consult details illustrating the connection of dissimilar materials for caulking and sealant work. Not all locations requiring caulking and sealants are detailed on the plans. Frequently the work is included in the technical specification section 07 90 00—Joint Protection (see Figure 12.15).

Taking-off and Pricing

Take off interior caulking separately from exterior joint sealants. Check the specifications for the location and type of caulking and sealants needed and note their respective applications within the takeoff. Caulking and joint sealants are taken off and priced by the LF. Separate them according to the size (width) of the bead, interior or exterior application, and type of material. Backer rod is taken off and listed by the LF according to size, composition, and application.

Many types of caulking and sealants are available, ranging from inexpensive latex caulking used by painters to expensive two-part elastomeric compounds used where flexibility and longevity are critical. Determining the quantity in gallons, quarts, or tubes often involves the manufacturer's specifications for LF per gallon. Figure 12.16 shows LF quantities per gallon of material based on different joint sizes. Consider whether scaffolding or power lifts are required to access windows or other features above ground level. Mobile lifts or boom lifts are most often employed for this work, as they have the required reach and mobility for accessing the work. Labor costs are typically priced based on the productivity of a single individual for the linear foot of completed joint.

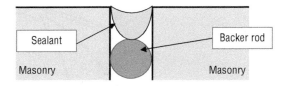

Figure 12.15 Illustration of Backer Rod and Sealant at Masonry Joint

Linear Feet per Full Gallon (231 cu. in.)		Width of Joint						
		1/4"	3/8"	1/2"	5/8"	3/4"	7/8"	1"
Depth	1/4"	308	205	154	123	102	88	—
of	3/8"	205	136	102	82	68	58	—
Joint	1/2"	154	102	77	61	51	44	38
	3/4"	102	68	51	43	34	30	26

Example: One full gallon is sufficient material to fill a joint 1/2" wide, 3/8" deep, and 102' long.

Cartridges: When figuring feet per cartridge for a particular joint size, divide linear feet shown above by 12.

Figure 12.16 Estimating Guide for Caulking

Stucco

Stucco, a cement plaster used to cover exterior wall and ceiling surfaces, is usually applied to a wood, metal lath, concrete, or masonry base. It is composed of Portland cement, lime, and sand, with water as the mixing agent, and is typically applied in a three-coat system. The *scratch coat* is the first, applied directly to the substrate. The *brown coat* is next, applied over the scratch coat that builds up the surface. The final coat is the *finish coat*. Stucco can be applied directly to the substrate, as in the case of masonry or concrete walls, or over a metal lath installed on a wood surface. Stucco provides a durable, weather-resistant surface that is virtually maintenance-free and impervious to moisture. The finish coat can be troweled in a variety of textures and then painted.

The completed stucco system requires special trim pieces to achieve square and plumb corners at the intersection of wall and ceiling surfaces. Other trims are also needed at the intersection of stucco with other materials, similar to J-bead. Expansion and control joints, used to control cracking, are also required at specific spacing.

Architectural drawings should be reviewed for the location of stucco surfaces, with special attention to exterior elevations, exterior reflected ceiling plans, and exterior sections and details. Consult the specifications for the particular mixing proportions of water, Portland cement, sand, and lime to achieve the required strength, as well as the texture and thickness of the finish coat. Specifications also note the type and location of the various trim pieces, such as outside corners, expansion joints, and metal lath (if applicable).

Taking-off and Pricing

Stucco is taken off by the SF by measuring the length and width of the surfaces to be covered. The SF is then converted to square yards (SY) for pricing. Quantities of metal lath can be determined from the SF area and converted to standard-sized sheets of the particular product. Allowances for overlap at the sides and ends of the individual sheets should also be included in the takeoff, based on the manufacturer's recommendations or the specifications. Trims, such as inside and outside corner beads, control joints, and J-beads, are taken off and priced by the LF and can be converted to the manufacturer's standard unit length (typically 8' or 10'). Convert SY of stucco to the individual components of Portland cement, sand, and lime by weight, and metal lath based on the proportions in the specifications for more accurate material pricing.

Labor for Stucco Applications

The stucco process is a crew task and should be priced based on the productivity of the entire crew. The principal trade is plastering. However, the crew may also include plasterer tenders to mix the stucco; to stock materials; to erect, dismantle, and relocate staging; and for cleanup. Assemble and price the proper crew based on the individual project. Productivity will be reduced if an elevation has a lot of openings or is generally cut up, in comparison to wide-open areas that several plasterers can work on simultaneously. Small walls between windows and entry soffits are often substantially more costly per SY than open areas, due to the amount of prep and cleanup required.

A predominance of plasters makes up the crew with laborers or tenders included to support the process. The size of the expected output per day may regulate the tenders required. Small areas with many openings may reduce the need for a second tender, while large open areas may require more frequent batching of the stucco mix due to increased productivity.

Because stucco uses water as a mixing agent, it is subject to freezing and may require temporary enclosures and heating during its curing period. In addition, only what can be applied in a timely manner is batched. (See discussion of Temporary Facilities and Controls in Chapter 6, "General Requirements.") Temporary scaffolding may also be necessary to access the work, including labor to erect, dismantle, and move sections of scaffolding. The estimator should also consider labor for cleanup of surrounding work areas as well as the mixer and tools.

Exterior Insulation and Finish System

An *exterior insulation and finish system* (EIFS, pronounced *ee-fiss*), sometimes referred to as "synthetic stucco," is an exterior siding material with the durability of stucco plus a thermal insulation value. It is composed of expanded polystyrene insulation board and a cementitious base coat applied in varying thicknesses, with a synthetic woven mesh that acts as reinforcement. The top, or finish coat is an acrylic material with the appearance of stucco, available in a variety of textures and colors. The insulation is adhered or mechanically fastened to the sidewall substrate of plywood, masonry, or gypsum sheathing. The base coat is troweled onto the insulation, similar to a conventional stucco system, and the mesh reinforcing is embedded in the base coat. The finish coat is applied after the base coat has had sufficient time to dry. There are a variety of competing products each with their own specific protocols and procedures.

Refer to the architectural drawings for plan view and elevations to find the dimensions required for determining quantities. Corresponding sections and details of the system will provide additional information for an accurate estimate. Special conditions, such as returns at windows and doors or the termination at dissimilar materials, also affect material and labor pricing. Check the reflected ceiling plan, which should show exterior soffits over entries or windows, as the product is often used as an exterior ceiling finish. Control and expansion joints are required per the manufacturer's recommendations to control expansion and contraction. The expanded polystyrene board is part of the EIFS and is estimated accordingly. Special or unique shapes are configured from the expanded polystyrene either on site or in a shop.

Expansion and control joints are typically shown on the exterior elevations. In the absence of expansion or control details on the drawings, consult the specifications for the recommended spacing. The specs should indicate the individual product as well as the manufacturer's standard application and installation procedures. EIFS includes a wide variety of products that are similar in composition, design, and performance. It is recommended the estimator become familiar with the specified product and its system prior to starting the takeoff or estimating the EIFS work.

Taking-off and Pricing

EIFS is taken off by the SF of surface area to be covered. Separating out the individual components, such as polystyrene insulation, base coat, reinforcing mesh, and finish coat, may be required for more accurate pricing, depending on the individual product, but as a rule this is not required. All components are taken off by the SF, and their individual material costs are added together to arrive at a total cost per SF. Base and finish coats may require conversion of SF quantities to gallons of material needed for the particular thickness specified. Control and expansion joints are taken off and priced by the LF and should be noted separately in the takeoff. List expanded polystyrene insulation according to its thickness (in inches) and application, such as soffits, fascias, walls, and ceilings. Special shapes of polystyrene should be noted as well, and whether the shape will be created on- or off-site. Also list finish coats with different textures, colors, or thicknesses separately. The estimator should deduct openings greater than 2 SF. Consider temporary heat and protection for cold weather installations and any scaffolding required to access the work.

Labor for EIFS Applications

Labor costs for installing EIFS can follow a similar procedure to stucco. In fact, the crews consist of the same trades and, in some cases, the same size crew. Productivity in SF per day is not comparable to that for stucco, as the process is different. Different proprietary systems will have individual processes that will impact productivity. The estimator should also include the labor to erect, dismantle, and move sections of scaffolding. The estimator should also consider labor for cleanup of surrounding work areas as well as the mixer and tools. (See Figure 12.17.)

Description	Labor-Hours	Unit
Field applied, 1″ EPS insulation	.136	SF
With 1/2″ cement board sheathing	.182	SF
2″ EPS insulation	.136	SF
With 1/2″ cement board sheathing	.182	SF
3″ EPS insulation	.136	SF
With 1/2″ cement board sheathing	.182	SF
4″ EPS insulation	.136	SF
With 1/2″ cement board sheathing	.182	SF
Premium finish, add	.032	SF
Heavy duty reinforcement add	0.44	SF
2.5#/SY metal lath substrate add	.107	SY
3.4#/SY metal lath substrate add	.107	SY
Color texture change	.032	SF
With substrate leveling base coat	.015	SF
With substrate sealing base coat	.007	SF

Figure 12.17 Installation Time in Labor-Hours for Exterior Insulation and Finish System

Source RSMeans Estimating Handbook, Third Edition. © 2009 John Wiley & Sons, Inc.

FIRE-STOP SYSTEMS AND SPRAYED-APPLIED FIREPROOFING

Fire-stop systems consist of an assembly of fire-resistive materials used to fill the space around penetrations through walls and floors. The practice is also called *fire-safing*. These penetrations are for piping, conduits, ductwork, and similar services to pass from room to room or floor to floor, in most cases through fire-rated wall or floor assemblies. Typically, sleeves are installed as the wall or floor assembly is being constructed or installed remedially in existing work. The sleeve is 1–2" larger in diameter than the pipe or conduit that will pass through it. Rectangular openings for ductwork are similarly oversized.

Once the pipe or duct has been installed, the oversized space at the perimeter of the sleeve must be filled with a fire-resistive material. The most common method is to fill most of the space with a rock wool or slag fiber insulating material, hand-packing it into the sleeve. It is held back from the edge of the sleeve approximately 1½ " at either end to allow space for the *intumescent filler*. An intumescent filler is a noncombustible material, similar in consistency to mortar or caulking, that resists the transfer of fire or toxic gases through the penetration. It solidifies once it dries and maintains its shape without shrinking or cracking. The product can be troweled, pumped, or poured into place. Other types of intumescent fillers are similar in appearance to caulking in a tube and are placed in much the same manner as regular caulking. Other applications include packing the firestop at the top of the partition or between the metal stud or masonry partition.

Taking-off and Pricing

Determining the quantities of fire-stop materials required depends on the individual product used. Most manufacturers provide printed data or online services for converting opening sizes into materials required. Consult the specifications for the required system and its application. Also consult the mechanical and electrical specifications to avoid duplication of fire-stop costs; it is frequently the responsibility of the individual trade to fire-stop penetrations for their work. This practice is not universal.

Quantities can be reported by each (EA) based on the size (diameter) and its thickness or depth of the sleeve. For fire-stopping at the top of a wall between the flutes of a metal deck, the unit of measure is per LF. Quantities of firestopping for walls and floors should be listed separately, as productivity is often different.

Labor for Fire-Stopping Installations

Labor costs are calculated based on the quantity of material that can be installed by a single individual per day. (Many manufacturers provide guidelines for their individual products.) Consider critical factors that affect productivity, such as accessibility or room to perform the work, staging requirements, or any formwork required to hold the material in place. Productivity for fire-stopping of a penetration can range from 8–20 units per day per individual installer. A single installer can install approximately 50–75 LF per day of fire-stopping materials at the top of a partition where it abuts fluted metal deck, again, individual conditions may cause production rates to vary. Fire-stopping can be labor intensive.

Spray-Applied Fireproofing

This fireproofing material is sprayed directly on a building's structural components, conforming to the contour of the element being protected. It is intended to achieve a fire endurance rating. Its main function is to insulate and resist the transfer of heat from fire to the structural members, thereby delaying deformation and failure. The most common type of sprayed fireproofing is *cementitious,* a composition of Portland cement or gypsum binders with inorganic fibers, fillers, aggregates, and additives. When mixed with water and spray-applied, it has a high bond strength to steel beams and decks, as well as bar joist. Consult the specifications for the materials to be used and the required thicknesses. Most specifications refer to a UL (Underwriters Laboratory) design or ASTM (American Society for Testing and Materials) testing number as a standard.

Taking-off and Pricing

Quantities are calculated as a volume in cubic feet (CF), based on thickness (inches converted to feet) multiplied by SF of surface area to be covered. The manufacturer's guideline is the best source for determining the actual quantity of the individual product. This trade requires both mixing and spraying equipment, and frequently scaffolding or lifts to access structural members or

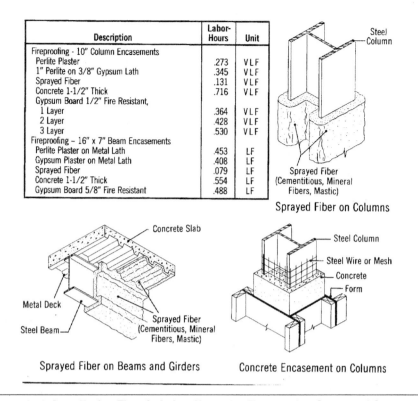

Description	Labor-Hours	Unit
Fireproofing - 10" Column Encasements		
Perlite Plaster	.273	VLF
1" Perlite on 3/8" Gypsum Lath	.345	VLF
Sprayed Fiber	.131	VLF
Concrete 1-1/2" Thick	.716	VLF
Gypsum Board 1/2" Fire Resistant,		
1 Layer	.364	VLF
2 Layer	.428	VLF
3 Layer	.530	VLF
Fireproofing – 16" x 7" Beam Encasements		
Perlite Plaster on Metal Lath	.453	LF
Gypsum Plaster on Metal Lath	.408	LF
Sprayed Fiber	.079	LF
Concrete 1-1/2" Thick	.554	LF
Gypsum Board 5/8" Fire Resistant	.488	LF

Figure 12.18 Installation Time in Labor-Hours for Fireproofing Structural Steel

decking above. The estimator can use online tables for calculating the square foot (SF) area of a variety of structural shapes per LF of the specific member.

Labor for Sprayed Fireproofing

Labor costs are based on the production of the crew and the equipment. There is typically a minimum of two people on a crew. Productivity varies by application and access to the area. Other cost considerations include accessibility, surrounding conditions, and staging requirements. Sprayed fireproofing can be extremely messy and warrants a liberal waste allowance for material. Costs for protection of adjacent work of other trades must also be included, along with detailed cleanup. Consider the impact of sprayed fireproofing work in multiple phases, if necessary, and touch-up of sprayed surfaces that may be damaged by other trades when hanging pipes, ducts, or conduits from the steel structure above. Figure 12.18 shows average labor-hours.

EXPANSION CONTROL

Section 07 95 00—Expansion Control includes expansion joints that occur in a variety of surfaces on a construction project. They are located in floors, walls, and roofs. They allow adjacent components of a structure to expand or contract with temperature without compromising the weathertight integrity of the surface or exposing a gap. Expansion control in this particular application differs from the expansion control that is part of another system such as EIFS or stucco.

Expansion Joints

Many expansion joints are part of the work it is incorporated within. For example, roof expansion joints are typically installed by the roofing contractor in the normal process of installing the roof. Similarly, expansion joints in masonry applications are installed by the masonry contractor. In addition, coordination with the contractor applying the joint sealants may be required to make the application weathertight.

Other types of expansion control include specific devices that are constructed of a wide variety of materials; aluminum, stainless steel, and copper to name a few. Individual components within the expansion joint may include neoprene to create a *bellows* that allows a larger range of movement.

Taking-off and Pricing

For most expansion joints, quantities are taken off and listed by the LF in both the horizontal and vertical applications. Quantities should be separated by application, composition, and type. The estimator is directed to become familiar with the individual application and product as this will have the greatest impact on costs. These devices are fabricated off site and then assembled on site. Individual products can be cast into concrete slabs or require steel angles for mounting in roof decks. Most installations require one or two people to complete, and trades can range from carpenters to cement finishers to roofers depending on the application.

SUMMARY

The work of Division 7—Thermal and Moisture Protection includes a variety of different and unique tasks that protect the structure from the elements. A thorough understanding of the individual product and how it is installed is essential in producing an accurate estimate for the work of this division.

Division 7 work starts with waterproofing and dampproofing of structural components below grade. While both are very different, they share a similar estimating process. Other Division 7 work includes insulation of varying types included in the exterior envelope of the structure. The insulation is protected by both vapor and air barriers at the exterior envelope of the building. While different, they follow similar procedures for estimating.

On the exterior surface of the structure is both siding and roofing. They differ significantly in how they are estimated and are subject to the individual type of roofing and siding and even the product. The estimator is urged to become familiar with the specified product for both siding and roofing.

Division 7 also includes Joint Protection more commonly known as caulking and sealants. This occurs both inside and outside the building and, while often overlooked in the estimate, is a key part to weatherproofing the structure. The work of this section can have few graphic details on the plans and is most often detailed in the technical specification section.

Stucco and a synthetic stucco called EIFS are included as a siding material in this division. It is taken off by the square foot or area and converted to square yards (SY) for pricing. Each has unique characteristics to be considered when estimating.

Near the end of Division 7 is firestopping and spray-applied fireproofing. While both contributing to the fire protection of the structure, they are estimated differently based on the product and application.

QUESTIONS/ PROBLEMS

1. Identify some graphic views and the work of Division 7 that an estimator may find on those graphic views.
2. Explain how iron oxide waterproofing works.
3. A 2" rigid insulation is required to be placed under the entire slab of a building that has an outside of foundation dimension of 100' × 180'. The wall is 1' thick and the slab is 4" thick. The insulation is also required to be continuous from the underside of slab to the top of footing which is 3'-9" below the top of slab. The slab and top of the wall are at the same elevation. Calculate the total area of 2" rigid insulation required.
4. If a laborer will place 500 SF of vertical insulation per day and 2000 SF of horizontal insulation per day, how many laborers will be required to complete the task in question #3 in a single workday.
5. A wood-framed roof has an 8:12 pitch on each side of the ridge. It is 60' long and 32' wide with a 12" overhang along the 60' lengths. Calculate the following: (a) SF of asphalt felt required; (b) squares of roof shingles; (c) quantity of 10' length of drip edge for fascia and ridge; and (d) LF of ridge cap. Allow 5% waste for each material.
6. If a crew of three roofers will install 16-SQ of shingles, including drip edge, cap, and underlayment per 8-hr day, how many days will the roof in Problem #5 take to complete.
7. Using the appropriate figure from the chapter, how many gallons of sealant will be required to fill 788 LF of ½" wide × ½" deep joint? Round your answer up to the nearest gallon. Cite the table used.
8. The net area of a building sidewall is 4,765 SF. If the plans call for ½" × 6" clapboard with 4" exposure, how many LF will be required with 4% waste? Round your answer to the nearest LF.
9. An EIFS crew consists of three plasters at $96.50 per hour and two plasterer tenders at $73.40 per hour. If they will install complete 550 SF of EIFS in a day, what is the cost per SF?
10. If the building in problem #5 has a flat ceiling with an R-38 insulation and the cost of the material is $75.99 per bundle containing 47.9 SF, what is the material cost and how many bundles are required?

13 | Openings

D ivision 8 is called *Openings* in the revised CSI MasterFormat® structure. It includes windows, doors and frames, finish hardware, glass and glazing, entrances, storefronts, and curtain walls, as well as skylights and louvers and vents. Hardware for this division refers to finish hardware in contrast to rough hardware and includes items such as hinges, locksets, passage sets, thresholds, weather stripping, door closers, and panic devices. Doors and windows are available in a multitude of sizes with various functions, insulating values, finishes, and glass types.

The size, location, quantity, and specific information for each door or window are included on the architectural drawings. Plan and elevation view drawings show their locations, operations, and quantities. In addition to the graphic views there are *schedules* that provide the information in a table format. This allows the reader to find all of the critical information in one location.

The specifications define door quality either by gauge of steel facing or species of wood veneer. Figure 13.1 illustrates some of the more common types of doors as seen in elevation view. Figure 13.2 illustrates common window types. Consult the details to clarify the type and specific operations of windows.

The work of Division 8 often has a significantly high cost per unit opening. It is in the best interest of accuracy that the estimator to have a contemporaneous quote for the materials of this division whenever possible.

ARCHITECTURAL DRAWINGS

The architectural drawings include both a window schedule and a door schedule. (Refer to the "Schedules" section of Chapter 1.) Schedules are laid out in block column form and list all the information concerning each item.

Estimating Building Costs for the Residential and Light Commercial Construction Professional, Third Edition. Wayne J. Del Pico.
© 2023 John Wiley & Sons, Inc. Published 2023 by John Wiley & Sons, Inc.

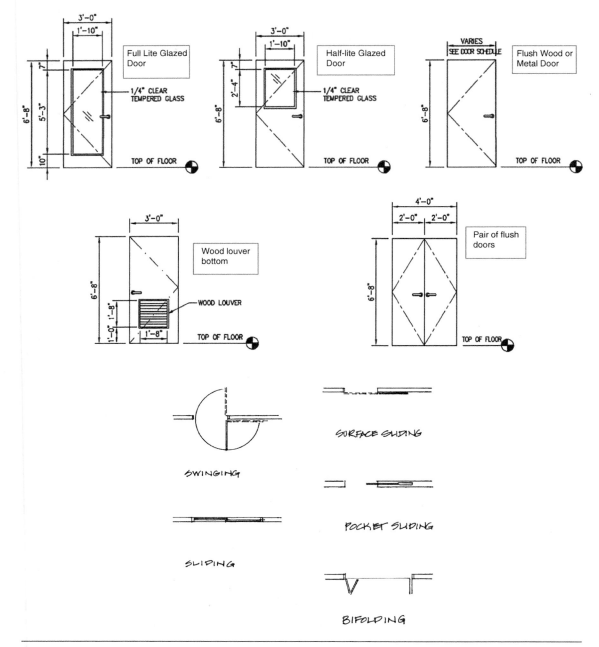

Figure 13.1 Door Types

Many projects have detailed drawings of the head, sill, and jambs of the doors, frames, or windows for the purpose of clarification and to provide adequate detail to estimate and eventually build the work.

Window Schedules

Windows schedules typically include the following:

- Designation (usually by letter)
- Size in width by height (W × H)
- Material composition of the sash and frame

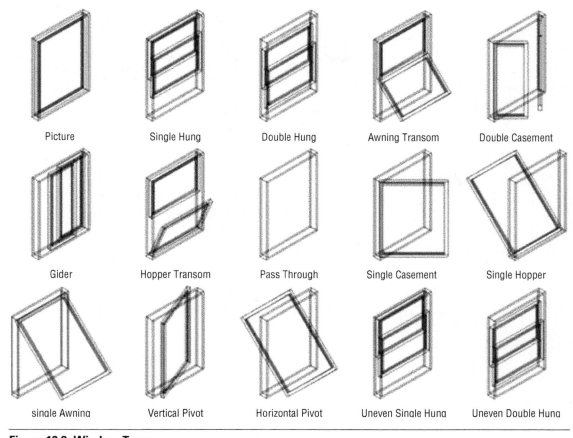

Picture   Single Hung   Double Hung   Awning Transom   Double Casement

Gider   Hopper Transom   Pass Through   Single Casement   Single Hopper

single Awning   Vertical Pivot   Horizontal Pivot   Uneven Single Hung   Uneven Double Hung

Figure 13.2 Window Types

- Manufacturer and model number
- Type or operation (e.g., fixed, double-hung, casement, or awning)
- Glazing requirements
- Jamb, head, and sill details for this window
- Remarks

The schedule sometimes lists the size of the rough opening for the installation of the window rather than the size of the window itself. Figure 13.3 is a typical window schedule with related details.

Door Schedules

Door schedules typically list each door opening by a specific number, or designation, called the *door mark*. This mark can also be used to identify the location of the door within the structure. Prefixes to the door mark typically indicate the floor on which the door is located. For example, B05 specifies door #5 in the basement level. Listed beside the mark in the schedule are the following details:

- The door size (width by height by thickness)
- Door type (flush, louvered, divided light, etc.)
- Material composition (steel, wood, or glass, etc.)

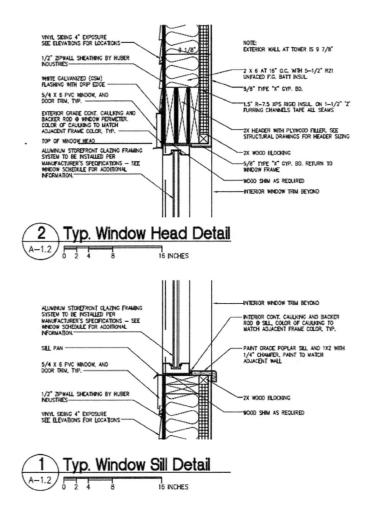

Figure 13.3 Window Schedule and Details

- Frame type and material composition
- Frame size (width by height) and thickness
- Fire rating requirements (if any)
- Louver or vision panels (if required)

- The hardware set for the door
- A "remarks" column for specific instructions (such as undercutting the door)

Figure 13.4 is a typical door schedule showing some of the corresponding details of the head, jamb, and sill.

Hardware Schedules

Doors require special hardware, called *finish hardware* that allow the door to operate. The finish hardware schedule is different from window and door schedules in that it does not always appear in column form on the architectural drawings.

DOOR SCHEDULE (x)

NUMBER	LOCATION	SIZE (W X H)	MANUFACTURER/ DOOR TYPE	MODEL #	LOCKSET	REMARKS
1	ENTRANCE DOOR	3'-6" X 6'-8" W/14" SL	THERMA-TRU	CC40 W/CC2020-LE	ENTRANCE	WITH 14"W SIDELITE U-FACTOR=0.14 SHGC=.01
2	GARAGE (OVERHEAD)	16'-0" X 8'-0"	OVERHEAD DOOR COMPANY	THERMACORE 194	–	WHITE FACTORY FINISH– PROVIDE WITH GARAGE DOOR OPENER
3	STORAGE EXTERIOR	3'-6" X 6'-8"	THERMA-TRU	S-100	ENTRANCE	–
4	FOUR SEASON ROOM EXTERIOR	3'-6" X 6'-8"	MASONITE (OR EQUAL)	FIBERGLASS	ENTRANCE- SINGLE ACTING	FULL LITE GLASS U-FACTOR=0.30 SHGC=0.18
5	BEDROOM 1 EXTERIOR	PAIR 2'-6" X 6'-8"	MASONITE (OR EQUAL)	FIBERGLASS	ENTRANCE- SINGLE ACTING	FULL LITE LOW-E GLAZING W/ BLINDS BETWEEN GLASS
6	BEDROOM 2 EXTERIOR	PAIR 2'-6" X 6'-8"	MASONITE (OR EQUAL)	FIBERGLASS	ENTRANCE- SINGLE ACTING	FULL LITE LOW-E GLAZING W/ BLINDS BETWEEN GLASS
7	BEDROOM 3 EXTERIOR	PAIR 2'-6" X 6'-8"	MASONITE (OR EQUAL)	FIBERGLASS	ENTRANCE- SINGLE ACTING	FULL LITE LOW-E GLAZING W/ BLINDS BETWEEN GLASS
8	BEDROOM 4 EXTERIOR	PAIR 2'-6" X 6'-8"	MASONITE (OR EQUAL)	FIBERGLASS	ENTRANCE- SINGLE ACTING	FULL LITE LOW-E GLAZING W/ BLINDS BETWEEN GLASS
9	BEDROOM 5 EXTERIOR	PAIR 2'-6" X 6'-8"	MASONITE (OR EQUAL)	FIBERGLASS	ENTRANCE- SINGLE ACTING	FULL LITE LOW-E GLAZING W/ BLINDS BETWEEN GLASS
10	FOYER CLOSET	PAIR 2'-0" X 6'-8"	SEE NOTES BELOW	FLUSH WOOD	SINGLE LEVER PULL W/ ROLLER LATCH #590 ROCKWOOD	–
11	DEN	3'-6" X 6'-8"	SEE NOTES BELOW	FLUSH WOOD	PASSAGE	–
12	GREAT ROOM / HALL	3'-0" X 6'-8" CASED OPENING	–	–	–	FLUSH METAL FRAME WITH WOOD CASING
13	BATH 3	3'-0" X 6'-8"	SEE NOTES BELOW	FLUSH WOOD	PRIVACY	W/ 1" UNDERCUT; INSTALL ROBE HOOK ON DOOR
14	CLOSET	3'-0" X 6'-8"	SEE NOTES BELOW	FLUSH WOOD	STORAGE	–
15	BASEMENT STAIRS	2'-8" X 6'-8"	SEE NOTES BELOW	FLUSH WOOD	STOREROOM	SEE NOTE BELOW

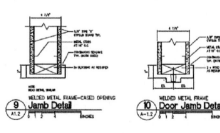

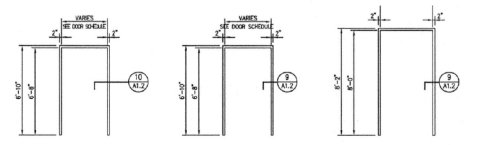

Figure 13.4 Door Schedule and Frame Details

It lists the items needed to outfit a particular door, called the *hardware set*, typically noted by a number. Each piece of hardware is listed by manufacturer, model number, size, and color or appearance, called the *finish*. The finish should be noted as part of the hardware takeoff, as it has a major impact on material pricing. The following is a sample description of a hardware set that might be encountered in a set of specifications.

Hardware Set #4

- 1½ pair 4½" Stanley CB Series Hinges
- Corbin 977L-9500 Series Mortise Lockset
- LCN 4010 CUSH Series Closer
- Ives 436 B Floor Stop
- Ives #20 Silencers

"Hardware Set #4" in this example may apply to several different doors on one project. Sometimes the door schedule may include a column that lists hardware sets that are applicable to particular doors. In the absence of such a column, the finish hardware section of the specifications will list the applicable doors under the individual hardware sets. Both practices are common and should be reviewed carefully for an accurate quantity of each hardware set.

As a general note to the estimator, when reviewing a window or door schedule, the estimator should also compare various plan views and exterior elevations to make sure that the quantities are correct. Window schedules frequently do not include quantities and only identify types. Door schedules on the other hand identify each individual door by the door mark on a floor-by-floor basis. Verifying quantities between plan views, elevations, and schedules is a good check and balance to ensure correct quantities.

Hardware schedules on residential projects are frequently missing. In this case, the schedule is sometimes substituted with an allowance for purchase and installation of the hardware. Refer to Chapter 2-Understanding the Specifications for more on Allowances.

HOLLOW METAL FRAMES AND DOORS

Hollow metal frames and doors are typically used for commercial projects that require durable, heavy-duty door and frame systems and where function supersedes aesthetics. Hollow metal components are sometimes referred to as steel doors and frames. They are available in a wide variety of sizes, fire- and sound-ratings, functions, and price ranges. Installation labor is predominantly, but not exclusively, a carpentry task. Metal door frames set in masonry opening are sometimes set by the bricklayer. The work can be an individual task, or it may require two individuals.

Hollow Metal Frames

Hollow metal frames are formed of 18-, 16-, or 14-gauge steel and are made to accommodate 1³⁄₈" and 1¾" wood or metal doors. They come in a variety of standard wall thicknesses, sometimes called *throat*, typically 4¾", 5¾", 6¾",

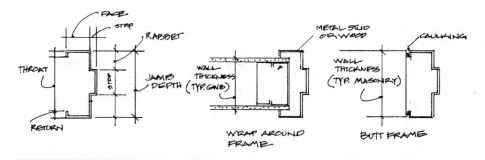

Figure 13.5 Hollow Metal Door Frames

and 8¾". They are available prefinished, galvanized, primed, or unfinished. Hollow metal frames can be installed in wood frame walls, masonry walls, metal stud and drywall walls, or a combination of all three materials. They are available in two standard levels of fabrication: *knockdown* (KD), where the frame is delivered disassembled into the two jambs and the headpiece and reassembled on-site, and the *welded assembly,* where the frame is welded at the factory at the corners to produce a true frame for site installation. Figure 13.5 illustrates some typical hollow metal door frame applications.

Taking-off and Pricing

Hollow metal door frames are taken off and priced by the individual frame or piece (EA). Takeoff quantities should be separated according to type (knockdown or welded), size (width by height), finish (raw or primed), gauge of the frame (metal thickness), and throat size (wall thickness). Any special fabrications required for installation, such as dimpled jambs for tee-anchors for masonry applications, should be noted. The height of the head (standard 2", or it can be 4" to match certain masonry coursing) should also be noted for pricing. Be sure to note any special preparation of or machining to the frame for hardware applications. Modifying frames in the field can be labor intensive and difficult at best. All machine prepping of the door frame should be done during fabrication to minimize on-site costs.

Labor to Install Metal Door Frames

Labor to install hollow metal door frames is priced by the individual frame as EA. The type of door frame (welded or knockdown), the size, installation type (drywall, wood, or masonry), and whether the frame is installed in the course of new work or as a retrofit application are all critical details for accurate labor pricing. Welded frames are installed as the wall or partition is being built, whereas KD frames are installed after the wall or partition has been drywalled or sheathed. Welded frames often require two individuals to erect and set, although a KD frame is an individual task. The labor to install hollow metal frames is often included as part of the work of the trade that is responsible for building the wall or partition. This is especially true of welded frames since they are integral to the wall construction.

An individual carpenter should install a minimum of two knockdown frames per hour, for double frames, slightly longer. For welded frames, installation times will vary by size and application and can be as much as twice that of a knockdown frame.

The specification section for masonry might require that the installation of hollow metal frames in all masonry walls and partitions be done by a bricklayer as the walls or partitions are being constructed. Note that many specifications require that door frames with fire ratings be filled solid with noncombustible material. Welded frames installed in masonry walls are filled with mortar as the wall or partition is constructed. Welded frames installed in drywall walls and partitions are frequently filled with plaster or fire-safing material. Due to the installation process of the frame in the drywall opening, this is a labor-intensive and often difficult task. For hollow metal welded frames installed in a masonry opening, the time-consuming task is the bracing of the frame to ensure that the frame is square, plumb, and it is not racked. Once that has been done, the bricklayers can proceed as normal.

Hollow Metal Doors

Hollow metal doors are constructed of 16-, 18-, or 20-gauge face sheets, configured for a $1\frac{3}{8}$" or $1\frac{3}{4}$" finished door thicknesses. They are available in a variety of styles, including flush, small vision panels, full or half glass, raised panel, and louvered. Building codes require doors and frames in certain locations to be fire-rated. A general rule is that if the wall is rated, the door within it is also rated. Typical fire-rating capacity labels are:

- C label for a $\frac{3}{4}$-hour rating
- B label for a $1\frac{1}{2}$- to 2-hour rating
- A label for a 3–hour rating

A door's *rating* is based on its physical composition and capacity to slow the transmission of fire. Labeled doors have a metal tag attached to the hinge edge of the door identifying its rating from the factory. Doors are subjected to testing by independent organizations, such as Underwriters Laboratories (UL). Other restrictions and qualifications also govern labeled doors and frames, such as glass size for vision panels and fire-rated louvers. Labeled doors and frames are considerably more expensive than other doors. Review the architectural drawings, particularly the floor plans and elevations, for door locations and quantities. The door schedule should identify any special fire-rating (label), undercutting, louvers, and hardware sets. Consult the drawings for rated wall types. Doors within the rated wall should match with a comparable rating.

Taking-off and Pricing

Metal doors are taken off and priced by the individual piece (EA), according to size, thickness, type, gauge of metal, and finish/style. Any special preparations for a deadbolt, openings for louvers or vision panels, or undercutting must be noted. Double door sets are noted as a pair (PR) in the takeoff and estimate, with the same qualifications. An individual door within a pair is referred to as a *leaf*. Figure 13.6 lists the common designations for steel doors.

Labor to Install Hollow Metal Doors

Hollow metal doors are typically installed by carpenters. The actual installation on the frame, referred to as *hanging the door,* requires installing the hinges first. This is typically a two-person task due to the awkwardness and weight of most steel doors. Rated or labeled doors can be heavier than most standard metal doors. The actual hanging of the door is typically included as part of the hardware package labor in an effort to save an estimating step. Labor should be provided to swing and adjust the door for smooth operation. (See the "Finish Hardware" section of this chapter. Also consult the "General Notes on Estimating Doors" section later in this chapter.)

WOOD DOORS AND FRAMES

Wood doors come in many types, materials, sizes, and thicknesses. They can be supplied separately for hollow metal frames, or they can be specified prehung. *Prehung* refers to a packaged unit, consisting of a finished door on a frame, with trim, and hinges. Wood doors can be classified in one of two categories. The first, *flush doors,* are flat slab doors that are either solid-core or hollow-core. Solid-core wood doors are made of particle board or mineral core composition with a wood veneer facing. They are used where increased fire resistance, sound insulation, and dimensional stability are specified.

Letter symbol	Description of door surface
F	Flush face steel surface
T	Textured face
E	Embossed face
TE	Textured and embossed face panels
L	Louvered top or bottom
LL	Louvered top and bottom
V	Vision lite
VL	Vision lite top and louvered bottom
N	Narrow (vision) lite
NL	Narrow lite top and louvered bottom
GL	Half glass top and louvered bottom
G	Half glass at top with varying lites
FG	Full glass
FL	Full louver
D	Dutch door

Figure 13.6 Standard Door Nomenclature

Hollow-core doors are used for interior applications and have a honeycombed cardboard/wood core with a wood veneer facing. They are less stable than solid-core doors and have no real thermal or sound-insulating value. They are used mainly in the residential market.

The second major classification is stile and rail doors. *Stile* and *rail* doors have vertical (stile) and horizontal (rail) members that provide the framework for wood, glass, or louver center panels. Wood doors are manufactured in widths ranging from 2' to 4', with smaller widths available in some styles. Heights range from 6'-6" to 9'-0", and thicknesses are 1¾", 1⅜", and 1⅛" for some residential-grade bifold and sliding closet doors.

Wood doors for high-end residential and commercial applications are available in a variety of wood veneers, ranging from oak and birch to exotic hardwoods. Consult the architectural drawings and door schedules (as noted for metal doors) for sizes, quantities, composition, veneer, and styles. Review the specifications for quality terms such as *book-matched* and *balanced*, which indicate matching grains within the door veneers.

Wood doors with exotic veneers can fluctuate in price significantly. It is recommended that the estimator obtain a contemporaneous price for the doors whenever possible, including machining (prepping for hardware) and factory finish.

Factory finishing of high-end wood doors is fairly common and almost always preferable.

Taking-off and Pricing

Wood doors are taken off and priced by the piece (EA). They are listed in the takeoff according to size (width by height by thickness); composition (solid- or hollow-core); species of wood; type of door (flush or stile and rail); and any special features, such as fire-rating label, vision panels, and preparation for special hardware. Wood frames are taken off and priced by the piece and specified according to size of the door they fit, wall thickness, species, and quality of wood (paint- or stain-grade), and any preparation to the frame, such as for hinges, locksets, or deadbolts. Prehung units are taken off and priced by the piece, EA, which in this case is the complete assembly. They are listed according to the size and type of the door itself, wall thickness, species and quality of door and frame wood, labels, vision panels, and any additional characteristics that would aid in accurate pricing. Double doors are taken off and priced by the pair (PR), with the aforementioned qualifications. Of special concern for estimating double doors is the hardware prep required for the active and inactive leaves of the doors.

Many high-end residential and commercial wood doors are available prefinished from manufacturers. Often, wood door specifications require that doors be prefinished by the manufacturer to maintain the warranty. Be sure to note this in the takeoff so that there is no duplication of finishing costs under the painting section. In addition, wood doors should always be machined and prepped for hardware off-site when this option is available. While doing this clearly adds to the cost of the door, it reduces on-site costs for labor, provides quality control not available in the field, and saves time.

Labor for Wood Doors and Frames

Wood door frames are installed by carpenters. The process includes measuring and prepping the opening, setting the frame in place, and ensuring the frame is plumb, true, and square. It is then shimmed, and permanently fastened to the rough opening. Lightweight residential interior prehung units are frequently installed by an individual carpenter, while exterior prehung entry units may require a crew of two, at least to set and plumb the door. Labor costs are based on carpenter labor-hours per door frame and the complexity of the frame and opening. Prehung units with the door and frame attached are also estimated by the carpenter labor-hours to install. Premachined, prefinished doors are hung as part of the installation of the finish hardware. Hanging the door encompasses installing the hinges on each door leaf, then attaching the door to the frame. In addition, the door should be adjusted so that it swings freely and the spacing on the strike side is uniform. This can often add time to the process, especially in renovation work where the partition or wall may be less than perfect.

For wood doors installed in metal door frames, the process is similar.

See "Hollow Metal Doors and Frames" earlier in this chapter. See Figure 13.7 for guidance on installation time for wood doors.

SLIDING AND SWINGING GLASS DOORS

Sliding glass doors, commonly referred to as *sliders,* consist of a stile and rail door with a full glass panel, sold as a single unit. Slider units range from 5' to 12" in width and from 6'-8" to 8'-0" in height. The stiles and rails can be wood or aluminum, or they can be covered with a metal or vinyl coating, referred to as *cladding,* at the exterior. The panels are insulated, tempered safety glass. Slider units are noted on the plan view and exterior elevation drawings and should also be shown on the door schedule. The process is similar with swinging glass doors commonly referred to as *French* or *patio* doors. Both types can have an assembly component to the labor or can be prehung. The estimator should refer to the specifications and door schedule for the information necessary for pricing such as special hardware, glazing, U-factors, solar-heat gain coefficients (SHGC), species of wood and factory finishes. Suppliers should be able to advise of the level of assembly required for a particular product.

Similar products called *window walls* are a series of glazed panels that slide along tracks to stack in a specific location. It gives the appearance of a wall open to the exterior. Window walls can fall under the classification of

Description	Labor-hours	Unit
Architectural, flush, interior, hollow-core		
Veneer face		
Up to 3'–0" × 7'–0"	0.941	EA
4'–0" × 7'–0"	1.000	EA
Plastic laminate face, hollow-core		
Up to 2'–6" × 6'–8"	1.000	EA
3'–0" × 7'–0"	1.067	EA
4'–0" × 7'–0"	1.143	EA
Particle-core, veneer face		
2'–6" × 6'–9"	1.067	EA
3'–0" × 6'–8"	1.143	EA
3'–0" × 7'–0"	1.231	EA
4'–0" × 7'–0"	1.333	EA
M.D.O. on hardboard face		
3'–0" × 7'–0"	1.333	EA
4'–0" × 7'–0"	1.600	EA
Plastic laminate face, solid-core		
3'–0" × 7'–0"	1.455	EA
4'–0" × 7'–0"	2.000	EA
Flush, exterior, solid-core, veneer face		
2'–6" × 7'–0"	1.067	EA
3'–0" × 7'–0"	1.143	EA
Decorator, hand-carved solid wood		
Up to 3'–0" × 7'–0"	1.143	EA
3'–6" × 8'–0"	1.600	EA
Fire door, flush, mineral-core		
B label, 1 hour, veneer face		
2'–6" × 6'–8"	1.143	EA
3'–0" × 7'–0"	1.231	EA
4'–0" × 7'–0"	1.333	EA
Plastic laminate face		
3'–0" × 7'–0"	1.455	EA
4'–0" × 7'–0"	1.600	EA
Residential, interior		
Hollow-core or panel		
Up to 2'–8" × 6'–8"	0.889	EA
3'–0" × 6'–8"	0.941	EA
Bifolding closet		
3'–0" × 6'–8"	1.231	EA
5'–0" × 6'–8"	1.455	EA
Interior prehung, hollow-core or panel		
Up to 2'–8" × 6'–8"	0.941	EA
3'–0" × 6'–8"	1.000	EA
Exterior, entrance, solid-core or panel		
Up to 2'–8" × 6'–8"	1.000	EA
3'–0" × 6'–8"	1.067	EA
Exterior, entrance, prehung		
Up to 3'–0" × 7'–0"	1.000	EA

Left Hand Reverse

Right Hand Reverse

Left Hand

Right Hand

Hand Designations

Figure 13.7 Installation Time in Labor-Hours for Interior Doors

RSMeans Estimating Handbook, Third Edition. © 2009 John Wiley & Sons, Inc.

Description	Labor-hours	Unit
Hollow metal doors, flush		
Full panel, commercial		
20 gauge		
2'–0" × 6'–8"	.800	EA
2'–6" × 6'–8"	.889	EA
3'–0" × 6'–8" or 3'–0" × 7'–0"	.941	EA
4'–0" × 7'–0"	1.067	EA
18 gauge		
2'–6" × 6'–8" or 2'–6" × 7'–0"	0.941	EA
3'–0" × 6'–8" or 3'–0" × 7'–0"	1.000	EA
4'–0" × 7'–0"	1.067	EA
Residential		
24 gauge, prehung		
2'–8" × 6'–8"	1.000	EA
3'–0" × 7'–0"	1.067	EA
Bifolding		
3'–0" × 6'–9"	1.000	EA
5'–0" × 6'–8"	1.143	EA
Steel frames		
16 gauge		
3'–0" wide	1.000	EA
6'–0" wide	1.143	EA
14 gauge		
4'–0" wide	1.067	EA
8'–0" wide	1.333	EA
Transom lite frames		
Fixed add	.103	EA
Movable	.123	EA

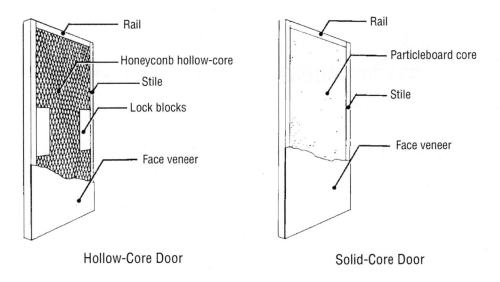

Hollow-Core Door   Solid-Core Door

Figure 13.7 *(Continued)*

specialty doors. The estimator is urged to research the individual product as the prep of the opening can require significant labor, along with the installation of the head and sill tracks. This type of application is always a multi-person crew installation.

Taking-off and Pricing

Sliding glass door units are taken off and priced by the piece (EA) or complete unit. *Piece*, in this example, refers to the entire unit or assembly of doors, hardware, and track. The slider should be specified in the takeoff by the manufacturer and model number, when available for accurate pricing. If the manufacturer and model number are not specified, the unit should be listed by the size (width by height), composition, wall thickness the unit will occupy, type of glazing, screens or grilles, special hardware, and the color of the finish (if applicable). SHGC and U-factors may also be specified. Swinging glass door units are qualified the same way, with a special notation about the number and location of operable panels.

Labor

There is a wide selection of sliding and swinging doors as well as window walls on the market, with dramatically different prices and quality. Some are shipped assembled and ready for installation, while others require assembly prior to installation. Assembly requirements should be noted in the takeoff and included as part of the installation costs. Installation typically requires a crew of two or more, although assembly can often be done by a single carpenter. The estimator may consider viewing videos online to watch the assembly and installation of specific products for more accurate understanding of the process.

SPECIAL DOORS

Special doors include folding doors, pocket doors, and surface sliding doors, along with some more unique special purpose doors.

Folding Doors

These are accordion or bifold doors that fold or stack against a wall or jamb. Folding doors over 150 square feet (SF) in area are referred to as *folding partitions*. Both folding doors and partitions are typically provided with tracks and all related hardware for installation. Available finishes include wood, fabric, and vinyl. Folding doors and partitions are often used to separate space within a room and, therefore, may require resistance to sound transmission, expressed as the door's sound transmission class (STC). The STC rating impacts costs; in general, the higher the sound-insulating value, the more expensive the door.

Larger folding doors or partitions may require a structural addition of either steel or wood beams above the door/partition to carry the load. While the beam is a direct result of the door/partition, it is rarely if ever, part of the work of the crew installing the door/partition. The addition of any structural members should be included in the appropriate division: 5 for steel and 6 for wood. The estimator is urged to become familiar with the details of the head track of the product specified.

Pocket Doors

These are installed within the framework of a wall so that, when opened, the door can be stored within the wall cavity. They are used in residential applications where special constraints preclude the use of swing or bifold doors. Pocket doors are available as a package unit with all necessary hardware and the track, in a variety of compositions, finishes, styles, and sizes. The estimator is urged to become familiar with the product specified.

Insulating Door Units

Steel insulating door units for residential entrances are composed of thin steel sheets over a wood-and-foam insulating core. Fiberglass units have the same internal composition with a fiberglass skin. The skin is often textured to resemble wood grain. They are typically provided prehung in a wood frame with an integral aluminum threshold, bored for locksets and/or deadbolts. Sizes range from 2'-8" to 3'-0" in width and 6'-6" or 6'-8" in height. Steel doors are provided primed for field-applied paint, and fiberglass units are unfinished, ready for field-applied stains or paints.

Storm Doors

These provide protection of the entrance from weather and allow air passage through the screen in warm weather. Aluminum storm doors are lightweight and prefinished, with interchangeable glass sashes and screens, set within a matching aluminum frame. The storm unit is surface-applied by screwing through flanges, attached to the frame jambs and head, on the surface of the exterior door trim. Handles, latches or locksets, and hydraulic closers are included as part of the unit. Storm door units are available in sizes to fit most exterior entry units, in a variety of colors and designs.

Taking-off and Pricing

Folding doors and partitions are taken off and priced by the square foot of the opening (SF). The takeoff description should note size (width by height), composition of the door and finishes, method of door operation (manual or electrical), STC classifications, fire rating (if applicable), and any special locking hardware. Folding partitions can be installed by each individual panel and may require some scaffolding for access and lifting equipment for support.

Pocket doors are taken off and listed by the piece (EA), according to size (width by height), composition (wood or plastic), and thickness of the wall. The takeoff designation should include all components necessary in the unit so that they can be priced as a single unit.

Steel and fiberglass insulating entry units are taken off and priced by the piece and listed according to size, wall thickness, function (in-swing or out-swing), design or model, exterior casing type, and the preparation for the door (boring for lockset or deadbolt). The piece designation includes the complete unit, less lockset and deadbolt hardware.

Storm door units are taken off and listed by the piece according to size, design, special options (e.g., insulated glazing or locksets), and color of finish. Size is often designated according to the size of the entry unit. They are priced by the individual piece or unit, with the designation each (EA). The individual piece is typically provided as a complete unit with fasteners, hardware, continuous hinge, weatherstripping and lockset.

Labor to Install Special Doors

Special doors are most often installed by carpenters. Costs are based on carpenter labor-hours and vary dramatically with the type of door and application. For most special doors, installation requires a minimum crew of two carpenters. Folding partitions often require two or more carpenters. Consult applicable manufacturers' literature on individual units for guidelines on assembly and installation. The estimator is urged to become familiar with the product specified and consider a subcontractor quote for the installation, especially with proprietary systems such as folding doors or partitions.

Pocket doors require some level of assembly prior to installation, and the amount of work depends on the manufacturer and quality of the unit.

See Figure 13.8 for guidance on installation times of special doors.

GENERAL NOTES ON ESTIMATING DOORS

Doors and frames are typically delivered to the job FOB (freight on board) and require handling from the truck to storage or immediate distribution. Handling and distribution can be costly for projects with a large quantity of doors and frames, such as a school or hotel. Review the specifications for any special handling or storage requirements, such as temperature and humidity control. Costs for labor to distribute doors, frames, and hardware from the storage area to the actual opening must be included in the labor portion of the estimate. Many collective bargaining agreements require that finished products,

Description	Labor-hours	Unit
Access panels & doors, metal		
12" × 12"	0.800	EA
18" × 18"	0.889	EA
24" × 24"	0.889	EA
36" × 36"	1.000	EA
Cold storage doors		
Single galvanized steel horizontal sliding		
5'–0" × 7'–0"		
Manual operation	8.000	EA
Power operation	8.421	EA
9'–0" × 10'–0"		
Manual operation	9.412	EA
Power operation	10.000	EA
Hinged lightweight 3'–0" × 6'–6"		
2" thick	8.000	EA
4" thick	8.421	EA
6" thick	11.429	EA
Biparting electric operated 6' × 8'	20.000	Opng.
For door buck framing & door protection add	6.400	Opng.
Galvanized batten door		
4'–0" × 7'–0"	8.000	Opng.
6'–0" × 8'–0"	8.889	Opng.
Fire door, 6' × 8'		
Single slide	20.000	Opng.
Double biparting	22.857	Opng.
Darkroom doors revolving		
2-way 36" diameter	5.161	Opng.
3-way 51" diameter	11.429	Opng.
4-way 49" diameter	11.429	Opng.
Hinged safety		
2-way 41" diameter	6.957	Opng.
3-way 51" diameter	11.429	Opng.
Pop-out safety		
2-way 41" diameter	5.161	Opng.
3-way 51" diameter	11.429	Opng.
Vault front door and frame		
32" × 78", 1 hour, 750 lbs.	10.667	Opng.
40" × 78", 2 hours, 1130 lbs.	16.000	Opng.
Day gate		
32" wide	10.667	Opng.
40" wide	11.429	Opng.

Figure 13.8 Installation Time in Labor-Hours for Special Doors

Source RSMeans Estimating Handbook, Third Edition. © 2009 John Wiley & Sons, Inc.

including doors and frames, be distributed by carpenters instead of laborers. Otherwise, distribution is most often performed by laborers. Distribution should be included and identified as a separate item in the labor portion of the estimate.

Occasionally, the *handing* or direction of a door's swing may affect the price (especially for exterior residential door units that swing out instead of the normal in-swing). Similar requirements for metal doors should be noted. It is not uncommon for all doors to need some minor adjustments. Allow labor time for adjusting the swing of the door after it has been hung. Because prehung units come with the casing attached or included, it is necessary to specify the type of casing for accurate pricing of the unit. Finishing of wood and metal doors will be discussed in Chapter 14, "Finishes." Accurate pricing of doors and frames often requires contemporaneous pricing by a material supplier. Door and frame takeoffs should be provided to material suppliers for up-to-date prices.

Folding partitions for commercial applications may need to be fire-rated to meet building codes. Review the specifications carefully for this requirement. Many of the folding partitions are covered in fabric, with different levels of fabric having an impact on the price. Larger folding partitions will require special structural details to accommodate the weight of the partition itself. As previously noted, this is not usually part of the scope of the partition manufacturer or installer; clarification of the limits of the work is essential for accurate pricing. Many specifications require a structural engineer's design and calculations, complete with a stamp, for the design and sizing of the structural member that will carry the partition. This may also be required by the permitting authority.

OVERHEAD AND COILING DOORS

Overhead doors, such as garage doors, are constructed of wood stiles and rails with hardboard flush inserts or thin steel face sheets over a steel frame and foam insulating core. Face sheets can be flush or may have an embossed design. Special designs are also manufactured to include glass lights within individual panels. Wood doors are either unfinished or primed for field painting. Metal doors are usually provided prefinished in the manufacturer's standard colors, white being most common, with an increased price for colors.

Standard door openings range from 8' to 18' in width and 7' to 12' in height. Custom sizes are also available but may constitute an additional cost. Overhead doors are typically provided with the necessary hardware for installation, as well as locking mechanisms for security. Other types of overhead doors, called *coiling doors* or *rolling shutters,* are taken off and priced per opening based on the square foot area of the coiling door. Material composition (steel, stainless steel, or aluminum) of the door and its operation will affect the price. Added costs for steel jambs and heads for mounting, hoods, or guides must be included. The estimator is urged to review door schedules, architectural plans, and elevations for information on overhead doors, and consult the specifications for special requirements or manufacturers. The operation of the doors – manual or electric – will also impact the cost. Safety features may require the coordination of another trade such as the electrician. The clear demarcation of the scope of work is important.

Taking-off and Pricing

Overhead doors are taken off and listed by the piece (EA) according to size (width by height), thickness of the door panels, composition and finish of the panels, door style or design, manufacturer's model number (if applicable), and special options such as glass lights, electrical operators, safety devices, and security mechanisms. Additional information, such as R-value or the amount of overhead clearance distance (from the head of the door to the ceiling above), may also be required for accurate pricing. Custom-sized overhead doors are taken off and qualified by the same units. Overhead doors are typically priced by the square foot area of each door. Special functions, such as door operation or high R-values, will affect the square foot cost.

Labor to Install Overhead Doors

All but the simplest of overhead doors are installed by specialty subcontractors. Always solicit a price for a furnished and installed overhead door to ensure accurate pricing. Installation costs for overhead doors are calculated per labor-hour per door. Labor costs are based on the productivity of the crew or individual, depending on the size of the door and its application. Single opening residential garage overhead doors are often installed by an individual carpenter, whereas double width overhead doors may require a crew of two carpenters. The estimator should also note that some type of lift may be required to handle and access the door. Labor costs for the electrical wiring portion of a motorized overhead door should be calculated in Division 26—Electrical.

ENTRANCES AND STOREFRONTS

These consist of metal framework surrounding fixed or operable windows and entrance doors. The framework is manufactured from aluminum alloy and can be extruded into a variety of shapes and sizes. Finishes are anodized in a number of colors (or clear) to prevent oxidation from exposure. Durable coatings are available in a range of colors. Colors can have a significant impact on cost.

Insulated or plain glazing panels manufactured off-site are installed within channels of the extruded framework to provide a weather-tight window or wall system. Specialty glazing can be used to obscure structural elements or insulating nonglazed panels may be incorporated into the design. Entrance doors of the same construction with hinges or pivots can be manually operated or motorized and may be purchased with panic and finish hardware. Figure 13.9 illustrates a typical storefront and details.

The precise location and width of the entrance and storefront system are noted on the architectural floor plans. Additional drawings showing sections through the head, jamb, and sills of the various windows are usually provided on a separate sheet, called *window details*. Elevation views are found on the exterior elevations. Review the specifications for the type of

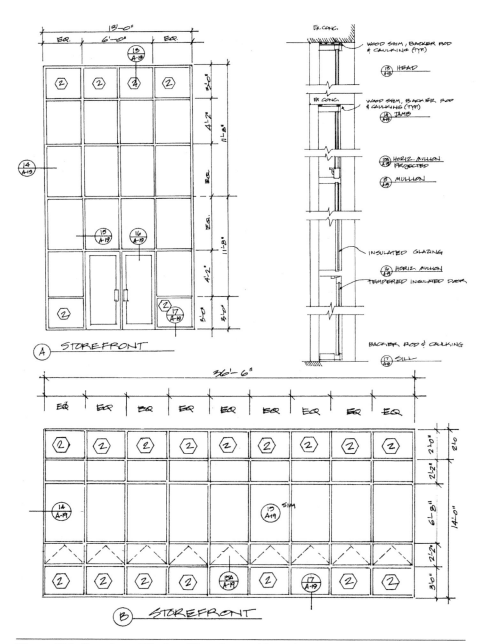

Figure 13.9 Storefront Window Elevation and Details

glass (laminated safety or tempered, colored or reflective), frame system, manufacturer, door and window systems and hardware. Samples and shop drawings for storefront systems are almost always a requirement which will add costs to the system.

Taking-off and Pricing

Each individual component of the glass and glazing system is taken off and priced separately for better accuracy. The aluminum framework is taken off and priced by the linear foot (LF), including all vertical and horizontal mullions, jambs, heads, and sills. Extruded sections should be separated according to size, thickness of the material (gauge), finish, and shape. Aluminum flashings at the sills, or head of the system, should be taken off by the LF. Flashings over 12" in width may be converted to SF.

Caulking and sealants at the perimeter of the unit and between metal components are taken off and priced by the LF. Be careful to avoid duplication of work that is specified under Section 07 90 00—Joint Protection. Frequently, specifications require that the storefront section in Division 8 include caulking and sealing of storefront work, in an effort to maintain one source of responsibility for weather-tightness of the system.

Glass and insulated glass panels are taken off and priced by the square foot (SF) of the area of the individual insulated glass panels that fit within an extruded frame system. These should be listed in the takeoff separately according to size (width by height), thickness of the insulating panel or pane of glass, type of glass (e.g., clear, tinted, or spandrel), and special treatment of glass (e.g., tempered or laminated safety glass, or security films). Insulated nonglazed panels are taken off and priced by the SF according to the size, thickness, facing or finish of the panel, and insulating material within the panel.

Windows are taken off and priced by the piece (EA) according to size (width by height), function (e.g., casement, awning, hopper, or sliding), hardware requirements, material and finish of the sash, and glass type. They can also be priced by the square foot.

Doors are taken off and priced by the piece (EA) according to size (width by height), thickness, stile and rail materials, hardware (such as hinges or pivots), panic devices, locking mechanisms, and type of glass. Finish door hardware for aluminum entrances may be specified under this section or can be included as part of Section 08 70 00—Finish Hardware. The estimator should review the hardware section carefully for correct scope and to avoid duplication.

Custom fabrication of glass panels and their framework require shop drawings showing the exact sizes of each fabrication. Field measurement and verification of the rough opening sizes are needed. The cost of shop drawings is considered part of the cost of the work and should be accounted for in the takeoff. Shop drawings are listed as a lump sum (LS).

Labor

There are two classifications of labor for storefront systems: shop fabrication and field installation labor. Both must be included in the estimate to be complete. Aluminum extrusions that make up the storefront system come from the manufacturer in stock lengths and are then cut to specific sizes and assembled in accordance with the shop drawings. Assembled window sashes and doors fit into the frame. Frames are delivered to the site unglazed and installed within the premeasured rough openings. Once secured in place, the frames are glazed; stops and gasketing are set; and doors, finish hardware, and window sashes are installed.

Labor for both the fabrication and the installation process are crew tasks. The individuals that perform the work are referred to as *glaziers*. Crews consist of two or more glaziers, depending on the size of the storefront. Fabrication labor should be kept separate from installation labor. Fabrication of glass panels within the frame is typically priced by the square foot area of the panel. Installation costs are based on labor-hours per square foot area of the frame, including doors and windows. Additional labor costs for installation of doors, finish hardware, and glazed and nonglazed panels must be included. Again, include labor costs for adjustments to door and window hardware once installed.

Depending on the size of the storefront system, scaffolding for access and a piece of equipment for support may be required. It is suggested the estimator solicit contemporaneous pricing for fabrication and installation from a subcontractor whenever possible.

WOOD AND PLASTIC WINDOWS

Windows are furnished as complete, factory-assembled units, including frame, sash, operating hardware, weather stripping, and glazing. Wood windows are constructed of kiln-dried, clear, straight-grain woods, usually western pine. Frames are treated with a water-repellent preservative and are available primed and ready for field-applied paint, factory-painted, aluminum-clad, or vinyl-clad. Plastic windows, featuring jambs, heads, sills, and sashes, are made of various grades of PVC. Insulated glass panels are set within plastic stiles and rails. Wood and plastic windows come in literally hundreds of sizes within the various types of operation. Combinations of various styles and sizes are often joined together in the factory or field to produce a desired appearance.

Review architectural floor plans and elevations for the locations of various window types. The quantities and type derived from the plan view should be cross checked with the elevations and the window schedule for confirmation of quantity. Unlike door schedules, window schedules often show type only and not a letter or number for each window on the project. Window schedules (see Figure 13.3) should be studied for the specific manufacturer, model, and size of the individual or combined unit. Occasionally, special details or sections through the unit may be used for clarification. The specifications should be studied for detailed information on the type of glazing, U-factors, SHGCs, finish of the window, cladding color (if applicable), treatment of glazing, and accessories, such as screens, security hardware, or grilles. Acceptable manufacturers and required warranties may also be listed in the specifications.

Taking-off and Pricing

Wood and plastic windows are taken off and priced by the individual unit or piece (EA). Quantities should be noted according to manufacturer, model or size, function (e.g., double-hung, casement, awning, or fixed sash), wall thickness, type of exterior finish, color of the finish, type of insulating glass, and exterior casing (for non-clad wood windows). Special options, such as grilles, screens, or blinds, can be taken off and priced separately or included as part of the actual window cost. For larger window units, such as bays or bows, many manufacturers offer the option of making normally fixed sashes operable, such as the center sashes in a four- or five-lite bow window. This constitutes an additional cost and should be noted in the takeoff for accurate pricing. Windows, especially custom windows, can have extremely long lead times. The estimator should inquire about lead time when pricing the windows. Pricing should include all shipping costs.

Many manufacturers fabricate windows to custom specifications, which can be expensive and require considerable lead times. To accurately price custom window fabrication, contact the manufacturer and obtain a direct quote, including any shipping costs. While most wood windows are manufactured for a normal wall thickness (2" × 4" construction), those constructed of thicker framing members (such as 2" × 6") require finish wood pieces to bring the jambs, head, and sill out flush to the interior finish surface. These are called *extension jambs* and are additional to the cost of the window.

In much the same manner as with a lumber list, windows are listed with all specifics, type of operation, and sizes and are forwarded to a vendor to obtain a quote.

Labor

Most wood and plastic window installations require two carpenters. Smaller units may be installed by an individual carpenter. For replacement windows set within an existing frame, the task can often be accomplished by an individual carpenter from the interior of the building. Labor costs are based on labor-hours per window or, alternately, the quantity of windows that can be installed by a crew in an eight-hour day. The estimator must consider scaffolding or lifts to access openings above the first floor. Installation labor is separate from framing or reconfiguring the rough opening part of the work. Framing and reconfiguring the rough opening is the work of Division 6. Installation is typically defined as setting the window frame and sash within the opening, shimming and leveling the frame within the opening, and fastening the window frame to the structural members. This can also include applying the membrane weather barrier along the flanges of the window at the exterior.

METAL WINDOWS

Metal windows are frequently used for commercial applications. They can be *stock units* (standard models that the manufacturer keeps in stock for fast delivery) or custom-fabricated to meet the specific needs of an individual project. Metal windows are comparable in operation, glazing, and most options to wood and plastic windows. Aluminum windows are premanufactured and available in a wide variety of operations, sizes, colors, and glazing specifics. The specifications should be studied for detailed information on the type of glazing, U-factors, SHGCs, finish of the window, cladding color (if applicable), treatment of glazing, and accessories, such as screens, security hardware, or grilles. Acceptable manufacturers and required warranties are also listed in the specifications.

Taking-off and Pricing

For stock units, metal windows are taken off and priced by the individual piece (EA). Trims and receiver channels at the perimeter of the window are taken off and priced by the LF of the perimeter. Custom-fabricated windows are taken off and priced by the SF of the window itself. It can also be taken off and priced by a unique unit of measure coined by the window industry to represent the product dimensions more accurately. This unique unit is called the *united inch*. The united inch measurement of a window is the sum of the width and height of a window in inches. Custom windows may be priced by all of the same considerations noted under the "Wood and Plastic Window" section, with the exception of the extension jambs. For stock and custom fabrications, shop drawings are often required, which can also add to the cost. Pricing should be quoted by the manufacturer to ensure accuracy and as with wood and plastic windows, lead times can be significant.

Large metal window units are sometimes assembled in the field. The estimator should inquire as to the amount of field assembly required. The estimator must consider scaffolding or lifts to access openings above the first floor.

Labor

Installation is typically performed by a multiperson crew, most often carpenters. Productivity is measured by unit-per-day installed by a specific crew. Labor costs per window unit can be assigned as labor-hours per unit. Note that, as in the case of entrance and storefront systems, metal window specifications may require caulking and sealing or to be flashed to the air barrier system. Review the specifications so that there are no duplications. Additional labor should be added for installing screens and hardware that do not come assembled from the factory. The installation is typically defined as setting the window frame and sash within the opening, shimming, and leveling the frame within the opening and fastening the window frame to the structural members. This can also include applying the membrane weather barrier along the flanges of the window at the exterior.

See Figure 13.10 for guidance on installation times for exterior windows.

Description	Labor-hours	Unit
Custom aluminum sash, glazing not included	.080	SF
Stock aluminum windows, frame, and glazing casement, 3'–1" × 3'–2"opening	1.600	EA
Combination storm and screen		EA
2'–0" × 3'–5"opening	.533	EA
4'–0" × 6'–9"opening	.640	EA
Projected, with screen		
3'–1" × 3'–2"opening	1.600	EA
4'–5" × 5'–3"opening	2.000	EA
Single hung, 2'–3"opening	1.600	
3'–4" × 5'–0"opening	1.778	EA
Sliding, 3'–2"opening	1.600	EA
5'–0" × 3'–0"opening	1.778	EA
8'–0" × 4'–0" opening	2.667	EA
Custom steel sash units, glazing & trim not incl.	.080	SF
Stock steel windows, frame, trim and glass incl.		
Double hung, 3'–2"opening	1.333	EA
Commercial projected, 3'–9" × 5'–5" opening	1.600	EA
6'–9" × 4'–1"opening	2.286	EA
Intermediate projected, 2'–9" × 4'–1" opening	1.333	EA
Custom wood sash, incl. double/triple glazing but not trim		
5'–0" × 4'–0"opening	3.200	EA
7'–0" × 4'–6"opening	3.721	EA
8'–6" × 5'–0"opening	4.571	EA
Stock wood windows, frame, trim and glass incl.		
Awning type, 2'–10" × 1'–10" opening	.800	
5'–0" × 3'–0"opening	1.000	EA
Bow bay, 8'–0" × 5'–0"opening	1.600	EA
12'–0" × 6'–0" opening	2.667	EA
Casement 2'–0" × 3'–0"opening	.800	EA
4'–0" × 4'–0" opening 2 leaf	.920	EA
6'–0" × 4'–0" opening 3 leaf	1.040	EA
10'–0" × 4'–0" opening 5 leaf	1.280	EA
Double hung, 2'–0" × 3'–0" opening	.800	EA
3'–0" × 5'–0" opening	1.000	EA
Picture window, 5'–0" × 4'–0" opening	1.445	EA
Roof window, complete		
2'–9" × 4'–0" opening	3.556	EA
Sliding, 3'–0" × 3'–0" opening	.800	EA
6'–0" × 5'–0" opening 2 leaf	1.000	EA

Figure 13.10 Installation Time in Labor-Hours for Exterior Windows/Sash

Source RSMeans Estimating Handbook, Third Edition. © 2009 John Wiley & Sons, Inc.

GENERAL NOTES ON WINDOWS

Specifications for commercial projects may require the window and storefront system installer to clean all glass (inside and out) prior to acceptance by the owner. Warranties on the glazing portion of windows and storefront systems, typically issued by the manufacturer, may require a site test or inspection by the manufacturer's technical representative. Depending on the manufacturer, this may have an associated cost. Check the specifications carefully for any "attic stock" or replacement glazing panels that must be provided as part of the base contract.

In much the same manner as noted in the discussion on doors in this chapter, projects with large quantities of windows require handling and distribution. The estimator is directed to consider added labor-hours for the unloading, handling, and distribution of windows on a project. In addition, many windows come prepackaged in cardboard boxes which can generate a considerable amount of recycling debris. This should also be considered.

In contrast to doors, many window installations require some level of access from the exterior. The estimator should analyze whether the window units can be installed from the interior of the building or if they must be installed from the exterior. Should the latter be required, the estimator may want to consider if the project will have scaffolding available by the GC or if it must be provided as part of the window installation. This may be noted in Division 1.

ROOF WINDOWS AND SKYLIGHTS

Many projects incorporate windows in the roof, called *skylights*. Skylights provide natural light and help reduce energy costs. Skylights or roof windows are furnished as complete, factory-assembled units, including frame, sash, operating hardware if applicable, weather stripping, and glazing. Skylight frames are constructed out of straight-grain wood or metal and are usually furnished with the appropriate flashing kits for the application. They can be operable, allowing the circulation of air, or fixed. Glazing can be flat or domed, insulating glass or non-yellowing polycarbonate. Hardware can be manual cranks as with windows or motor operated and may even be Wi-Fi enabled. They are available with precipitation sensors that will close the window when it rains. Every feature on the skylight adds to the price.

Light tubes are cylindrical tubes from the roof to the ceiling of an interior space. They provide natural light to dark interior areas such as hallways but use a smaller area of roof and ceiling than a skylight.

The estimator is directed to study the plans and specifications carefully for the products and quantities included. The internet is an excellent source of information on specific product details.

Taking-off and Pricing

For stock units, roof windows, light tubes, and skylights are taken off and priced by the individual piece (EA). Trims and flashing kits at the perimeter of the opening are taken off and priced by the LF of the perimeter or by the individual piece depending on the product. Pricing should be quoted by the manufacturer to ensure accuracy. Products should be listed in the takeoff by size (length by width), operation, glazing type, energy ratings, operating hardware, and accessories such as screen and shades.

Labor

Installation is typically performed by a multiperson crew. Productivity is measured by unit-per-day installed by a specific crew. Labor costs per skylight unit can be assigned as labor-hours per unit. It should be noted that in contrast to windows, which are typically installed by a single trade – a carpenter – skylights, roof windows, and light tubes also require roofers to flash and weather-tight the product to the roof. The estimator should consider the specific application, as cutting into an existing roof is more time consuming than installing the feature prior to the roof work, as in new construction. Framing the roof opening, whether for a new structure or a renovation, is typically considered a separate task for estimating purposes. The same is true about cutting and patching an existing roof or flashing in a new skylight or light tube.

Figure 13.11 provides guidance for the installation of skylights.

	Labor-Hours	Unit
Domed unit skylights		
Plastic domes, flush or curb mounted, ten or more units		
Nominal size under 10 SF, double	.246	SF
Single	.200	SF
10 SF to 20 SF, double	.102	SF
Single	.081	SF
20 SF to 30 SF, double	.081	SF
Single	.069	SF
30 SF to 65 SF, double	.069	SF
Single	.052	SF
Ventilating insulated plexiglass dome with curb mounting, 36 × 36	2.667	EA
52 × 52	2.667	EA
28 × 52	3.200	EA
36 × 52	3.200	EA
Field-fabricated, factory type, aluminum and wire glass	.267	SF
Insulated safety glass with aluminum frame	.200	SF
Sandwich panels, fiberglass, for walls, 1-9/16" thick, to 250 SF	.160	SF
250 SF and up	.121	SF
As above, but for roofs, 2¾" thick, to 250 SF	.108	SF
250 SF and up	.097	SF
Metal-framed skylights		
Prefabricated glass block with metal frame		
Minimum	.121	SF
Maximum	.200	SF

Figure 13.11 Installation Time in Labor-Hours for Skylights

Source RSMeans Estimating Handbook, Third Edition. © 2009 John Wiley & Sons, Inc.

FINISH HARDWARE

The following is a list of the most common U.S. Code Symbol numbers and their respective finishes:

- *US P—Primed paint coat*
- *US 3—Polished brass*
- *US 4—Satin brass*
- *US 9—Polished bronze*
- *US 10—Satin bronze*
- *US 10B—Satin bronze, oil-rubbed*
- *US 14—Polished nickel*
- *US 15—Satin nickel*
- *US 20—Statuary (light) bright bronze*
- *US 26—Polished chrome*
- *US 26D—Satin chrome*
- *US 28—Satin aluminum, anodized*
- *US 32—Polished stainless steel*
- *US 32D—Satin stainless steel*

Finish hardware is hardware that is applied to a door to allow it to function as desired. The most common examples of finish hardware are hinges, locksets, latch sets, closers, stops, deadbolts, thresholds, weather stripping, and panic devices. Most hardware is available in a variety of finishes, designated by the U.S. Code Symbol Designation.

Consult the architectural drawings and door schedule, which typically follow two basic formats for noting finish hardware. The first and most common method is to designate the particular hardware set for each door by a number, such as Hardware Set #2, as discussed at the beginning of this chapter. The individual components of the specific set number are identified by make and model and quantity per leaf.

The second method uses a series of columns included in the Door Schedule, each of which is headed by a particular item of finish hardware. The row lists the door mark and corresponding hardware items with a number that defines the specific item in the specifications and in some cases the quantity of each piece.

Taking-off and Pricing

Hardware items are taken off by counting the individual units and are listed as EA. The individual pieces can be grouped into hardware sets for pricing. Items should be separated according to the manufacturer, series model, type, finish, and any other means of identification specified, such as Federal Specification Series Designation or American National Standards Institute (ANSI) series number. Note that the finish on hardware can have a direct impact on its cost and may require significant lead time. Some of the less-expensive finishes are readily available. Others are special orders and may require long lead times for delivery. Consult the supplier for delivery schedules so that temporary hardware items are included in the estimate, if necessary.

Many construction projects require temporary cores for locksets during construction, with permanent ones installed at turnover. This is typically noted in the specifications and may be an additional cost. Most doors supplied to project sites have been prepped for the hardware off-site.

Occasionally, doors are mortised and bored for hardware on-site. Be sure to verify how the doors will be supplied prior to pricing the hardware installation.

Pricing hardware material is a specialized discipline requiring advanced knowledge and experience beyond that of the average estimator. A simple addition of a number or letter to a model can have a dramatic impact on an item's function and price. For projects with detailed finish hardware requirements, solicit pricing and lead times from suppliers in advance of the bid date.

Labor

Finish hardware is typically installed by finish carpenters. Occasionally, the installation portion of the finish hardware is specified under Section 06 20 00—Finish Carpentry. Finish hardware installation requires first-class workmanship and is measured by the productivity of the individual carpenter.

Review the documents to determine whether the doors will be provided premachined, which will impact the costs of the finish hardware installation. Other concerns include receiving, cataloging, and distributing the hardware to the individual locations where they will be installed. On larger commercial projects with many doors, such as hotels, schools, and apartment buildings, these costs can be labor intensive and must be included in the estimate. Labor costs are calculated as labor-hours per hardware set. The cost in labor-hours must be determined for each set, then multiplied by the quantity of each. As an example, Hardware Set #4 below is a tally of the estimated time in hours it takes to install each component of the hardware set.

Labor-hours to install Hardware Set #4	
1½ pairs 4½" Stanley CB Series Hinges (1½ PR)	0.50 hours
Hang and swing door	0.30 hours
Corbin 977L-9500 Series Mortise Lockset (1 EA)	1.00 hours
LCN 4010 CUSH Series Closer (1 EA)	1.40 hours
Ives 436 B Floor Stop (1 EA)	0.30 hours
Ives #20 Silencers (3 EA)	0.20 hours
Adjust hardware and door	0.50 hours
Subtotal for labor-hours for HS #4	4.20 hours

If the 4.20 labor-hours for each Hardware Set #4 is multiplied by the takeoff quantity of 12 for Hardware Set #4, then:

$$HS \#4 - 12 \text{ sets} \times 4.20 \text{ labor-hours per set} = 50.40 \text{ labor-hours}$$

The 50.40 labor-hours could then be multiplied by the billing rate of $86.50 per labor-hour to arrive at the costs for installing all Hardware Sets #4:

$$50.40 \text{ labor-hours} \times \$86.50 \text{ per labor-hour} = \$4,359.60$$

The labor-hours could alternately be calculated for *all* hardware sets on the project before pricing is done. This method can save some estimating time and reduce the level of detail. Remember to calculate individual labor-hours to the nearest 5-minute increment (0.083 labor-hours) whenever possible. Allow time for sorting, reading instructions or hardware schedules, and distribution to the opening. Rounding should be at the final summary of all labor-hours for the entire hardware package. Figure 13.12 provides some guidance for calculating installation labor-hours on specific types of hardware. This figure should be consulted as a guide only, and a contractor's own historical data should always be used first.

Description	Labor-hours	Unit
Astragals, 1/8" × 3"	.089	LF
Spring hinged security seal with cam	.107	LF
Two-piece overlapping	.133	LF
Automatic openers, swing doors, single	20.000	EA
Single operating, pair	32.000	Pair
Sliding doors, 3" wide including track and		
hanger, single	26.667	Opng.
Biparting	40.000	Opng.
Handicap opener button, operating	5.333	Pair
Bolts, flush, standard, concealed	1.143	EA
Bumper Plates, 1½" × ¾" U Channel	.200	LF
Door closer, adjustble backcheck, 3-Way mount	1.333	EA
Doorstops	.250	EA
Kickplate	.533	EA
Lockset, nonkeyed	.667	EA
Keyed	.800	EA
Dead locks, heavy-duty	.889	EA
Entrance locks, deadbolt	1.000	EA
Mortise lockset, nonkeyed	.889	EA
Keyed	1,000	EA
Panic device for rim locks		
Single door, exit only	1.333	EA
Outside key and pull	1.600	EA
Bar and vertical rod, exit only	1.600	EA
Outside key and pull	2.000	EA
Push-pull plate	.667	EA
Weather stripping, window double-hung	1.111	Opng.
Doors, wood frame, 3' × 7'	1.053	Opng.
6' × 7'	1.143	Opng.
Metal frame, 3' × 7'	2.667	Opng.
6' × 7'	3.200	Opng.

Figure 13.12 Installation Time in Labor-Hours for Finish Hardware

Be sure to note the labor-hours for retrofit situations on remodeling jobs, where existing hardware is removed and replaced, on a door-by-door basis. Aligning and adjusting hardware on this type of application can be painstakingly slow. Verify that doors with closers, or specialty features are adjusted to meet the requirements of the Americans with Disabilities Act (ADA). Adjustment times should be added to the labor cost.

GLASS AND GLAZING

All glass and mirrors that are not specifically part of a storefront, entry unit, or premanufactured windows make up the category of glass and glazing. Most glass used in construction today is called *float* glass, which refers to its manufacturing process. Float glass undergoes *annealing,* which is a heating treatment that increases the toughness of the finished product. Specific examples of glass and glazing include glass for vision panels within doors, fire-rated glass for rated doors, glass for sidelights and transoms, glass for interior borrowed lights, and nonframed mirrors. Glass materials under this section can be classified into one of the following types for estimating purposes:

- **Insulated glass.** A composite of multiple pieces (panes) of glass with a sealed or gas-filled space between them, which helps reduce the loss of heat or cooling through the unit. Units are custom-made to specific sizes and thicknesses for individual applications. A variety of different types of glass can be used for the individual panes, which can affect the price.
- **Tempered glass.** Heat-treated annealed glass that has a high strength. When tempered glass breaks, it ruptures into cube-shaped fragments. This fragmenting helps reduce the incidence of serious injury that would normally occur with pointed fragments.
- **Laminated glass.** A "sandwich" unit of multiple panes of glass with a high-strength, transparent plastic film between each layer. It is available in various thicknesses and is frequently referred to as *safety glass*. Safety glass will crack under sufficient impact but will remain intact.
- **Wired glass.** Contains wire within the thickness of the glass sheet and is used for protective applications, such as fire door vision panels. Its purpose is to maintain the shape and integrity of the glass in the event of a fire. The glass will crack from the heat but maintain its shape due to the wire.
- **Coated glass.** Float glass that has a thin reflective metallic coating. Some coatings reduce glare, some reduce heat loss, and others are heat-absorbing. The majority of coated glass is used for energy conservation applications.
- **Mirrors.** Sheets of float glass, usually $\frac{1}{8}$" or $\frac{1}{4}$" in thickness, to which a reflective coating has been applied. The coating is then sealed to protect against moisture and handling damage. Mirrors that have been cut from larger pieces require dressing of the edges. The most common form of edgework is a plain polished edge, although beveled edges are sometimes specified for a particular application. Mirrors can be installed using adhesives applied to the back, or they can be supported by special tracks and hardware. Trim pieces may also be required to cover vertical and horizontal seams or inside and outside corners.
- **Plastic glazing.** Plastic glazing materials are either polycarbonate or acrylic plastic. Both materials are durable, shatterproof, and crack-resistant thermoplastics. They are lighter and resist impact better. Both can be coated with abrasion resistant chemicals.

All types of glass, plastics, and mirrors are shown on the exterior and interior elevations of architectural drawings for wall applications and on reflected ceiling plans for ceiling applications. Plan view architectural drawings should be cross-referenced to verify quantities, sizes, and location. The specifications should be reviewed for the type, thickness, and method of application, as well as for the associated trim pieces or setting materials needed. Consult the door schedules for glass in vision panels and borrowed lights.

Taking-off and Pricing

All glass, plastics, and mirror materials are taken off and priced by the SF and should be separated in the takeoff according to size, type, color, thickness, edge treatment, and method of installation for accurate pricing. The estimator may elect to take off the quantity of edge treatment for mirrors separately, by measuring the perimeter of the individual pieces and listing treatment by the LF. Adhesives for mirrors are taken off by the SF and extended to the gallon or manufacturer's typical sales unit for pricing. SF quantities are determined by the individual product's coverage. Setting blocks are calculated by the perimeter of the individual glass size.

Cutting special shapes or boring holes in glass or mirrors for custom applications can be expensive and should be noted in the takeoff for accurate pricing. Some types of glass can be drilled or cut only in the factory.

Labor

Most glass and mirrors are installed by *glaziers*. Glaziers constitute the trade that works with glass and mirrors. Most glazing is done by a single individual or a multiperson crew when the glazed pieces are too large for an individual to handle. Special equipment may also be required to handle the pieces. Labor is calculated by the daily output of the crew for the particular task. As with most labor costs, crew costs can be calculated by summing the individual hours of each glazier and the divide by the daily output of the task to arrive at a cost per unit. The estimator should allow time for set up and distribution of the product, as well as any scaffolding or lifts needed for work above grade.

LOUVERS AND VENTS

The last section of Division 8 is called Louvers and Vents. It includes all products, fixed and operable, that allow the flow of air between the exterior and interior through the building envelope. *Louvers* are defined as wall openings with horizontal slats that admit light and air. They can be fixed-blade or adjustable, providing the ability to regulate the volume of air being admitted. Operation can be controlled manually or electrically. Operation can be automatic, as in the case of fresh-air intake louvers for generators, or the louvers can be opened at a specific temperature.

Louvers can be made of wood, vinyl, plastic, or metal and are typically located in vertical surfaces such as exterior walls. Louvers can also be found in doors to allow air to be exchanged between interior spaces.

Vents also allow the passage of air through the exterior envelope but are most often found in ceilings or soffits. In contrast to louvers, there are no blades to prevent the passage of water. Vents can also be used to allow water pressure to equalize on both sides of a wall, as in the case of a flood vent.

Taking-off and Pricing

Quantities of both louvers and vents are taken off and listed according to species, size (width by height), wall thickness, operation, composition, and location in the structure. Louvers can be listed by the square foot of opening (SF) or free air area. Vents can be listed by each (EA) as in the case of button vents, or by the linear foot (LF) as in soffit vents.

The estimator should review the plans carefully for quantity and location. Specifications should be reviewed to determine if it is part of another system, as in the case of the aforementioned HVAC example, or a stand-alone application.

Labor

Louvers and vents can be installed by a variety of trades, most often depending on the surface in which they are to be placed. For example, small vents and louvers in masonry walls are typically installed by bricklayers. Larger louvers that are operable in masonry walls may be installed by the HVAC contractor as part of a fresh-air system. Gable-end louvers in wood frame construction are installed by a carpenter during the siding process.

Installation is typically performed by an individual for small units or a multiperson crew for larger units. Productivity is measured by units-per-day installed by a specific crew. Labor costs per unit can be assigned as labor-hours or as a cost per square foot, SF.

SUMMARY

Division 8—Openings encompasses doors, frames, hardware, windows, skylights, and louvers. There is a wide range of costs for products in Division 8. For most of the products in this division, a full understanding of the product and its accessories is essential to accurately price both the material components and the installation labor. The plans should illustrate the work of Division 8 in multiple views so that the estimator can quantify and then confirm those quantities using separate views.

Door and window schedules are used to determine sizes of the item as well as special features. The work of this division can often represent a large amount of the total construction estimate. Many windows have U-factor and SHGC numbers as part of their specification. This can impact cost significantly. Specialty items such as storefront entrances and window wall units often require some assembly once on site; this may also be true of individual windows. Installation may require lifts or scaffolding for access to work above grade or the windows may be installed from the inside. The means and methods of installation should be considered carefully.

Division 8 work at exterior walls is closely related to the work of Division 7—Thermal and Moisture Protection. The caulking and sealing of windows and door frames to adjacent surfaces can often be specified as part of the work of installing the window or door frame. Maintaining the integrity of the air barrier by sealing the window perimeter may also be included. The estimator is urged to review the specifications carefully for scope responsibility.

For large skylights or roof windows, a piece of equipment may be required to set the product on the roof.

Much of the work of CSI Division 8—Openings involves specialty suppliers or subcontractors. Whenever possible, secure pricing from potential suppliers or subcontractors to ensure that it is accurate and timely.

QUESTIONS/ PROBLEMS

1. Schedules are frequently used in this division to identify doors, windows, or hardware. Select one (1) of these categories and explain what a reader would expect to see in that schedule.
2. Explain the difference between a knockdown metal door frame and a welded door frame from an estimating perspective.
3. Identify the characteristics that an estimator should consider when pricing a wood door.
4. How is the labor estimated for a wood, metal, or plastic window?
5. Describe the estimating process for a specific hardware set.
6. Define the unit of measure called the *united inch*. What product is it typically applied to?
7. How do U-factors and solar heat gain coefficients (SHGCs) impact the cost of a window unit?
8. Using the information below and the hourly rate of $88.50 for a carpenter, what is the cost of installing Hardware Set #2?

Labor-hours to install Hardware Set #2	
1½ pairs 4½" Stanley HD Series Hinges (1½ PR)	0.60 hours
Hang and swing door	0.25 hours
Corbin 123L-5500 Series Mortise Lockset (1 EA)	1.10 hours
LCN 4010 CUSH Series Closer (1 EA)	1.80 hours
Ives 436 B Floor Stop (1 EA)	0.45 hours
Ives #20 Silencers (3 EA)	0.15 hours
Adjust hardware and door	0.40 hours
Subtotal for labor-hours for HS #2	0.00 hours

9. Explain why it is preferable for a door to be premachined and finished by the manufacturer.
10. Why is it important to secure subcontractor pricing for materials and labor for specialty doors?

14 | Finishes

CSI MasterFormat®—Division 9 includes a variety of interior finish work, such as drywall and metal stud partitions, plaster, tile, acoustical ceilings, wood and resilient flooring, carpeting, and painting. Although this work does not typically represent a major segment of the total project cost, it does account for a large amount of time in the schedule. Finishes have the largest impact on the aesthetic value of the structure, and time is often required to produce the necessary quality, which is always expected to be first class.

Finish work is detailed on the architectural drawings. Quantities are derived from floor plans, reflected ceiling plans, interior elevations, and the related details that support the design. The details and sections often provide the additional information for accurately defining the work, as well as critical dimensions for changes in finishes. Architectural drawings allow the estimator to count, measure, or calculate quantities of finishes to be priced.

Room Finish Schedules, also found in the architectural set of drawings, provide more information about the finishes of floors, walls, and ceilings. They display information in a table format for easy reference. Each room is assigned a number, and the specific finishes for each surface (walls, floors, base, and ceilings) are listed. Some finish schedules list each wall separately, referring to it by its compass location (north, south, east, or west). This allows the architect to call out different finishes on each wall if desired. Finish schedules are most often found on commercial projects with a large number of rooms.

While the drawings are essential for determining finish quantities, the specifications are necessary for determining the quality and individual

Estimating Building Costs for the Residential and Light Commercial Construction Professional, Third Edition. Wayne J. Del Pico.
© 2023 John Wiley & Sons, Inc. Published 2023 by John Wiley & Sons, Inc.

characteristics of each specified product. Specifications also define the quality of workmanship and acceptable standards of installation. Special installation methods or techniques are also outlined.

Due to the large variety of finish products available, it is critical that the estimator become familiar with the product specified. Simple changes in a model number or color of a finish material can have a significant cost impact on both material and labor. The internet can be a valuable source of information, where the estimator can learn about new products and installation techniques from a manufacturer's website or video.

For estimating purposes, Division 9 work can be classified into one of the following general categories:

- Plaster systems
- Gypsum wallboard systems
- Metal stud framing and furring
- Tile: floor and wall
- Ceiling systems
- Flooring
- Acoustical treatment
- Painting and wallcoverings

PLASTER SYSTEMS

Once the main choice for wall and ceiling surfaces, plaster has seen a steady decline in popularity since the introduction of its competitor, gypsum drywall. Plaster still has an appeal with residential contractors, however, due to its more durable finish. The plaster system, commonly called *lath and plaster*, consists of a rigid substrate, *lath*, and a coating of plaster. Lath has evolved from wood strips to perforated metal sheets, then to SHEETROCK®, and finally to its current form, a gypsum base, called *blueboard*. Blueboard is fastened to the wood or metal framing as a base for the application of plaster. Most plaster in residential/light-commercial construction is applied using a thin (⅛") coat, called *skim coat plaster*.

The gypsum base is a gypsum-core board faced with a multilayer, specially treated paper that allows the skim coat or *veneer* plaster to bond to the gypsum base. Veneer plasters are designed to be applied in one- or two-coat systems for a strong, abrasion-resistant surface. Common blueboard sheets are 4' in width by 8', 10', or 12' in length. Other sizes are also available on special order. Standard thicknesses are ⅜", ½", and ⅝". The gypsum base is applied by screwing or nailing it to the framing. The plaster base is also manufactured for fire-rated applications and with a vapor backing to retard the transmission of moisture. Veneer plaster is sold dry in 50- or 80-lb. bags. When mixed with water to a pastelike consistency, it can be troweled onto the surface of the base. Coverage is based on the individual product, the specified thickness, and the type of finish.

An alternate type of plaster system, called a *conventional three-coat* system, requires the first two coats to be mixed with sand, perlite, or vermiculite aggregates in varying proportions. Three separate coats are applied, called *scratch, brown,* and *finish coats.* Each coat is allowed to dry prior to the application of the next. Conventional plaster systems are extremely labor intensive and therefore costly, and, as a result, they have suffered a reduction in popularity.

In addition to these two methods of plastering, there are unique or specialty plastering techniques that are replications of period plastering. One such technique is called *Venetian* plastering. It is intended to give the surface a finish that replicates those of antiquity. Another type is cement plastering, a hybrid of plastering and stucco.

Taking-off and Pricing

Materials for both types of lath and plaster work are taken off by the square foot (SF) of surface area and converted to sales units for pricing. For accurate material pricing, the individual components, such as base, plaster, trims, and joint reinforcing tape, can be priced separately, then combined to arrive at the price of the system.

Skim-coat plaster systems are priced by the square foot (SF). Conventional plaster systems are priced by the square yard (SY). The takeoff should separate wall and ceiling applications due to labor pricing considerations. Openings less than 4 SF in area are not deducted, while openings without plaster *reveals,* or returns, are deducted in full. Reveals at openings should be calculated at 1.5 times their actual size. Special configurations, such as arches, pilasters, columns, decorative features, and special patterns, should be separated in the takeoff and priced on an individual basis. Allow one fastener for each SF of plaster base. Multiple layers follow the same procedure. The size and type of each fastener is determined by the thickness of the lath and the corresponding requirement in the specifications.

Specialty plastering techniques are also taken off and priced as described here. Pricing for these particular techniques varies with current trends and the abilities of local artisans. Some can include additional products such as silica for sparkles, or simply a different stroke of the trowel.

Labor

Regardless of the system, plastering is a crew task, except for very small patches. Multiple trades comprise the crew. Carpenters hang the gypsum base, or metal lath. The tenders (laborers that tend to plasterers) mix the plaster, erect and dismantle scaffolding, distribute materials, stock the plasterers with mix, and clean up. Plasterers cover the base with each coat to the desired finished texture. Plastering, similar to masonry and finish carpentry, is considered a *craft,* and only high-quality or first-class workmanship is acceptable.

GYPSUM WALLBOARD SYSTEMS

Gypsum wallboard, more commonly known as *drywall,* is a manufactured product of powdered gypsum mixed with water and sandwiched between two layers of treated paper. It has become the system of choice in recent years. Because of its gypsum and mineral core, drywall, like plaster base, does not support combustion. It is manufactured in the same thicknesses and sheet sizes as plaster base. It can be used for a number of applications, including for fire ratings, moisture resistance, mold resistance, and foil-backed for retarding vapor transmission. Other special products, such as 1"-thick shaft-wall liners for use in cavity-wall applications (see "Metal Stud Framing and Furring" later in this chapter) and gypsum sheathing for exterior curtain walls, also fall within the gypsum drywall category.

Drywall is installed with screws on metal or wood framing in the same manner as plaster base. Taping and finishing conceals the joints and results in a smooth surface ready for paint or other finishes. Drywall ceiling installations can be covered with a textured finish to achieve an acoustical finish.

Various types of drywall have different color paper facing for identification purposes.

Taking-off and Pricing

Drywall is taken off and priced by the square foot (SF). Individual components can be converted to typical sales units, such as sheets, then priced separately in order to arrive at a SF price. The procedure for taking off and pricing drywall is the same as for plaster base. Multiple layers of drywall for special fire-rated assemblies should be separated in the takeoff, since the level of finish differs at each layer. Taping and finishing are also listed in the takeoff by the SF of area to be finished. The level of finish required should also be noted (for example, fire-coat only, or full tape to Level 4 or Level 5 finish) as this can significantly impact the final price.

Textured finishes for ceilings are listed in the takeoff separately by the SF. Different application methods, such as spray or handwork, should also be separated. Height above the floor is also a factor that should be noted for accurate estimating.

Consult the architectural drawings, plans, elevations sections, and details. In addition, small cross-sectional views of the different walls and partitions used in the project, called *wall types,* are instrumental in identifying the components required. They illustrate the different components of the particular assembly. Like lath and plaster work, drywall should be priced separately based on the location and application. Costs for items such as scaffolding to access the work or special equipment for spraying must also be included in the estimate. Deductions in openings follow the same rules as with lath and plaster.

Consult the specification for the level of taping and finish required. Finishing of drywall can range from Level 0 (unfinished) to Level 5 (highest-quality full surface finish). Each level of finish represents a very different requirement in labor-hours and, therefore, cost.

Similar to blueboard, distribution of drywall sheets is cumbersome and labor intensive. The estimator should consider how the board will be loaded into the project. This may require the use of a crane or the board supplier's boom truck. It may also require hand carrying for small quantities. Regardless, there is likely to be a distribution cost for labor.

Labor

Hanging drywall is a crew task and is most often done by carpenters. It is very similar to hanging blueboard, with the exception of orientation of the board. Production is based on the crew's combined output. Taping and finishing are done by *tapers*. Taping, however, is an individual task measured by the production of an individual. Consider additional costs for scaffolding to finish drywall located above reach from the floor and the costs associated with handling and distribution of the drywall. Labor costs can be calculated as a cost per SF for both hanging the board and taping the surface.

It is common practice to carry an allowance for drywall touch-up after the surfaces have been primed. Figure 14.1 lists labor-hours for installation of various wallboard systems on wood partitions.

Description	Labor-Hours	Unit
Metal Lath Diamond Expanded		
2.5 lb. per SY	.094	SY
3.4 lb. per SY	.100	SY
Gypsum Lath		
3/8" Thick	.094	SY
1/2" Thick	.100	SY
Gypsum Plaster		
2 Coats	.381	SY
3 Coats	.460	SY
Perlite or Vermiculite Plaster		
2 Coats	.435	SY
3 Coats	.541	SY
Wood Fiber Plaster		
2 Coats	.556	SY
3 Coats	.702	SY
Drywall Gypsum Plasterboard Including Taping		
3/8" Thick	.015	SF
1/2" or 5/8" Thick	.017	SF
For Thin Coat Plaster Instead of Taping Add	.013	SF
Prefinished Vinyl-Faced Drywall	.015	SF
Sound-deadening Board	.009	SF
Walls in Place		
2" x 4" Studs with 5/8"		
Gypsum Drywall Both Sides Taped	.053	SF
2" x 4" Studs with 2 Layers Gypsum Drywall		
Both Sides Taped	.078	SF

Figure 14.1 Installation Time in Labor-Hours for Wallboard Systems

GYPSUM SHEATHING AND UNDERLAYMENTS

In addition to regular gypsum board applied to the interior of a framed wall, a similar product called *gypsum sheathing* is used on the exterior of the metal stud assembly. Gypsum sheathing is applied to the exterior of the metal stud frame to serve as a sheathing material in much the same way that plywood is applied to wood studs. It is manufactured with a water-resistant core and water-repellant facing paper or reinforced mat. It is noncombustible and is used as a substrate or underlayment for exterior siding materials. Gypsum sheathing is attached with screws similar to drywall.

Other underlayments include a tile backer board for walls and floors. There are a variety of products for this application, but all are generally designed to keep moisture out of the cavity wall. Some are a blend of concrete and gypsum with a reinforcing mat; others are lighter in weight and constructed of composite materials.

Taking-off and Pricing

All underlayment products follow the same takeoff procedure as drywall. Final square footage is converted to whole sheets, where the size is dependent on the product. The estimator is directed to read the specifications carefully for any required joint treatments such as taping with a mastic-based tape. This can be time consuming and add substantially to the cost of labor.

The estimator should include associated costs such as scaffolding or lifts for exterior applications on walls, ceilings, and soffits.

Labor

Gypsum sheathing can be installed by an individual carpenter or by a crew, depending on the application. Production rate is based on the crew's combined output or on that of the individual. Typically, gypsum sheathing applied to exterior walls is a crew task, while underlayment used as a backer for ceramic floor and wall tile is installed by an individual. Labor costs can be calculated as a cost per SF. Stocking gypsum sheathing or underlayments follows the same procedure and considerations as with drywall.

METAL STUD FRAMING AND FURRING

A popular choice for partitions and ceilings on commercial projects is metal stud framing and furring because it does not support combustion and is lighter to handle. The materials covered in this chapter are limited to non-load-bearing applications. For load-bearing applications, consult Chapter 10, "Metals."

Non-load-bearing metal framing and furring are manufactured from cold-rolled galvanized metal and are available in a variety of sizes, thicknesses (gauge), and shapes. The most common sizes of metal stud and track are $1^5/_8$", $2^1/_2$", $3^5/_8$", 4", and 6" with metal thicknesses of 20 and 25 gauge. Standard lengths range from 8' to 16' in 2' increments. Longer lengths of the larger sizes can be special-ordered. Metal stud framing components are fastened with framing screws, while furring can be attached with screws or power-actuated fasteners.

Metal stud partitions are framed in much the same way as wood partitions. Channel-shaped runners, called *tracks,* are positioned at the top and bottom of the partition and are similar to wood plates in wood framing. The tracks are anchored to the floor and overhead structure with screws or, in the case of concrete or steel, a power-actuated fastener. Metal studs are installed perpendicular to the track by fastening the flanges of the stud and track together with self-tapping screws. The studs are located within the track at the on center spacing. Figure 14.2 illustrates a typical non-load-bearing metal stud partition assembly.

Special steel framing components for use in cavity shaft and fire-rated construction are manufactured in 20-, 22-, 24-, and 25-gauge thicknesses and are galvanized to resist corrosion. Their components are similar to regular metal stud framing but include support for 1"-thick shaft wall liner panels. Components are available in 2½", 4", and 6" widths and lengths from 8' to 28', depending on the shape and width specified. Some of the more common shapes encountered in shaft wall construction are C-H studs installed between abutting liner panels, E studs used to cap panels at vertical intersections of walls, and J and C runners used as tracks at the top and bottom of cavity wall construction. Cavity wall framing materials are all non-load-bearing. The estimator should become familiar with the specified system prior to taking off the components. Figure 14.3 shows the different types of studs and track used in cavity wall construction.

In addition to metal framing, shaft wall construction has specialty board called *shaft wall liner.* Shaft wall construction uses a gypsum liner panel that is friction-fit within the channels of the C-H, J, and E framing components.

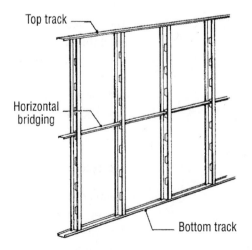

Non−Load-Bearing Steel Studs

Figure 14.2 Non-Load-Bearing Steel Studs

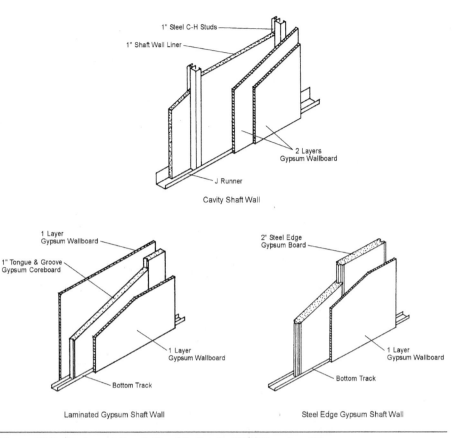

Figure 14.3 Studs and Track Used in Cavity Wall Construction

Panels are 1" thick and 24" in width and are available in reinforced and mold-resistant types. Shaft wall liner does not require taping and finishing as with regular gypsum board, however the exposed drywall may.

Other specialty members, such as cold-rolled channels (CRCs) for use as main runners in ceiling suspension systems, are manufactured in ¾", 1½", and 2" sizes. The material is 16-gauge cold-rolled galvanized steel and is available in 16' and 20' lengths. It is typically suspended by a wire tied to the structure above. CRCs are spaced at a maximum of 4'-0" on center longitudinally, with a hat channel fastened transversely to the CRC at 12", 16", or 24" on center by wire or special clips. Figure 14.4 shows different types of metal-framed ceiling systems.

Taking-off and Pricing

Metal studs and track are taken off by the linear foot (LF). The track and studs are separated by gauge, type, size, and length of studs. Quantities are determined by dividing the length of the partition (in feet) by the stud spacing (also in feet) and adding additional studs for the end and for intersecting partitions. Doubled-up metal studs at the door and window jambs, heads, and sills are also needed. Although there are no structural headers required at

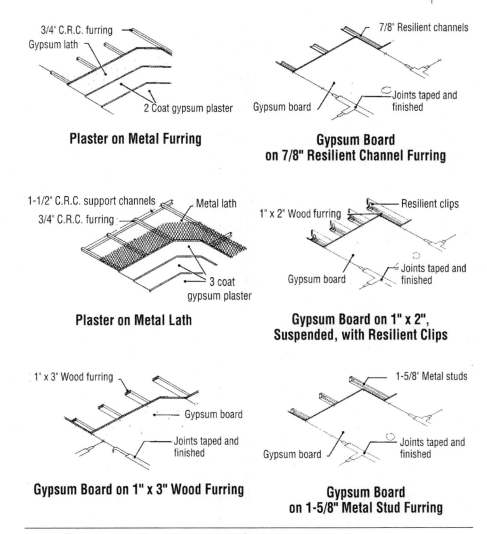

3/4" C.R.C. furring
Gypsum lath
2 Coat gypsum plaster

Plaster on Metal Furring

7/8" Resilient channels
Gypsum board
Joints taped and finished

**Gypsum Board
on 7/8" Resilient Channel Furring**

1-1/2" C.R.C. support channels
Metal lath
3/4" C.R.C. furring
3 coat gypsum plaster

Plaster on Metal Lath

Resilient clips
1" x 2" Wood furring
Gypsum board
Joints taped and finished

**Gypsum Board on 1" x 2",
Suspended, with Resilient Clips**

1" x 3" Wood furring
Gypsum board
Joints taped and finished

Gypsum Board on 1" x 3" Wood Furring

1-5/8" Metal studs
Gypsum board
Joints taped and finished

**Gypsum Board
on 1-5/8" Metal Stud Furring**

Figure 14.4 Types of Metal-Framed Ceiling Systems

non-load-bearing partitions, a piece of track is required to terminate the drywall at the head of the door and window and at the windowsill.

Furring channels are taken off by the LF and separated according to type, size, gauge, and application (attached to wood, masonry, concrete, etc.).

CRCs are taken off by the LF according to size and location. To determine the LF quantity, divide the length of the ceiling (in feet) by the on-center spacing (in feet) and add one (1) piece to start. Multiply the result by the length of the individual pieces.

Cavity wall framing members are taken off by the LF and separated according to size, type, and length of the individual piece. The quantity of J and C runners is determined by measuring the length of the partition at the top and bottom. The number of C-H studs can be found by dividing the length of the partition (in feet) by the spacing (in feet) and adding additional studs for the ends or for corners. Additional quantities of track should also be considered for openings in cavity walls for doors, as with standard metal partitions previously noted.

Shaft wall liner panels are taken off by the SF and converted to the nearest full panel, along with an allowance for waste. Note that these panels, as with most drywall products, are heavy and require labor for distribution.

The quantity of screws for framing is typically calculated at four screws per stud, in the absence of a detailed specification for fastening. Power-actuated fasteners are typically expensive and are calculated by the LF of track, the top and bottom of the partition, and the total length of furring channels by the on-center spacing as dictated by the specifications. A general rule for fastening track is one fastener per 2 LF of track.

Labor

Metal framing and furring are installed by multiple carpenters. Typical crews consist of pairs of carpenters working together. Production is measured by the crew's daily output. A laborer to handle and distribute materials may also be part of the crew when the work requires it. Production is measured by the LF of framed partition. This includes installing the track (top and bottom), studs, and special framing members at door or window openings. Productivity can also be based on the SF of framed partition.

Additional costs for equipment, such as staging or lifts, including erection, dismantling, and delivery charges, must be included for accurate pricing. Framing partitions around ductwork, piping, conduits, and the like will substantially reduce productivity, as will numerous window or door openings. Take this into account when calculating labor-hours.

Similar to drywall, metal stud framing components may need to be stocked above grade level. Figure 14.5 provides a guideline for determining labor-hours for framing and furring.

The estimator should also consider the labor cost for laying out partitions. Typically, a small crew of one or two persons arrives on site in advance of the production crew to do the layout of the partitions. This allows the production crew to focus on framing metal stud partitions. Labor to lay out the partitions is wholly dependent on the size and complexity of the project. The layout crew arrives in advance to transfer dimensions from the plans to the slab (or floor deck). Lines are chalked and clear paint is sprayed over the lines to keep them visible. The layout crew's goal is to stay just far enough ahead of the framing so that they do not catch up. The same process can be applied to overhead metal-framed drywall soffits and coffers.

TILE

Tile, manufactured from clay, porcelain, or stone, is available in an ever-changing variety of sizes, shapes, colors, textures, patterns, finishes, and thicknesses. Tile provides a hard, durable, and virtually maintenance-free

Description	Labor-hours	Unit
Non-load-bearing stud, galv. 25 gauge		
1⅝" wide		
16" on center	.019	SF
24" on center	.016	SF
2½" wide		
16" on center	.020	SF
24" on center	.016	SF
20 gauge, 1⅝" wide		
16" on center	.018	SF
24" on center	.016	SF
2½" wide		
16" on center	.019	SF
24" on center	.016	SF
25 gauge or 20 gauge, 3⅝" or 4" wide		
16" on center	.020	SF
24" on center	.017	SF
6" wide		
16" on center	.022	SF
24" on center	.018	SF
Load-bearing stud, galv. or painted, 18 gauge		
2½" wide		
16" on center	.019	SF
24" on center	.016	SF
3⅝" or 4" wide		
16" on center	.020	SF
24" on center	.017	SF
6" wide		
16" on center	.022	SF
24" on center	.018	SF
Load-bearing stud, galv. or painted		
16 gauge, 2½" wide		
16" on center	.020	SF
24" on center	.017	SF
3⅝" or 4" wide		
16" on center	.021	SF
24" on center	.018	SF
6" wide		
16" on center	.024	SF
24" on center	.019	SF

Figure 14.5 Installation Time in Labor-Hours for Steel Stud Partition Systems

surface for both interior and exterior use. It is considered an excellent choice where wear and tear and longevity are concerned. In addition to standard tile pieces, often referred to as *field tile,* special shapes and trim pieces are also available. Some of the more common wall trims are bullnose, cove base, inside and outside corner pieces, custom transition shapes, and accessories such as toilet paper holders, towel bars, and soap dishes.

The characteristics of the tile material itself, installation method, and setting materials influence the overall cost of the work. Tile can be set with water-resistant premixed adhesives – the most common method of tile installation in residential applications. Another common method is using dry-set Portland cement mortar, commonly referred to as *thin-set mortar.* Thin-set mortar is available in bags of varying sizes and is mixed with water to obtain a toothpaste-like consistency. Specialty types of thin-set mortars for flexibility, exterior applications, and epoxy-based thin-sets are also available and in wide use.

Tile comes in individual pieces or back-mounted sheets for faster installation. It can be set with small spaces between the pieces for a gridlike appearance. The size of the spacing, called the *grout joint,* depends on the size, type, location, and specified design of the tile. Some ceramic tile is manufactured with preset spacing, as with back-mounted sheets; others are spaced by the tile setter. Once the tile has set, the joints between the tile are grouted to provide a continuous surface. The excess is cleaned off with a sponge or by other means to leave the surface clean.

Review the architectural drawings, plans, elevations, and reflected ceiling drawings for the location of tile work. Sections and details may add the necessary perspective to accurately price the work. Sections can be used to confirm wall tile heights shown on elevations, and details can provide information on the substrate or trim pieces required to complete the installation. Room Finish Schedules should also be consulted as a check and balance against other architectural drawings for locations of tile work.

A comprehensive review of the technical specifications is also essential for the specific proprietary information on the products, setting method, and grout. Without this information, accurate pricing of the tile work will be compromised.

As the varieties of tile are numerous, for the purpose of this text and estimating tile work in general, discussion is limited to two classifications: manufactured tile (ceramic and quarry tile) and natural tile (marble and granite tile).

Manufactured Tile

Ceramic and quarry tiles are manufactured from clay, porcelain, or similar materials and baked in a kiln to a permanent hardness. Ceramic tile can be glazed or unglazed. Glazed tile can have an impervious glassy facial finish available in a multitude of colors, shapes, and sizes. Special trim pieces, such as bull-nose tile, are used as transition pieces to terminate the tile work at adjacent surfaces. Other special trims include pieces that are used at the intersection of wall and floor tile, called *coves* or *cove base,* are available with inside and outside corners. Ceramic accessories, such as soap dishes, toilet paper holders, and towel bars, are also manufactured to match the field tile.

Ceramic mosaic tiles are small, with a face area of less than 6 square inches. They are mounted on pre-spaced backing sheets for ease of installation and are manufactured glazed or unglazed in various colors, patterns, or designs. Ceramic mosaic tile is used primarily for floor or decorative wall applications. Figure 14.6 shows the typical use and location of field and trim ceramic tile pieces.

Figure 14.6 Uses of Ceramic Tile

Another extremely hard, durable form of tile, called *quarry tile,* is manufactured for use primarily as a floor tile. It is long wearing and resembles brick pavers. It is available with abrasive chips embedded in the wearing surface for nonskid applications. Trim pieces similar to ceramic wall tile trims are also available to match.

Taking-off and Pricing

Tile and accessories are taken off separately, using the following guidelines:

- **Field tile for both ceramic and quarry tile:** Taken off and priced by the SF area. Quantities should be separated according to finish (glazed or unglazed) and location (floor or wall).
- **Accent tile:** Listed separately by the SF or by the piece, EA.
- **Trim pieces:** By the LF, converted to the number of pieces required.
- **Inside and outside corners:** By the individual piece (EA).

Deductions for door and window openings should be made in full. Returns, or reveals, at doors and window openings should be included. The quantity of tile in a SF area can be converted to boxes of tile based on the individual sales unit. Tile-setting materials, such as thin-set or adhesive, are calculated by the SF of tile to be set. These quantities are then converted to typical sales units based on the manufacturer's coverage. Required bonding additives are determined by the manufacturer's formula of dry materials to liquid. This is then converted to the appropriate sales unit, most often gallons. Calculating grout follows the same procedure for setting materials. The SF area and size of the grout joint (width and depth) are required. Manufacturers typically provide tables for calculating the amount of grout required for specific joint sizes. Identify the grouting and cleaning phase separately in the estimate from the actual tile-setting process.

It is prudent to include a percentage for material waste based on the application and the anticipated number of cuts. A factor of 3% to 8% is within the typical range, although waste is calculated on a project-by-project basis.

Labor for Setting Tile

Ceramic and quarry tile setting is done by tile layers or setters, with assistance from laborers, sometimes referred to as *tile setters' tenders*. Productivity is measured by the crew's daily output. Crews can range from an individual tile setter in small applications, such as a residential bathroom, to several tile setters and helpers for commercial projects. The quality of workmanship is always estimated as first class due to the highly aesthetic nature of tile. This is true even for utilitarian applications. Special designs within the field of tile may reduce productivity, so additional layout time should also be allowed for in the estimate. Applying adhesive to the substrate and setting the tile on the wall or floor are considered a single process. The labor required to grout and clean the tile is also estimated as a separate task, often because it occurs after the tile has set.

The cost of cutting tile with a hand cutter must be included as part of the setting costs. Harder or more dense types of tiles, such as quarry tiles and those that require unusual cuts, may need an electric or gas water-cooled saw with a diamond blade. Projects with a large number of cuts may even require an individual setter or helper dedicated to making cuts. This cost, along with the cost of the saw and blade replacement resulting from wear and tear, must be included in the estimate.

Delays, or time lapses, between setting and grouting tile, allow for the materials to cure and harden. Be sure to account for the costs in downtime for tile setters. The cost of cleaning agents should also be included.

Because of the weight and bulk of tile, shipping and handling costs can be significant. Secure the delivery price whenever possible, especially if the tile originates outside the continental United States.

Natural Tile

This type of tile has been cut and polished from natural stone such as marble and granite. The popularity of natural tile has risen exponentially as a result of the increased varieties and availability. Marble and granite tile, cut from larger blocks into uniform sizes, commonly come in 12" × 12" (nominal) squares in ½" thicknesses. The face or exposed surface is factory-polished to a high gloss. Other types of natural tile, such as slate for floor and wall applications, are less uniform in size and thickness and suggest a more rustic appearance. Slate tile is packaged by square footage. The setting materials used for ceramic and quarry tiles can also be used for stone tiles, although specialty products for certain types of marble tiles may be required. Trim pieces are not typically manufactured for stone tiles. Consequently, the external corners may require field-mitering. Grouting joints between natural tiles is accomplished in the same manner as in ceramic and quarry tile. Grouted surfaces can be cleaned with water and light detergents, in contrast with acid for ceramic and quarry tile work.

Taking-off and Pricing

Natural tiles are taken off by the SF, extended to the quantity of individual pieces required for material pricing. They should be separated according to application or location (wall or floor), size, type of stone, and setting method. Cutting or mitering can be labor intensive in large quantities and must be done with a wet saw. The estimator should include a percentage for material waste based upon the application and the anticipated number of cuts. A factor of 3% to 8% is within the typical range, although waste is calculated on a project-by-project basis.

Full mortar beds consisting of a 1" subbase are fairly rare due to the cost. The estimator must research the specifications carefully to determine what is required as a subbase for setting the tile on.

(Refer to the "Manufactured Tile" section for more information on cutting tile.)

Labor to Set Natural Tiles

As with manufactured tiles, natural tiles are installed by tile setters and helpers. Crews rarely consist of less than two people due to the nature of the work. Review the considerations in the ceramic tile labor section earlier in this chapter since most are applicable to natural tile. Other considerations include costs for sealing the surface and special epoxy setting materials for installing certain colored marble tiles. Sealing of natural surfaces can be expensive in both material and labor costs. Refer to the specifications and additional literature on the specific product. Installation productivity can be significantly reduced with certain types of epoxy setting adhesives, due to the small coverage per unit.

For all tile work, factor in waste due to tile damage, breakage, or mishandling. As with all natural products, tiles can have color and pattern variations in an individual tile that may render the tile unusable despite no apparent flaws. This should be considered when estimating waste.

Specifications for all tile work may require the contractor to provide additional materials to be turned over to the owner upon completion. This is commonly called *attic stock,* and it allows the owner to make repairs and replacement of tile outside the warranty. Attic stock is often specified as a percentage of each style, color, or type incorporated in the project. Common percentages range from 2% to 10%, depending on the size of the project and type of tile.

Figure 14.7 provides a guideline to determine labor-hours for setting various types of tiles.

Description	Labor-Hours	Unit
Flooring Cast Ceramic 4" x 8" x 3/4" Pressed	.160	SF
Hand-Molded	.168	SF
8" x 3/4" Hexagonal	.188	SF
Heavy-Duty Industrial Cement Mortar Bed	.200	SF
Ceramic Pavers 8" x 4"	.168	SF
Ceramic Tile Base, Using 1" x 1" Tiles, 4" High,		
Mud-Set	.195	LF
Thin-Set	.125	LF
Cove Base, 4-1/4" x 4-1/4" High, Mud-Set	.176	LF
Thin-Set	.125	LF
6" x 4-1/4" High, Mud-Set	.160	LF
Thin-Set	.117	LF
Sanitary Cove Base, 6" x 4-1/4" High, Mud-Set	.172	LF
Thin-Set	.129	LF
6" x 6" High, Mud-Set	.190	LF
Thin-Set	.137	LF
Bullnose Trim, 4-1/4" x 4-1/4", Mud-Set	.195	LF
Thin-Set	.125	LF
6" x 4-1/4" Bullnose Trim, Mud-Set	.190	LF
Thin-Set	.129	LF
Floors, Natural Clay, Random or Uniform,		
Thin-Set, Color Group 1	.087	SF
Color Group 2	.087	SF
Porcelain Type, 1 Color, Color Group 2,		
1" x 1"	.087	SF
2" x 2" or 2" x 1", Thin-Set	.084	SF
Conductive Tile, 1" Squares, Black	.147	SF
4" x 8" or 4" x 4", 3/8" Thick	.133	SF
Trim, Bullnose, Etc.	.080	LF
Specialty Tile, 3" x 6" x 1/2", Decorator Finish	.087	SF
Add For Epoxy Grout, 1/16" Joint,		
1" x 1" Tile	.020	SF
2" x 2" Tile	.020	SF
Pregrouted Sheets, Walls, 4-1/4", 6" x 4-1/4",		
and 8-1/2" x 4-1/4", SF Sheets,		
Silicone Grout	.067	SF
Floors, Unglazed, 2 SF Sheets		
Urethane Adhesive	.089	SF
Walls, Interior, Thin-Set, 4-1/4" x 4-1/4" Tile	.084	SF
6" x 4-1/4" Tile	.084	SF
8-1/2" x 4-1/4" Tile	.084	SF
6" x 6" Tile	.080	SF

Figure 14.7a Installation Time in Labor-Hours for Floor and Wall Tile Systems

WATERPROOF MEMBRANE FOR TILE

For many floor and wall tile applications, the specifications require that the substrate be waterproofed prior to the installation of the tile. This prevents water that is spilled from penetrating the substrate and damaging floors below or a cavity wall. Many of the products are self-curing liquid rubber polymers with or without a reinforcing sheet that create a waterproof barrier once they

Description	Labor-Hours	Unit
Decorated Wall Tile, 4-1/4"x 4-1/4"		
Minimum	.059	EA
Maximum	.089	EA
Exterior Walls, Frostproof, Mud-Set,		
4-1/4" x 4-1/4"	.157	SF
1-3/8" x 1-3/8"	.172	SF
Crystalline Glazed, 4-1/4" x 4-1/4", Mud-Set,		
Plain	.160	SF
4-1/4" x 4-1/4", Scored Tile	.160	SF
1-3/8" Squares	.172	SF
For Epoxy Grout, 1/16" Joints, 4-1/4" Tile,		
Add	.020	SF
For Tile Set in Dry Mortar, Add	.009	SF
For Tile Set in Portland Cement Mortar, Add	.055	SF
Regrout Tile 4-1/2" x 4-1/2", or Larger, Wall	.080	SF
Floor	.064	SF
Ceramic Tile Panels Insulated, Over 1000		
Square Feet,		
1-1/2" Thick	.073	SF
2-1/2" Thick	.073	SF
Glass Mosaics 3/4" Tile on 12" Sheets,		
Color Group 1 and 2 Minimum	.195	SF
Maximum (Latex Set)	.219	SF
Color Group 3	.219	SF
Color Group 4	.219	SF
Color Group 5	.219	SF
Color Group 6	.219	SF
Color Group 7	.219	SF
Color Group 8, Gold, Silvers and Specialties	.250	SF
Marble Thin-Gauge Tile, 12" x 6", 9/32", White		
Carara	.250	SF
Filled Travertine	.250	SF
Synthetic Tiles, 12" x 12" x 5/8",		
Thin-Set, Floors	.250	SF
On Walls	.291	SF
Metal Tile Cove Base, Standard Colors,		
4-1/4" Square	.053	LF
4-1/8" x 8-1/2"	.040	LF
Walls, Aluminum, 4-1/4" Square, Thin-Set,		
Plain	.100	SF
Epoxy Enameled	.107	SF
Leather on Aluminum, Colors	.123	SF
Stainless Steel	.107	SF
Suede on Aluminum	.123	SF
Plastic Tile Walls, 4-1/4" x 4-1/4",		
.050" Thick	.064	SF
.110" Thick	.067	SF

Figure 14.7b *(Continued)*

have cured. They require minimal curing (drying) times and often can be tiled over in the same day. It is recommended the estimator study the specifications carefully for not only the product, but any special testing of the membrane required prior to setting the tile.

Description	Labor-Hours	Unit
Quarry Tile Base, Cove or Sanitary, 2" or 5" High, Mud-Set		LF
1/2" Thick	.145	LF
Bullnose Trim, Red, Mud-Set, 6" x 6" x 1/2" Thick	.133	LF
4" x 4" x 1/2" Thick	.145	LF
4" x 8" x 1/2" Thick, Using 8" as Edge	.123	LF
Floors, Mud-Set, 1000 SF Lots, Red,		SF
4" x 4" x 1/2" Thick	.133	SF
6" x 6" x 1/2" Thick	.114	SF
4" x 8" x 1/2" Thick	.123	SF
Brown Tile, Imported, 6" x 6" x 7/8"	.133	SF
9" x 9" x 1-1/4"	.145	SF
For Thin-Set Mortar Application, Deduct	.023	SF
Stair Tread and Riser, 6" x 6" x 3/4", Plain	.320	SF
Abrasive	.340	SF
Wainscot, 6" x 6" x 1/2", Thin-Set, Red	.152	SF
Colors Other Than Green	.152	SF
Windowsill, 6" Wide, 3/4" Thick	.178	LF
Corners	.200	EA
Terra-Cotta Tile on Walls, Dry Set, 1/2" Thick		
Square, Hexagonal or Lattice Shapes, Unglazed	.059	SF
Glazed, Plain Colors	.062	SF
Intense Colors	.064	SF

Figure 14.7c *(Continued)*

Taking-off and Pricing

Takeoff is done by calculating the area to be waterproofed in SF and then converting this figure to the nearest sales unit of the product being applied. Coverage is based on the number of coats and the individual product being used. Material costs can vary substantially depending on the product. The estimator is directed to research the specified product carefully for accurate pricing. Do not forget to include base areas (vertical wall area adjacent to the floor) when calculating floor areas, as most floor applications require a minimum vertical return at the floor.

Labor for Waterproofing Membranes

Waterproofing membranes can be installed by a specialty trade called a *waterproofer* or by the tile installer, depending on the product. Productivity is measured by either an individual waterproofer or a waterproofing crew for larger areas. Unit costs for the labor portion of the installation is the total cost of the crew for the workday divided by that crew's daily output. Productivity will vary with the interruptions to the flat surface such as corners and jogs in the surface or penetrations of conduits or pipes through the surface.

ACOUSTICAL CEILING SYSTEMS

Designed to absorb and reduce sound transmission while still providing access above the ceiling, acoustical ceiling tiles are available in a variety of sizes, colors, patterns, and textures. For estimating purposes, most acoustical ceiling systems can be classified in one of three groups:

- Acoustical ceiling systems attached directly to ceiling substrates
- Suspended acoustical ceiling tiles with a concealed spline
- Acoustical ceiling tiles installed in an exposed suspension system

Acoustical ceiling tiles are manufactured in increments of 12" in each direction, with 12" × 12", 24" × 24", and 24" × 48" the most common. Other nonacoustic ceiling tiles can be installed within a suspension system, or *grid*. Decorative, thin metal panels, either embossed or plain, in a variety of metal finishes are available. Vinyl-coated gypsum panels are manufactured for ceilings that require washable surfaces, such as kitchens and food-processing areas.

Suspension systems consist of aluminum or steel main runners of light, intermediate, or heavy-duty construction with snap-in cross tees in 1' to 5' lengths. Main runners are hung from the supporting structure with tie wire. Small metal angles, called *wall angles,* are attached to the perimeter wall or vertical surface at the ceiling height to complete the grid at the vertical surface. Components of the suspension system are available in colored or metallic plating.

Review the architectural floor and reflected ceiling plans to determine the locations and quantities of components. Cross-referencing the room finish schedule can help confirm locations and provide additional information, such as the type of ceiling, height of ceiling above the finished floor, and any special fire-rated assemblies. Quantities are determined by measuring the length and width of the room. Additional information, including the "length of the hang" of suspension systems, can be obtained from building cross sections.

Consult the specifications for the actual products and their various characteristics for accurate materials pricing. Critical characteristics, such as color, STC (sound transmission classification) rating, surface finish, fire rating, and seismic requirements, are major price determinants. Note that some tiles are color treated only on the surface and others have color all through the tile. This will impact the material price. Also review the specifications for attic stock requirements.

Taking-off and Pricing

Acoustical ceiling systems can be taken off in one of two ways: by the SF area of the room or by the individual components of the system. The latter is recommended for more accurate pricing, as it accounts for the SF area of the actual acoustical ceiling tile or panel. Dimensions of the room should be rounded to the nearest 1'-0" for 12" tile and nearest 2'-0" for 2' × 2' or 2' × 4' tiles. This can then be converted to the number of pieces or units required. Suspension components, such as the main runners, cross tees, wall angles, and hanging wire, should be taken off by the LF and converted to the individual piece. Hold-down clips, if required, are calculated by the individual piece or by total area. All quantities should be separated according to size, color, type, manufacturer, model, texture, and finish. Some acoustical tiles are *directional,* where the pattern on the exposed face has a direction and cut pieces from one side may not be used on the opposite side. This can contribute significantly to waste. The description in the takeoff should provide as much information as possible for accurate pricing. Seismic applications require additional components for the installation. The estimator is urged to review specific seismic requirements for the application.

Labor to Install Acoustic Ceilings

Acoustical ceiling systems are installed by carpenters. Productivity can be measured by the output of an individual carpenter or by the combined efforts of a crew by completed SF area per day. Large, open areas should be distinguished in the takeoff from small, confined areas, as different productivities will result. Another factor that affects productivity is the support structure from which the ceiling will hang. Tying hanging wire to open-web bar joists is less labor intensive than anchors drilled or "shot" into the concrete deck above. The length of the ceiling hang is also a determining factor. The longer the hang, the slower the process. Other considerations include staging or lifts required to access the work. Be sure to include erecting, dismantling, delivery, and rental costs if required.

Fire-rated systems or those with hold-down clips can significantly reduce productivity, as do sloped or splayed ceilings and soffits. These features all require additional time to install. Special tiles that are recessed in the grid, called *reveal edge* tiles, are sometimes cut to fit a grid opening. The cut edge must be grooved to replicate the reveal edge. This process, called *kerfing* the tile, is done with a knife by hand on-site and may, with sufficient quantity, require a separate individual to maintain the progress of the project. Figure 14.8 provides guidelines for installation labor-hours of different types of acoustical ceiling systems.

Distribution and handling of acoustical ceiling materials can be significant. Labor should be included for stocking materials if the quantity is sufficient. Consider added labor for cleanup and disposal of debris, if required in the contract.

Estimating acoustical ceilings systems can be separated into two steps: hanging the grid and installing the tile. Both are calculated by the daily cost of the crew divided by their output in SF. For parametric estimates both steps can be combined into a completed cost per SF, grid and tile.

Description	Labor-Hours	Unit
Ceiling Tile Stapled, Cemented, or Installed on Suspension System, 12" x 12" or 12" x 24", Not Including Furring		
Mineral Fiber, Plastic-Coated	.027	SF
Fire-Rated, 3/4" Thick, Plain-Faced	.027	SF
Plastic-Coated Face	.027	SF
Aluminum-Faced, 5/8" Thick, Plain	.027	SF
Metal Pan Units, 24-ga. Steel, Not Incl. Pads,		
Painted, 12" x 12"	.023	SF
12" x 36" or 12" x 24", 7% Open Area	.024	SF
Aluminum, 12" x 12"	.023	SF
12" x 24"	.022	SF
Stainless Steel, 12" x 24", 26-ga., Solid	.023	SF
5.2% Open Area	.024	SF
Suspended Acoustic Ceiling Boards Not Including Suspension System		
Fiberglass Boards, Film Faced, 2' x 2' or 2' x 4',		
5/8" Thick	.012	SF
3/4" Thick	.016	SF
3" Thick, Thermal, R11	.018	SF
Glass-Cloth-Faced Fiberglass, 3/4" Thick	.016	SF
1" Thick	.016	SF
1-1/2" Thick, Nubby Face	.017	SF
Mineral Fiber Boards, 5/8" Thick, Aluminum-Faced, 24" x 24"	.013	SF
24" x 48"	.012	SF
Plastic-Coated Face	.020	SF
Mineral Fiber, 2-Hour Rating, 5/8" Thick	.012	SF
Mirror-Faced Panels, 15/16" Thick	.016	SF
Air Distributing Ceilings, 5/8" Thick, FRD		
Water Felted Board	.020	SF
Eggcrate, Acrylic, 1/2" x 1/2" x 1/2" Cubes	.016	SF
Polystyrene Eggcrate	.016	SF
Luminous Panels, Prismatic	.020	SF
Perforated Aluminum Sheets, .024" Thick, Corrugated, Painted	.016	SF
Mineral Fiber, 24" x 24" or 48", reveal edge,		
Painted, 5/8" Thick	.013	SF
3/4" Thick	.014	SF
Wood Fiber in Cementitious Binder,		
2' x 2' or 4',		
Painted, 1" Thick	.013	SF
2" Thick	.015	SF
2-1/2" Thick	.016	SF
3" Thick	.018	SF
Access Panels, Metal, 12" x 12"	.400	EA
24" x 24"	.800	EA

Figure 14.8 Installation Time in Labor-Hours for Ceiling Systems

This method is often used when an estimator is checking or approximating a subcontractor's quote.

As with most finishes, the estimator must consider the cleanup of packaging and waste generated by the work.

FLOORING

Flooring includes an almost-infinite number of different and perpetually changing products. For estimating purposes, flooring can be classified in three general categories:

- Wood, including natural and synthetic products
- Resilient, including individual tiles and sheet goods and base
- Carpet, including carpet tiles

Material costs for flooring products range from economy to luxury grade. It is important that the estimator become familiar with the characteristics of the specified products to accurately price both materials and installation. Special concerns that have a cost impact include preparing the substrate, maintaining temperature and humidity conditions, washing/waxing or vacuuming the finish product, and protecting the installed work when complete. The estimator should study the architectural drawings and specifications in detail to determine the full scope of flooring work. In the absence of detailed specifications, the estimator must consider costs that relate to complying with industry standards or manufacturers' requirements. This is a scope of work that frequently requires attic stock for all types of flooring incorporated in the project.

Wood Flooring

Wood flooring comes in a wide variety of styles, grades, species, and patterns for interior use. It is manufactured in solid and laminated planks or parquet. *Laminated flooring* – high-grade wood veneers laminated over a lesser-grade base material – is sold prefinished. Solid wood flooring can be prefinished or unfinished and is classified by grade and quality. Common hardwood flooring materials are milled from oak, ash, maple, beech, walnut, cherry, and mahogany. Other exotic hardwoods include teak, ebony, zebrawood, and rosewood. Popular softwood flooring materials include fir, pine, cedar, and spruce.

Unfinished wood flooring requires repeated machine sanding with increasingly finer-grit sandpaper, followed by the application of a finish coating. Finish coatings can vary from polyurethanes to tung oil and may require more than one application. The unfinished wood flooring may also require an additional step of staining the wood. Review the architectural drawing set carefully, with special attention to floor plans and corresponding flooring details that illustrate the transition between different types of flooring. Room finish schedules confirm locations and types of flooring materials. Specifications provide the wood type, species, size, and pattern. The grade and quality of the material should also be identified in the wood floor specifications section. For unfinished materials, the specifications should also define the steps in the sanding process, the finishing product, and the number of coats. This is extremely important to accurately price the materials and labor for the particular system. In the absence of a formal specification, as is often the case in the residential sector, consult the manufacturer's literature for recommendations in order to maintain the warranty.

Taking-off and Pricing

Wood flooring is taken off by the SF of floor area. It should be identified according to type (plank, strip, or parquet), finish (prefinished or unfinished), species (oak, maple, ash, etc.), size (width and thickness), and method of application (adhesive, face, or blind-nailed). Any patterns should also be noted. Grade and quality of unfinished wood flooring materials are critical in determining accurate quantities and pricing.

Waste is directly related to grade and quality of the wood. The lesser the grade, the more waste to be included in the takeoff as a result of natural imperfections. Prefinished materials are more closely regulated during the quality control process and, as a result, generally have less waste. Patterns and designs influence quantities and waste as well. Waste will vary significantly.

The SF quantity of unfinished flooring materials is often converted to *board foot measure* (BFM) for pricing. (See Chapter 11, "Wood, Plastics, and Composites.") Board foot quantities depend on the profile and thickness of the individual species and type of flooring. The estimator should differentiate between large, open areas and small rooms or closets that will impact productivity. Fasteners, adhesives, vapor barriers, and building paper are taken off by the SF and converted to the sales unit of the individual product.

TAKEOFF AND PRICING FOR REFINISHING

The quantity of flooring to be refinished is calculated by the SF of surface area to be refinished. The species of wood and the various stages of sanding should also be noted. Wood flooring with heavy surface damage may require additional passes with the sander to eliminate marks. Finishing can be separated in the takeoff from sanding for more accurate pricing, but this is not mandatory. Finishing should be defined by the product and the number of coats required. The SF area multiplied by the number of individual coats can help determine the gallons of product needed, based on the manufacturer's recommended coverage. Whenever possible, the estimator is urged to physically view the condition of the flooring to be refinished.

Labor to Install Wood Flooring

Wood flooring can be installed by carpenters or floor installers. While the installation of prefinished wood flooring has become extremely homeowner-friendly, use of unfinished wood flooring still remains more complicated and the domain of the professional flooring contractor. Depending on the size of the project, crews can consist of a single individual or several flooring installers and laborers to distribute the flooring material. Productivity is based on the individual flooring installer. Patterns and designs generally reduce productivity, as do small, confined work areas. Blending new flooring into existing flooring is also particularly time consuming. It often requires the removal of existing flooring and the labor-intensive process of weaving new into old. Include any additional costs for handling and distributing the flooring, as well as equipment costs. Pneumatic nail guns have become increasingly popular for flooring installation and, as a result, have increased productivity in general.

Some full-thickness wood flooring projects may require laser-cut designs in the wood. These often require a shop drawing and then the factory laser cutting and overlay of a different species of wood. The designs can be simple or intricate, such as vines or flowers. It is recommended that the estimator obtain a contemporaneous quote for the laser cutting and overlay work.

Figure 14.9 provides guidelines for estimating installation labor-hours for wood flooring.

LABOR FOR REFINISHING FLOORING

Labor to refinish floors, including sanding and coating, is calculated and priced by the SF. Incidental costs of sandpaper, rollers, and brushes for applying the finish and equipment must be included. Refinishing of floors is typically a crew task, where one individual sands the field of the floor and a second individual sands the edges along vertical wall surfaces. Refinishing floors (versus new finishing) may require protection of adjacent work and the associated costs. Cleanup, vacuuming, hand sanding, and touch-up also must be addressed in the labor portion of the estimate. Productivity rates per crew can range between 400 and 1,000 SF per day, depending on the condition and species of the wood. Once the sanding is completed and vacuumed, the floor is coated. After the initial coat is allowed to dry, the floor is screened (lightly sanded) between each coat and recoated as required by the project. Most specifications require a minimum of two coats.

For refinishing projects, the estimator should consider the replacement of damaged areas of the floor. This can include single pieces of flooring or larger sections where damage cannot be sanded out of the floor. This should be considered a separate task and price. Correctly removing and replacing damaged wood flooring can be labor intensive.

Resilient Flooring

Resilient flooring is designed for areas where flooring durability, low maintenance, and longevity are primary concerns. Materials include asphalt tiles, vinyl composition tiles (VCT), cork tiles, and rubber tiles. Sheet materials include rubber, vinyl, and polyvinyl chloride. Rubber and vinyl accessories for resilient flooring include bases, thresholds, transition strips, stair treads and risers, and stair nosings. Resilient flooring materials are installed with adhesives, though some are manufactured with a self-adhering backing.

Consult the floor plans for information on the quantities and accessories, and the room finish schedules for locations and limits of flooring within

Description	Labor-Hours	Unit
Wood Floors Fir, Vertical Grain, 1" x 4", Not Including Finish	.031	SF
Gym Floor, in Mastic, Over 2-Ply Felt, #2 and Better 25/32"-Thick Maple, Including Finish	.080	SF
33/32"-Thick Maple, Including Finish	.082	SF
For 1/2" Corkboard Underlayment, Add	.011	SF
Maple Flooring, Over Sleepers, #2 and Better Including Finish, 25/32" Thick	.094	SF
33/32" Thick	.096	SF
For 3/4" Subfloor, Add	.023	SF
With Two 1/2" Subfloors, 25/32" Thick	.116	SF
Maple, Including Finish, #2 and better, 25/32" Thick, on Rubber Sleepers, with Two 1/2" Subfloors	.105	SF
With Steel Spline, Double Connection to Channels	.110	SF
Portable Hardwood, Prefinished Panels	.096	SF
Insulated with Polystyrene, Add	.048	SF
Running Tracks, Sitka Spruce Surface	.129	SF
3/4" Plywood Surface	.080	SF
Maple Strip, Not Including Finish	.047	SF
Oak Strip, White or Red, Not Including Finish	.047	SF
Parquetry, Standard, 5/16" Thick, Not Including Finish, Minimum	.050	SF
13/16" Thick, Select Grade, Minimum	.050	SF
Maximum	.080	SF
Custom Parquetry, Including Finish, Minimum	.080	SF
Maximum	.160	SF
Prefinished White Oak, Prime Grade, 2-1/4" Wide	.047	SF
3-1/4" Wide	.043	SF
Ranch Plank	.055	SF
Hardwood Blocks, 9" x 9", 25/32" Thick	.050	SF

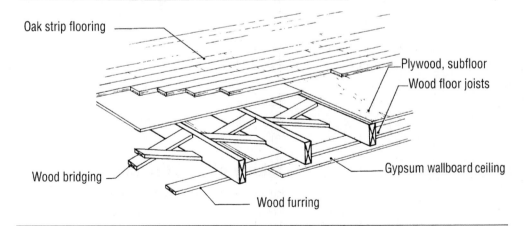

Figure 14.9a Installation Time in Labor-Hours for Wood Flooring

individual rooms. Review the technical specifications section for critical information necessary to accurately price materials and installation. Product data can aid the estimator in accurately pricing the materials and accessories. Other unique considerations that have a cost impact – such as

Description	Labor-Hours	Unit
Acrylic Wood Parquet Blocks,		
12" x 12" x 5/16", Irradiated, Set in Epoxy	.050	SF
Yellow Pine, 3/4" x 3-1/8", T & G, C and		
Better, Not Including Finish	.040	SF
Refinish Old Floors, Minimum	.020	SF
Maximum	.062	SF
Sanding and Finishing, Fill, Shellac, Wax	.027	SF
Wood Block Flooring End Grain Flooring,		
Natural Finish, 1" Thick	.029	SF
1-1/2" Thick	.031	SF
2" Thick	.033	SF
Wood Composition Gym Floors		
2-1/4" x 6-7/8" x 3/8", on 2" Grout Setting		
Bed	.107	SF
Thin-Set, on Concrete	.064	SF
Sanding and Finishing, Add	.040	SF

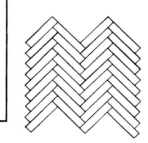

 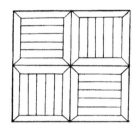

Random Patterns of Parquet Flooring

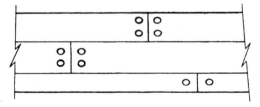

Strip Plank Flooring with Plugs

Figure 14.9b *(Continued)*

protection of completed flooring, washing, and waxing, and preparation of the substrate – are identified within the specifications.

Taking-off and Pricing

Resilient flooring is taken off by the SF by calculating the area of the floor. Resilient tiles are priced by the SF and can be converted to sales units of the individual product. Sheet goods quantities are extended from the SF to the SY for pricing. All takeoff quantities should be separated according to size, type, thickness, method of installation, location, or any other identifying characteristics. Accessories – such as base, transitions strips, and thresholds – are taken off by the LF, while stair treads, risers, and nosings of a repetitive size are taken off by the piece (EA). Adhesives are calculated by converting the square footage of flooring to units (most often gallons) of the individual product. Coverage varies with the subfloor material and condition as well as the manufacturer's recommendations.

Other considerations include allowances for waste, which vary with layout and patterns. Flooring tiles laid in a diagonal pattern, for example, require additional quantities. Most specifications call for layout to start at the center of the room so that the cuts on opposite walls are equal. The estimator may choose to increase the measurements of the room to the nearest full tile on either side. Because of the gauge (thinness) and flexibility of resilient products, defects in the subfloor easily telegraph through the flooring. Consequently, the subfloor materials and the quality of its surface preparation are very important. Subfloor preparation, even for new construction, is required and should be included in the estimate. The quantity is expressed by the SF of subfloor to be prepared. The specifications should be reviewed carefully for the requirement of wood or latex underlayment. There is a considerable cost difference. Latex underlayment is mixed from a powder to a paste-like consistency and then troweled over the damaged substrate. It is allowed to dry, then screened to remove imperfections and finally vacuumed to leave a clean surface for the application of flooring. Wood underlayment is typically 4' × 8' sheets of varying thicknesses that are nailed or screwed to the wood substate to allow a smooth clean surface.

Labor to Install Resilient Flooring

Resilient floor tile and sheet goods are installed by floor layers and priced by the SF and SY, respectively. The productivity of a crew composed of multiple floor layers is measured by the total output in SF per day. Installation of base, treads, risers, and nosing is priced by the individual piece (EA) or linear foot (LF). Costs for prep work should be separate in the takeoff from actual installation and are calculated by the SF of subfloor that can be prepped in a day. Floor prep costs are extremely subjective and vary dramatically depending on the conditions of the substrate. Floor prep costs are most accurate when the subfloor can be viewed prior to pricing. Other labor costs, such as washing and waxing or protection of newly installed floors, are calculated based on the SF area. Since most resilient flooring products are heavy and bulky, include costs for delivery, stocking, and distribution throughout the building. Flooring materials are one of the products that most often require attic stock. Quantities can vary from 1% to 5% of each product depending on size of each. The estimator is urged to review the specifications carefully for attic stock quantities.

Figure 14.10 provides a guideline for calculating labor.

Description	Labor-Hours	Unit
Resilient Asphalt Tile, on Concrete, 1/8" Thick		
Color Group B	.015	SF
Color Group C and D	.015	SF
For Less Than 500 SF, Add	.002	SF
For Over 5,000 SF, Deduct	.007	SF
Base, Cove, Rubber or Vinyl, .080" Thick		
Standard Colors, 2-1/2" High	.025	LF
4" High	.025	LF
6" High	.025	LF
1/8" Thick, Standard Colors,		
2-1/2" High	.025	LF
4" High	.025	LF
6" High	.025	LF
Corners, 2-1/2" High	.025	EA
4" High	.025	EA
6" High	.025	EA
Conductive Flooring, Rubber Tile,		
1/8" Thick	.025	SF
Homogeneous Vinyl Tile, 1/8" Thick	.025	SF
Cork Tile, Standard Finish, 1/8" Thick	.025	SF
3/16" Thick	.025	SF
5/16" Thick	.025	SF
1/2" Thick	.025	SF
Urethane Finish, 1/8" Thick	.025	SF
3/16" Thick	.025	SF
5/16" Thick	.025	SF
1/2" Thick	.025	SF
Polyethylene, in Rolls, No Base Incl.,		
Landscape Surfaces	.029	SF
Nylon Action Surface, 1/8" Thick	.029	SF
1/4" Thick	.029	SF
3/8" Thick	.029	SF
Golf Tee Surface with Foam Back	.033	SF
Practice Putting, Knitted Nylon Surface	.033	SF
Polyurethane, Thermoset, Prefabricated in Place, Indoor		
3/8" Thick for Basketball, Gyms, etc.	.080	SF

Figure 14.10a Installation Time in Labor-Hours for Resilient Flooring

Carpeting

A popular flooring material for both residential and commercial construction, carpet provides a comfortable, sound-absorbing, and attractive finish surface. It is manufactured from a variety of different materials, including nylon, wool, acrylic, polyester, and rayon, as well as recycled materials. Products are specified by weight, type, pile thickness, and density. Special carpets for commercial applications, such as fire-rated materials, stain protected, and anti-static carpet are also available. Commercial-grade carpeting is manufactured in rolls and tiles, but residential grade is available mainly in rolls.

Description	Labor-Hours	Unit
Stair Treads and Risers		
Rubber, Molded Tread, 12" Wide, 5/16"		
Thick, Black	.070	L F
Colors	.070	L F
1/4" Thick, Black	.070	L F
Colors	.070	L F
Grit Strip Safety Tread, Colors 5/16" Thick	.070	L F
3/16" Thick	.067	L F
Landings, Smooth Sheet Rubber,		
1/8" Thick	.067	S F
3/16" Thick	.067	S F
Nosings, 1-1/2" Deep, 3" Wide, Residential	.057	L F
Commercial	.057	L F
Risers, 7" High, 1/8" Thick, Flat	.032	L F
Coved	.032	L F
Vinyl, Molded Tread, 12" Wide, Colors,		
1/8" Thick	.070	L F
1/4" Thick	.070	L F
Landing Material, 1/8" Thick	.040	S F
Riser, 7" High, 1/8" Thick, Coved	.046	L F
Threshold, 5-1/2" Wide	.080	L F
Tread and Riser Combined, 1/8" Thick	.100	L F

Figure 14.10b *(Continued)*

The size of the roll varies with type, manufacturer, quality, and model. Rolls are generally available in 12' and 15' widths. Occasionally, 9' and 18'-wide rolls are available in limited patterns. Carpet tiles are available in 18" × 18", 24" × 24", and 36" × 36" sizes with self-adhering backing. Carpeting can be installed on cushion material called *padding*. This is typically a residential-type installation.

Accessories for carpeting include metal or rubber edging and specialty carpeting for stair runners. Most carpeting is installed using one of two methods: directly gluing the carpet to the floor substrate or using pad underlayments and fastening the carpet at the perimeter of the room. Some carpet tiles in commercial applications have an easy-release self-adhesive.

The architectural drawings, floor plans, and room finish schedules provide the location and limits of specific types of carpeting. Review the specs for manufacturer, model, size (roll width), color, weight, and method of installation for correct pricing. Other special requirements, such as seaming, patterns, vacuuming, and protection, are also detailed in the specifications. Patterns or borders can reduce productivity substantially.

Taking-off and Pricing

Carpeting and padding are taken off by calculating the area of the room in SF, extended to the pricing unit of SY, where 1 SY equals 9 SF. Carpet tile quantities are calculated the same way, but instead of extending the quantities to SY, they are converted to the required number of pieces or SF. Carpet and padding quantities should be separated for more accurate pricing. Accessories such as tack strip for the edge of the carpet and pad installation, is taken off by calculating the perimeter of the room where the carpet is installed. Separate different types of carpet according to manufacturer, model, type, style, and method of installation for accurate pricing. Patterns or combinations of different carpeting should also be noted, as they are sometimes labor intensive. Protection or cleaning requirements of completed work should be taken off separately.

The estimator should read the specifications carefully for carpet that may require "off-gassing" prior to installation. This is a process typically done by the distributor or supplier to allow the VOCs from manufacturing to escape before installation. It can add to the material cost of the product.

Many specifications have strict seaming requirements and may call for the layout of "drops" of carpeting or seaming drawings for approval. Seams should be minimized during installation, as they will increase the waste of the carpet if remnants cannot be used elsewhere. For example, if a room measuring 10'-3" × 20'-3" has an area of 23 SY, but a 12'-wide carpet is used, one piece 12' × 21' or 28 SY must be purchased. The difference represents 22% waste. Carpeting with special designs or repeating patterns requires careful calculation to allow for matching the pattern at seams. A classic example is carpet that simulates an Oriental or Persian rug.

Direct glue carpeting, much like resilient flooring, requires a smooth substrate on which to install the carpet. For most carpet installations, where the carpet is directly glued to the wood or concrete substrate, the surface must be patched with some type of underlayment. This can be sheets of plywood underlayment nailed or screwed to the existing surface or a liquid underlayment that is applied by trowel. Both processes are expensive; however, the latter approach can be labor intensive and require multiple applications. Latex or liquid underlayment is a fine powder mixed with a liquid fortifying agent and then troweled by hand onto the surface of the damaged area. Once dry it is screened by machine to remove the trowel marks or imperfections. This is a significant cost, especially in renovation work where the substrate may be damaged due to demolition.

While not typically part of the carpet installation itself, carpet installers often take care of fixing the substrate to maintain its integrity before installing the carpet. Whenever possible, the pricing process should include a site visit to assess the condition of the substrate. The underlayment should be quantified and estimated separately from the carpet work itself.

Carpet, in all its various forms, is frequently required for attic stock. The estimator should review the specifications carefully for the percentage of each type of carpet required for attic stock.

Labor to Install Carpet

Regardless of the method of installation, carpet is typically installed by carpet installers or floor layers. The crew most often consists of two individuals to handle the awkward and bulky carpet rolls. Productivity is measured in SY installed per crew per day. Large, open areas and small, confined spaces should be priced separately as productivities will vary. Installations with patterns or "repeats" should be estimated separately, as well as borders and carpet base. Figure 14.11 provides labor-hour guidelines for the installation of various types of carpeting and applications.

The estimator should include labor to clean up, vacuum, and protect the carpet, if required, in the specifications. The estimator must also consider

Description	Labor-Hours	Unit
Carpet Commercial Grades, Cemented		
Acrylic, 26 oz., Light to Medium Traffic	.216	SF
28 oz., Medium Traffic	.229	SF
35 oz., Medium to Heavy Traffic	.242	SF
Nylon, Non Antistatic, 15 oz., Light Traffic	.229	SF
Nylon, With Antistatic, 17 oz., Light to Medium Traffic	.211	SF
20 oz., Medium Traffic	.216	SF
22 oz., Medium Traffic	.216	SF
24 oz., Medium to Heavy Traffic	.229	SF
26 oz., Medium to Heavy Traffic	.229	SF
28 oz., Heavy Traffic	.229	SF
32 oz., Heavy Traffic	.242	SF
42 oz., Heavy Traffic	.258	SF
Needle Bonded, 20 oz., No Padding	.143	SF
Polypropylene, 15 oz., Light Traffic	.143	SF
22 oz., Medium Traffic	.182	SF
24 oz., Medium to Heavy Traffic	.182	SF
26 oz., Medium to Heavy Traffic	.182	SF
28 oz., Heavy Traffic	.205	SF
32 oz., Heavy Traffic	.205	SF
42 oz., Heavy Traffic	.216	SF
Scrim Installed, Nylon Sponge-Back Carpet		
20 oz.	.242	SY
60 oz.	.267	SY
Tile, Foam-Backed, Needle Punch	.014	SF
Tufted Loop or Shag	.014	SY
Wool, 30 oz., Medium Traffic	.229	SY
Wool, 36 oz., Medium to Heavy Traffic	.242	SY
Sponge Back, Wool, 36 oz., Medium to Heavy Traffic	.143	SY
42 oz., Heavy Traffic	.229	SY
Padding, Sponge Rubber Cushion, Minimum	.108	SY
Maximum	.123	SY
Felt, 32 oz. to 56 oz., Minimum	.108	SY
Maximum	.123	SY
Bonded Urethane, 3/8" Thick, Minimum	.094	SY
Maximum	.107	SY
Prime Urethane, 1/4" Thick, Minimum	.094	SY
Maximum	.107	SY

Figure 14.11 Installation Time in Labor-Hours for Carpeting

distributing the carpet and related materials to the space. For commercial applications this may require that the carpet is hand carried to the space or in more extreme situations, a window may need to be removed and the carpet hoisted by crane to the space. Off-gassing done on site may require space and additional labor to expose the carpet to the air prior to installation.

ACOUSTICAL TREATMENT

Many projects require control of sound generated in one space from traveling to an adjacent space. This type of acoustic control requires an *acoustic treatment*. Acoustic treatment can be as simple as sound-attenuation batts or blankets within the partitions enclosing the space or sound-deadening board applied to the wall, or as sophisticated as noise-canceling systems. For the scope of this text, discussions are limited to the most common practices found in the commercial and residential applications – specifically, sound insulation and sound-deadening boards. Noise-cancelling systems are a unique scope of work best left to the specialty subcontractor to estimate.

Sound insulation is installed within the cavity space of partitions or above ceilings. It is denser than thermal insulation but is installed the same way. Sound-attenuation insulation is reserved for partitions and ceilings that are on the interior versus thermal insulation used in exterior walls.

Sound-deadening board is typically available in 4' × 8' sheets ½" thick with a 15- to 20-lb.-per-cubic-foot density. It is applied to wood or metal studs with screws as an underlayment to the drywall. It can also be installed on floors between the subfloor and finished floors. Both applications provide a reduction in sound transmission.

Other types of acoustical treatment include poured or placed underlayment, discussed in Chapter 8, "Gypsum Cement Underlayment."

Taking-off and Pricing

Acoustical insulation is taken off and quantified by the SF by measuring length × width. Sound-deadening board is also taken off and quantified by the SF. Both quantities can be converted to the nearest sales unit to approximate materials costs more accurately. For acoustic treatment requiring Gypsum Cement Underlayment, refer to Chapter 8, "Concrete."

Labor for Acoustical Treatment

Most often, sound insulation is installed by carpenters just prior to installing the drywall or floor underlayment. However, acoustic insulation can be installed by the same crew that installs the thermal insulation. Productivity can be calculated by a single individual or by the crew. A single skilled worker should install between 600 and 1,000 SF per day of sound-attenuation batts in walls. Productivity on ceiling or floor installations ranges between 800 and 900 SF per day per worker.

Sound-deadening board is typically installed by carpenters on the drywall crew. One carpenter should install 500 to 700 SF per day.

PAINTING

Finishing and painting are often required to protect surfaces from wear and deterioration and to provide a decorative appearance. The work typically includes some level of preparation of the surface to receive the paint or finish, and it may involve sanding, filling in holes, and removing dust or oils. More extensive preparation can involve stripping old finishes, applying wood conditioner, or applying primers to neutralize the substrate, as in the case of galvanized finishes. Painting and finishing products run the full spectrum from latex and alkyd or oil-based paints and stains to epoxy coatings and urethanes. Application methods include spraying, rolling, and brushing. All three may be used on the same project or even in the same process.

Floor plans, interior elevations, and reflected ceiling plans can help define the interior painting work. Interior elevations frequently illustrate wood trims that require finishes, and exterior elevations will provide dimensions for exterior siding and trims. Room finish schedules should also be reviewed for individual locations and any limitations of painting or finishing work. Consult the mechanical and electrical drawings for the painting of pipes, conduits, and unfinished equipment as a means of identification. Roof drawings can be helpful in determining the quantities of rooftop equipment or gas piping to be painted.

The estimator is urged to study the technical specifications carefully for the location and summary of finishing work, type of products to be used, application methods, number of coats, and level of preparation required. Colors, finishes, and any patterns or borders should also be noted, as any type of detail work will reduce productivity.

Also note that some painting may involve the removal of existing finishes in their entirety. This may constitute enclosing the work and either sand blasting or chemically removing the finish. The residual material may contain contaminates such as lead. Please refer to Chapter 7, "Existing Conditions," for additional information on estimating this type of work.

Taking-off and Pricing

For estimating purposes, separate all of the work into two main classifications, exterior and interior, in the takeoff. While they share similar methods, techniques, and materials, some characteristics are unique to each type. The principal difference is that exterior work is affected by weather and temperature, and typically requires some type of scaffolding, ladders, or lifts to access work above grade level.

Exterior

Exterior work includes painting or staining exterior trims, siding, soffits, shutters, columns, doors and frames, windows, decks, porches, and rooftop equipment. Other, less common items to be painted include metal railings, flagpoles, bollards, gates, and grating.

Take off exterior components separately, using the following guidelines:

- **Trims and soffits less than 12" in width:** By the LF.
- **Trims and soffits larger than 12" in width:** Extended to SF for pricing.
- **Siding:** By the SF of surface area to be painted. Openings for windows and doors less than 10 SF are not deducted.
- **Shutters, columns, doors, frames, and windows:** By the piece (EA). Descriptions in the takeoff should include the size of each item, number of sides (doors and shutters), number of coats, and any special preparatory work.
- **Deck and porch surfaces:** By the SF.
- **Railings and balusters:** By the LF. Balusters can also be taken off by the individual piece, EA.
- **Exterior stairs:** By the individual components. Treads and risers by the piece EA, and stringers, skirts, or cheek boards by the LF.
- **Small rooftop equipment:** By the individual piece, EA. Larger units can be taken off by the SF of surface area on the unit.

Note the method of application (brush, spray, or roller) for each takeoff item, as well as the number of coats and whether the work is stain, paint, or epoxy. Surface preparation should always be listed separately. Special techniques, such as *back priming* – the process of priming all surfaces of trims before installation, even if not exposed – should be taken off and priced separately.

Interior

Interior work includes painting or staining walls, ceilings, doors and frames, windows, standing and running trims, wood shelving, casework, wainscoting, stair parts, and concrete floors. All work should be listed separately in the takeoff and described according to type, item to be painted, coating (paint or stain), number of coats, color, product, and method of application.

Generally, coverage of paints and finishes varies with the surface being finished (its porosity), the number of coats (the first coat covers less due to absorption into the surface), and the product itself. Consider the manufacturer's recommended coverage for the individual product and adjust, if necessary, for the porosity of the surface being painted. Also, review the specifications for the requirement of "attic stock" paint. It is not uncommon, especially with commercial projects, for the painter to be required to provide 1 gallon of unopened paint of each color to be turned over to the owner at the end of the project.

Wall and ceiling painting is taken off and priced by the SF of surface area to be painted. Wall area is calculated by multiplying the perimeter of the room by its height. Openings larger than 4 SF are deducted in full; openings smaller than 4 SF are negligible. Ceiling area is calculated by multiplying the length by the width of the surface. Walls and ceilings are listed separately in the takeoff for pricing purposes. Interior wood trims, such as baseboards, cornices, chair rails, door and window casings, railings, and stair cheek or skirt boards, are taken off and priced by the LF. Stair components, such as balusters, treads, risers, and newel posts, are taken off and priced by the individual piece and, again, are listed separately in the takeoff according to type. Painting of windows and accessories, such as grilles, is taken off and priced by the piece (EA). Interior doors and frames are taken off and priced by the individual unit (EA) or can be separated by frame and door leaves. Remember to differentiate between single and double door units in the takeoff.

Casework, cabinetry, and shelving are typically finished off-site during the manufacturing process. However, when these items are field finished, they are taken off and priced by the SF of surface area to be finished or by the individual piece for repetitive items of the same size. Take off wainscoting and wood ceilings by the SF of surface area, and be sure to note the type of work, such as raised panel wainscoting or tongue-and-groove ceilings.

Concrete floors are taken off and priced by the SF of floor area to be painted. The area is computed by multiplying the length by the width of the space. Verify whether preparation of the floor is necessary, such as sandblasting, acid-etching, or washing.

Painting of pipes and conduits is taken off and priced by the LF. Quantities should be separated according to pipe diameter. Most piping requires cleaning and priming. The solvents required to remove the oils and coatings from the pipes should be included as part of the preparatory work.

Labor to Paint

Productivity of painting can be measured by the individual painter or by a crew, depending on the task. Since painting and finishing of walls and trims has a high aesthetic value, it often proceeds at a slow pace and requires greater care. This is especially true of interior painting or finishing in high-end residential work.

Ordinary workmanship, such as used in painting industrial or warehouse buildings, would most likely not be suitable for residential or commercial applications, such as a home, restaurant, or library, where first-class workmanship is required. When estimating painting work, consider the class of workmanship required by the specifications or by the nature of the task.

Productivity and cost vary dramatically with the method of application of the product. It is important to select the proper method for the particular task. Large, open areas and new unoccupied areas are ideal for paint spraying. Paint rolling is necessary when the area is small or obstacles would interfere with the path of spray, or in the case of a renovation, where spray particles would contaminate the HVAC system or surrounding surfaces. Detailed work on trims and cutting-in around other features, such as casework or cabinetry, are done by brush. The method of application is critical to pricing the labor portion of the work and should be noted in the takeoff. New or renovation projects may require taping and masking off of surfaces to remain unpainted. This can be labor-intensive and should be included as part of the prep.

The quality and longevity of a paint job frequently depends on the level of preparation. Preparation time and costs vary from surface to surface. Repainting work that has a sound surface in good condition often requires only minimal time to correctly prep. Conversely, steel that has rusted or wood with flaking, scaling, or peeling paint may require considerable time to prepare the surface. For renovation work, the estimator should examine the existing substrate before preparing the estimate. Figure 14.12 provides labor-hour guidelines for the application of various coatings. Be sure to account for staging, ladders, platforms, and any other equipment necessary to access the work. Drop cloths and polyethylene sheeting should also be added to the materials, as these are often used and discarded when the work is complete. Productivity always decreases as a result of frequent moves of a ladder or setting up scaffolding or work platforms.

Most specifications require that the painter do minor touch-up work after the installation of work of other trades. Since this can be labor intensive and costly, consult historical data on similar projects to allow adequate labor-hours. Most materials used in touch-up work are left over from the main work.

HIGH-PERFORMANCE COATINGS

High-performance coatings can be classified as any painting type product that involves a high level of preparation or application procedure. Examples include epoxy coatings, dryfall (sweep-up) coatings, high-heat-resistant coatings, and urethane coatings. While these products have their greatest appeal in the industrial construction market, they are occasionally found in the light-commercial/residential market.

Estimators should research the individual product specified carefully as application procedures and cost differences can occur from manufacturer to manufacturer. Seek any guidance available from the manufacturer for coverage, drying time between coats, preparation, cleanup, and hazard protection to applicators, as these have an impact on costs. Information on individual products is available on the internet along with video

Description	Labor-Hours	Unit
Cabinets and casework		
Primer coat, oil base, brushwork	.012	SF
Paint, oil base, brushwork	.012	SF
1 coat	.012	SF
Stain, brushwork, wipe-off	.012	SF
Shellac, 1 coat, brushwork	.012	SF
Varnish, 3 coats, brushwork	.025	SF
Doors and windows, interior latex		
Doors, flush, both sides, incl. frame & trim		
Roll and brush primer	.800	EA
Finish coat, latex	.800	EA
Primer & 1 coat latex	1.143	EA
Primer & 2 coats latex	1.600	EA
Spray, both sides, primer	.400	EA
Finish coat, latex	.400	EA
Primer & 1 coat latex	.727	EA
Primer & 2 coats latex	1	EA
Doors, French, both sides, 10–15 lite, incl. frame & trim		
Roll & brush primer	1.333	EA
Finish coat, latex	1.333	EA
Primer & 1 coat latex	2.667	EA
Primer & 2 coats latex	4	EA
Doors, louvered, both sides, incl. frame & trim		
Roll & brush primer	1.143	EA
Finish coat, latex	1.143	EA
Primer & 1 coat latex	2	EA
Primer & 2 coats latex	2.667	EA
Spray, both sides, primer	.400	EA
Finish coat, latex	.400	EA
Primer & 1 coat latex	.727	EA
Primer & 2 coats latex	1	EA
Doors, panel, both sides, incl. frame & trim		
Roll & brush primer	1.333	EA
Finish coat, latex	1.333	EA
Primer & 1 coat latex	2.667	EA
Primer & 2 coats latex	3.200	EA
Spray, both sides, primer	.800	EA
Finish coat, latex	.800	EA
Primer & 1 coat latex	1.600	EA

Figure 14.12a Application Time in Labor-Hours for Painting

demonstrations of the application of the coating. The estimator should keep in mind that many of these products are specialized and may require a certified installer to maintain the warranty. The estimator may want to solicit a proposal for the work.

Description	Labor-Hours	Unit
Primer & 2 coats latex	2	EA
Windows, per interior side, based on 15 SF		
1 to 6 lite		
Brushwork, primer	.615	EA
Finish coat, enamel	.615	EA
Primer & 1 coat enamel	1	EA
Primer & 2 coats enamel	1.333	EA
7 to 10 lite		
Brushwork, primer	.727	EA
Finish coat, enamel	.727	EA
Primer & 1 coat enamel	1.143	EA
Primer & 2 coats enamel	1.600	EA
12 lite		
Brushwork, primer	.800	EA
Finish coat, enamel	.800	EA
Primer & 1 coat enamel	1.333	EA
Primer & 2 coats enamel	1.600	EA
Fences		
Chain link or wire metal, one side, water base		
Roll & brush, first coat	.008	SF
Second coat	.006	SF
Spray, first coat	.004	SF
Second coat	.003	SF
Picket, water base		
Roll & brush, first coat	.009	SF
Second coat	.008	SF
Spray, first coat	.004	SF
Second coat	.003	SF
Stockade, water base		
Roll & brush, first coat	.008	SF
Second coat	.007	SF
Spray, first coat	.004	SF
Second coat	.003	SF
Floors		
Concrete paint, latex		
Brushwork		
first coat	.008	SF
second coat	.007	SF
third coat	.006	SF
Roll		
first coat	.003	SF
second coat	.002	SF
third coat	.002	SF

Figure 14.12b *(Continued)*

Takeoff for high-performance coatings follows the same procedure as for regular painting. Classify work by product as well as quantity. Also note that many of these products require that the work be done when there are no other trades on site due to smell or work overhead. This may require that the work be done after normal hours or on weekends.

Description	Labor-Hours	Unit
Spray		
first coat	.003	SF
second coat	.002	SF
third coat	.002	SF
Acid stain and sealer		
Stain, 1 coat	.012	SF
2 coats	.014	SF
Acrylic sealer, 1 coat	.003	SF
2 coats	.006	SF
Floors, conc./wood, oil base, primer/sealer coat, brushwork		
Stain, wood floor, brushwork, 1 coat	.004	SF
Roller	.003	SF
Spray	.003	SF
Varnish, wood floor, brushwork	.004	SF
Roller	.003	SF
Spray	.003	SF
Grilles, per side, oil base, primer coat, brushwork		
Spray	.007	SF
Paint 2 coats, brushwork	.025	SF
Spray	.012	SF
Gutters and downspouts, wood		
Downspouts, 4", primer	.013	LF
Finish coat, exterior latex	.013	LF
Primer & 1 coat exterior latex	.020	LF
Primer & 2 coats exterior latex	.025	LF
Pipe, 1"–4" diameter, primer or sealer coat, oil base, brushwork	.013	LF
Spray	.007	LF
Paint 2 coats, brushwork	.021	LF
Spray	.012	LF
13"–16" diameter, primer or sealer coat, brushwork	.052	LF
Spray	.030	LF
Paint 2 coats, brushwork	.082	LF
Spray	.052	LF
Trim, wood, including puttying under 6" wide		
Primer coat, oil base, brushwork	.012	LF
Paint, 1 coat, brushwork	.012	LF
3 coats	.025	LF

Figure 14.12c *(Continued)*

WALLCOVERINGS

There is a wide selection of materials that can be applied as decorative treatment to interior walls in residential and commercial projects. Wallcoverings are manufactured, printed, or woven, in fabric, paper, vinyl, leather, suede, cork, wood veneers, and foils. They are available in a variety of weights, backings, and quality.

Wallcoverings are installed with adhesives, and some require special pastes. Preparation of the wall surface is necessary to ensure proper bonding of the

Description	Labor-Hours	Unit
Over 6" wide, primer coat, brushwork	.012	LF
Paint, 1 coat, brushwork	.012	LF
3 coats	.025	LF
Cornice, simple design, primer coat, oil base,		
Brushwork	.012	SF
Paint, 1 coat	.012	SF
Ornate design, primer coat	.023	SF
Paint, 1 coat	.023	SF
Balustrades, primer coat, oil base, brushwork	.015	SF
Paint, 1 coat	.015	SF
Trusses and wood frames, primer boat, oil base,		
brushwork	.010	SF
Spray	.013	SF
Paint, 2 coats	.016	SF
Spray	.013	SF
Stain, brushwork, wipe off	.013	SF
Varnish, 3 coats, brushwork	.029	SF
Siding, exterior		
Steel siding, oil base, paint 1 coat,		
brushwork	.008	SF
Spray	.004	SF
Paint 2 coats, brushwork	.012	SF
Spray	.006	SF
Stucco, rough, oil base, paint 2 coats,		
brushwork	.012	SF
Roller	.010	SF
Spray	.005	SF
Texture 1–11 or clapboard, oil base, primer		
Coat, brushwork	.012	SF
Spray	.004	SF
Paint 2 coats, brushwork	.020	SF
Spray	.006	SF
Stain 2 coats, brushwork	.017	SF
Spray	.006	SF
Wood shingles, oil base primer coat,		
brushwork	.012	SF
Spray	.004	SF
Paint 2 coats, brushwork	.020	SF
Spray	.007	SF
Stain 2 coats, brushwork	.017	SF
Spray	.006	SF

Figure 14.12d *(Continued)*

wallcovering and includes minor sanding and repairing of defects in the wall and applying *sizing,* or primer, for proper adhesion.

Review the architectural drawings, with particular attention to floor plans and elevations. Room finish schedules will provide the location of wallcoverings within specific rooms. Reflected ceiling drawings or room finish schedules should be reviewed for the vertical heights of the walls. The technical specifications should be consulted for information on individual

Description	Labor-Hours	Unit
Wall coatings		
High build epoxy, 50 mil, minimum	.021	SF
Maximum	.084	SF
Laminated epoxy with fiberglass, minimum	.027	SF
Maximum	.055	SF
Sprayed perlite or vermiculite, 1/16" thick,		
Minimum	.003	SF
Maximum	.013	SF
Vinyl plastic wall coating, minimum	.011	SF
Maximum	.033	SF
Urethane on smooth surface, 2 coats, minimum	.007	SF
Maximum	.012	SF
Ceramic-like glazed coating, cementitious,		
Minimum	.018	SF
Maximum	.023	SF
Resin base, minimum	.013	SF
Maximum	.024	SF

Figure 14.12e *(Continued)*

products, including model and manufacturer, surface preparation, and adhesives. Special requirements in commercial applications, such as fire-treated wallcoverings that do not support combustion, should also be noted in the specs. Waste on wallcovering applications can range from a low of 10% for a product with a basic pattern to as much as 60% for wallcoverings with intricate patterns.

Taking-off and Pricing

Wallcoverings are taken off by the SF of surface area, calculated by multiplying the LF of the wall by its height. Deductions are taken for windows, doors, and openings in excess of 4 SF. The SF area is then converted to wallcovering rolls, which typically contain 36 SF. The total number of rolls can be determined and then priced. Wallcoverings are generally produced in double and triple roll units called *bolts*. Commercial wallcoverings are manufactured in widths ranging from 21" to 54" and in lengths up to 100-yard bolts. For an accurate quantity, the linear footage of the walls to be covered can be divided by the width of the wallcovering (in feet), which results in the number of strips. Calculate the number of strips per bolt by dividing the length of the bolt by the height of the hang. Adjustments in the number of strips per bolt will be required based on the repetition of the pattern, if applicable. The number of strips required (rounded up to the nearest whole number) is divided by the number of strips per bolt (rounded down to the nearest whole number) to determine the number of bolts required.

Quantities of adhesives or paste are calculated by dividing the total SF area to be covered by the manufacturer's recommended coverage per sales unit (typically, the gallon). Coverage may be expressed in rolls.

Labor to Install Wallcoverings

Wallcoverings are typically installed by paperhangers and painters, and productivity is measured by the daily output of each worker. For larger projects, it is not uncommon for a paperhanger to have a helper for trimming and pasting. Due to the high aesthetic value of wallcoverings, the quality of workmanship must always be estimated as first class. Calculate the cost of wall preparation, including sizing, separate from the actual cost of hanging the wallcovering. Layout and cleanup should also be included as part of the cost of the work. While the surface area of the wall is increased to include waste and repeating patterns, remember to calculate labor based on the actual area of the walls to be covered. Figure 14.13 provides guidelines for wallcovering installation labor-hours.

Description	Labor-Hours	Unit
Wall Covering		
Aluminum Foil	.029	SF
Copper Sheets, .025" Thick		
Vinyl Backing	.033	SF
Phenolic Backing	.033	SF
Cork Tiles, Light or Dark, 12" x 12"		
3/16" Thick	.033	SF
5/16" Thick	.034	SF
1/4" Basketweave	.033	SF
1/2" Natural, Nondirectional Pattern	.033	SF
Granular Surface, 12" x 36"		
1/2" Thick	.021	SF
1" Thick	.022	SF
Polyurethane-Coated, 12" x 12"		
3/16" Thick	.033	SF
5/16" Thick	.034	SF
Cork Wallpaper, Paperbacked		
Natural	.017	SF
Colors	.017	SF
Flexible Wood Veneer, 1/32" Thick		
Plain Woods	.080	SF
Exotic Woods	.084	SF
Gypsum-based, Fabric-backed, Fire-Resistant for Masonry Walls		
Minimum	.010	SF
Maximum	.013	SF
Acrylic, Modified, Semi-rigid PVC		
.028" Thick	.048	SF
.040" Thick	.050	SF
Vinyl Wall Covering, Fabric-backed		
Lightweight	.013	SF
Medium Weight	.017	SF
Heavy Weight	.018	SF
Grass Cloths with Lining Paper		
Minimum	.020	SF
Maximum	.023	SF

Figure 14.13 Installation Time in Labor-Hours for Wallcoverings

Installation of wall coverings may require scaffolding or platforms for access in much the same manner as painting. The estimator may consider using the scaffolding or lift for multiple trades.

GREEN OR SUSTAINABLE PRODUCTS

Many of the products of Division 9—Finishes available in the residential/light-commercial marketplace are offered in green or sustainable versions. The estimator is directed to be extremely vigilant in ensuring that the product being offered meets the specification requirements, especially in the case of achieving a predetermined standard such as those set by the Leadership in Energy and Environmental Design (LEED) program or the National Association of Home Builders (NAHB). Green or sustainable products tend to be more costly to purchase and, in some cases, to install. The estimator is directed to research extensively new products for guidelines on pricing installation and any related labor costs such as cleanup.

SUMMARY

Always pay careful attention to the plans and specifications when estimating finishes in general. Their aesthetic value cannot be overstated – often one of the most important and most noticed features to the owner. Accurate estimating of both materials and labor costs is a byproduct of understanding the requirements set forth in the documents.

Drywall or gypsum wall board and plaster systems provide the means to identify and delineate spaces within the structure or complete the exterior wall system. They are both taken off and estimated by the square foot of surface area. They also act as the substrate for the application of paint, tile, wood trims, etc. Quantities derived for drywall surfaces can often be used for the calculation of painting, tile, or wallcoverings. Partition types should be reviewed for composition, especially materials such as acoustical treatment and wood blocking. There may also be a layout cost to the work.

Ceiling systems can be "hard," constructed of metal stud and drywall or acoustical suspension systems. In both cases, the quantity and pricing are dependent on the systems and the area calculated in square feet. The composition of the system – fire-rated, seismic applications, or some other unique characteristic for the acoustical ceiling – can be a determining factor. For ceiling and soffit work, typically some type of work platform is required.

Ceramic tile can be a wall or floor finish is typically a two-step process; set the tile and then grout the tile. Both steps are priced by the square area of tile with consideration to the setting or grouting method. The estimator is urged to solicit material quotes for the tile and trim pieces as they can vary significantly, based on size, color, origin, and so on.

Flooring comes in a variety of materials and types. The key to accurate pricing is to become familiar with the specified product and the recommended installation procedure. In addition, floor prep such as latex or wood underlayment can be required. Most flooring is taken off and priced by the square foot, SF with the exception of carpet sheet goods which is priced by the square yard, SY. Carpet can have repeat patterns that often dictate the waste percentage.

Painting and high-performance coatings are taken off by the area to be coated. Interior and exterior applications are separated due to ambient conditions and requirement for access above grade. Surface preparation for both old and new work is often labor intensive and time consuming. Wall coverings are calculated by the area to be covered with an allowance for repeat patterns and waste. They are then converted to the appropriate sales unit. Surface prep for wall coverings can also be labor intensive, but necessary.

For much of Division 9 work the specification of material can be tied to a green or sustainable standard. The estimator is urged to review this requirement carefully.

QUESTIONS/ PROBLEMS

1. Calculate the amount of ⅝" gypsum wallboard required for a partition that is 276' long by 12' high with two layers of board each side full height. Provide your answer in square feet, SF.

2. If the two carpenters on the crew will hang 1,500 SF per day (total), how many days will the hang in problem #1 take? Round your answer to the full day.

3. The partition in problem #1 requires 3" sound-attenuation batts within the metal studs. If the cost for materials and labor is $1.91 per SF, what is the cost of the acoustical treatment? Assume the installers will use the lift on site.

4. The lobby to a mall has 6" × 6" quarry tile in an area 136' × 42'-8". If the tile material cost $4.22 per SF, thin set at $0.83 per SF, and grout at $1.22 per SF, what is the material cost for the work with 5% waste and a 6% sales tax?

5. The tile laying crew (one tile setter and one helper) will set 900 SF per day. How long should the tile in problem #4 take to set? (Round to the full day.) If the tile setter is billed at $86.50 per hour and the helper is billed at $60.25 per hour what is the labor cost per SF for the job?

6. There are seven (7) offices along the outside wall of an office space. They are 11'-6" wide and total 138'-0" long. Calculate the amount of carpet to be installed (net area) and the amount of carpet materials to be carried in the estimate if the carpet is sold 12'-0". Include 5% waste.

7. The following trim will need to be back-primed with paint (all 4-sides); 220' of 1" × 6", 280' of 1" × 8", 135' of 1" × 10", and 96' of 1" × 12". Calculate the surface area to be painted. If a single gallon of primer will cover 220 SF of surface area (one coat), how many gallons are required to prime the wood trim with two coats?

8. The exterior elevation of a house shows eight (8) windows, each with a pair of shutters. Historical data for the painting company reveals that it takes 0.3 hrs. to remove a single shutter from the house, 0.85 hrs. to

prep a shutter, 1.2 hrs. to paint each shutter with two coats (both sides), and 0.25 hrs. to rehang each shutter. If the painter's billing rate is $66.50 per hr., what is the total cost to complete the shutters on this elevation?

9. Using the partition in problem #1, calculate the material cost to prime and paint two coats of interior paint on the exposed surface of the GWB full height each side. Primer coverage is 180 SF per gallon at a cost of $22.00 per gallon, first coat of finish latex coverage is 220 SF per gallon, and the second coat of finish latex coverage is 270 SF per gallon. The finish latex paint cost $31.50 per gallon.

10. The interior painting in problem #9 will be rolled. A single painter will roll primer at 900 SF per day, first coat of finish at 750 SF per day, and second coat of finish at 1,000 SF per day. How many days rounded to the full day will it take to complete the paint in problem #9 if there are two painters on the crew? At the billing rate of $66.50 per hr. what is the labor cost for the painting?

15 Specialties

CSI MasterFormat® Division 10—Specialties includes items that are manufactured off-site and shipped prefinished, and often preassembled, for easy on-site installation. Specialty items are generally more common in commercial than residential projects, though not exclusively. This division includes a wide variety of familiar items. Division 10—Specialties is categorized in groups, as the following outline shows.

To begin any takeoff for Division 10—Specialties items or tasks, the estimator should start with a careful review of the technical specifications and create a list of the items or tasks on the project. The list should be separated according to interior and exterior specialties. Once the estimator is familiar with the items, he or she can go to the appropriate drawings for the quantities and confirmation.

A brief checklist and their respective CSI number are outlined in the sections below.

INTERIOR SPECIALTIES

A. 10 10 00 Information Specialties
- Visual Display Units and Display Cases
- Chalkboards, Tackboards, and Markerboards
- Directories
- Signage: Interior and Exterior
- Telephone Specialties and Informational Kiosks

B. 10 20 00 Interior Specialties
- Compartments and Cubicles
- Partitions and Service Walls
- Wall and Door Protection
- Toilet Bath and Laundry Accessories

Estimating Building Costs for the Residential and Light Commercial Construction Professional, Third Edition. Wayne J. Del Pico.
© 2023 John Wiley & Sons, Inc. Published 2023 by John Wiley & Sons, Inc.

C. 10 30 00 Fireplaces and Stoves
- Manufactured Fireplaces
- Fireplace Specialties
- Stoves

D. 10 40 00 Safety Specialties
- Emergency Access and Information Cabinets
- Emergency Aid Specialties
- Fire Protection Specialties

E. 10 50 00 Storage Specialties
- Lockers
- Postal Specialties
- Storage Assemblies
- Wardrobe and Closet Specialties

EXTERIOR SPECIALTIES

F. 10 70 00 Exterior Protection
- Canopies and Awnings
- Sun Protection
- Protective Covers
- Rolling Shutters and Storm Panels
- Flagpoles
- Manufactured Exterior Specialties
- Cupolas, Weathervanes, and Steeples

G. 10 80 00 Other Specialties
- Bird, Insect, and Rodent Control Devices
- Grilles and Screens
- Security Mirrors and Domes
- Flags and Banners
- Scales
- Monument Signs

Division 10 has often been referred to as the "catchall" division. The items included in this section are often added after the finishes on the project are complete and, therefore, occur in the latter part of the schedule. Specialty items can be installed by a contractor's own forces or subcontracted to a specialty subcontractor.

Most of the work will be depicted on the architectural drawings. All plan views, elevations, and schedules should be reviewed carefully for reference notes and illustrations. Occasionally, typical details will be presented in latter sheets of the architectural drawings. There are also instances in which the specific work items of Division 10 are a single note on the plan, or under remarks in the Room Finish Schedule. Classic examples include fire extinguishers and cabinets, toilet partitions and accessories, signage, and metal lockers. The work of Division 10 is not relegated exclusively to the interior of the building; therefore, be sure to consult the site drawings for items such as site signage and flagpoles. The architect does not always provide the graphic detail for Division 10 work that is frequently found with other items, but more commonly relies on the language of the specifications to define the scope of work. Review the specifications for items not shown on the drawings. Often, designers provide the name of a vendor or distributor and contact information for a particular product. This is a fairly good indicator that substitutions will most likely not be accepted. Coordination with drawings within other disciplines is also recommended to verify quantities and sizes.

Many of the items in this section come furnished and delivered only. Therefore, it is essential to obtain current pricing for each product. This is often as simple as searching the specified manufacturer's website or soliciting a price. Websites often provide illustrations or information that allow you to determine if assembly is required or if there are other unique features of the product that affect pricing.

The estimator is recommended to keep interior and exterior specialties separate in the estimate, as one is affected by weather or may require hoisting or a lift for access.

TAKING-OFF AND PRICING

The most common takeoff method is by the individual piece, or each (EA). Some specialty items such as fire extinguishers, toilet accessories, signage (both interior and exterior), prefabricated fireplaces, lockers, and flagpoles can be repetitive and are best estimated by counting and assigning a lump-sum price. Other items, such as chalkboards and tackboards, toilet partitions, and grilles, are priced by the square foot (SF) area of the individual piece. Be sure to consider equipment costs for setting items that are too large to handle by hand. For example, most large flagpoles require some type of crane to install. This cost must be included in the estimate.

Also consider coordination with other trades such as providing power to lighted signage, wood blocking for anchorage, or on-site painting. Knowing the level of assembly required when a product arrives on-site can often mean the difference between success and failure in predicting an accurate cost. The estimator is directed to the manufacturer's website to find the most accurate information on assembly guidelines.

Be sure to consider learning-curve productivity for self-performed work under this section if the installers do not have historical cost data models on these items for reference.

Labor to Install Division 10 Items

Labor for this division is most often estimated by the piece (EA). For example, the labor cost for installing interior signage can be calculated by the quantity of signs that would be installed in 1 labor-hour by a single worker. Alternate methods include the number of labor-hours per unit, as in the case of installing a prefabricated fireplace. Installation costs for chalkboards and tackboards can be calculated by the individual piece for boards of the same size, or by the SF area for boards of varying sizes or dimensions. This is more of a matter of preference than actual standard estimating procedure.

The work of Division 10 cannot be classified into one individual trade or crew size. It is often a challenge to predict who will actually do the work. One thing is for sure: Skilled labor is required for a correct installation of Division 10 work. Tasks such as installing chalkboards and tackboards, setting toilet partitions, and installing flagpoles are better executed by multiperson crews due to the size, weight, and other physical attributes of these products. Many tasks, such as installing bathroom accessories or room signage, are best executed by an individual tradesperson.

Other special considerations for accurately estimating Division 10 work include the cost for unwrapping or uncrating items and disposing of shipping packaging. While this may not be a major concern for a single prefabricated fireplace, it does constitute a cost when estimating multi-unit dwellings, such as condominium and townhome projects. Projects that are phased often pose other challenges. For example, many manufacturers prefer to price their product with a single drop shipment versus per phase. If storage, rehandling, and distribution costs are required per phase, be sure to include them in the estimate.

Takeoff for Division 10 work can be used as a checklist against other items in separate divisions. For example, if grab bars in toilet compartments require wood blocking within framed wood or metal stud walls, the estimator can check the quantity of wood blocking against the total number of grab bars as a way of verifying both tasks.

SUMMARY

Although Division 10—Specialties does not typically amount to a large portion of the total estimated cost of a project, there are often many items included from a variety of locations. While specialties may be shown on the drawings, occasionally they are not. In this case, the specifications may be the determining factor for both quantity and quality.

Division 10 includes such an enormous quantity and variety of different types of specialties that it would be impractical to attempt to provide labor-hours for installation costs within this chapter. For additional information on installation costs, consult *RSMeans Estimating Handbook, Third Edition*, or the individual manufacturer's product literature.

Determining the actual trade required to install a particular item can often pose a challenge, but all require skilled labor. Consider lifts for access to the work or cranes for the hoisting of the product into place. Separate interior from exterior tasks when creating a checklist. In addition, always consider assembly time and cost if required as well as disposal costs for the debris generated by the packaging.

QUESTIONS/ PROBLEMS

1. The project has twelve (12) marker/tackboards all 4'-0" high and 5–10' long, 3–12' long and 4–8' long. If it takes 8 hours for a crew of two (2) carpenters to install all 12, what are the labor-hours required per SF?

2. If the two carpenters on the crew in problem # 1 are billed at $86.50 per hr, what is the cost per SF?

3. What sources should an estimator consider searching to determine how a bank of metal lockers is assembled?

4. Historical data indicates that it takes one (1) carpenter 8 hours to install all of the toilet accessories for a six (6) -fixture toilet room (3 water closets and 3 lavs). If there are four (4) more toilet rooms with four (4) fixtures each, what is the amount of labor hours required by the carpenter to complete the installation of the accessories in the remaining four toilet rooms? (Assume same accessories per fixture.)

5. What is the designer's purpose for including the name and contact information for the vendor of a specific product within a technical specification section?

16 | Equipment, Furnishings, Special Construction, and Conveying Equipment

While equipment, furnishings, special construction, and conveying systems occupy four separate divisions in CSI MasterFormat® (Divisions 11–14), they have been combined into one chapter because of their limited applicability to residential/light commercial construction. Much of the work is subcontracted due to its highly specialized nature and unique applications. Examples include Division 11's equipment for bank, library, medical, or ecclesiastical applications, as well as elevators and wheelchair lifts in Division 14—Conveying Systems. Other items, such as rugs, furniture, window treatments, and artwork in Division 12—Furnishings, are frequently contracted separately by the owner and only coordinated by the general contractor during installation. Each particular scenario requires careful evaluation and acknowledgment within the estimate. The discussion of these divisions in this chapter concerns only their most common applications and estimating guidelines for the average residential/light commercial project.

DIVISION 11—EQUIPMENT

CSI Division 11—Equipment includes a wide range of equipment such as central vacuum cleaning systems, residential kitchen appliances, audio-visual equipment, and any related specialized equipment. All buildings are initially designed with a specific function in mind – for example, libraries, churches, medical or dental offices, retail spaces, or restaurants, all of which require specialized equipment to allow them to perform their individual trade or business. Occasionally, systems or equipment are contracted directly by the owner under a separate agreement from the contract with the general contractor, but more often it is the responsibility of the general contractor to provide this equipment installed, tested, and fully functional as part of the base contract. Many contracts even provide for owner training. A checklist is

Estimating Building Costs for the Residential and Light Commercial Construction Professional,
Third Edition. Wayne J. Del Pico.
© 2023 John Wiley & Sons, Inc. Published 2023 by John Wiley & Sons, Inc.

always helpful when determining the scope of Division 11. The work of Division 11—Equipment is classified in the following categories:

A. 11 10 00 Vehicle and Pedestrian Equipment
- Vehicle Service Equipment
- Parking Control Equipment
- Loading Dock Equipment
- Pedestrian Control Equipment
- Security Control Equipment

B. 11 20 00 Commercial Equipment
- Retail and Service Equipment
- Banking Equipment
- Hospitality Equipment
- Office Equipment
- Postal, Packaging, and Shipping Equipment

C. 11 30 00 Residential Equipment
- Residential Appliances
- Retractable Stairs
- Unit Kitchens

D. 11 40 00 Foodservice Equipment
- Commercial Kitchen Equipment
- Food Prep Equipment
- Walk-in Coolers and Freezers
- Ice Machines
- Food Service Cleaning and Disposal Equipment

E. 11 50 00 Educational/Scientific Equipment
- Library Equipment
- Audio-Visual Equipment
- Laboratory Equipment
- Vocational Shop Equipment
- Planetarium Equipment
- Exhibit Equipment

F. 11 60 00 Entertainment and Recreation Equipment
- Broadcast, Theater, and Stage Equipment
- Musical Equipment
- Athletic Equipment
- Recreational Equipment
- Play Field Equipment

G. 11 70 00 Healthcare Equipment
- Medical Sterilizing Equipment
- Examination and Treatment Equipment
- Patient Care Equipment
- Dental Equipment
- Optical Equipment
- Operating Room Equipment
- Radiology Equipment

- Mortuary Equipment
- Therapy Equipment

H. 11 80 00 Facility Maintenance and Operation Equipment
- Facility Maintenance Equipment
- Facility Solid Waste Handling Equipment

I. 11 90 00 Other Equipment
- Religious Equipment
- Agricultural Equipment
- Horticultural Equipment
- Veterinary Equipment
- Arts and Crafts Equipment
- Security Equipment
- Detention Equipment

Much of the equipment in Division 11 is delivered in the final phase of the project and installed with minimum impact to the schedule, while other items are installed in various phases of construction and require the general and specialized equipment contractors to maintain close contact as the project progresses. For these types of equipment, the estimator must calculate the costs of interfacing the various systems or equipment with the structure itself. Consider the cost related to leaving a portion of the structure (roof or wall) open until such time as the equipment can be loaded into the building and installed.

For example, consider bank vault doors. While the actual cost of the equipment portion of the work may be outside the experience of the average general contractor's estimator, calculating the cost of the interface or preparatory work is not. Many estimators attempt to establish budgets based on historical cost data from previous projects. However, since most general contractors subcontract this work, detailed costs of materials, labor, and equipment may not be available. The only information may be the subcontract amount, plus any authorized change orders. This information should be used to establish budgets as "plugs" in the estimate until firm subcontractors' quotes are available. Other options include subcontractor quotes received over a period of time, which can establish a unit cost for the work. This is referred to as a *parametric* estimate and will be discussed in a later chapter.

When soliciting a bid from a specialized subcontractor, it is the estimator's responsibility to carefully define the scope of work to be priced with reference to the relevant CSI MasterFormat® section of work within the project documents. This includes coordination between mechanical and electrical disciplines for work that may be needed to complete the specialized equipment package. As an example, consider the coordination required for restaurant kitchen equipment. There is gas, electrical, and plumbing requirements for virtually every piece of equipment within the kitchen. Some equipment, such as exhaust hoods over cooking lines, requires coordination with roof work, ductwork, structural reinforcement, and even the fire alarm system. Making sure that all coordinated work is included is no insignificant task. Once the subcontractor's proposal has been submitted, be sure to carefully review and qualify the work.

Most of the work of Division 11 can be found on the architectural drawings. All plan views, elevations, and schedules should be reviewed for reference notes and illustrations. On some projects where the scope of work is extensive, separate drawings in the bid set include specialty drawings to coordinate this work with the mechanical and electrical scopes as well as adjacent work. The architect does not always provide the graphic detail for Division 11 work that is frequently found with other items and may rely on the language of the specifications and the expertise of the specialty contractor to fill in the missing details. Due to the highly specialized nature of the design of these systems, the architect will often refer to the manufacturers' product literature for pricing information. Review the Division 11 specification section for items not shown on the drawings. Consult drawings from other disciplines, such as the mechanical and electrical drawings, as many pieces of equipment require building utilities to operate.

Residential appliances in Division 11 include ranges, refrigerators, dishwashers, and laundry equipment. Most vendors provide pricing on residential appliances furnished and delivered to the site only, excluding installation. Appliances can also be purchased, delivered, and set in place if specifically requested. Be sure to price appliances through a supplier that can provide up-to-date quotes – often as simple as searching the specified manufacturers' website and soliciting a price. The actual hookup, or *tie-in,* labor and material costs are the responsibility of other trades, such as the plumbing and electrical contractors. Coordinate the various trades to ensure that all tie-in costs are included.

Taking-off and Pricing

Simple Division 11 items, such as residential appliances, are taken off and priced by the individual piece (EA). Other, more complicated equipment, such as athletic equipment and all of the related components that complete a system – including installation – can be priced as a lump sum (LS). Assembly can be a factor in some of the tasks, especially larger items that are shipped in pieces. Consider any equipment or lifts that may be required for those items above grade.

Labor for Division 11 Items

Estimating installation labor for Division 11 work follows the takeoff unit costs, by the individual piece (EA). For residential appliances, a multiperson crew is required. Labor costs are calculated by the number of labor-hours per each appliance or the installation as a whole. The work of Division 11 cannot be classified into one individual trade or crew size, as it depends on the task at hand. Often, the crew is a mix of skilled and unskilled labor.

Special considerations for accurately estimating this work include costs of unwrapping or uncrating items and disposing of shipping packaging. While this may not be a major concern for a single refrigerator, it does impose a financial impact when estimating multiple appliances for multiunit dwellings, such as for apartment projects. If storage, rehandling, and distribution costs are required per phase, they must also be included in the estimate. In addition, consider costs for assembly once on-site, testing, and training of the owner's personnel, which may be part of the cost.

Figures 16.1 through 16.5 are provided as guidance for estimating the costs of labor to install some of the more common items of Division 11.

Maintenance Equipment	Crew Makeup	Daily Output	Labor-Hours	Unit
VACUUM CLEANING Central, 3-inlet, residential	1 Skilled Worker	.90	8.890	Total
Commercial		.70	11.430	Total
5-inlet system, residential		.50	16.000	Total
7-inlet system		.40	20.000	Total
9-inlet system		.30	26.670	Total

Figure 16.1 Installation Time for Maintenance Equipment

Source RSMeans Estimating Handbook, Third Edition. © 2009 John Wiley & Sons, Inc.

Library Equipment	Crew Makeup	Daily Output	Labor-Hours	Unit
LIBRARY EQUIPMENT Bookshelf, mtl., 90" high, 10" shelf				
Single face	1 Carpenter	12	.667	LF
Double face		12	.667	LF
Carrels, hardwood, 36" × 24", minimum		5	1.600	EA
Maximum		4	2.000	EA
Charging desk, built-in, with counter, plastic laminated top		7	1.140	LF

Figure 16.2 Installation Time for Library Equipment

Source RSMeans Estimating Handbook, Third Edition. © 2009 John Wiley & Sons, Inc.

Projection Screens	Crew Makeup	Daily Output	Labor-Hours	Unit
PROJECTION SCREENS Wall or ceiling hung,				
glass beaded, manually operated, economy	2 Carpenters	500	.032	SF
Intermediate		450	.036	SF
Deluxe		400	.040	SF
Electric operated, glass beaded, 25 SF, economy		5	3.200	EA
Deluxe		4	4.000	EA
50 S.F., economy		3	5.330	EA
Deluxe		2	8.000	EA
Heavy duty, electric operated, 200 SF		1.50	10.670	EA
400 S.F.		1	16.000	EA
Rigid acrylic in wall, for rear projection, 1/4" thick	2 Glaziers	30	.533	SF
1/2" thick (maximum size 10' × 20')	"	25	.640	SF

Figure 16.3 Installation Time for Projection Screens

Source RSMeans Estimating Handbook, Third Edition. © 2009 John Wiley & Sons, Inc.

Parking Control Equipment	Crew Makeup	Daily Output	Labor-Hours	Unit
PARKING EQUIPMENT Traffic, detectors, magnetic	2 Electricians	2.70	5.930	EA
Single treadle		2.40	6.670	EA
Automatic gates, 8' arm, one way		1.10	14.550	EA
Two way		1.10	14.550	EA
Fee indicator, 1" display		4.10	3.900	EA
Ticket printer and dispenser, standard		1.40	11.430	EA
Rate computing		1.40	11.430	EA
Card control station, single period		4.10	3.900	EA
4 period		4.10	3.900	EA
Key station on pedestal		4.10	3.900	EA
Coin station, multiple coins	↓	4.10	3.900	EA

Figure 16.4 Installation Time for Parking Control Equipment

Source RSMeans Estimating Handbook, Third Edition. © 2009 John Wiley & Sons, Inc.

Food Service Equipment	Crew Makeup	Daily Output	Labor-Hours	Unit
APPLIANCES Cooking range, 30" freestanding, 1 oven				
Minimum	2 Building Laborers	10	1.600	EA
Maximum		4	4.000	EA
2 oven, minimum		10	1.600	EA
Maximum		4	4.000	EA
Built-in, 30" wide, 1 oven, minimum	2 Carpenters	4	4.000	EA
Maximum		2	8.000	EA
2 oven, minimum		4	4.000	EA
Maximum		2	8.000	EA
Freestanding, 21"-wide range, 1 oven, minimum	2 Building Laborers	10	1.600	EA
Maximum	"	4	4.000	EA
Countertop cook tops, 4-burner, standard				
Minimum	1 Electrician	6	1.330	EA
Maximum		3	2.670	EA
As above, but with grille and griddle attachment				
Minimum		6	1.330	EA
Maximum		3	2.670	EA
Induction cooktop, 30" wide		3	2.670	EA
Microwave oven, minimum		4	2.000	EA
Maximum	↓	2	4.000	EA
Combination range, refrigerator and sink, 30" wide				
Minimum	1 Electrician 1 Plumber	2	8.000	EA
Maximum		1	16.000	EA
60" wide, average		1.40	11.430	EA
72" wide, average		1.20	13.330	EA
Office model, 48" wide		2	8.000	EA
Refrigerator and sink only	↓	2.40	6.670	EA
Combination range, refrigerator, sink, microwave oven, and ice maker	1 Electrician 1 Plumber	.80	20.000	EA
Compactor, residential size, 4-to-1 compaction				
Minimum	1 Carpenter	5	1.600	EA
Maximum	"	3	2.670	EA
Deep freeze, 15 to 23 CF, minimum	2 Building Laborers	10	1.600	EA
Maximum		5	3.200	EA
30 CF, minimum		8	2.000	EA
Maximum	↓	3	5.330	EA
Dishwasher, built-in, 2 cycles, minimum	1 Electrician 1 Plumber	4	4.000	EA
Maximum		2	8.000	EA
4 or more cycles, minimum		4	4.000	EA
Maximum	↓	2	8.000	EA
Dryer, automatic, minimum	1 Carpenter 1 Helper	3	5.330	EA
Maximum	↓	2	8.000	EA
Garbage disposer, sink type, minimum	1 Electrician 1 Plumber	5	3.200	EA
Maximum	↓	3	5.330	EA

Figure 16.5a Installation Time for Food Service Equipment

Source RSMeans Estimating Handbook, Third Edition. © 2009 John Wiley & Sons, Inc.

Food Service Equipment	Crew Makeup	Daily Output	Labor-Hours	Unit
Appliances, Heater, electric, built-in, 1,250-watt				
Ceiling type, minimum	1 Electrician	4	2.000	EA
Maximum		3	2.670	EA
Wall type, minimum		4	2.000	EA
Maximum		3	2.670	EA
1,500-watt wall type with blower		4	2.000	EA
3,000-watt		3	2.670	EA
Hood for range, 2-speed, vented, 30" wide				
Minimum	1 Carpenter .5 Electrician .5 Sheet Metal Worker	5	3.200	EA
Maximum		3	5.330	EA
42" wide, minimum		5	3.200	EA
Maximum		3	5.330	EA
Icemaker, automatic, 13lbs./day	1 Plumber	7	1.140	EA
51lbs./day		2	4.000	EA
Refrigerator, no frost, 10 to 12 CF, minimum	2 Building Laborers	10	1.600	EA
Maximum		6	2.670	EA
14 to 16 CF, minimum		9	1.780	EA
Maximum		5	3.200	EA
18 to 20 CF, minimum		8	2.000	EA
Maximum		4	4.000	EA
21 to 29 CF, minimum		7	2.290	EA
Maximum		3	5.330	EA
Sump pump cellar drainer, 1/3 HP, minimum	1 Plumber	3	2.670	EA
Maximum		2	4.000	EA
Washing machine, automatic, minimum		3	2.670	EA
Maximum		1	8.000	EA
Water heater, electric, glass-lined, 30-gallon				
Minimum	1 Electrician 1 Plumber	5	3.200	EA
Maximum		3	5.330	EA
80-gallon		2	8.000	EA
Maximum		1	16.999	EA
Water heater, gas, glass-lined, 30-gallon, minimum	2 Plumbers	5	3.200	EA
Maximum		3	5.330	EA
50-gallon, minimum		2.50	6.400	EA
Maximum		1.50	10.670	EA
Water softener, automatic, to 30 grains/gallon		5	3.200	EA
To 75 grains/gallon		4	4.000	EA
Vent kits for dryers	1 Carpenter	10	.800	EA
KITCHEN EQUIPMENT Bake oven, single deck	1 Plumber 1 Plumber Apprentice	8	2.000	EA
Double deck		7	2.290	EA
Triple deck		6	2.670	EA
Electric convection, 40" × 45" × 57"	2 Carpenters 1 Building Laborer .5 Electrician	4	7.000	EA

Figure 16.5b *(Continued)*

DIVISION 12—FURNISHINGS

Division 12—Furnishings is frequently handled under a separate contract directly with the owner. The items most commonly included within the general contract for this division are window treatments and manufactured casework. Window treatments include vertical or horizontal blinds, roller shades, and drapery rods. Manufactured casework is defined as almost any mass-produced cabinetry, countertops, or casework for a specific use that would be routinely found at a big-box home improvement retailer. Again, this work is often estimated as furnished and installed, since it can be easily subcontracted.

Food Service Equipment	Crew Makeup	Daily Output	Labor-Hours	Unit
KITCHEN EQUIPMENT Broiler, without oven				
Standard	1 Plumber	8	2.000	EA
	1 Plumber Apprentice			
Infrared	2 Carpenters	4	7.000	EA
	1 Building Laborer			
	.5 Electrician			
Cooler, reach-in, beverage, 6' long	1 Plumber	6	2.670	EA
	1 Plumber Apprentice			
Dishwasher, commercial, rack-type				
10 to 12 racks/hour	1 Plumber	3.20	5.000	EA
	1 Plumber Apprentice ↓			
Semiautomatic 38 to 50 racks/hour		1.30	12.310	EA
Automatic 190 to 230 racks/hour	2 Plumbers	.70	34.290	EA
	1 Plumber Apprentice			
235 to 275 racks/hour	↓	.50	48.000	EA
8,750 to 12,500 dishes/hour		.20	120.000	EA
Fast-food equipment, total package, minimum	6 Skilled Workers	.08	600.000	EA
Maximum	↓	.07	686.000	EA
Food mixers, 20 quarts	2 Carpenters	7	4.000	EA
	1 Building Laborer			
	.5 Electrician			
60 quarts	↓	5	5.600	EA
Freezers, reach-in, 44 CF	1 Plumber	4	4.000	EA
	1 Plumber Apprentice			
68 CF		3	5.330	EA
Fryer, with submerger, single		7	2.290	EA
Double		5	3.200	EA
Griddle, 3' long		7	2.290	EA
4' long		6	2.670	EA
Ice cube maker, 50 lbs./day		6	2.670	EA
500 lbs./day	↓	4	4.000	EA
Kettles, steam-jacketed, 20 gallons	2 Carpenters	7	4.000	EA
	1 Building Laborer			
	.5 Electrician			
60 gallons	↓	6	4.670	EA
Range, restaurant-type, 6 burners and				
1 oven, 36"	1 Plumber	7	2.290	EA
	1 Plumber Apprentice			
2 ovens, 60"		6	2.670	EA
Heavy-duty, single 34" oven, open top		5	3.200	EA
Fry top		6	2.670	EA
Hood fire protection system, minimum		3	5.330	EA
Maximum		1	16.000	EA
Refrigerators, reach-in type, 44 CF		5	3.200	EA
With glass doors, 68 CF	↓	4	4.000	EA

Figure 16.5c *(Continued)*

Food Service Equipment	Crew Makeup	Daily Output	Labor-Hours	Unit
KITCHEN EQUIPMENT Steamer, electric, 27 KW	2 Carpenters	7	4.000	EA
	1 Building Laborer			
	.5 Electrician			
Electric, 10 KW or gas 100,000	.5 Electrician	5	5.600	EA
Rule of thumb: Equipment cost based on				
kitchen work area				
Office buildings, minimum	2 Carpenters	77	.364	SF
	1 Building Laborer			
	.5 Electrician			
Maximum		58	.483	SF
Public eating facilities, minimum		77	.364	SF
Maximum		46	.609	SF
Hospitals, minimum		58	.483	SF
Maximum	↓	39	.718	SF
WINE VAULT Redwood, air-conditioned, walk-in type				
6'-8" high, incl. racks, 2' × 4' for 156 bottles	2 Carpenters	2	8.000	EA
4' × 6' for 614 bottles		1.50	10.670	EA
6' × 12' for 1,940 bottles	↓	1	16.000	EA

Figure 16.5d *(Continued)*

Other items, especially on commercial projects, include artwork, rugs, interior plantings, and furniture and seating, as in the case of a restaurant.

Commercial projects often have a *Furniture, Fixtures, and Equipment* scope that is identified with a separate plan. This is abbreviated as FF&E and in many cases the owner purchases and furnishes to the site the products, and the contractor installs the package. Consider the tables, chairs, and bar stools for a restaurant. The estimator should review the scope carefully to determine the extent of the work. Often, the scope can include unloading, storage, unpacking, assembly, setting in place and disposal of the packing. Other can have an installation requirement such as artwork hung on the wall.

Developers of residential properties often furnish the model with a wide variety of furniture and accessories to enhance the appeal of the space. This process is called *staging* and contains many of the items in this division. While this is not within the typical domain of the contractor, it can be. It is almost always a labor-only task.

Consult the architectural drawings in plan view and elevation to determine window-opening sizes. Window schedules are also helpful. Quantities can be determined from reviewing plan views in conjunction with elevations. It is essential that you review the window treatment specification section for products and special characteristics necessary for accurate pricing. Fire-rated and noncombustible window treatments, for example, have a tremendous impact on pricing.

Manufactured casework frequently has a schedule defining the individual components that constitute the work. Repetitive projects such as apartments, medical offices, classrooms, or hotel suites are most often the application for manufactured casework. The work of Division 12—Furnishings can be classified in the following categories:

A. 12 10 00 Art
   - Murals
   - Wall Decorations
   - Sculptures
   - Art Glass & Religious Art
B. 12 20 00 Window Treatments
   - Window Blinds
   - Curtain and Drapes
   - Interior Shutters
   - Window Shades
   - Window Treatment Operating Hardware
   - Interior Daylighting Devices
C. 12 30 00 Casework
   - Manufactured Metal Casework
   - Manufactured Wood Casework
   - Manufactured Plastic Casework
   - Specialty Casework
   - Countertops

D. 12 40 00 Furnishings and Accessories
- Office Accessories
- Table Accessories
- Portable Lamps
- Bath Furnishings
- Bedroom Furnishings
- Furnishing Accessories
- Rugs and Mats

E. 12 50 00 Furniture
- Office Furniture
- Seating
- Retail Furniture
- Hospitality Furniture
- Detention Furniture
- Institutional Furniture
- Industrial Furniture
- Residential Furniture
- Systems Furniture

F. 12 60 00 Multiple Seating
- Fixed Audience Seating
- Portable Audience Seating
- Stadium and Arena Seating
- Booths and Tables
- Multiuse Fixed Seating
- Telescoping Stands
- Pews and Benches
- Seat and Table Assemblies

G. 12 90 00 Other Furnishings
- Interior Planters and Artificial Plants
- Interior Public Space Furnishings

Taking-off and Pricing

Window treatments, such as blinds and shades, are taken off by the area of the opening they occupy, in square feet (SF), and are listed in the takeoff as width by height. Drapery rods are taken off by the width of the opening and listed in the takeoff in linear feet (LF). For motorized window treatment hardware, items are listed in the takeoff as each (EA). The estimator is advised to coordinate the electrical component of the work in the electrical portion of the estimate. Note that window treatments for odd-shaped or custom windows may have substantially higher costs per SF than the more common rectangular shaped treatments.

Manufactured casework is taken off by the piece (EA) and listed according to model number (if available), application (base or wall cabinet), size (width × depth × height) in inches, and any defining characteristics. Do not forget to include the costs of assembly, door and drawer hardware, distribution, and adjustments. Note that many manufacturers offer a diverse range of functional accessories for cabinets and drawers that can substantially increase the base cost.

Most other items within Division 12 are taken off by the individual piece and are listed as EA. The takeoff should describe the item or task sufficiently, including any assembly required to allow for accurate pricing. Keep in mind that some of the work of Division 12 can be installation and handling only. For example, it is not uncommon for the owner to purchase the Division 12 item and have it delivered to the site. The contractor then takes responsibility for the assembly, handling, distribution, and installation of the items. These costs may be negligible when considering two or three walk-off mats at entry doors; however, the costs can be substantial if one were to consider the seating and tables for a large restaurant.

One last consideration is the disposal of the packaging and debris generated by the items in Division 12, especially manufactured casework and furniture. Many of these items are shipped considerable distances and as a result are well protected to avoid damage during shipping. There can be a substantial amount of debris.

Labor for Division 12 Items

Estimating labor for Division 12 work can be done in a variety of ways. The most common method is to follow the takeoff units and estimate labor costs by the individual piece (EA). For example, the labor cost for installing window shades could be calculated by the quantity of shades that can be installed in 1 labor-hour by a single individual. The difference in labor costs between installing a 36" wide shade versus a 42" wide shade is negligible. Most estimators price installation by the piece rather than by size, although alternate methods include quantifying the number of labor-hours per unit, as for larger vertical blinds. Productivity is measured and based on the production of an individual, rather than a crew. Installation costs for other items follow the same procedure.

Other items in this division may require a crew to assemble and set in place, or to install. Again, the estimator would be wise to consult the individual manufacturer for as much information as possible. Items such as detention or institutional furniture make up a unique market and may be best served by a subcontractor price.

Figures 16.6 through 16.12 provide some guidance for the installation costs of the more common items of Division 12.

Artwork	Crew Makeup	Daily Output	Labor-Hours	Unit
ARTWORK Framed				
Photography, minimum	1 Carpenter	36	.222	EA
Maximum		30	.267	EA
Posters, minimum		36	.222	EA
Maximum		30	.267	EA
Reproductions, minimum		36	.222	EA
Maximum		30	.267	EA

Figure 16.6 Installation Time for Artwork

Source RSMeans Estimating Handbook, Third Edition. © 2009 John Wiley & Sons, Inc.

Hospital Casework	Crew Makeup	Daily Output	Labor-Hours	Unit
CABINETS				
Hospital, base cabinets, laminated plastic	2 Carpenters	10	1.600	LF
Enameled steel		10	1.600	LF
Stainless steel		10	1.600	LF
Cabinet base trim, 4" high, enameled steel		200	.080	LF
Stainless steel		200	.080	LF
Counter top, laminated plastic, no backsplash		40	.400	LF
With backsplash		40	.400	LF
For sink cutout, add		12.20	1.310	EA
Stainless steel counter top		40	.400	LF
Nurses station, door type, laminated plastic		10	1.600	LF
Enameled steel		10	1.600	LF
Stainless steel		10	1.600	LF
Wall cabinets, laminated plastic		15	1.070	LF
Enameled steel		15	1.070	LF
Stainless steel		15	1.070	LF
Kitchen, base cabinets, metal, minimum		30	.533	LF
Maximum		25	.640	LF
Wall cabinets, metal, minimum		30	.533	LF
Maximum		25	.640	LF
School, 24" deep		15	1.070	LF
Counter height units		20	.800	LF
Wood, custom fabricated, 32" high counter		20	.800	LF
Add for counter top		56	.286	LF
84" high wall units		15	1.070	LF

Figure 16.7 Installation Time for Hospital Casework

Source RSMeans Estimating Handbook, Third Edition. © 2009 John Wiley & Sons, Inc.

DIVISION 13— SPECIAL CONSTRUCTION

Division 13 includes a variety of unique and highly specialized work, hence the name Special Construction. In the residential/light commercial market, there are a few tasks, including swimming pools, sauna and steam rooms, interior and exterior fountains, ice rinks, clean rooms, and greenhouses, in Division 13. All require specialized training – both to execute and to estimate.

The work of Division 13—Special Construction can be classified in the following categories:

A. 13 10 00 Special Facility Components
- Swimming Pools
- Fountains: Interior and Exterior
- Aquariums
- Amusement Park Structures and Equipment
- Specialty Element Construction

Blinds	Crew Makeup	Daily Output	Labor-Hours	Unit
BLINDS, INTERIOR Solid colors				
Horizontal, 1" aluminum slats, custom, minimum	1 Carpenter	590	.014	SF
Maximum		440	.018	SF
2" aluminum slats, custom, minimum		590	.014	SF
Maximum		440	.018	SF
Stock, minimum		590	.014	SF
Maximim		440	.018	SF
2" steel slats, stock, minimum		590	.014	SF
Maximum		440	.018	SF
Custom, minimum		590	.014	SF
Maximum		400	.020	SF
Alternate method of figuring:				
1" aluminum slats, 48" wide, 48" high	1 Carpenter	30	.267	EA
72" high		29	.276	EA
96" high		28	.286	EA
72" wide, 72" high		25	.320	EA
96" high		23	.348	EA
96" wide, 96" high		20	.400	EA
Vertical, 3" to 5" PVC or cloth strips, minimum		460	.017	SF
Maximum		400	.020	SF
4" aluminum slats, minimum		460	.017	SF
Maximum		400	.020	SF
Mylar mirror-finish strips, to 8" wide, minimum		460	.017	SF
Maximum		400	.020	SF
Alternate method of figuring:				
2" aluminum slats, 48" wide, 48" high	1 Carpenter	30	.267	EA
72" high		29	.276	EA
96" high		28	.286	EA
72" wide, 72" high		25	.320	EA
96" high		23	.348	EA
96" wide, 96" high		20	.400	EA
Mirror finish, 48" wide, 48" high		30	.267	EA
72" high		29	.276	EA
96" high		28	.286	EA
72" wide, 72" high		25	.320	EA
96" high		23	.348	EA
96" wide, 96" high		20	.400	EA
Decorative printed finish, 48" wide, 48" high		30	.267	EA
72" high		29	.276	EA
96" high		28	.286	EA
72" wide, 72" high		25	.320	EA
96" high		23	.348	EA
96" wide, 96" high		20	.400	EA

Figure 16.8a Installation Time for Window Treatments – Blinds

Source RSMeans Estimating Handbook, Third Edition. © 2009 John Wiley & Sons, Inc.

- Tubs and Pools (Hot Tubs and Therapeutic Pools)
- Ice Rinks
- Kennels and Animal Shelters

B. 13 20 00 Special-Purpose Rooms
- Controlled Environment Rooms (Clean Rooms, Hyperbaric Rooms, Cold Storage Rooms, etc.)
- Office Shelters and Booths
- Planetariums
- Special Activity Rooms (Saunas and Steam Baths)
- Fabricated Rooms (Storm Shelters)
- Vaults
- Athletic and Recreational Special Construction (Safety Netting, Indoor Soccer Boards)

Blinds	Crew Makeup	Daily Output	Labor-Hours	Unit
BLINDS, INTERIOR Wood folding panels with moveable louvers				
7" × 20" each	1 Carpenter	17	.471	PR
8" × 28" each		17	.471	PR
9" × 36" each		17	.471	PR
10" × 40" each		17	.471	PR
Fixed louver type, stock units, 8" × 20" each		17	.471	PR
10" × 28" each		17	.471	PR
12" × 36" each		17	.471	PR
18" × 40" each		17	.471	PR
Insert panel type, stock, 7" × 20" each		17	.471	PR
8" × 28" each		17	.471	PR
9" × 36" each		17	.471	PR
10" × 40" each		17	.471	PR
Raised panel type, stock, 10" × 24" each		17	.471	PR
12" × 26" each		17	.471	PR
14" × 30" each		17	.471	PR
16" × 36" each		17	.471	PR

Figure 16.8b *(Continued)*

Shades	Crew Makeup	Daily Output	Labor-Hours	Unit
SHADES Basswood, roll-up, stain finish, 3/8" slats	1 Carpenter	300	.027	SF
7/8" slats		300	.027	SF
Vertical side slide, stain finish, 3/8" slats		300	.027	SF
7/8" slats		300	.027	SF
Mylar, single layer, non-heat-reflective		685	.012	SF
Double-layered, heat-reflective		685	.012	SF
Triple-layered, heat-reflective		685	.012	SF
Vinyl-coated cotton, standard		685	.012	SF
Lightproof decorator shades		685	.012	SF
Vinyl, lightweight, 4-gauge		685	.012	SF
Heavyweight, 6-gauge		685	.012	SF
Vinyl laminated fiberglass, 6-gauge, translucent		685	.012	SF
Lightproof		685	.012	SF
Woven aluminum, 3/8" thick, lightproof and fireproof		350	.023	SF
Insulative shades		125	.064	SF
Solar screening, fiberglass		85	.094	SF
Interior insulative shutter, stock unit, 15" × 60"		17	.471	PR

Figure 16.9 Installation Time for Window Treatments – Shades

Source RSMeans Estimating Handbook, Third Edition. © 2009 John Wiley & Sons, Inc.

C. 13 30 00 Special Structures
   - Fabric Structures (Air-Supported Structures)
   - Space Frames
   - Geodesic Structures
   - Fabricated Engineered Structures (Greenhouses, Solariums, Sunrooms, and Conservatories)
   - Rammed Earth Construction
   - Towers

D. 13 40 00 Integrated Construction
   - Building Modules (Hospital Modules, Detentions Cell Modules, etc.)
   - Modular Mezzanines
   - Facility Protection
   - Sound, Vibration, and Seismic Control
   - Radiation Protection

Furniture	Crew Makeup	Daily Output	Labor-Hours	Unit
DORMITORY				
Desktop, built-in, laminated plastic, 24" deep				
Minimum	2 Carpenters	50	.320	LF
Maximum		40	.400	LF
30" deep, minimum		50	.320	LF
Maximum		40	.400	LF
Dressing unit, built-in, minimum		12	1.330	LF
Maximum	↓	8	2.000	LF
LIBRARY				
Attendant desk, 36" × 62" × 29" high	1 Carpenter	16	.500	EA
Book display, "A" frame display, both sides		16	.500	EA
Table with bulletin board	↓	16	.500	EA
Book trucks, descending platform,				
Small, 14" × 30" × 35" high	1 Carpenter	16	.500	EA
Large, 14" × 40" × 42" high		16	.500	EA
Card catalogu, 30-tray unit		16	.500	EA
60-tray unit	↓	16	.500	EA
72-tray unit	2 Carpenters	16	1.000	EA
120-tray unit	↓	16	1.000	EA
Carrels, single face, initial unit	1 Carpenter	16	.500	EA
Additional unit	↓	16	.500	EA
Double face, initial unit	2 Carpenters	16	1.000	EA
Additional unit		16	1.000	EA
Cloverleaf	↓	11	1.450	EA
Chairs, sled base, arms, minimum	1 Carpenter	24	.333	EA
Maximum		16	.500	EA
No arms, minimum		24	.333	EA
Maximum		16	.500	EA
Standard leg base, arms, minimum		24	.333	EA
Maximum		16	.500	EA
No arms, minimum		24	.333	EA
Maximum	↓	16	.500	EA
Charge desk, modular unit, 35" × 27" × 39" high				
Wood front and edges, plastic laminate tops				
Book return	1 Carpenter	16	.500	EA
Book truck port		16	.500	EA
Card file drawer, 5 drawers		16	.500	EA
10 drawers		16	.500	EA
15 drawers		16	.500	EA
Card and legal file		16	.500	EA
Charging machine		16	.500	EA
Corner		16	.500	EA
Cupboard		16	.500	EA
Detachable end panel		16	.500	EA
Gate		16	.500	EA
Knee space		16	.500	EA
Open storage		16	.500	EA
Station charge		16	.500	EA
Work station		16	.500	EA
Dictionary stand, stationary		16	.500	EA
Revolving	↓	16	.500	EA

Figure 16.10a Installation Time for Furniture

Source RSMeans Estimating Handbook, Third Edition. © 2009 John Wiley & Sons, Inc.

E. 13 50 00 Special Instrumentation
- Stress Instrumentation
- Seismic Instrumentation
- Meteorological Instrumentation

Taking-off and Pricing

Takeoff quantities in Division 13—Special Construction can have very diverse units of measure. Swimming pools, fountains, hot tubs, and aquariums can be quantified in the takeoff by the gallons of water they hold and listed as GL. Ice rinks can be classified by their size in SF, kennels by the number of cages or

Furniture	Crew Makeup	Daily Output	Labor-Hours	Unit
LIBRARY (cont.)				
Exhibit case, table style, 60" × 28" × 36"	1 Carpenter	11	.727	EA
Globe stand		16	.500	EA
Magazine rack		16	.500	EA
Newspaper rack		16	.500	EA
Tables, card catalog reference, 24" × 60" × 42"		16	.500	EA
24" × 60" × 72"		16	.500	EA
Index, single tier, 48" × 72"		16	.500	EA
Double tier, 48" × 72"	↓	16	.500	EA
Study, panel ends, plastic lam. surfaces 29" high				
36" × 60"	2 Carpenters	16	1.000	EA
36" × 72"		16	1.000	EA
36" × 90"		16	1.000	EA
48" × 72"	↓	16	1.000	EA
Parsons, 29" high, plastic laminate top,				
wood legs and edges				
36" × 36"	2 Carpenters	16	1.000	EA
36" × 60"		16	1.000	EA
36" × 72"		16	1.000	EA
36" × 84"		16	1.000	EA
42" × 90"		16	1.000	EA
48" × 72"		16	1.000	EA
48" × 120"		16	1.000	EA
Round, leg or pedestal base, 36" diameter		16	1.000	EA
42" diameter		16	1.000	EA
48" diameter		16	1.000	EA
60" diameter	↓	16	1.000	EA
RESTAURANT Bars, built-in, front bar	1 Carpenter	5	1.600	LF
Back bar	↓	5	1.600	LF

Figure 16.10b *(Continued)*

pens, and special-purpose rooms by their SF of floor space, SF FLR. Each can have its own distinct pricing structure.

It is recommended that the estimator prepare an estimate for comparison purposes based on any available historical data or published data. Costs can be distilled to specific parameters such as SF, SF FLR, VLF, GAL, or even by the number of occupants the space will hold. These parameters are then used to produce a *parametric* estimate or budget and are compared to solicited prices from subcontractors that will furnish and install all work complete. (See Chapter 28, "Conceptual Estimating," for more information on parametric estimating.)

The estimator is strongly encouraged to research the specific product to gain as much information as possible, such as shipping costs, assembly requirements, testing and calibration costs, and even equipment required for installation. Make sure all coordinated work is included within the appropriate division, such as electrical, mechanical, excavation, and backfill, to complete the installation of the Special Construction work. The estimator's main responsibility for this work is to ensure that the specialty subcontractor's proposal is all-inclusive.

Labor for Division 13 Items

Estimating labor for Division 13 work can be done in a variety of ways. The most common method is to follow the takeoff units and estimate labor costs

Booths and Tables	Crew Makeup	Daily Output	Labor-Hours	Unit
BOOTHS				
Banquet, upholstered seat and back, custom				
Straight, minimum	2 Carpenters	40	.400	LF
Maximum		36	.444	LF
"L" or "U" shape, minimum		35	.457	LF
Maximum	↓	30	.533	LF
Upholstered outside finished backs for				
single booths and custom banquets				
Minimum	2 Carpenters	44	.364	LF
Maximum	"	40	.400	LF
Fixed seating, one-piece plastic chair and				
plastic laminate tabletop				
Two-seat, 24" × 24" table, minimum	2 Carpenters 2 Building Laborers Power Tools	30	1.070	EA
Maximum		26	1.230	EA
Four-seat, 24" × 48" table, minimum		28	1.140	EA
Maximum		24	1.330	EA
Six-seat, 24" × 76" table, minimum		26	1.230	EA
Maximum		22	1.450	EA
Eight-seat, 24" × 102" table, minimum		20	1.600	EA
Maximum	↓	18	1.780	EA
Freestanding, wood fiber core with				
plastic laminate face, single booth				
24" wide	2 Carpenters	38	.421	EA
48" wide		34	.471	EA
60" wide		30	.533	EA
Double booth, 24" wide		32	.500	EA
48" wide		28	.571	EA
60" wide	↓	26	.615	EA
Upholstered seat and back				
Foursome, single booth, minimum	2 Carpenters	38	.421	EA
Maximum		30	.533	EA
Double booth, minimum		32	.500	EA
Maximum	↓	26	.615	EA
Mount in floor, wood fiber core with				
plastic laminate face, single booth				
24" wide	2 Carpenters 2 Building Laborers Power Tools	30	1.070	EA
48" wide		28	1.140	EA
60" wide		26	1.230	EA
Double booth, 24" wide		26	1.230	EA
48" wide		24	1.330	EA
60" wide	↓	22	1.450	EA

Figure 16.11 Installation Time for Booths and Tables

Source RSMeans Estimating Handbook, Third Edition. © 2009 John Wiley & Sons, Inc.

Description	Crew Makeup	Daily-Output	Labor-Hours	Unit
Cabinets, kitchen base, 24" × 24" × 35" high	2 Carpenters	22.30	.717	EA
Kitchen wall, 12" × 24" × 30" high		20.30	.788	
Casework frames				
Base cabinets, 36" high, two-bay, 36" wide	1 Carpenter	2.20	3.636	EA
Bookcases, 7' high, two-bay, 36" wide		1.60	5.000	
Coatracks, 7' high, two-bay, 48" wide		2.75	2.909	
Wall-mounted cabinets, 30" high, two-bay, 36" wide		2.15	3.721	
Wardrobe, 7' high, 48" wide		1.70	4.706	
Cabinet doors				
Glass panel, hardware frame, 18" wide, 30" high	1 Carpenter	29.00	.276	EA
Hardwood, raised panel, 18" wide, 30" high		14.00	.571	

Figure 16.12a Installation Time in Labor-Hours Manufactured Casework

Source RSMeans Estimating Handbook, Third Edition. © 2009 John Wiley & Sons, Inc.

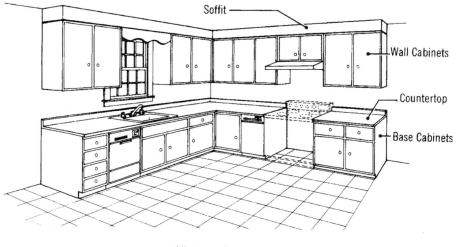

Kitchen Cabinets

Figure 16.12b *(Continued)*

by specific task in the process. This requires a thorough understanding of each step in the process. Most of the tasks in Division 13 are crew tasks. That crew may comprise multiple trades, depending on the task in the process.

Figures 16.13 through 16.17 are provided as guidance for estimating the costs of labor to install some of the more common items of Division 13.

DIVISION 14—CONVEYING EQUIPMENT

Elevators, wheelchair lifts, escalators, and industrial conveying systems are included in Division 14. All are highly specialized equipment that require licensing and are heavily regulated by government agencies. Accurately estimating the work requires an in-depth understanding of the various components, their assembly, and even proprietary information about the individual manufacturer's product. As in the case of all of the work covered in this chapter, the estimator should prepare a budget estimate as a place marker, or "plug" for use in comparing prospective bids. Any historical data can be

Athletic Rooms	Crew Makeup	Daily Output	Labor-Hours	Unit
SPORT COURT				
Rule of thumb for components:				
Walls	3 Carpenters	.15	160.000	Court
Floor	↓	.25	96.000	Court
Lighting	2 Electricians	.60	26.670	Court
Handball, racquetball court in existing building				
Minimum	3 Carpenters 1 Building Laborer Power Tools	.20	160.000	Court
Maximum	↓	.10	320.000	Court
Rule of thumb for components: walls	3 Carpenters	.12	200.000	Court
Floor		.25	96.000	Court
Ceiling	↓	.33	72.730	Court
Lighting	2 Electricians	.60	26.670	Court

Figure 16.13 Installation Time for Athletic Rooms
Source RSMeans Estimating Handbook, Third Edition. © 2009 John Wiley & Sons, Inc.

Saunas	Crew Makeup	Daily Output	Labor-Hours	Unit
SAUNA Prefabricated, incl. heater & controls, 7' high				
6' × 4'	2 Carpenters 1 Building Laborer .5 Electrician	2.20	12.730	EA
6' × 5'		2	14.000	EA
6' × 6'		1.80	15.560	EA
6' × 9'		1.60	17.500	EA
8' × 12'		1.10	25.450	EA
8' × 8'		1.40	20.000	EA
8' × 10'		1.20	23.330	EA
10' × 12'		1	28.000	EA
Door only, with tempered insulated glass window	2 Carpenters	3.40	4.710	EA
Prehung, incl. jambs, pulls, & hardware		12	1.330	EA

Figure 16.14 Installation Time for Saunas

Source RSMeans Estimating Handbook, Third Edition. © 2009 John Wiley & Sons, Inc.

Steam Baths	Crew Makeup	Daily Output	Labor-Hours	Unit
STEAM BATH Heater, timer & head, single, to 140 CF	1 Plumber	1.20	6.670	EA
To 300 CF		1.10	7.270	EA
Commercial size, to 800 CF		.90	8.890	EA
To 2,500 CF		.80	10.000	EA
Multiple baths, motels, apartment, 2 baths	1 Plumber 1 Plumber Apprentice	1.30	12.310	EA
4 baths		.70	22.860	EA

Figure 16.15 Installation Time for Steam Baths

Source RSMeans Estimating Handbook, Third Edition. © 2009 John Wiley & Sons, Inc.

Greenhouses	Crew Makeup	Daily Output	Labor-Hours	Unit
GREENHOUSE Shell only, stock units				
not incl. 2' stud walls, foundation, floors				
heat or compartments				
Residential type, freestanding, 8'-6" long				
7'-6" wide	2 Carpenters	59	.271	SF FLR
10'-6" wide		85	.188	SF FLR
13'-6" wide		108	.148	SF FLR
17'-0" wide		160	.100	SF FLR
Lean-to type, 3'-10" wide		34	.471	SF FLR
6'-10" wide		58	.276	SF FLR
Geodesic hemisphere, 1/8" plexiglass glazing				
8' diameter	2 Carpenters	2	8.000	EA
24' diameter		.35	45.710	EA
48' diameter		.20	80.000	EA

Figure 16.16 Installation Time for Greenhouses

Source RSMeans Estimating Handbook, Third Edition. © 2009 John Wiley & Sons, Inc.

organized into a database that can be used as a reliable prediction of future costs. Even bids from subcontractors that were not used can be a source of cost information. Review the project bid documents carefully to identify and coordinate the work of other trades.

It is recommended that the estimator research the individual product, if possible, for an understanding of any coordinated trades that have related

Swimming Pools	Crew Makeup	Daily Output	Labor-Hours	Unit
SWIMMING POOL ENCLOSURE Translucent, freestanding, not incl. foundations, heat, or light				
Economy, minimum	2 Carpenters	200	.080	SF HOR
Maximum		100	.160	SF HOR
Deluxe, minimum		100	.160	SF HOR
Maximum		70	.229	SF HOR
SWIMMING POOL EQUIPMENT Diving stand, stainless steel, 3 meter		.40	40.000	EA
1 meter		2.70	5.930	EA
Diving boards, 16' long, aluminum		2.70	5.930	EA
Fiberglass		2.70	5.930	EA
Gutter system, stainless steel, with grating, stock, contains supply and drainage system	1 Welder Foreman 1 Welder 1 Equip. Oper. (light) 1 Gas Welding Mach.	20	1.200	LF
Integral gutter and 5' high wall system, stainless steel		10	2.400	LF
Ladders, heavy-duty, stainless steel, 2-tread	2 Carpenters	7	2.290	EA
4-tread		6	2.670	EA
Lifeguard chair, stainless steel, fixed		2.70	5.930	EA
Lights, underwater, 12 volts, with transformer 300-watt	1 Electrician	1	8.000	EA
110-volt, 500-watt, standard		1	8.000	EA
Low water cutoff type		1	8.000	EA
Slides, fiberglass, aluminum handrails & ladder 5'-0", straight	2 Carpenters	1.60	10.000	EA
8'-0", curved		3	5.330	EA
10'-0", curved		1	16.000	EA
12'-0", straight with platform		1.20	13.330	EA
Hydraulic lift, movable pool bottom, single ram Under 1,000 SF area	1 Labor Foreman (ins.) 2 Building Laborers 1 Struc. Steel Worker .5 Electrician	.03	1200.000	EA
Four ram lift, over 1,000 SF	1 Carpenter, 1 Helper	.02	1800.000	EA
Removable access ramp, stainless steel	2 Building Laborers	2	8.000	EA
Removable stairs, stainless steel, collapsible		2	8.000	EA

Figure 16.17 Installation Time for Swimming Pool Equipment

Source RSMeans Estimating Handbook, Third Edition. © 2009 John Wiley & Sons, Inc.

work. The work of Division 14—Conveying Equipment can be classified in the following categories:

A. 14 10 00 Dumbwaiters
  • Manual Dumbwaiters
  • Electric Dumbwaiters
  • Hydraulic Dumbwaiters
B. 14 20 00 Elevators
  • Electric Traction Elevators
  • Hydraulic Elevators
  • Rack and Pinion Elevators
  • Limited Use/Limited Application Elevators
  • Custom Elevator Cabs and Doors
  • Elevator Equipment and Controls
C. 14 30 00 Escalators and Moving Walks
  • Escalators
  • Moving Walks
  • Moving Ramps
D. 14 40 00 Lifts
  • People Lifts
  • Wheelchair Lifts

- Platform Lifts
- Sidewalk Lifts
- Vehicle Lifts
- Material Lifts

E. 14 70 00 Turntables
- Industrial Turntables
- Hospitality Turntables
- Exhibit Turntables
- Entertainment Turntables

F. 14 80 00 Scaffolding
- Suspended Scaffolding
- Rope Climbers
- Elevating Platforms
- Powered Scaffolding

G. 14 90 00 Other Conveying Equipment
- Facility Chutes (Coal, Trash, Laundry, and Escape Chutes)
- Pneumatic Tube Systems
- Slide Pole Systems

Taking-off and Pricing

Takeoff quantities in Division 14—Conveying Equipment can have very diverse units of measure. Elevators and dumbwaiters can be quantified in the takeoff by the number of stops or floors and listed as STPS. Pneumatic tube systems can be classified by their size in diameter and the linear footage, both vertical and horizontal. Each can have its own distinct pricing structure.

It is recommended that the estimator prepare an estimate for comparison purposes based on any available historical data or published data. Costs can be distilled to specific parameters such as CARS, STPS, LF, and so on. These parameters are then used to produce a *parametric* estimate or budget and are compared to solicited prices from subcontractors that will furnish and install all work complete. The estimator is urged to seek a subcontractor's quote for inclusion within their bid. (See Chapter 28, "Conceptual Estimating," for more information on parametric estimating.)

Labor for Division 14 Items

Estimating labor for Division 14 work can be done in a variety of ways. The most common method is to follow the takeoff units and estimate labor costs by specific task in the process. This requires a thorough understanding of each step in the process. Most of the tasks in Division 14 are crew tasks. The crew may consist of multiple trades, depending on the task in the process. For elevators, escalators, and wheelchair lifts, specific regulations govern the work. Most require licensed elevator constructors to perform and/or supervise the work. Figures 16.18 through 16.22 are provided as guidance for estimating the costs of labor to install some of the more common items of Division 14.

Elevators	Crew Mavkeup	Daily Output	Labor-Hours	Units
ELECTRIC TRACTION FREIGHT ELEVATORS				
Electric freight, base unit, 4,000 lb., 200fpm, 4-stop, std. fin.	2 Elevator Constructors	.05	320	EA
For travel over 40 VLF, add		7.25	2.207	VLF
For number of stops over 4, add		.27	59.259	Stop
HYDRAULIC FREIGHT ELEVATORS				
Hydraulic freight, base unit, 2,000 lb., 50 fpm, 2-stop, std. fin.	2 ElevatorConstructors	.10	160	EA
For travel over 40 VLF, add		7.25	2.207	VLF
For number of stops over 2, add		.27	59.259	Stop
ELECTRIC TRACTION PASSENGER ELEVATORS				
Electric pass., base unit, 2,000 lb, 200 fpm, 4-stop, std. fin.	2 Elevator Constructors	.05	320	EA
For travel over 40 VLF, add		7.25	2.207	VLF
For number of stops over 4, add		.27	59.259	Stop
Electric hospital, base unit, 4,000 lb., 200 fpm, 4-stop, std. fin.		.05	320	EA
For travel over 40 VLF, add		7.25	2.207	VLF
For number of stops over 4, add		.27	59.259	Stop
HYDRAULIC PASSENGER ELEVATORS				
Hyd. Pass., base unit, 1,500 lb.,100 fpm, 2-stop, std. fin.	2 Elevator Constructors	.10	160	EA
For travel over 12 VLF, add		7.25	2.207	VLF
For number of stops over 2, add		.27	59.259	Stop
Hydraulic hospital, base unit, 4,000 lb.,100 fpm, 2-stop, std. fin.		.10	160	EA
For travel over 12 VLF, add		7.25	2.207	VLF
For number of stops over 2, add		.27	59.259	Stop

Figure 16.18 Installation Time for Elevators

Source RSMeans Estimating Handbook, Third Edition. © 2009 John Wiley & Sons, Inc.

Electric Dumbwaiters	Crew Makeup	Daily Output	Labor-Hours	Unit
DUMBWAITERS 2-stop, electric, minimum	2 Elevator Constructors	.13	123.000	EA
Maximum	↓	.11	145.000	EA
For each additional stop, add		.54	29.630	Stop

Figure 16.19 Installation Time for Electric Dumbwaiters

Source RSMeans Estimating Handbook, Third Edition. © 2009 John Wiley & Sons, Inc.

Manual Dumbwaiters	Crew Makeup	Daily Output	Labor-Hours	Unit
DUMBWAITERS 2-stop, hand, minimum	2 Elevator Constructors	.75	21.333	EA
Maximum	↓	.50	32.000	EA
For each additional stop, add		.75	21.333	Stop

Figure 16.20 Installation Time for Manual Dumbwaiters

Source RSMeans Estimating Handbook, Third Edition. © 2009 John Wiley & Sons, Inc.

Chutes	Crew Makeup	Daily Output	Labor-Hours	Unit
CHUTES Linen or refuse, incl. sprinklers				
Aluminized steel, 16 gauge, 18" diameter	2 Sheet Metal Workers	3.50	4.570	Floor
24" diameter		3.20	5.000	Floor
30" diameter		3	5.330	Floor
36" diameter		2.80	5.710	Floor
Galvanized steel, 16 gauge, 18" diameter		3.50	4.570	Floor
24" diameter		3.20	5.000	Floor
30" diameter		3	5.330	Floor
36" diameter		2.80	5.710	Floor
Stainless steel, 18" diameter		3.50	4.570	Floor
24" diameter		3.20	5.000	Floor
30" diameter		3	5.330	Floor
36" diameter		2.80	5.710	Floor
Linen bottom collector, aluminized steel		4	4.000	EA
Stainless steel		4	4.000	EA
Refuse bottom hopper, aluminized steel				
18" diameter		3	5.330	EA
24" diameter		3	5.330	EA
36" diameter		3	5.330	EA
Package chutes, spiral type, minimum		4.50	3.560	Floor
Maximum	↓	1.50	10.670	Floor

Figure 16.21 Installation Time for Chutes

Source RSMeans Estimating Handbook, Third Edition. © 2009 John Wiley & Sons, Inc.

Tube Systems	Crew Makeup	Daily Output	Labor- Hours	Unit
PNEUMATIC TUBE SYSTEM Single tube, 2 stations, 100' long, stock, economy,				
3" diameter	2 Steamfitters	.12	133.000	Total
4" diameter	2 Steamfitters	.09	177.000	Total
Twin tube, two stations or more, conventional system				
2½" round	2 Steamfitters	62.50	.256	LF
3" round		46	.348	LF
4" round		49.60	.323	LF
4" × 7" oval		37.60	.426	LF
Add for blower		2	8.000	System
Plus for each round station, add		7.50	2.130	EA
Plus for each oval station, add		7.50	2.130	EA
Alternate pricing method: base cost, minimum		.75	21.330	Total
Maximum		.25	64.000	Total
Plus total system length, add, minimum		93.40	.171	LF
Maximum		37.60	.426	LF
Completely automatic system, 4" round,				
15 to 50 stations		.29	55.170	Station
51 to 144 stations		.32	50.000	Station
6" round or 4" × 7" oval, 15 to 50 stations		.24	66.670	Station
51 to 144 stations		.23	69.570	Station

Figure 16.22 Installation Time for Pneumatic Tube Systems

Source RSMeans Estimating Handbook, Third Edition. © 2009 John Wiley & Sons, Inc.

SUMMARY

While most general contractors subcontract the work of these divisions, it is recommended that the estimator become familiar with the scope of work and establish budget costs as guidelines for reviewing subcontractors' proposals. It is also important to verify that the scope of work being priced is defined accurately and completely. Whenever possible the general contractor estimator is urged to solicit, qualify, and include within their bid, firm fixed prices for the work of these divisions.

Divisions 11, 12, 13, and 14 include such an enormous quantity and variety of different types of specialties that it would be impractical to attempt to provide labor-hours for installation costs for all of the work within this chapter. It is suggested that the estimator understand the limits of the scope for each of these divisions so there is no omission or duplication of cost.

For additional information on installation costs, consult *RSMeans Estimating Handbook, Third Edition,* or the manufacturer's specific product literature.

QUESTIONS/ PROBLEMS

1. What is the typical unit of measure for window treatments? How is it obtained?
2. What is the typical unit of measure for residential appliances? What should be included in the labor costs?
3. What is the suggested unit of measure for special purpose rooms such as a sauna?
4. How would an estimator develop the labor portion of an estimate for manufactured wood casework?
5. Explain in detail how an estimator might develop a budget estimate or plug from historical cost data for an elevator.

17 | Fire Suppression

Division 21 of CSI MasterFormat® is dedicated to fire protection and suppression systems. The work is typically performed by trades called sprinkler fitters or pipefitters and by firms with specialized training, often with licenses, insurances, and permits separate from those of the general contractor. Taking off and estimating fire suppression systems requires a working knowledge of the system or trade and, often, specialized education and training not normally within the realm of the general contractor's estimating experience. Nevertheless, the estimator should be able to develop sound, realistic budgets for comparison purposes and to evaluate fire suppression contractors' pricing on bid day.

While reviewing mechanical drawings and specifications, the estimator should become familiar with the scope of work required, which provides the basis for a more thorough review of subcontractor pricing.

Division 21 work is typically shown on the fire protection drawing(s) in the mechanical drawings section of the project plans. However, not all projects have a sprinkler drawing(s) included within the set. Many projects include only an outline technical specification for the type of sprinkler system required. This outline specification, typically referencing NFPA requirements, is used by the sprinkler subcontractors to design the project system and subsequently bid on the work.

Fire protection drawings are typically designated by the prefix SP, FP, or even F. Review all drawings in the bid set, including architectural and civil drawings, for related work that may be shown on other drawings, as well as for detailed information on dimensions and measurements for room sizes, floor-to-floor heights, location of services entering the building, and coordination with other work within the particular area.

Estimating Building Costs for the Residential and Light Commercial Construction Professional,
Third Edition. Wayne J. Del Pico.
© 2023 John Wiley & Sons, Inc. Published 2023 by John Wiley & Sons, Inc.

Fire protection drawings with schedules are helpful in determining types and quantities of materials for takeoff. (See the "Schedules" section in Chapter 1.) Specialized details, such as riser diagrams showing the configuration and components of piping systems, may be included.

It should also be noted that some projects have only an outline specification for the sprinkler. In many jurisdictions, the fire sprinkler contractor is required to do the detailed design and calculations.

Consult the specification sections for the products, methods, and techniques of installation, as well as the related work of other trades that will affect pricing. A review of Division 1—Temporary Facilities, of the specs may be necessary to determine what, if any, special requirements, such as hoisting or scaffolding, may be included.

This chapter covers the basic procedures and methods for takeoff and pricing, limited to the fire protection/suppression systems normally encountered in residential and light commercial construction using water as a suppressing agent. For extinguishing systems that use chemicals as a suppressing agent, the process is similar.

FIRE SUPPRESSION SYSTEMS

The *International Fire Code* (IFC) requires fire suppression sprinkler systems in buildings based on their occupancy and use. The design engineer will use the occupancy group to determine the system required. The three main components of a fire protection system are detection, alarm, and suppression. Suppression of fire may be accomplished by fire standpipe systems, automatic sprinkler systems, or a combination of the two. There are several different classifications of automatic sprinkler systems designed for specific firefighting applications. For budgeting purposes, most fire protection work can be reduced to a cost per square foot (SF) of the space being protected or cost per sprinkler head. For project plans complete with detailed sets of drawings, the estimator can count heads, sections of pipe, and even fittings. For most projects however, this level of detail is not required as the general contractor will be soliciting prices from subcontractors doing the work. (See Figure 17.1.)

Fire suppression standpipe systems require hose stations placed at various locations within the structure, according to International Fire Code design and local building code requirements. Standpipe systems are divided into three classes: Class I, Class II, and Class III, with Class I being the most common. In addition to the various classes of standpipe there are also different types. They are automatic dry and wet pipe, manual dry and wet pipe, and semiautomatic dry type. Wet systems are filled with water and dry systems are filled with air or gas. The design engineer will be the determining party as to which type, and class is required for the project.

Most fire protection systems in residential and light commercial construction are classified into one of two groups for estimating purposes: wet pipe sprinkler systems and dry pipe sprinkler systems. In wet pipe systems, water is constantly under pressure in each head and the system. These sprinklers can

Sprinkler systems may be classified by type as follows:

1. **Wet Pipe System.** A system employing automatic sprinklers attached to a piping system containing water and connected to a water supply so that water discharges immediately from sprinklers opened by a fire.

2. **Dry Pipe System.** A system employing automatic sprinklers attached to a piping system containing air under pressure. When the pressure is released from the opening of sprinklers, the water pressure opens a valve known as a "dry pipe valve." The water then flows into the piping system and out the opened sprinklers.

3. **Preaction System.** A system employing automatic sprinklers attached to a piping system containing air that may or may not be under pressure. Preaction systems have heat-activated devices that are more sensitive than the automatic sprinklers themselves, installed in the same areas as the sprinklers. Actuation of the heat-responsive system, as from a fire, opens a valve, which permits water to flow into the sprinkler piping system and to be discharged from any sprinklers which may be open.

4. **Deluge System.** A dry system connected to a water supply through a valve which is opened by the operation of a heat-responsive system (installed in the same areas as the sprinklers). When this valve opens, water flows into the piping system and discharges from all attached sprinklers.

5. **Combined Dry Pipe and Preaction Sprinkler System.** A system employing automatic sprinklers attached to a piping system containing air under pressure. A supplemental heat-responsive system of generally more sensitive characteristics than the automatic sprinklers themselves is installed in the same areas as the sprinklers. Operation of the heat-responsive system, as from a fire, actuates tripping devices, which open dry pipe valves simultaneously and without loss of air pressure in the system. Operation of the heat-responsive system also opens approved air exhaust valves at the end of the feed main, which facilitates the filling of the system with water (which usually precedes the opening of sprinklers). The heat-responsive system also serves as an automatic fire alarm system.

6. **Limited Water Supply System.** A system employing automatic sprinklers and conforming to automatic sprinkler standards, but supplied by a pressure tank of limited capacity.

7. **Chemical Systems.** Systems using FM200, carbon dioxide, dry chemical, or high-expansion foam as selected for special requirements. The agent may extinguish flames by chemically inhibiting flame propagation or suffocate flames by excluding oxygen, interrupting chemical action of oxygen uniting with fuel or sealing and cooling the combustion center.

8. **Firecycle System.** Firecycle is a fixed fire protection sprinkler system utilizing water as its extinguishing agent. It is a time delayed, recycling, preaction type, which automatically shuts the water off when the heat is reduced below the detector operating temperature. Water is turned on again when that temperature is exceeded. The system senses a fire condition through a closed-circuit electrical detector, which controls water flow to the fire automatically. Batteries supply up to 90-hour emergency power supply for system operation. The piping system is dry (until water is required) and is monitored with pressurized air. Should any leak in the system piping occur, an alarm will sound, but water will not enter the system until heat is sensed by a Firecycle detector.

Area coverage sprinkler systems may be laid out and fed from the supply in any one of several patterns. It is desirable, if possible, to utilize a central feed and achieve a shorter flow path from the riser to the farthest sprinkler. This permits use of the smallest sizes of pipe possible, with resulting savings.

Figure 17.1 Type of Automatic Sprinkler Systems

Source RSMeans Estimating Handbook, Third Edition. © 2009 John Wiley & Sons, Inc.

be used in areas that are not subject to freezing. Dry pipe sprinkler systems are used where freezing is a concern, and they feature compressed air that restrains water at the supply sources until needed. Dry pipe sprinkler systems tend to be more expensive due to added equipment and appurtenances.

Other types of fire protection systems include preaction, deluge, and fire cycle systems, all of which have their own unique equipment to complete the system. Sprinkler systems may need to function in conjunction with other systems in the building, such as fire alarms. Review subcontractor proposals and ensure that all components, including those being provided by other trades, have been accounted for in the estimate. Gaps in the scope of work being priced between trades can result in missed or duplicated costs in the estimate.

Study the fire protection drawings for locations of sprinkler heads; sizes and materials of piping; and location of valves, fittings, appurtenances, and water sources. Special equipment, such as compressors for dry systems, backflow preventers, and alarm bells, may be shown on riser diagrams or details. Architectural floor and reflected ceiling plans provide dimensions and ceiling heights. The specifications detail the materials to be used, method of installation, and required compliances of the various governing agencies. Related work by other trades is also listed in this section and includes wired connection of flow or tamper switches to the fire alarm system or power wiring for a dry pipe system compressor. The estimator should also review the documents to determine which trade is responsible for coring or drilling floors or partitions to allow the passage of the piping. Consequently, these cored holes may require sleeves and even fire safing if the partition or floor is rated. (See Figure 17.2.)

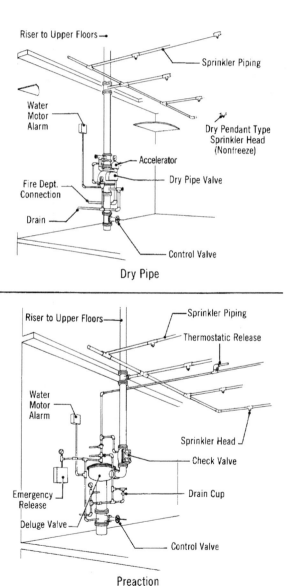

Figure 17.2 Graphics of Dry Pipe and Pre-action Systems

Class	Design-Use	Pipe Size Minimums	Water Supply Minimums
Class I	2-1/2" hose connection on each floor All areas within 30' of nozzle with 100' of hose Fire department trained personnel	Height to 100', 4" diam. Heights above 100', 6" diam. (275' max. except with pressure regulators 400' max.)	For each standpipe riser 500 GPM flow For common supply pipe allow 500 GPM for first standpipe plus 250 GPM for each additional standpipe (2,500 GPM max. total), 30 min. duration, 65 PSI at 500 GPM
Class II	1-1/2" hose connection with hose on each floor All areas within 30' of nozzle with 100' of hose Occupant personnel	Height to 50', 2" diam. Height above 50', 2-1/2" diam.	For each standpipe riser 100 GPM flow For multiple riser common supply pipe 100 GPM 30 min. duration, 65 PSI at 100 GPM,
Class III	Both of above. Class I valved connections will meet Class III with addition of 2-1/2" by 1-1/2" adapter and 1-1/2" hose	Same as Class I	Same as Class I

Figure 17.3 Standpipe Systems

Source RSMeans Estimating Handbook, Third Edition. © 2009 John Wiley & Sons, Inc.

Although systems are generally similar in residential and light-commercial construction, the grades and types of materials can be significantly different. In general, the delivery system for residential work is CPVC piping. CPVC is a polyvinylchloride pipe designed to work under pressure. Black iron or steel piping is most commonly used in the commercial industry. In the absence of a definitive specification, reference the applicable NFPA code or the authority having jurisdiction over the project.

Not all fire suppression systems use water as the main ingredient to fight the fire. Chemical systems are often used where water is not the appropriate agent or can cause more damage. Consider a grease fire in a restaurant kitchen or an electrical fire in a computer center. Clearly, water would not be the proper suppression agent in either case. Chemical suppression systems can have dry or wet chemical agents. Not all of the chemical systems are outside of the residential/light commercial industry.

Another component to the fire protection system includes a *standpipe system* or vertical pipes that rise between floors to provide water for fire hoses or fire department connections. Figure 17.3 provides design criteria for the different classes of standpipe systems.

Taking-off and Pricing

All takeoff quantities should be listed separately according to type and classification of system – wet, dry, chemical, etc. In addition to materials, special items such as flow tests, shop drawings, permits and fees, and special staging or rigging equipment for the installation of larger-diameter pipes and valves should be included within the estimate. Additional takeoff guidelines are as follows:

- **Shop drawings and calculation:** Can be listed by the sheet of drawing or by the square area of space being protected. This can also include flow tests to determine the pressure available at the site.
- **Riser and distribution piping material for standpipe and automatic sprinkler systems:** By the linear foot (LF), separated according to type of pipe (steel, black iron, or CPVC), size (diameter), and method of connection (grooved joint or threaded).
- **Fittings:** By the piece (EA) according to type (elbow, tee, reducer, etc.), size (diameter), method of connection, and material composition (steel, black iron, CPVC, etc.).
- **Valves and special appurtenances:** By the individual piece (EA) according to type, manufacturer, model, size, use or application, and other identifying information needed for pricing. Includes gauges, couplings, flanges, water motor alarms and bells, fire department connections, backflow preventers, control panels or devices, and storage cylinders.
- **Sprinkler heads and discharge nozzles:** By the piece (EA) according to type (pendant, upright, or sidewall), temperature range, manufacturer, and model number.
- **Trim pieces/escutcheons:** By the piece (EA) according to type, finish, model, and application.

In most cases, the sprinkler contractor's work begins at the interior of the building where the fire protection water service line enters the structure. Site piping and related excavation and backfill are typically the responsibility of the site contractor, although the sprinkler contractor in some jurisdictions may do or supervise this work. Review the specifications regarding the exact scope of work included.

Other components such as hoses, hose racks, hose reels, and cabinets for standpipe systems are part of the estimate. Miscellaneous components such as extra heads, tools, and even fire department key boxes must be added if specified. These can have a wide variety of takeoff units from EA to LF.

Be sure to include nonproduction costs such as testing water pressure, reviewing plans with local government agencies, commissioning and testing, owner training, as-built drawings, and the review/approval processes by insurance underwriters, if applicable.

Not all fire suppression systems are attached to the building structure only. Consider the rack/shelving system in a big-box retail center. This system often has its own fire sprinkler system within it to deliver the water more directly to the source of the fire. The same can be true for the dry chemical systems under the kitchen hood in restaurants. The important consideration is that in both of these applications there is the interface with the main suppression system or other building systems. Dry chemical systems under the kitchen hood are often tied into a gas valve delivering gas to the cooking appliance under the hood. As soon as the suppression system deploys, it shuts the gas valve to prevent fuel flow.

Labor-Fabrication and Installation

Fabrication and installation of fire protection systems are done by pipe fitters. Commercial projects in particular require special tools, licenses, and equipment, and many of the components are not available on the open market. As a means of increasing installation productivity, most grooved steel pipe or threaded black iron pipe is fabricated off-site from a *cut list* generated from shop drawings. A specialty subcontractor to the sprinkler subcontractor cuts and threads or grooves all pipe to specific lengths based on the cut list. The materials are then marked and delivered to the site for installation by the sprinkler subcontractor. This reduces or eliminates the need for on-site fabrication.

Crews consisting of multiple pipefitters assemble prefabricated pipe, starting from the largest-diameter distribution pipes to smaller branch piping. Production is measured by the crew's output, most often in LF of pipe per day. In addition to the actual assembly of piping, installation of hangers, bracing, and seismic restraint of pipe assemblies also consumes labor-hours. As with most trades, productivity can be affected by a variety of circumstances, such as extremes in temperature. Large, open floor plans proceed more quickly than small, confined spaces, such as attics or crawl spaces. Working from staging or lifts reduces productivity and may even require an additional person to hand materials up from the ground. Labor costs for the testing and commissioning of systems should also be included in the estimate. Figures 17.4 through 17.8 provide details on labor-hour guidelines for installing components of various fire suppression systems.

OTHER FIRE SUPPRESSION COMPONENTS

In addition to standpipe and automatic sprinkler fire suppression systems, there are other components required to make the system work. These include things like fire pumps and water storage tanks for fire suppression water. Fire pumps and accessories are used to increase water supply to floors above grade in a building or to pump water from underground storage tanks into the fire suppression system.

Water storage tanks come in a variety of sizes and composition. They can be steel tanks placed in the upper levels or roof of a multifloor building. They can also be concrete and buried below grade outside the structure they serve.

With both storage tanks and fire pumps, the estimator must consider what other trades and scope of work are involved. For underground storage tanks, excavation, backfill and a concrete anchoring pad may be required. For pumps, equipment might be needed for hoisting and rigging to set the pump in place and also the electrical wiring to run the pump or a generator run by fossil fuel.

Description	Labor-Hours	Unit
Steel Pipe Labor-hours to Install Black, Schedule #10, Grooved Joint or Plain End with a Mechanical Joint Coupling and Pipe Hanger Every Ten Feet		
2" Pipe Size	.186	LF
2-1/2" Pipe Size	.262	LF
3" Pipe Size	.291	LF
3-1/2" Pipe Size	.302	LF
4" Pipe Size	.327	LF
5" Pipe Size	.400	LF
6" Pipe Size	.522	LF
8" Pipe Size	.585	LF
10" Pipe Size	.706	LF
12" Pipe Size	.800	LF
Black or Galvanized Schedule #40 Grooved Joint or Plain End with a Mechanical Joint Coupling and Pipe Hanger Every Ten Feet		
3/4" Pipe Size	.113	LF
1" Pipe Size	.127	LF
1-1/4" Pipe Size	.138	LF
1-1/2" Pipe Size	.157	LF
2" Pipe Size	.200	LF
2-1/2" Pipe Size	.281	LF
3" Pipe Size	.320	LF
3-1/2" Pipe Size	.340	LF
4" Pipe Size	.356	LF
5" Pipe Size	.432	LF
6" Pipe Size	.571	LF
8" Pipe Size	.649	LF
10" Pipe Size	.774	LF
12" Pipe Size	.889	LF
Fittings for Use with Grooved Joint or Plain End Steel Pipe Elbows 90° or 45°		
3/4" Pipe Size	.160	EA
1" Pipe Size	.160	EA
1-1/4" Pipe Size	.200	EA
1-1/2" Pipe Size	.242	EA
2" Pipe Size	.320	EA
2-1/2" Pipe Size	.400	EA
3" Pipe Size	.485	EA
4" Pipe Size	.640	EA
5" Pipe Size	.800	EA
6" Pipe Size	.960	EA
8" Pipe Size	1.143	EA
10" Pipe Size	1.333	EA
12" Pipe Size	1.600	EA

Flanged joint

Grooved joint steel pipe

Grooved joint coupling

Mechanical joint elbow 90° — plain end pipe

Figure 17.4a Installation Time in Labor-Hours for Steel Pipe

Taking-off and Pricing

All takeoff quantities should be listed separately according to component and operation. In addition to materials, items such as special permits and fees for hauling oversize tanks over the road, and special hoisting or rigging equipment for the installation of tanks, pumps, piping, and valves should be included within the estimate. Additional takeoff guidelines are as follows.

Description	Labor-Hours	Unit
Tees		
3/4" Pipe Size	.211	EA
1" Pipe Size	.242	EA
1-1/4" Pipe Size	.296	EA
1-1/2" Pipe Size	.364	EA
2" Pipe Size	.471	EA
2-1/2" Pipe Size	.593	EA
3" Pipe Size	.727	EA
4" Pipe Size	.941	EA
5" Pipe Size	1.231	EA
6" Pipe Size	1.412	EA
8" Pipe Size	1.714	EA
10" Pipe Size	2.000	EA
12" Pipe Size	2.400	EA
Labor-hours to Install Black or Galvanized Schedule #40 Threaded with a Coupling and Pipe Hanger Every Ten Feet. The Pipe Hanger is Oversized to Allow for Insulation.		
1/2" Pipe Size	.127	LF
3/4" Pipe Size	.131	LF
1" Pipe Size	.151	LF
1-1/4" Pipe Size	.180	LF
1-1/2" Pipe Size	.200	LF
2" Pipe Size	.250	LF
2-1/2" Pipe Size	.320	LF
3" Pipe Size	.372	LF
3-1/2" Pipe Size	.400	LF
4" Pipe Size	.444	LF
Fittings for Use with Steel Pipe. Threaded Fittings, Cast Iron, 125 lb. or Malleable Iron Rated at 150 lb. Elbows, 90° or 45°		
1/2" Pipe Size	.533	EA
3/4" Pipe Size	.571	EA
1" Pipe Size	.615	EA
1-1/4" Pipe Size	.727	EA
1-1/2" Pipe Size	.800	EA
2" Pipe Size	.889	EA
2-1/2" Pipe Size	1.143	EA
3" Pipe Size	1.600	EA
3-1/2" Pipe Size	2.000	EA
4" Pipe Size	2.667	EA

Tee — plain end pipe

Threaded and coupled steel pipe

45° elbow — malleable iron

Tee — cast iron

Figure 17.4b *(Continued)*

- **Fire pumps and related piping:** By the individual pump (EA) according to size (horsepower), capacity (gallons per minute), and fuel source (diesel or electric). Include the control panels and piping for the fire pumps.
- **Water storage tanks and related piping:** By the gallons of water (GAL) the storage tank will hold, its composition (steel or concrete), and its location (rooftop or inground). The units of measure for piping and testing of the piping from the tank to the fire suppression system can vary.

Description	Labor-Hours	Unit
Valve, Gate, Iron Body, Flanged OS and Y, 125 lb. 4" Diameter	5.333	EA
Valve, Swing Check, w/Ball Drip, Flanged, 4" Diameter	5.333	EA
Valve, Swing Check, Bronze, Thread End, 2-1/2" Diameter	1.067	EA
Valve, Angle, Bronze, Thread End, 2" Diameter	.727	EA
Valve, Gate, Bronze, Thread End, 1" Diameter	.421	EA
Alarm Valve, 2-1/2" Diameter	5.333	EA
Alarm, Water Motor, with Gong	2.000	EA
Fire Alarm Horn, Electric	.308	EA
Pipe, Black Steel, Threaded, Schedule 40,		
4" Diameter	.444	LF
2-1/2" Diameter	.320	LF
2" Diameter	.250	LF
1-1/4" Diameter	.180	LF
1" Diameter	.151	LF
Pipe Tee, 150 lb. Black Malleable		
4" Diameter	4.000	EA
2-1/2" Diameter	1.778	EA
2" Diameter	1.455	EA
1-1/4" Diameter	1.143	EA
1" Diameter	1.000	EA
Pipe Elbow, 150 lb. Black Malleable		
1" Diameter	.615	EA
Sprinkler Head, 135° to 286°, 1/2" Diameter	.500	EA
Sprinkler Head, Dry Pendant 1" Diameter	.571	EA
Dry Pipe Valve, w/Trim and Gauges, 4" Diameter	16.000	EA
Deluge Valve, w/Trim and Gauges, 4" Diameter	16.000	EA
Deluge System Monitoring Panel 120 Volt	.444	EA
Thermostatic Release	.400	EA
Heat Detector	.500	EA
Firecycle Controls w/Panel, Batteries, Valves and Switches	32.000	EA
Firecycle Package, Check and Flow Control Valves, Trim, 4" Diameter	16.000	EA
Air Compressor, Automatic, 200 gal. Sprinkler System 1/3 HP	6.154	EA

Figure 17.5 Installation Time for Automatic Sprinkler Systems

Source RSMeans Estimating Handbook, Third Edition. © 2009 John Wiley & Sons, Inc.

SUMMARY

As most general contractors subcontract fire suppression work, it is not always essential to have a detailed unit price estimate for the work. However, when done correctly, a detailed estimate is almost always more accurate. What is required is an understanding of the system and its components so that the estimator can produce a parametric estimate to be used in comparison to the solicited subbids for the work. The estimator must consider the full scope of the project, including all design, testing, scaffolding or lifts, and the coring of masonry or concrete partitions and floors. The estimator's goal should be to ensure there are no gaps or missing tasks in the estimate. This can often be done with a checklist during the review process. Historical data from projects with similar criteria or published cost data are excellent sources of costs for developing a parametric estimate.

Description	Labor-Hours	Unit
FM200 System, Filled, Including Mounting Bracket		
26-lb. Cylinder	2.000	EA
44-lb. Cylinder	2.286	EA
63-lb. Cylinder	2.667	EA
101-lb. Cylinder	3.200	EA
196-lb. Cylinder	4.000	EA
Electro/Mechanical Release	4.000	EA
Manual Pull Station	1.333	EA
Pneumatic Damper Release	1.000	EA
Discharge Nozzle	.570	EA
Control Panel Single Zone	8.000	EA
Control Panel Multizone (4)	16.000	EA
Battery Standby Power	2.810	EA
Heat Detector	1.000	EA
Smoke Detector	1.290	EA
Audio Alarm	1.194	EA

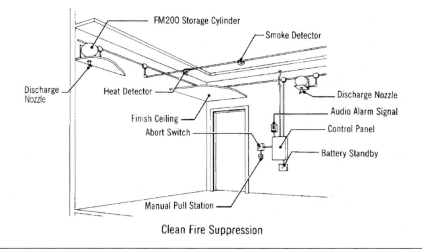

Clean Fire Suppression

Figure 17.6 Installation Time for FM200 Fire Suppression Systems
Source RSMeans Estimating Handbook, Third Edition. © 2009 John Wiley & Sons, Inc.

QUESTIONS/ PROBLEMS

1. Provide a brief explanation as to difference between a wet pipe and a dry pipe sprinkler system from an estimator's perspective.
2. Prepare a checklist of possible interfacing trades for a fire pump on the upper level of a building.
3. Explain in detail how an estimator would prepare a parametric estimate for a wet pipe sprinkler system for a project.
4. What is a typical unit of measure for a water storage tank?
5. Provide a list of some of the nonproduction costs associated with a fire suppression system.

Description	Labor-Hours 4″	6″	8″	Unit
Black Steel Pipe	.444	.774	.889	LF
Pipe Tee	4.000	6.000	8.000	EA
Pipe Elbow	2.667	3.429	4.000	EA
Pipe Nipple, 2-1/2″	1.000	1.000	1.000	EA
Hose Valve, 2-1/2″	1.140	1.140	1.140	EA
Pressure-Restricting Valve, 2-1/2″	1.140	1.140	1.140	EA
Check Valve with Ball Drip	5.333	8.000	10.667	EA
Siamese Inlet	3.200	3.478	3.478	EA
Roof Manifest with Valves	3.333	3.478	3.478	EA

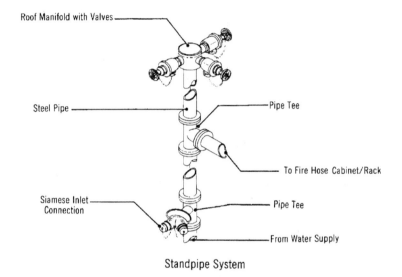

Standpipe System

Figure 17.7 Installation Time for Fire Standpipes

Source RSMeans Estimating Handbook, Third Edition. © 2009 John Wiley & Sons, Inc.

Description	Labor-Hours	Unit
Fire Pumps with Controller and Fittings Electric		
4″ Pipe Size 100 HP 500 GPM	51.619	EA
6″ Pipe Size 250 HP 1000 GPM	88.889	EA
8″ Pipe Size 300 HP 2000 GPM	114.286	EA
10″ Pipe Size 450 HP 3500 GPM	133.333	EA
Diesel		
4″ Pipe Size 111 HP 500 GPM	53.333	EA
6″ Pipe Size 255 HP 1000 GPM	80.000	EA
8″ Pipe Size 255 HP 2000 GPM	100.000	EA
10″ Pipe Size 525 HP 3500 GPM	160.000	EA

Figure 17.8 Installation Time in Labor-Hours for Pumps

Source RSMeans Estimating Handbook, Third Edition. © 2009 John Wiley & Sons, Inc.

18 | Plumbing

Division 22 of CSI MasterFormat® and is dedicated to plumbing systems. Plumbing work is typically performed by trades or firms with specialized training, insurance, plumbing licenses and permits separate from those of the general contractor. Taking off and estimating plumbing systems requires a working knowledge of the particular trade or system and, often, specialized education and training not normally within the realm of the general contractor's estimating experience. Nevertheless, the estimator should be able to develop sound, realistic budgets for comparison purposes and evaluate plumbing contractors' pricing on bid day. While reviewing plumbing drawings and specifications, the estimator will become familiar with the scope of work involved, which provides the basis for a more thorough review of subcontractor pricing.

Division 22 work is typically shown on the plumbing drawing(s) in the mechanical drawings section of the project plans. (See the discussion of mechanical drawings in Chapter 1.) Plumbing drawings are labeled with the letter "P" for plumbing. Review all drawings in the bid set, including architectural and civil, for related work that may be shown on other drawings, as well as for detailed information on dimensions and measurements for room sizes, floor-to-floor heights, location of services entering the building, and coordination with other work within the particular area. Plumbing drawings often provide schedules of fixtures or equipment that are helpful in determining types and quantities of materials for takeoff. (See the "Schedules" section of Chapter 1.) Schedules can show a wide variety of items that have been arranged in tables for the convenience of the estimator. Plumbing schedules can be dedicated to toilet fixtures, gas-burning appliances, venting, water supply piping, or even pipe insulation.

Estimating Building Costs for the Residential and Light Commercial Construction Professional, Third Edition. Wayne J. Del Pico.
© 2023 John Wiley & Sons, Inc. Published 2023 by John Wiley & Sons, Inc.

Specialized details, such as riser diagrams showing the configuration and components of piping systems, are often included and are helpful in visualizing the piping layout in a diagrammatic manner.

Study the specification sections for the products, methods, and techniques of installation, as well as the related work of other trades that affect pricing. A review of the specs for Division 1—Temporary Facilities may be necessary to determine what, if any, special requirements, such as hoisting or staging, testing, storage, or cleanup may be included.

CSI MasterFormat® uses the following organizational structure for Division 22.

- 22 07 00 Plumbing Insulation
- 22 08 00 Plumbing Commissioning
- 22 10 00 Plumbing Piping
- 22 30 00 Plumbing Equipment
- 22 40 00 Plumbing Fixtures
- 22 50 00 Pool and Fountain Plumbing Systems
- 22 60 00 Gas and Vacuum Systems for Laboratory and Healthcare Facilities

This chapter covers the basic procedures and methods for takeoff and pricing, limited to the plumbing systems normally encountered in residential and light commercial construction.

PLUMBING SYSTEMS

All buildings that will be regularly occupied by people require some type of plumbing – from simple toilets in warehouses to elaborate bathrooms in upscale residences, and to sophisticated plumbing systems in hospitals and restaurants. As with fire protection, plumbing work requires special knowledge and training for both the estimator and the tradespeople doing the work. It requires licensed plumbers and a separate permit and inspection process.

Like many other types of drawings, the plumbing drawings are designed around the architectural layout. Carefully review the plumbing drawings, as well as the architectural plans since critical dimensions are defined on the architectural plans. Some plumbing plans provide *riser* diagrams, or vertical layouts of plumbing pipes, to be used in determining piping quantities. Fixture and equipment schedules are also helpful in determining quantities and types of plumbing fixtures and equipment to be furnished and/or installed. Other drawings within the bid set, such as special equipment plans (e.g., kitchen equipment plans for restaurants), should be studied for equipment furnished by others, and installed or connected under the plumbing contract. Some plumbing drawings show under-slab and above-slab piping on the same sheet. This requires graphic symbols to illustrate the difference between locations. It may be helpful to review plumbing legends so that all symbols are understood prior to starting the estimate. Figure 18.1

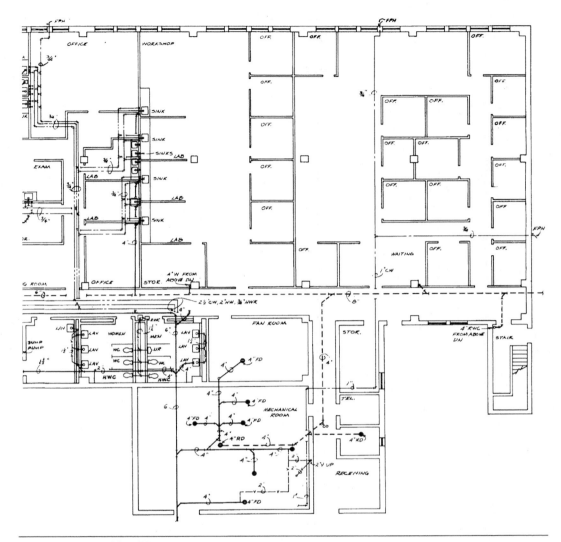

Figure 18.1 Plumbing Plan

is an example of a plumbing plan showing waste, vent, and water piping on the same plan. Figure 18.2 illustrates an example of a riser diagram for soil and vent piping for the preceding plumbing plan. Figure 18.3 is the riser diagram for the hot and cold water piping for the plan in Figure 18.1.

Special drawings with enlarged plan views and details for clarification of certain aspects of work are common, as in the example of the domestic water heater in Figure 18.4. Details or diagrams offer clarification only and often are not drawn to scale.

Consult Division 22—Plumbing and its various subsections of the specifications for the scope of work, product information, and acceptable methods of installation, as well as the related work in other sections. For the purpose of takeoff and pricing, plumbing work can be broken down into subsystems or categories, described in the following sections.

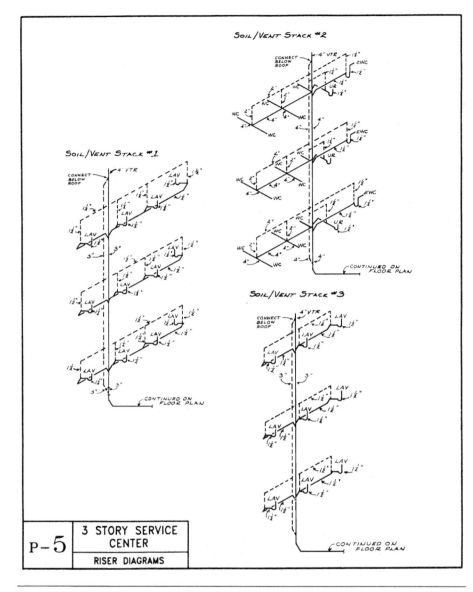

Figure 18.2 Riser Diagrams for Soil/Vent Piping

Plumbing Labor

It should come as no surprise that the work of the plumbing section is performed by plumbers and plumber's helpers. The plumbing scope of work can be done by an individual plumber or by a crew with some combination of plumbers and plumber's helpers, sometimes called *apprentices*. Generally speaking, installing heavy, cast-iron pipe above or below ground is typically a crew task, whereas installation of fixtures and trim is performed by an individual. However, there are exceptions to this generalization.

Insulation of water piping or roof leaders may be done by a separate trade or subcontractor to the plumber called an *insulator*.

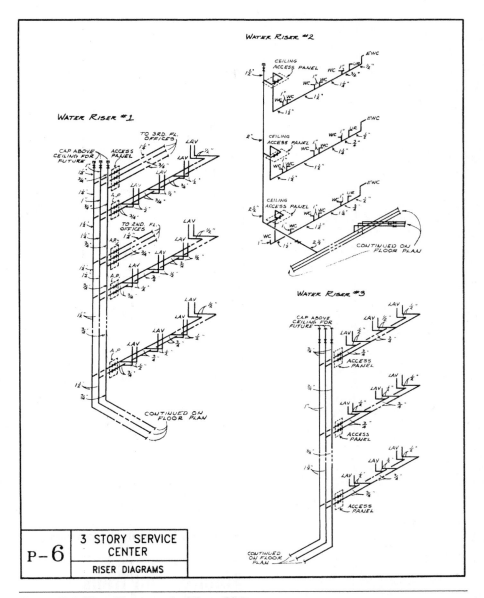

Figure 18.3 Riser Diagrams for Water Piping

Installation of all types and sizes of piping is priced by the linear foot (LF) and based on the amount of pipe that a crew can install in a single day. As with all above-grade piping, productivity is affected by the number of fittings, direction changes in the pipe, and accessibility in the workspace.

Productivity can be affected by a wide variety of issues, from access to temperature. Estimates of labor-hours must consider actual conditions in order to be accurate. Figure 18.5 details labor considerations resulting from special conditions.

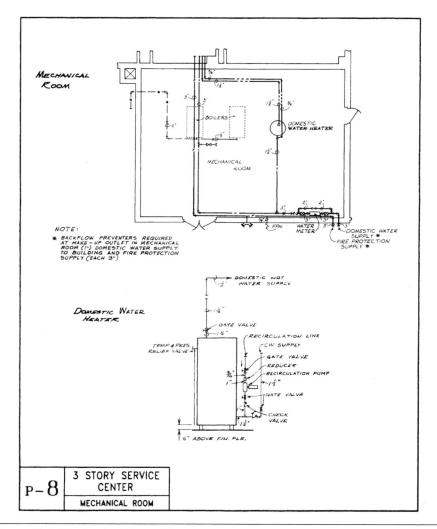

Figure 18.4 Mechanical Room Plan

Fixtures and Trims

Plumbing *fixtures* include water closets (toilets), urinals, shower stalls, tubs, and lavatories. *Trims* are the valves, faucets, trip levers, and drains that are added to the fixtures in the finish stage to complete their installation. The estimator would be wise to consider other incidentals to the installation, such as the flexible water supply lines or the wax ring for toilets. Although costs for these items are less than they are for the trims, they are still essential to include for a comprehensive estimate.

Taking-off and Pricing

Fixtures and trims are taken off and priced by counting the individual pieces and listing them as EA. Quantities of fixtures may be used in determining the quantities of trims. For example, each lavatory requires a faucet and drain, each tub requires a shower/tub valve and trip lever drain, and each water

Special Condition	Labor Increase
Add to labor for elevated installation (Above floor level)	
10' to 14.5' high	10%
15' to 19.5'	20%
20' to 24.5'	25%
25' to 29.5'	35%
30' to 34.5	40%
35' to 39.5'	50%
40' and higher	55%
Add to labor for crawl space	
3' high	40%
4' high	30%
Add to labor for multi-story building	
Add per floor floors 3 thru 19	2%
Add per floor for floors 20 and up	4%
Add to labor for working in existing occupied buildings	
Hospital	35%
Office building	25%
School	20%
Factory or warehouse	15%
Multi dwelling	15%
Add to labor, miscellaneous	
Cramped shaft	35%
Congested area	15%
Excessive heat or cold	30%
Labor factors, the above are reasonable suggestions, however each project should be evaluated for its own peculiarities.	
Other factors to be considered are:	
Movement of material and equipment through finished areas	
Equipment room	
Attic space	
No service road	
Poor unloading/storage area	
Congested site area/heavy traffic	

Figure 18.5 Labor Factors Due to Special Conditions

Source RSMeans Estimating Handbook, Third Edition. © 2009 John Wiley & Sons, Inc.

closet requires a toilet seat. Fixtures and their corresponding trims are listed according to type, manufacturer, model, color, or finish, and special features, such as ADA compliance. To reduce the chance of error, the estimator may opt to separate fixture and trim quantities according to specific bathrooms or toilets, or by the floor for multilevel commercial toilet rooms. This allows the takeoff to be completed for each bathroom or toilet before proceeding to the next one. See Figure 18.6 for guidance on labor-hours for plumbing fixtures and trim.

Equipment

Equipment for plumbing work includes such items as water heaters, hot water boilers, water storage tanks, interior grease interceptors, and sump pumps, to name but a few. Hookup of equipment supplied by others, such as garbage disposals, gas ranges and ovens, refrigerators with automatic icemakers, and dishwashers, must be included. Special devices, such as washing machine outlets and venting kits for water heaters, are necessary for a comprehensive estimate.

Description	Labor-Hours	Unit
For Setting Fixture and Trim		
Bathtub	3.636	EA
Bidet	3.200	EA
Dental Fountain	2.000	EA
Drinking Fountain	2.500	EA
Lavatory		
Vanity Top	2.500	EA
Wall-Hung	2.000	EA
Laundry Sinks	2.667	EA
Prison/Institution Fixtures		
Lavatory	2.000	EA
Service Sink	5.333	EA
Urinal	4.000	EA
Water Closet	2.759	EA
Combination Water Closet and Lavatory	3.200	EA
Shower Stall	8.000	EA
Sinks		
Corrosion-Resistant	5.333	EA
Kitchen, Countertop	3.330	EA
Kitchen, Raised Deck	7.270	EA
Service, Floor	3.640	EA
Service, Wall	4.000	EA
Urinals		
Wall-Hung	5.333	EA
Stall Type	6.400	EA
Wash Fountain, Group	9.600	EA
Water Closets		
Tank Type, Wall-Hung	3.019	EA
Floor-Mount, One Piece	3.019	EA
Bowl Only, Wall-Hung	2.759	EA
Bowl Only, Floor-Mount	2.759	EA
Gang, Side by Side, First	2.759	EA
Each Additional	2.759	EA
Gang, Back to Back, First Pair	5.520	Pair
Each Additional Pair	5.520	Pair
Water-Conserving Type	2.963	EA
Watercooler	4.000	EA

Figure 18.6 Installation Time in Labor-Hours for Plumbing Fixtures

Source RSMeans Estimating Handbook, Third Edition. © 2009 John Wiley & Sons, Inc.

Taking-off and Pricing

Equipment is taken off and priced by counting the individual piece (EA), and it is listed according to type, manufacturer, model, size, and capacity. Equipment that requires only hookup should be listed separately by the piece with the same qualifications as noted here, although each piece of equipment scheduled for hookup can be estimated by the number of labor-hours required. It should clearly designate installation only. Special devices needed for the complete installation should be noted separately, with reference to the equipment for which it is required – for example, gas regulators and shutoff valves. Some equipment provided under the plumbing contract may require the work of other trades, such as power wiring of water heaters and faucets, power wiring of draft inducers, and special flues for gas-burning appliances such as water heaters. Penetrations through roofing systems and flashing of roof vents may also be required. The estimator should consider developing a checklist for all costs associated with the hookup of a piece of equipment to check against other sections of the estimate.

Below-Grade Sanitary Waste and Vent Piping

Below-grade sanitary waste piping carries waste from the individual plumbing fixture and above-grade waste piping in the structure to outside the building. Below-grade vent piping ties into vertical risers that vent the system to the exterior. A variety of fittings and devices are used, in addition to the piping itself, including couplings, elbows, tees, tee-wyes, cleanouts, traps, and floor drains, to mention just a few. Fittings, devices, and piping are available in a number of different materials, from PVC to cast iron. Review the specifications for the type of materials for specific applications. Plumbing plans help determine quantities, size and type of piping, fittings, and devices that will be used in the below-grade application. In some applications, especially residential, it is common for the plumbing crew to do their own excavation and backfill for below grade piping. This should be verified if not clearly specified within the documents.

Taking-off and Pricing

Fittings and devices are taken off and priced by counting the individual piece (EA) and listing them separately according to type (elbow, tee, cleanout, etc.), material composition (PVC, service-weight cast iron, extra-heavy-weight cast iron, etc.), method of connection to piping (lead and oakum, hubless, neoprene joint, or PVC cement), and size (diameter). Less-accurate alternative methods include allowing a percentage of the costs of the pipe for the fittings. This approach is not recommended, except for preliminary budgeting.

Piping is taken off and priced by the LF and should be listed separately according to type, size, application (waste or vent), and method of installation. Linear feet of individual types and sizes of pipe can be converted to actual lengths of pipe required. If cast-iron soil pipe is of the lead-joint design, it will be necessary to determine the total amount of lead, oakum (jute packing), and gas (propane) to complete the installation. This can be done after the fittings and piping takeoff has been completed so that the total number of joints is known. Cast-iron soil pipe is manufactured in 5' and 10' lengths (for lead, 5' is used). The amount of lead per joint is determined by the diameter of the pipe. Figure 18.7 lists the amount of lead (in pounds) per joint to caulk cast-iron pipe.

The following example shows how to determine quantities of lead, oakum, and gas for cast-iron soil pipe, assuming the following quantities of service-weight soil pipe and fittings have been taken off:

> 20 LF—4" *service-weight soil pipe*
>
> 3 EA — 4-1/8" *wye*
>
> 1 EA—6" × 4" *wye*

The procedure is as follows. (Note that each hub, rather than the opening, is the basis for the number of joints.)

Pipe & Fitting Diams. Inches	Lead Ring Depth Inches	Service Weight		Extra Heavy Weight	
		Cu. Ins.	Wt. Lbs.	Cu. Ins.	Wt. Lbs.
2	1	2.81	1.15	2.91	1.19
3	1	3.90	1.60	4.17	1.71
4	1	4.98	2.04	5.25	2.15
5	1	6.06	2.49	6.24	2.56
6	1	7.15	2.93	7.42	3.04
8	1.25	15.06	6.17	15.49	6.35
10	1.25	18.90	7.75	19.34	7.93
12	1.25	25.53	10.47	26.02	10.67
15	1.5	43.09	17.67	43.38	17.8

Figure 18.7 Lead Required to Caulk Cast-Iron Soil Pipe Joints

20 LF of 4" (5' length) = four 4" joints × 2.04 lb./joint = 8.16 lb.

3 EA—4-1/8" bends

3 × (1/4" joint) = 3 × 2.04 = 6.12 lb.

1 EA—6" × 4" wye 1 × (1—6" joint + 1—4" joint) = (1 × 2.93) + (1 × 2.04) = 4.97 lb. Total lead = 19.25 lb.

Oakum is typically estimated at one-tenth the weight of lead. Therefore, 19.25 lb. divided by 10 = 1.92 lb. of oakum. Gas consumption is approximated at one (instopropane) cylinder per 200 lb. of lead.

If cast-iron soil pipe is of the neoprene joint clamp type (hubless), or if the waste piping is PVC, count the number of joints to arrive at the quantity of clamps, gaskets, or couplings.

Below-grade sanitary waste piping may include excavation and backfill for the placement of the piping – typically the responsibility of the site or general contractor. However, this is not a foregone conclusion. If excavation and backfill are part of the plumbing scope for work for below-grade, this takeoff and pricing should be kept separate from the piping installation. Excavation and backfill by equipment can be taken off and estimated following the guidelines provided in Chapter 23, "Earthwork." For excavation and backfill using hand tools, the estimator is better served by calculating the labor-hours required for both phases in the process.

The estimator is urged to consider the type of soil if known. Review the technical specifications, including Division 1—General Requirements to determine the exact scope of work in order to avoid costly duplications or omissions for costs such as lifts, rigging of equipment, or testing of completed installations.

Above-Grade Sanitary Waste and Vent Piping

Above-grade sanitary waste provides piping for the flow of waste from upper floors of the building to the below-grade system, where it will exit the structure. Vent piping allows the escape of gases generated by the waste through the upper level of the structure, typically the roof. The piping and fittings used above grade are similar to those used below grade and are determined by the specifications and the plans. In addition, be sure to consider hangers and supports for supporting or hanging piping in both vertical and horizontal applications. Special conditions such as seismic bracing may also be required. Check the specifications for references to seismic codes that may apply. For cast-iron waste and vent above grade, mechanical means may be required to lift larger diameter pipe and fitting into place. This should be included in the estimate as part of the installation costs.

The estimator should review any riser diagrams on the plans for waste and vent piping locations and configurations. Remember that riser diagrams are diagrammatical in nature and not to scale. They do, however, offer a three-dimensional perspective that will help the estimator with directional changes and the fittings to accomplish those changes. Sometimes typical or generic riser diagrams are offered as a guide for the estimator to use.

Taking-off and Pricing

Procedures for takeoff and pricing follow those of the below-grade system. Fittings are counted and listed as EA according to types and sizes. Piping is taken off and priced by the LF and listed according to type, size, applications (waste or vent), and method of installation. While lead and oakum joints are less prevalent above grade, they are still used. The procedures for calculating lead, oakum, gas and/or gaskets, clamps, and couplings are the same as for the below-grade system. Additional items for the support of piping are taken off and priced by the individual piece and listed as EA, according to size, application, and type. Because pipe hangers and supports are omitted from the drawings for clarity, it is necessary to complete the piping estimate for horizontal run and riser to determine the quantity of each. Refer to the specifications for the required intervals of hangers (on-center spacing) and supports to determine the number needed. Always add a small percentage as extra that may be wasted or lost. Review specialty sections in the plumbing or mechanical sections for any seismic restraint or vibration isolators, as these can affect material and labor costs.

The takeoff and pricing procedure for above-grade waste and vent piping should be altered slightly to accommodate the fact that above-grade piping is in two planes: horizontal runs and branches in the horizontal plane, and risers and drops to fixtures and equipment in the vertical plane. Many plans do not illustrate vertical rise and drops, but only indicate them with piping symbols. The estimator is urged to carefully review the vertical path of piping before starting the quantity takeoff.

Riser diagrams and building sections provide dimensions showing floor-to-floor heights. With this information, the length of risers and the approximate length of drop pieces to fixtures can be calculated. Horizontal piping and risers are taken off separately to reduce the chance for error in referencing between multiple drawings. Special staging or rigging may be required to install some of the heavier cast-iron pipes in above-grade applications and must be included in the estimate. This work may require a lift to access the underside of the floor above in order to make connections and tie into the below-grade system. Miscellaneous clamps and hangers along with metal strut channel to anchor the piping to a wall or column should also be included.

Rarely is above-grade waste and vent piping in cast-iron a task for an individual plumber. It requires a crew of two as a minimum. However, with PVC and ABS piping, it can be done by an individual due to the lighter weights of the pipe and fittings.

Below-Grade Storm System Piping

The storm drainage system is another piping system that receives clear water from roof drains, (on flat-roof buildings), cooling or condensate water, or other clear water within the structure. It conveys the drainage to the building's storm sewer or disposal system on site. It is always separate from the sanitary waste system. There are several fittings and components unique to this system, such as roof drains and overflow drains and fittings. The installation often follows a similar process to the below-grade sanitary system.

Taking-off and Pricing

The takeoff and pricing procedure, as well as the units of measure, are the same as those for the below-grade sanitary waste and vent piping system. Many of the species of pipe and fittings are also the same. Although similar, the storm system takeoff and prices should be kept separate from the sanitary system, if for no other reason than it is a different subsection of *22 10 00 Plumbing Piping*. Some of these lines (pipes) can be rather large in diameter and require a multiperson crew and equipment to set, especially if the pipe is cast iron. Similar to below-grade waste and vent piping, there is frequently excavation and backfill required to direct the flow and get the pipe below the slab. Most often, storm drain systems are found on commercial projects, and as a result the sitework subcontractor does the excavation and backfill. This is not always a foregone conclusion; it is more a matter of what is customary in an area. The estimator should review the specifications to determine who will be responsible for the excavation and backfill of the below-grade system.

Above-Grade Storm System Piping

Above-grade storm system piping consists of roof drains, leaders, and horizontal offsets that tie into the below-grade storm system at the floor level. It often includes the overflow drains and piping that is part of the work but separate from the main system. The estimator would be prudent to review the roofing section of the specifications for the limitations of the plumbing work with regard to the roof drains. Many roofing specifications call for components of the roof drains to be installed by the roofer to ensure watertightness. Additionally, roof drains may require structural support in the form of steel angles under the roof deck, spanning between structural members. This is typically the responsibility of the structural steel fabricator/erector and only coordinated with the roof drain work.

Taking-off and Pricing

The takeoff, pricing procedure, and units of measure are the same as those for the above-grade sanitary waste and vent piping systems. Pipe is by the linear foot (LF), and fittings are by the piece (EA). Again, although similar, keep these two systems separate in the estimate. It is also best practice to also keep horizontal pipe and offsets for above-grade storm systems (that will be above finished ceilings) separate in the takeoff, since they are normally insulated to prevent condensation. This separation allows the estimator to reference the horizontal pipe quantities to calculate the quantity of insulation later in the estimate. Again, this work may require a lift to access the underside of the roof deck above in order to make connections and tie into the below-grade system. Miscellaneous clamps and hangers, along with metal strut channel to anchor the piping to a wall or column, should be included. Figure 18.8 provides guidelines for labor-hours to install various applications of cast piping and fittings for plumbing systems.

Hot and Cold Water Piping

Hot- and cold-water piping is a part of virtually every plumbing scope. This includes piping, fittings, valves, control devices, and all the related appurtenances for conveying potable water to plumbing fixtures and equipment – often referred to as *domestic water piping* (excluding piping for fire protection systems). The domestic water supply typically starts with distribution of the cold-water at the point where the water service enters the building and is distributed to the various fixtures and equipment within the structure. The hot water supply usually starts at the hot water heater and is distributed to the various fixtures within the building requiring hot water.

Standard piping materials include types L and K copper tubing and, in limited applications, brass, galvanized steel pipe, and PVC. Cross-linked polyethylene plastic, also known as XLPE (PEX) piping, has become extremely popular in residential and light commercial domestic water applications. The most common method of joining copper pipe and fittings is by solder joint, or rolled-groove pressure fittings for copper tubing, threaded fittings for steel and brass pipe, and cement joint fittings for PVC. Fittings include elbows, tees, 45° and 22.5° bends, couplings, and reducing

Description	Labor-Hours	Unit
Cast-Iron Soil Pipe Service Weight, Single-Hub with Hangers Every 5 Feet, Lead and Oakum Joints Every 10 Feet		
2" Pipe Size	.254	LF
3" Pipe Size	.267	LF
4" Pipe Size	.291	LF
5" Pipe Size	.316	LF
6" Pipe Size	.329	LF
8" Pipe Size	.542	LF
10" Pipe Size	.593	LF
12" Pipe Size	.667	LF
Push on Gasket Joints Every 10 Feet		
2" Pipe Size	.242	LF
3" Pipe Size	.254	LF
4" Pipe Size	.281	LF
5" Pipe Size	.304	LF
6" Pipe Size	.320	LF
8" Pipe Size	.516	LF
10" Pipe Size	.571	LF
12" Pipe Size	.653	LF
Cast-Iron Soil Pipe Fittings Hub and Spigot Service Weight		
Bends or Elbows		
2" Pipe Size	1.000	EA
3" Pipe Size	1.140	EA
4" Pipe Size	1.230	EA
5" Pipe Size	1.330	EA
6" Pipe Size	1.410	EA
8" Pipe Size	2.910	EA
10" Pipe Size	3.200	EA
12" Pipe Size	3.560	EA
Tees or Wyes		
2" Pipe Size	1.600	EA
3" Pipe Size	1.780	EA
4" Pipe Size	2.000	EA
5" Pipe Size	2.000	EA
6" Pipe Size	2.180	EA
8" Pipe Size	4.570	EA
10" Pipe Size	4.870	EA
12" Pipe Size	5.330	EA

Eighth bend

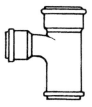

Sanitary tee

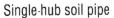

Single-hub soil pipe

Figure 18.8a Installation Time in Labor-Hours for Cast Iron

fittings. PEX piping or tubing is flexible and requires its own proprietary fittings, manifolds, appurtenances, and valves, but is significantly easier from a labor perspective.

Control devices include a variety of valves, such as check, globe, gate, ball, and butterfly. Other special devices include such items as backflow preventers, relief valves, pressure-reducing valves, shock absorbers, vacuum breakers, and frost-proof hose bibs. The list can be extensive.

Description	Labor-Hours	Unit
Push-on Gasket Joints		
Bends or Elbows		
2" Pipe Size	.800	EA
3" Pipe Size	.941	EA
4" Pipe Size	1.070	EA
5" Pipe Size	1.140	EA
6" Pipe Size	1.260	EA
8" Pipe Size	2.670	EA
10" Pipe Size	2.910	EA
12" Pipe Size	3.200	EA
Tees or Wyes		
2" Pipe Size	1.330	EA
3" Pipe Size	1.600	EA
4" Pipe Size	1.780	EA
5" Pipe Size	1.850	EA
6" Pipe Size	2.180	EA
8" Pipe Size	4.000	EA
10" Pipe Size	4.870	EA
12" Pipe Size	5.330	EA
Clean-outs		
Floor Type		
2" Pipe Size	.800	EA
3" Pipe Size	1.000	EA
4" Pipe Size	1.333	EA
5" Pipe Size	2.000	EA
6" Pipe Size	2.667	EA
8" Pipe Size	4.000	EA
Clean-out Tee		
2" Pipe Size	2.000	EA
3" Pipe Size	2.222	EA
4" Pipe Size	2.424	EA
5" Pipe Size	2.909	EA
6" Pipe Size	3.200	EA
8" Pipe Size	6.400	EA
Drains		
Heelproof Floor Drain		
2" to 4" Pipe Size	1.600	EA
5" and 6" Pipe Size	1.778	EA
8" Pipe Size	2.000	EA
Shower Drain		
1-1/2" to 3" Pipe Size	2.000	EA
4" Pipe Size	2.286	EA
Cast-Iron Service Weight Traps		
Deep Seal		
2" Pipe Size	1.143	EA
3" Pipe Size	1.333	EA
4" Pipe Size	1.455	EA

Clean-out, floor type

Clean-out tee

Heelproof floor drain

Shower drain

Deep seal trap

Figure 18.8b *(Continued)*

Taking-off and Pricing

Fittings, valves, and control devices are taken off and priced by counting the individual pieces (EA) and listing them according to type, material composition, size, and application. Water piping is taken off and priced by the LF, following a procedure similar to that for above-grade sanitary piping. The estimator should start by taking off mains and branches, then risers and drops. The linear foot (LF) quantities can be converted to individual lengths of pipe. All piping quantities should be listed according to type, grade, and size (diameter). For smaller-diameter copper pipe, such as ½" to 1", it is an impractical use of estimator time to count individual fittings; an allowance of one fitting per 10' of run may be more useful. For piping applications with more changes of direction in a shorter length, the increment can be reduced to 7' or 8' per fitting.

Description	Labor-Hours	Unit
P Trap		
2" Pipe Size	1.000	EA
3" Pipe Size	1.143	EA
4" Pipe Size	1.231	EA
5" Pipe Size	1.333	EA
6" Pipe Size	1.412	EA
8" Pipe Size	2.909	EA
10" Pipe Size	3.200	EA
Running Trap with Vent		
3" Pipe Size	1.143	EA
4" Pipe Size	1.231	EA
5" Pipe Size	2.182	EA
6" Pipe Size	3.000	EA
8" Pipe Size	3.200	EA
S Trap		
2" Pipe Size	1.067	EA
3" Pipe Size	1.143	EA
4" Pipe Size	1.231	EA
No Hub with Couplings Every 10 Feet OC		
1-1/2" Pipe Size	.225	LF
2" Pipe Size	.239	LF
3" Pipe Size	.250	LF
4" Pipe Size	.276	LF
5" Pipe Size	.289	LF
6" Pipe Size	.304	LF
8" Pipe Size	.464	LF
10" Pipe Size	.525	LF
No Hub Couplings*		
1-1/2" Pipe Size	.333	EA
2" Pipe Size	.364	EA
3" Pipe Size	.421	EA
4" Pipe Size	.485	EA
5" Pipe Size	.545	EA
6" Pipe Size	.600	EA
8" Pipe Size	.970	EA
10" Pipe Size	1.230	EA

P trap

Running trap with vent

S trap

No hub coupling

Figure 18.8c *(Continued)*

Pipe hangers and supports are taken off and priced by dividing the total LF of water piping in each size category by the specified intervals, as noted in the specifications or by code requirements. Solder, flux, and gas are used for joining copper water pipe. Copper pipe and fittings are joined by soft (nonlead) solder.

Solder, flux, and gas are difficult items to estimate, but by using the chart in Figure 18.9, one can arrive at a relatively accurate amount of each. The number of joints required for each size of fitting and device must be counted to determine a total number of joints in each size category.

Plumbing work can have a variety of miscellaneous items or appurtenances that are part of the estimate, especially when it comes to water piping. These must be included for a thorough estimate.

For XLPE piping the work follows a similar process however, the hot- and cold-water quantities are kept separate as the tubing is color coded. The system has its own fittings, valves, elbows, and so on in much the same way

Estimated Pounds of Soft Solder to Make 100 Joints.

Size	⅜"	½"	¾"	1"	1¼"	1½"	2"
Pounds	.5	.75	1.0	1.4	1.7	1.9	2.4
Size	2½"	3"	3½"	4"	5"	6"	8"
Pounds	3.2	3.9	4.5	5.5	8.0	15.0	32.0

Two oz. of flux will be required for each pound of solder. One tank of PRESTO gas will be required for every 500 joints.

Figure 18.9 Estimated Pounds of Soft Solder Required to Make 100 Joints

as copper pipe. Sizes of pipe (diameters) are separated for accurate pricing of materials and labor. Some specialty pieces, such as manifolds, are unique to the system. Special tools for flaring, cutting, prepping, disconnecting, etc. are also needed, but the system does not require solder, gas, or any of the typical items required when using copper. There are also limited diameters with XLPE piping. Renovation work can often be a mix of existing copper and new XLPE piping. The estimator is urged to watch internet videos of PEX pipe installations to clarify any questions with the process.

Pipe Insulation

Local plumbing, building, and energy codes require pipe to be insulated to reduce heat loss and prevent condensation. Pipe insulation is available in a wide variety of sizes and compositions for different applications and is manufactured in both rigid and flexible forms, with or without fittings. Fiberglass pipe insulation can require that the joints between pieces be taped and in some cases that the pipe be *jacketed*. Jacketing pipe involves encasing the pipe insulation with a hard but flexible covering for protection of the insulation. Most fiberglass pipe insulation is specified by the diameter of pipe it will cover and the wall thickness of the insulation. In addition to fiberglass pipe insulation there is polyethylene foam and rubber foam insulation. Lengths of the individual pieces range from 3' to 6', depending on type and manufacturer.

Taking-off and Pricing

Once the piping and fittings have been taken off, the estimator can calculate the quantity of insulation based on the takeoff quantity of piping. Diameters, wall thickness, type, and whether it has a jacketing should be noted for accurate pricing. Insulation for individual fittings (elbows, tees, valves, etc.) is taken off and priced by the piece (EA).

Insulation for piping runs should be listed in the takeoff according to the diameter and length of the pipe, LF to be covered, and the type of insulation (fiberglass or foam). Labor is based on the amount of insulation, in linear feet (LF), an individual insulator or plumber will install in a workday. Pricing should include any lifts or work platforms to access work above grade. Insulation is typically light in weight and as a result is most often an individual task.

Natural Gas System Piping

Piping for natural gas starts at the entrance of the gas service to the building and is distributed to the various gas-fueled appliances within the building, such as water heaters, furnaces, boilers, ranges, clothes dryers, and rooftop HVAC units. Piping materials are typically black steel pipe with malleable iron-threaded fittings. Fittings for gas piping are similar to those for other piping systems and include elbows, bends, unions, and tees. Special pieces such as flexible connectors are also used to connect the supply pipe to the equipment. Valves for the control of the flow of gas within the pipe are called *gas cocks*, and they are typically made of brass.

Study the plumbing plans and Division 22 specifications for the location, size, and type of pipe, and for the appliances to be connected. Division 23— HVAC specifications, HVAC plans, and any specialty plans such as kitchen equipment drawings should be reviewed for coordination and to ensure that all items are accounted for. Architectural and roof (mechanical) plans can be used to calculate the horizontal runs of gas pipe required to supply the units. Flexible connectors and valves quantities can also be determined from these drawings.

Taking-off and Pricing

The procedure for taking off and pricing gas piping is similar to that used for above-grade sanitary waste and vent piping or water piping. Fittings and valves are taken off and priced by the individual piece (EA) and listed according to type and size (diameter). Fittings occur most often at a change in direction of the pipe. Piping is taken off and priced by the LF, according to type and size (diameter). Since gas pipe and fittings are joined by a threaded connection, lead, oakum, or solder is not required. Pipe hangers and supports follow the same procedure as for water piping. For gas piping on a roof, special wood or synthetic blocking, called *dunnage*, is used to support the pipe slightly above the roof. The pipe is clamped to the dunnage at specific intervals, typically 10'.

For gas piping 3" in diameter or greater, the piping seams and fittings are required to be welded instead of threaded. This should be verified by local gas codes or the project specifications. For diameters less than 3", each cut will need to be threaded. Cut and thread quantities can be counted to assign a labor cost to each one. Larger lengths may require a second crew person to help with the cut-and-thread operation.

Special devices for regulating the pressure of gas supplied to appliances (such as for ranges or commercial ovens) may be required and must be included in the estimate. Flexible gas connectors for connecting movable appliances to a stationary gas supply may also need to be included in the estimate. Flexible connectors may be noted in the specification section dedicated to seismic and vibration restraint or in the gas specification itself.

Some local codes may require a separate permit for gas work. This usually constitutes an additional fee over and above that charged for the plumbing portion of the work.

Gas piping is frequently required to be painted or have labels stenciled on it for identification purposes. Painting can be the responsibility of the painting contractor, or it can be the responsibility of the gas piping installer. This should be noted in the specifications for the gas piping or painting.

Miscellaneous Items

Plumbing work can have a variety of miscellaneous costs or appurtenances that are part of the estimate. These must be included for a complete estimate. One of the first items to consider is the cost of the plumbing permit. Different municipalities have different methods for calculating the permit costs. Some base the permit fee on a percentage of the cost of the work. Others base the fee on a specific dollar amount per fixture or appliance. This applies to gas permits as well.

Access panels installed in the finish surface of walls or ceilings may be included as part of the plumbing work. They may also be a furnished item, installed by others. These are taken off and priced according to size, type, manufacturer, model, and location and are listed by the piece (EA). Note that access panels in fire-rated walls or ceiling assemblies are required to be fire-rated and some are required to be lockable.

Small brass or plastic tags with numbers, called *valve tags*, are often required to identify specific valves. In addition to the tags, a printed log is provided to the owner so that each valve with a tag can be identified according to its purpose. Labels identifying the direction of flow and whether the valve is for hot or cold water are also common. Identification systems should be included as a cost in the estimate.

The cost for flushing, sterilizing, and pressure testing is also commonly associated with a new plumbing system in a building. This cost is included within the plumbing scope but may be required to be done by an outside testing lab that can produce a report.

Painting of piping is often included within the Section 09900—Painting scope of work, although not exclusively. The estimator is directed to review the plumbing and painting section of the specifications to determine the trade that is responsible for painting the piping.

Cutting, coring, installing sleeves, and fire-stopping for plumbing piping can also be part of the plumbing scope of work. The estimator is directed to review the plumbing specifications carefully for responsibility for these items, as they are often a source of dispute. The estimator can be directed to Division 7, Section 07 80 00—Fire and Smoke Protection for the actual specifications.

Record drawings called *as-builts,* which illustrate the actual location of plumbing piping and appurtenances, are often part of the plumbing closeout requirements. As-builts can be simple redline updates on the plumbing drawings or revised electronic files that create new line drawings in CAD (computer-assisted design). The estimator is directed to research and include all costs associated with as-builts within the estimate. For many smaller plumbing contractors in the commercial market, as-builts in a CAD format will require that they solicit pricing for the work. This should be included in the cost of the work.

For renovation work, there is often a "make-safe" process whereby the plumber cuts and caps waste or water lines to remain and removes those scheduled for demolition. The process can often include dropping the work scheduled for demolition to the floor for pickup later or by a separate crew. This can also include the removal of old, antiquated equipment such as water heaters scheduled for replacement. Demolition work should be estimated separately from new work.

At the completion of the work, many commercial projects require that the systems withing the plumbing scope of work be commissioned. *Commissioning* is the process of evaluating the installation of the work in accordance with the design. Today's project designs are acutely in tune with energy conservation. Once the work is complete, the commissioning process confirms that the work is installed in accordance with the design. Occasionally, the commissioning process is under the domain of the design professional or the owner. However, on some projects the individual trade is responsible for the commissioning costs, done by an independent agency. The estimator must examine the specifications carefully for commissioning requirements, as they can be extremely expensive.

One final step in the closeout of the plumbing scope of work is the training of the owner's personnel who will maintain the systems once the project has been handed over to the owner. Operations and Maintenance (O&M) manuals are a compilation of the cut sheets and various literature for each piece of equipment that is turned over to the owner at closeout. This can be time consuming to prepare, review, and digitize. This is more common in the commercial market than residential. Many specifications require that the plumbing contractor video the training and present the video file as part of closeout. The complexity of the training is commensurate with the project. Regardless, there is a cost associated with training.

PARAMETRIC ESTIMATING

Most general contractors will never perform the work of the plumbing subcontractor for a variety of reasons, the least of which is the inability to accurately estimate the costs. This should not prevent the GC's estimator from developing a *parametric* estimate for comparison of plumbing bids. Parametric estimates, as the name would imply, are based on specific parameters such as square area or the cost of a project on a per-fixture basis. This is a very common practice in residential and light commercial projects. The GC uses prior projects of a similar nature to generate a cost per SF for just the plumbing scope of work. The cost can also be reduced to a cost per fixture or appliance, or even per bathroom. The parametric estimate can be used to compare to the plumbing subcontractor's quotes as a means of evaluating the bid. Parametric estimating will be further explained in Chapter 28, "Conceptual Estimating."

SUMMARY

One of the main premises of quantity takeoff is that the information derived from the contract documents is accurate. For example, a footing 1' high × 2' wide × 54' long contains 4 cubic yards (CY) of concrete, regardless of who does the takeoff. In order to do an accurate takeoff, detailed plans and specifications would be required, including an individual design for plumbing. For commercial projects, this is required by law for most states, but for residential construction it is often left to the individual plumbing contractor to design and install a system that will perform its function in accordance with local codes and standards.

It is not essential to itemize the takeoff and estimate for plumbing systems in the same manner as for other aspects of the project, such as carpentry, painting, or roofing. Listing the basic criteria helps establish a budget estimate for plumbing. Historical data from projects with similar criteria can then be compared parametrically with actual quotes from subcontractors to arrive at reliable budget costs. Additional costs include staging or platforms to access the work, which must be included as part of the costs of performing the work.

As with fire suppression and HVAC costs, budget estimates for plumbing work should be done purely for comparison or conceptual purposes, as most general contractors do not perform their own plumbing work. Only firm, well-qualified plumbing bids should be included within the GC's estimate scope.

For the estimator that prefers a detailed material, labor, and equipment estimate, similar to other scopes of work, the project must be broken down into its various systems – waste and vent piping; above- and below-grade stormwater systems; above- and below-grade domestic water piping; and insulation, fixtures, and the connection of appliances or equipment provided by others. Most piping is taken off and priced by the linear foot, LF with fittings and appurtenances being estimated by the individual piece, EA, noting its function and location for accuracy. Gas piping follows a similar process, but with some unique tasks such as threading and welding.

Miscellaneous items such as permits, valve tags, labeling, painting, or coring and sleeves must also be considered on a piece-by-piece basis. Make safe work may be required for renovation projects. Finally, as with all scopes of work, access and cleanup is normally part of the individual subcontractor's price and must be accounted for in the estimate.

**QUESTIONS/
PROBLEMS**

1. Provide a brief explanation of parametric estimating and how it can be used to generate a plumbing estimate for comparison with subcontractor bidders.

2. For below-grade systems such as waste and stormwater, explain what other costs may be associated with the work.

3. What part of the technical specification section 22 00 00—Plumbing would an estimator review to determine the submittals required? What section for cleanup of debris generated by the plumbing work?

4. Define the process of "make-safe" for plumbing?

5. Using the appropriate table from the chapter, calculate the cost of installation (setting fixture and trim) for a wall-hung urinal if the hourly billing rate for the plumber is $110.75 per hour. Cite the figure used.

19 | Heating, Ventilating, and Air-Conditioning (HVAC)

CSI MasterFormat® Division 23—Heating, Ventilating, and Air-Conditioning, better known by its acronym, *HVAC*, is the work of conditioning the space. Whether the work is to heat, cool, or just change the air (ventilate) within the space, it is included within Division 23. The work is typically performed by trades or firms with specialized training, with licenses, insurances, and permits separate from those of the general contractor. Taking off and estimating HVAC systems requires a working knowledge of the particular trade or system and, often, specialized education and training not normally within the realm of the general contractor's estimating experience. Nevertheless, the estimator should be able to develop sound, realistic budgets for comparison purposes to evaluate HVAC contractors' proposals on bid day. While reviewing HVAC drawings and specifications, the estimator will become familiar with the scope of work involved, which provides the basis for a more thorough review of subcontractor pricing.

Division 23 work is typically shown on the HVAC or *mechanical* drawings. (See the discussion of mechanical drawings in Chapter 1.) Mechanical drawings are labeled with the following prefixes: "M" for mechanical and "H" or "HVAC" for heating, ventilating, and air-conditioning. Review all drawings in the bid set, including architectural drawings, for related work that may be shown on other drawings, as well as for detailed information on dimensions and measurements for room sizes, floor-to-floor heights, location of services entering the building, and coordination with other work within the particular area. Mechanical drawings with schedules are helpful in determining types and quantities of materials such as rooftop units (RTUs), fans, and coils, for takeoff. (See the "Schedules" section of Chapter 1.) Specialized details, such as riser diagrams showing the configuration and components of refrigerant piping systems, are sometimes included.

Estimating Building Costs for the Residential and Light Commercial Construction Professional,
Third Edition. Wayne J. Del Pico.
© 2023 John Wiley & Sons, Inc. Published 2023 by John Wiley & Sons, Inc.

Consult the specification sections for the products, methods, and techniques of installation, as well as the related work of other trades that affect pricing. A review of Division 1—Temporary Facilities of the specifications may be necessary to determine what, if any, special requirements, such as hoisting or staging, may be included.

This chapter covers the basic procedures and methods for takeoff and pricing, limited to the HVAC systems normally encountered in residential and light commercial construction.

HVAC SYSTEMS

Heating, ventilating, and air-conditioning systems, commonly referred to as HVAC, include the various components that provide heating, cooling, and fresh air (ventilation) to the occupied space of the building or residence. One of the most common methods involves gas- or oil-fired furnaces that supply warm air through a series of supply and return-air ductwork. This same system of ductwork can be used to supply cooled air in warmer months.

Alternate methods of heating employ gas- or oil-fired boilers that force hot water through a system of radiant baseboard installed in individual rooms. This is referred to as a *hydronic heating system*.

Study the mechanical plans for the layout, locations, and sizes of ductwork and fin tube radiation baseboard. Special mechanical plans and details that illustrate the components of boilers and rooftop HVAC units may also be included. Architectural drawings should be reviewed for dimensions and coordination with architectural features, such as ceilings, windows, and doors. Division 23 specifications for the HVAC work also list the specific materials, manufacturer, and model of the heating and cooling units, as well as the various appurtenances required for a complete system.

For purposes of takeoff and estimating, HVAC work can be divided into two general categories: ducted systems for the distribution of heated or cooled air and radiant heating systems for forced hot water.

Both takeoff and estimating for HVAC systems, like other Division 23 work, require a specialized knowledge of the individual systems. The discussion within this chapter has been categorized into major categories for estimating. The various disciplines of labor involved in HVAC work are discussed in the following section.

HVAC Labor

The work of HVAC systems can be very diverse and, therefore, requires some specialized talents. The HVAC scope of work can be done by several different trades, for example:

- Sheet metal worker for the ductwork
- Pipe fitter or plumber for refrigerant or hydronic piping
- Insulators for the refrigerant/hydronic piping and ductwork

- Hoisting and rigging specialists to set rooftop equipment
- Controls technician for automatic temperature controls
- Calibration technicians for computerized systems
- Welders for heavy gauge welded steel ductwork
- Boilermakers for installing/assembling hot water boilers

Each of these trades has designated helpers or apprentices. HVAC crews can consist of numerous combinations depending upon the operation in progress.

In general, installing large or awkward runs of ductwork or fittings requires multiple individuals and some type of hoisting equipment. Insulators can work as single individuals or multiperson crews, hoisting/rigging is a multiperson task, and calibrating computerized systems can be done by a single individual or a multiperson crew.

Productivity can be affected by a wide variety of issues, from access to temperature. Estimates of labor-hours must consider actual conditions in order to be accurate. Tight or confined spaces such as boiler rooms or attics reduce productivity, whereas wide-open areas found in office buildings can enhance productivity. Production is measured by the combined activity of the crew over the workday. Renovations labor-hours for a task may be more than for new, due to the *cut and patch to match* scenario common with fitting new work to existing work.

Ducted Systems

Ducted systems include a wide variety of heating and cooling systems with metal or fiberglass ductwork for the supply and return of conditioned air to and from spaces within a building. The heated or cooled air can be supplied from a single, self-contained unit or from separate components at various locations within or outside the structure. Ductwork and equipment shown on the project drawings are engineered to suit the specific application based on the design criteria. In much the same manner as piping for water and waste systems, ductwork is installed in a series of main trunks and branches to specific areas as required. Different "fittings" or transition pieces, allow ducts to change direction, circumvent obstacles, or reduce in size, as required by the particular application.

Special devices that control the flow of air within the ductwork are called *dampers*. Dampers can be manual or automatic. They are typically on the main supply ductwork at the beginning of the branch duct. Round or rectangular outlets that diffuse the air delivered to the space are called *diffusers*. They are most often located at the ceiling level but can be installed on walls as well. Similar devices that have a grille and damper for regulating airflow at the device are called *registers*. Louvered or perforated panels at the inlet to return-air ducts are called *grilles*. Grilles are most often at the ceiling or at least near the top of the wall.

Controls that regulate the temperature of the space, called *thermostats*, signal to the equipment the need for more or less heat or cooling. Thermostatic controls can be simple rotary dial, digital, or Wi-Fi enabled. They can control the temperature in one space or can be tied to other sensors to average the temperature in multiple spaces. Thermostats can also be more sophisticated devices that allow the user to control and monitor temperatures from outside the space via wi-fi or sense when the space is unoccupied thereby preventing wasted heating or cooling. Figure 19.1 illustrates gas- and oil-fired warm air ducted systems, respectively.

Taking-off and Pricing

The takeoff process often begins at the heating/cooling device and moves through the distribution ductwork until it reaches the intended space. It starts with larger ducts and fittings and moves to smaller. Metal ductwork is taken off by the linear (LF) and listed according to type, size, and application (supply or return). It can be converted to weight in pounds (lb.) based on its thickness or gauge, for pricing (discussed later in this section). In addition to horizontal mains, vertical risers and drops are also necessary for the distribution of air between multiple floors. Flexible ducts for short runs to diffusers are taken off and priced by the LF. Most specifications have distinct limitations on the length of flexible ducts used.

Ductwork takeoffs can be done in a tabular format, showing each size, length, location in the project, and gauge of the metal. It can be noted as supply or return, insulated or plain, and if it has sound liner. This allows for an easier flow of calculations for all of the other components with the ductwork. The same is true for piping. Miscellaneous items such as sealant, tie wraps, duct tape, or reflective tape are also necessary. These are typically accounted for as a small percentage (2–3%) of the cost of the ductwork itself.

Since the supply ductwork and fittings are usually insulated, it is helpful to separate the quantities of each in the takeoff. Fittings, transition pieces, reducers, collars for the connection of flexible ducts, and dampers are taken off and priced by the individual piece (EA) and listed according to type, size, and application. Devices installed in the finished space, such as registers, grilles, thermostats, and diffusers, are taken off and priced by the individual piece (EA) and listed according to type, size, manufacturer, model, and finish.

Equipment takeoff for furnaces, air-conditioning condensing units, evaporators, electric coils, and heat pumps is by the individual piece (EA) and listed according to type, size or capacity, manufacturer, model, series, and any other special identifying criteria. Components to complete the system may include flues for the furnace, control wiring for the thermostat, testing and balancing, and filters, which are taken off individually and listed as each (EA), lump sum (LS), or whatever units best represent the scope of work. Here a

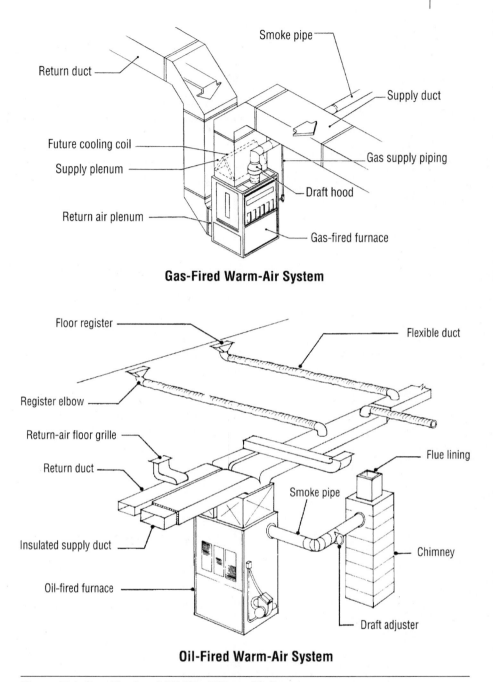

Gas-Fired Warm-Air System

Oil-Fired Warm-Air System

Figure 19.1 Gas-Fired and Oil-Fired Warm Air Systems

checklist can be a good way to avoid omissions. Once the list of all necessary components is complete, prices should be solicited from suppliers.

Consult the mechanical drawings and specifications to determine the exact scope of HVAC work. Items such as power wiring to the heating or cooling equipment, furnishing and installation of oil tanks for oil-fired systems, and gas or fuel piping for fossil-fuel-fired furnaces or boilers are not typically part of the HVAC contractor's work.

For commercial projects with rooftop equipment, hoisting or crane services are necessary. Check the specifications to verify who will provide the crane or hoisting of the equipment and note this information in the takeoff. It may even require a permit for the space on the public right-of-way occupied by the crane during the pick if it cannot be located on private property.

Testing and balancing of completed duct systems is mandatory for most commercial projects and must be done by an independent contractor at the expense of the HVAC contractor. Reports are typically required to confirm that the design criteria have been met and may also be a condition of occupancy with the local official. Testing and balancing should be included as a separate cost in the HVAC estimate. Prices should be solicited from a qualified test and balance contractor for the most accurate pricing and included within the bid as a subcontractor.

Sheet metal ductwork, fabricated from galvanized steel sheets, is often converted to weight (lb.) for the pricing of the raw material (sheets). The calculation for this conversion considers the gauge (thickness) and weight per square foot (SF) of the material being used. Consult Figure 19.2 to obtain the weight per LF of ductwork for various types of sheet metal.

The total weight in pounds of ductwork should be increased by 10% to 15% for waste, hangers, and clips. The total weight in pounds can be priced prior to the fabrication. Like most commodities, sheet metal fluctuates with the international price of steel. Many estimators get the most up-to-date price for steel on the day of the bid. Figure 19.3 provides information for estimating installation costs for ductwork and related work.

Pipe and Duct Insulation

Many designs, as well as local energy codes, require pipe to be insulated to reduce heat loss and prevent condensation. Pipe insulation is available in a wide variety of sizes and compositions for different applications and is

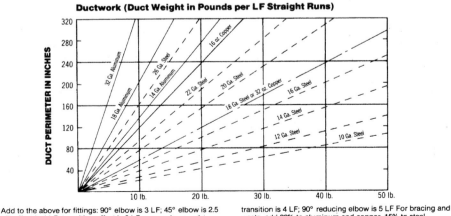

Ductwork (Duct Weight in Pounds per LF Straight Runs)

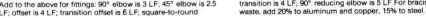

Add to the above for fittings: 90° elbow is 3 LF; 45° elbow is 2.5 LF; offset is 4 LF; transition offset is 6 LF; square-to-round

transition is 4 LF; 90° reducing elbow is 5 LF For bracing and waste, add 20% to aluminum and copper, 15% to steel.

Figure 19.2 Duct Weights

Description	Labor-Hours	Unit
Ductwork		
Fabricated Rectangular, Includes Fittings, Joints, Supports		
Allowance for Flexible Connections, No Insulation		
Aluminum, Alloy 3,003-H14, Under 300 lbs.	.320	lb.
300 to 500 lbs.	.300	lb.
500 to 1,000 lbs.	.253	lb.
1,000 to 2,000 lbs.	.200	lb.
2,000 to 10,000 lbs.	.185	lb.
Over 10,000 lbs.	.166	lb.
Galvanized Steel, Under 400 lbs.	.102	lb.
400 to 1,000 lbs.	.094	lb.
1,000 to 2,000 lbs.	.091	lb.
2,000 to 5,000 lbs.	.087	lb.
5,000 to 10,000 lbs.	.084	lb.
Over 10,000 lbs.	.080	lb.
Stainless Steel, Type 304, Under 400 lbs.	.145	lb.
400 to 1,000 lbs.	.130	lb.
1,000 to 2,000 lbs.	.120	lb.
2,000 to 10,000 lbs.	.107	lb.
Over 10,000 lbs.	.102	lb.
Flexible, Vinyl-Coated Spring Steel or Aluminum, Pressure to 10" (WG) UL-181		
Noninsulated, 3" Diameter	.040	LF
4" Diameter	.044	LF
5" Diameter	.050	LF
Ductwork, Flexible, Noninsulated		
6" Diameter	.057	LF
7" Diameter	.067	LF
8" Diameter	.080	LF
9" Diameter	.089	LF
10" Diameter	.100	LF
12" Diameter	.133	LF
14" Diameter	.200	LF
16" Diameter	.267	LF
Insulated, 4" Diameter	.047	LF
5" Diameter	.053	LF
6" Diameter	.062	LF
7" Diameter	.073	LF
8" Diameter	.089	LF
9" Diameter	.100	LF
10" Diameter	.114	LF
12" Diameter	.160	LF
14" Diameter	.200	LF
16" Diameter	.267	LF
18" Diameter	.356	LF
20" Diameter	.369	LF

Figure 19.3a Installation Time in Labor-Hours for Ductwork

Source RSMeans Estimating Handbook, Third Edition. © 2009 John Wiley & Sons, Inc.

manufactured in both rigid and flexible forms, with or without fittings. Fiberglass pipe insulation can require that the joints between pieces and fittings be taped and in some cases that the pipe be *jacketed*. Jacketing pipe means encasing the pipe insulation with a hard but flexible covering for protection of the insulation.

Supply-air ductwork is typically insulated to reduce heat or cooling loss as the air travels through the duct. Thermal insulation is installed on the outside of the ductwork. Other types of insulation reduce sound transmission as the air moves through the ductwork. This is called a *sound liner* and is installed on the interior for the duct.

Description	Labor-Hours	Unit
Fiberglass, Aluminized Jacket, 1-1/2" Blanket		
4" Diameter	.047	LF
5" Diameter	.053	LF
6" Diameter	.062	LF
7" Diameter	.073	LF
8" Diameter	.089	LF
9" Diameter	.100	LF
10" Diameter	.114	LF
12" Diameter	.160	LF
14" Diameter	.200	LF
16" Diameter	.267	LF
18" Diameter	.356	LF
Rigid Fiberglass, Round, .003" Foil Scrim Jacket		
4" Diameter	.052	LF
5" Diameter	.058	LF
6" Diameter	.067	LF
7" Diameter	.073	LF
8" Diameter	.089	LF
9" Diameter	.100	LF
10" Diameter	.114	LF
12" Diameter	.160	LF
14" Diameter	.200	LF
16" Diameter	.267	LF
18" Diameter	.356	LF
20" Diameter	.369	LF
22" Diameter	.400	LF
24" Diameter	.436	LF
26" Diameter	.480	LF
28" Diameter	.533	LF
30" Diameter	.600	LF
Rectangular, 1"-Thick, Aluminum-Faced, No Additional Insulation Required	.069	SF surf.

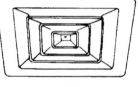

Supply Diffuser

Figure 19.3b *(Continued)*

Taking-off and Pricing

Once the piping and fittings have been taken off, the estimator can calculate the quantity of insulation. Insulation for individual fittings (elbows, tees, valves, etc.) is taken off and priced by the piece (EA).

Insulation for piping runs should be listed in the takeoff according to the diameter and length of the pipe (LF) to be covered and the type of insulation (fiberglass or closed cell), and wall thickness or R-value.

After the ductwork portion of the takeoff has been completed, calculate the quantity of insulation needed. To do so, the total surface area of the various-sized ducts and fittings must be determined. The easiest method considers the fittings as ductwork and measures through the fittings when doing a

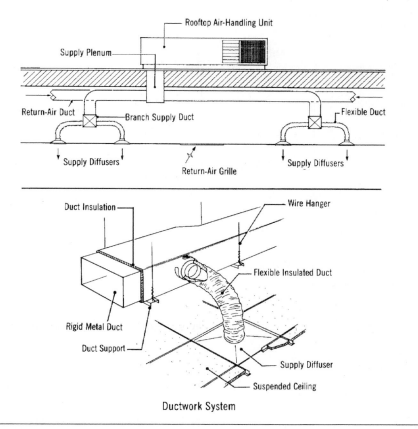

Ductwork System

Figure 19.3c *(Continued)*

takeoff from the plan. Many of these calculations can be avoided by using a chart or table from the internet for converting various sizes to square feet.

Consider as an example:

Assume a takeoff quantity of 43' of 16" × 12" supply duct to be insulated:

$$Surface\ area = (16" + 12") \times 2 = 4.67\ SF\ per\ LF\ of\ duct$$

$$43\ LF \times 4.67\ SF/LF = 200.8\ SF$$

$$Adding\ 15\%\ for\ lapping\ and\ sealing = 1.15 \times 200.8 = 231\ SF$$

Sound lining in the interior of the duct is taken off and priced by the SF using the same procedure as for duct thermal insulation without the added percentage for lapping and sealing.

Forced Hot Water Systems (Hydronic Heating Systems)

An alternative method of heating is oil- or gas-fired hot water (or steam) boilers, manufactured in cast iron, steel, or copper, which are available preassembled (packaged) or in sections for field assembly. This is frequently referred to as *wet heat* or more technically a *hydronic heating system*.

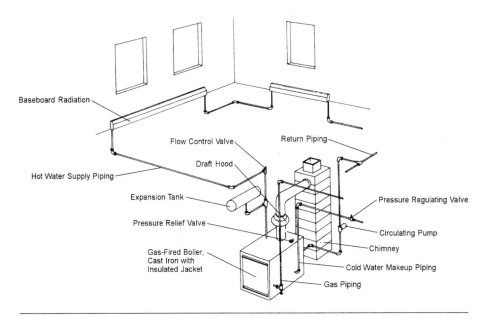

Figure 19.4 Forced Hot Water Heating System

Boilers heat water under pressure to almost boiling (or steam). The water (or steam) travels through a closed-loop system and radiates the heat caused by the hot water (or steam) to a device that delivers it to the location being heated. Those devices are called *fin-tube radiant baseboard* for water and *radiators* for steam. Figure 19.4 illustrates a forced hot water heating system.

In addition to the boiler itself, appurtenances required to complete the system include expansion tanks, pressure-relief and pressure-regulating valves, zone valves, circulators, pipe and fittings, flow control valves, oil burners (for oil-fired systems), operating controls, and fin-tube radiant baseboard.

Taking-off and Pricing

The following takeoff quantities apply to forced hot water systems:

- **Fin-tube radiant baseboard:** Taken off and priced by the LF (as is cast-iron baseboard). Quantities of each should be kept separate and listed according to type, manufacturer, model, or rating (BTU output), and finish of the protective enclosure.
- **Boilers and Furnaces:** Taken off and priced by the individual unit and quantified as EA; listed according to size (BTU rating), type of fuel used, construction (steel, cast iron), manufacturer, model or series, and level of assembly required. Quotes are often solicited from manufacturers or vendors for contemporaneous pricing.
- **Appurtenances (expansion tanks, circulators, zone valves, pressure valves, draft hoods, breeches, flues, and oil tanks):** Taken off and priced by the piece (EA) and listed according to manufacturer, model, function or type, size, and any other identifying features. The estimator must consider any pieces for connecting the various appurtenances.

- **Piping to and from radiant baseboard:** Taken off and priced by the LF and listed according to size and material (steel, copper, PEX, etc.). This should include the fittings at each end of the fin tube to complete the loop.
- **Fittings and valves:** Taken off and priced by the individual piece (EA) according to type, size, material, and method of joining. This same procedure is used for taking off looped radiant systems.

For much of the work noted above, a checklist can be used to ensure that all components are captured for the estimate. Similar to ductwork takeoffs, piping can be summarized in a tabular format allowing for better control of the different sizes and quantities within the estimate. Remember that related work of other trades may be necessary for a complete system, including power wiring, gas piping, control wiring, roof penetrations, coring concrete or masonry, and piping of oil tanks. Piping that passes through fire-rated walls or floors may require fire-stopping and sleeves (refer to Chapter 12—Division 7—Thermal and Moisture Protection). Review the specifications carefully to determine the exact scope of work that is included. Figure 19.5 provides information for estimating installation costs for forced hot water heating systems.

HVAC Equipment

In addition to the categories already discussed, there is a wide variety of equipment used in the HVAC scope of work. Some of the more common items include:

- Air-handling units (AHUs) for moving air
- Rooftop units (RTUs) for heating and cooling from a central rooftop location
- Packaged condensers and compressors for air-conditioning systems
- Electric heat pumps with cooling coils
- Split system components and line set
- Heat exchangers for efficient heat transfer
- Economizers
- Exhaust and intake air fans to remove or provide air
- Variable air volume units (VAV boxes) to zone airflow
- Air-filtering or cleaning units
- Unit heaters for individual locations
- Cabinet unit heaters and ceiling unit heaters

Taking-off and Pricing

HVAC equipment is typically taken off by the individual piece and listed in the takeoff as EA. The completed list is sent to a supplier of HVAC equipment for pricing. The same list is used for determining labor costs associated with the installation of each piece. Again, a checklist is often a professional way to

Description	Labor-Hours	Unit
Hot Water Heating System, Area to 2,400 SF		
Boiler Package, Oil-Fired, 225 MBH	17.143	EA
Oil Piping System	4.584	EA
Oil Tank, 550 Gallon, with Black Steel		
Fill Pipe	4.000	EA
Supply Piping, 3/4" Copper Tubing	.182	LF
Supply Fittings, Copper	.421	LF
Baseboard Radiation	.667	LF

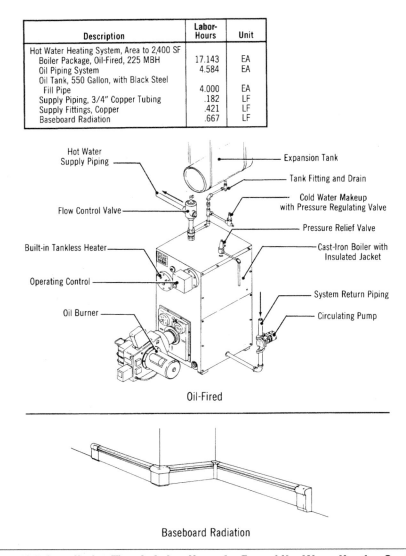

Oil-Fired

Baseboard Radiation

Figure 19.5 Installation Time in Labor-Hours for Forced Hot Water Heating Systems

Source RSMeans Estimating Handbook, Third Edition. © 2009 John Wiley & Sons, Inc.

ensure that all components are included in the estimate. Many specifications also refer to a piece of equipment by its *SEER rating (Seasonal Energy Efficiency Ratio)*. This is a ratio that identifies the energy efficiency of the equipment. Usually, higher efficiency translates to a higher price. Labor-hour guidelines for some of the more common items mentioned are shown in Figures 19.6 through 19.9.

HVAC Controls

All HVAC equipment requires some type of automatic temperature control (ATC) system to operate. Some systems are very sophisticated and are part of an integrated automation system, thus requiring a highly specialized subcontractor to install, test, and program. Others are simple thermostatic

Description	Labor-Hours	Unit
Electric Fired, Steel, Output		
60 MBH	20.000	EA
500 MBH	36.293	EA
1,000 MBH	60.000	EA
2,000 MBH	94.118	EA
3,000 MBH	114.286	EA
7,000 MBH	117.778	EA
Gas Fired, Cast Iron, Output		
80 MBH	21.918	EA
500 MBH	53.333	EA
1,000 MBH	64.000	EA
2,000 MBH	99.250	EA
3,000 MBH	133.333	EA
7,000 MBH	320.000	EA
Oil Fired, Cast Iron, Output		
100 MBH	26.667	EA
500 MBH	64.000	EA
1,000 MBH	76.000	EA
2,000 MBH	114.286	EA
3,000 MBH	139.130	EA
7,000 MBH	360.000	EA
Scotch Marine Packaged Units, Gas or Oil Fired, Output		
1,300 MBH	80.000	EA
3,350 MBH	152.000	EA
4,200 MBH	170.000	EA
5,000 MBH	192.000	EA
6,700 MBH	223.000	EA
8,370 MBH	230.000	EA
10,000 MBH	250.000	EA
20,000 MBH	380.000	EA
23,400 MBH	484.000	EA

Figure 19.6a Installation Time in Labor-Hours for Boilers

Source RSMeans Estimating Handbook, Third Edition. © 2009 John Wiley & Sons, Inc.

controls with either line voltage or low-voltage wiring. Estimating HVAC controls is often the work of a specialty subcontractor to the HVAC contractor in commercial projects and even in some of the more complex residential systems.

Taking-off and Pricing

Thermostats are taken off and listed by the individual piece (EA). They should be separated according to operation (low-voltage hard wired or Wi-Fi enabled) and application (heating, cooling, or both). Installation costs depend on the system and quantity of the devices as well as ancillary devices such as temperature sensors. Once a complete list of all control devices has been taken off, it can be sent to a supplier of HVAC controls for pricing. The same list is used for determining labor costs associated with the installation of each piece.

Control wire for each device can be taken off and listed by the LF based on the type of wire and its application. Do not forget to include vertical runs of wire between floors and possibly raceways (conduits) that may be required.

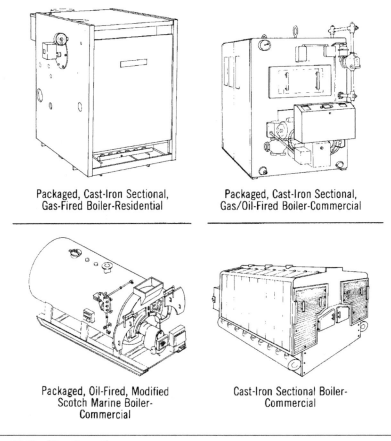

Packaged, Cast-Iron Sectional, Gas-Fired Boiler-Residential

Packaged, Cast-Iron Sectional, Gas/Oil-Fired Boiler-Commercial

Packaged, Oil-Fired, Modified Scotch Marine Boiler-Commercial

Cast-Iron Sectional Boiler-Commercial

Figure 19.6b *(Continued)*

For Wi-Fi enabled systems with a central control such as a computer or tablet the estimator must include the programming, software, and hardware costs. These can be substantial, not to mention additional programming costs as systems are integrated in the system.

For commercial projects with systems that are integrated with the HVAC controls, it is often best to solicit pricing for providing a complete package by a representative of the specified system. For most estimators without the background in controls, estimating this work can be dubious due to unfamiliarity with the system or the inability to access pricing of the various components for proprietary reasons. For larger commercial projects, controls vendors may be aware of the project through independent means and may have prepared pricing for the equipment needed.

Miscellaneous Items

A variety of miscellaneous items or appurtenances can be part of the estimate for HVAC work. These must be included for a thorough estimate.

Ductwork passing through a fire-rated wall or ceiling requires *fire dampers*. Fire dampers have a fusible link that melts in the high heat of a fire and releases a spring-loaded damper to close off the ductwork from the passage

Description	Labor-Hours	Unit
Fan Coil Unit, Freestanding		
Finished Cabinet, 3-Row Cooling		
or Heating Coil, Filter		
200 CFM	2.000	EA
400 CFM	2.667	EA
600 CFM	2.909	EA
1,000 CFM	3.200	EA
1,200 CFM	4.000	EA
4,000 CFM	8.571	EA
6,000 CFM	16.000	EA
8,000 CFM	30.000	EA
12,000 CFM	40.000	EA
Direct Expansion Cooling		
Coil, Filter		
2,000 CFM	5.333	EA
3,000 CFM	5.333	EA
4,000 CFM	9.231	EA
8,000 CFM	34.286	EA
12,000 CFM	40.000	EA
16,000 CFM	53.333	EA
20,000 CFM	63.158	EA
Central Station Unit, Factory		
Assembled, Modular, 4-, 6-, or		
8-Row Coils, Filter and Mixing Box		
1,500 CFM	13.333	EA
2,200 CFM	14.545	EA
3,800 CFM	20.000	EA
5,400 CFM	30.000	EA
8,000 CFM	40.000	EA
12,100 CFM	52.174	EA
18,400 CFM	72.727	EA
22,300 CFM	82.759	EA
33,700 CFM	126.316	EA
52,500 CFM	200.000	EA
63,000 CFM	246.154	EA

Figure 19.7a Installation Time in Labor-Hours for Air-Handling Units

Source RSMeans Estimating Handbook, Third Edition. © 2009 John Wiley & Sons, Inc.

of fire or smoke. Fire dampers are taken off and priced according to size, rating, type, manufacturer, model, and location and are listed by the piece (EA).

Renovation work may require new and existing ductwork to be connected together. Demolition can disturb dust accumulated in existing ductwork, so many renovation projects call for duct cleaning. This is a process requiring special equipment, and the work can be labor intensive. Whenever possible, the estimator should solicit pricing for the duct cleaning scope of work.

Small brass or plastic tags, similar to those used in plumbing work with numbers, called *valve tags,* may be required to identify specific valves in the heating or cooling assembly. Accompanying the tags is a printed log provided to the owner so that each valve with a tag can be identified according to its purpose. Labels identifying the direction of flow and whether it is for supply or return are also common. Identification systems should be included as a cost in the estimate.

Costs for flushing, sterilizing, and pressure testing are also commonly associated with a new hydronic heating system. They are included within the HVAC scope but may be required to be done by an outside testing lab

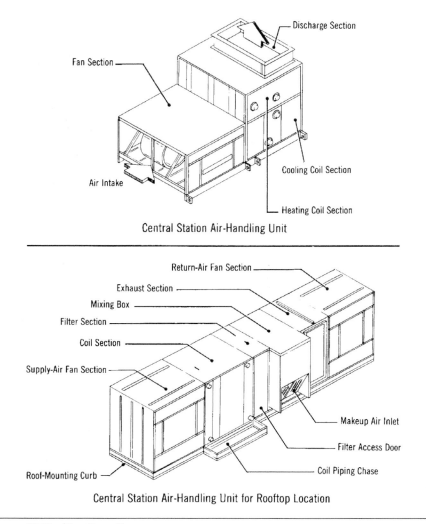

Central Station Air-Handling Unit

Central Station Air-Handling Unit for Rooftop Location

Figure 19.7b *(Continued)*

that can produce a report. In addition, chemicals such as ethylene glycol can be introduced to the water in the closed-loop systems to reduce the temperature at which the water will freeze. This service may likewise be required to be performed by an outside agency.

Start-Up and Testing constitute the factory start-up of large pieces of equipment such as RTUs, AHUs, and boilers. This task is typically performed by a manufacturer's certified technician, who can provide a documented report confirming that the equipment was installed appropriately and will operate as designed. This is not to be confused with commissioning of the HVAC system.

Painting of piping is often included within the *Section 09 90 00—Painting* scope of work, although not exclusively. The estimator is directed to review the HVAC and painting sections of the specifications to determine which trade is responsible for painting of the hydronic piping.

Description	Labor-Hours	Unit
Single Zone, Electric Cool, Gas Heat		
2-Ton Cooling, 60 M BTU/Hr. Heating	17.204	EA
4-Ton Cooling, 95 M BTU/Hr. Heating	26.403	EA
5-Ton Cooling, 112 M BTU/Hr. Heating	28.571	EA
10-Ton Cooling, 200 M BTU/Hr. Heating	35.982	EA
15-Ton Cooling, 270 M BTU/Hr. Heating	42.032	EA
20-Ton Cooling, 360 M BTU/Hr. Heating	47.976	EA
30-Ton Cooling, 540 M BTU/Hr. Heating	68.376	EA
40-Ton Cooling, 675 M BTU/Hr. Heating	91.168	EA
Multizone, Electric Cool, Gas Heat, Economizer		
15-Ton Cooling, 360 M BTU/Hr. Heating	52.545	EA
20-Ton Cooling, 360 M BTU/Hr. Heating	60.038	EA
25-Ton Cooling, 450 M BTU/Hr. Heating	71.910	EA
28-Ton Cooling, 450 M BTU/Hr. Heating	79.012	EA
30-Ton Cooling, 540 M BTU/Hr. Heating	85.562	EA
37-Ton Cooling, 540 M BTU/Hr. Heating	113.000	EA
70-Ton Cooling, 1,500 M BTU/Hr. Heating	198.000	EA
80-Ton Cooling, 1,500 M BTU/Hr. Heating	228.000	EA
90-Ton Cooling, 1,500 M BTU/Hr. Heating	256.000	EA
105-Ton Cooling, 1,500 M BTU/Hr. Heating	290.000	EA

Figure 19.8a Installation Time in Labor-Hours for Packaged Rooftop Air Conditioner Units

Source RSMeans Estimating Handbook, Third Edition. © 2009 John Wiley & Sons, Inc.

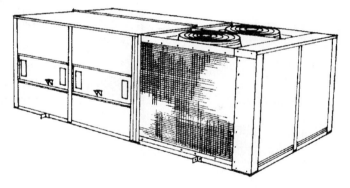

Packaged Rooftop Air Conditioner

Figure 19.8b *(Continued)*

Cutting, coring, installing sleeves, and fire-stopping for HVAC ductwork and piping can also be part of the HVAC scope of work. The estimator is directed to review the HVAC specifications carefully.

Supports and anchors are often required for the installation of HVAC equipment and piping. This can include drilling and setting anchor bolts into concrete or masonry to set the piece of equipment. Other supports such as strut (perforated steel channel) to attach pipe or duct can also be required. For commercial projects, hangers and supports can be specified, but often only shown in typical details on the plans. They are always steel or a metal alloy. This includes everything from riser clamps to clevis hangers with saddles for insulation. Read the specifications carefully and then allow

Description	Labor-Hours	Unit
Fans		
Belt Drive, In-Line Centrifugal		
3,800 CFM	5.882	EA
6,400 CFM	7.143	EA
10,500 CFM	8.333	EA
15,600 CFM	12.500	EA
23,000 CFM	28.571	EA
28,000 CFM	50.000	EA
Direct-Drive Ceiling Fan		
95 CFM	1.000	EA
210 CFM	1.053	EA
385 CFM	1.111	EA
885 CFM	1.250	EA
1,650 CFM	1.538	EA
2,960 CFM	1.818	EA
Direct-Drive Paddle Blade Fan		
36", 4,000 CFM	3.333	EA
52", 7,000 CFM	5.000	EA
Direct-Drive Roof Fan		
420 CFM	2.857	EA
675 CFM	3.333	EA
770 CFM	4.000	EA
1,870 CFM	4.762	EA
2,150 CFM	5.000	EA
Belt-Drive Roof Fan		
1,660 CFM	3.333	EA
2,830 CFM	4.000	EA
4,600 CFM	5.000	EA
8,750 CFM	6.667	EA
12,500 CFM	10.000	EA
21,600 CFM	20.000	EA
Direct-Drive Utility Set		
150 CFM	3.125	EA
485 CFM	3.448	EA
1,950 CFM	4.167	EA
2,410 CFM	4.545	EA
3,328 CFM	6.667	EA
Belt-Drive Utility Set		
800 CFM	3.333	EA
1,300 CFM	4.000	EA
2,000 CFM	4.348	EA
2,900 CFM	4.762	EA
3,600 CFM	5.000	EA
4,800 CFM	5.714	EA
6,700 CFM	6.667	EA
11,000 CFM	10.000	EA
13,000 CFM	12.500	EA
15,000 CFM	20.000	EA
17,000 CFM	25.000	EA
20,000 CFM	25.000	EA

Figure 19.9a Installation Time in Labor-Hours for Fans

Source RSMeans Estimating Handbook, Third Edition. © 2009 John Wiley & Sons, Inc.

a quantity based on the spacing required, plus some extra for on-site conditions. The count should be accurate as these supports are often more expensive than hangers in PVC.

Cranes or rigging equipment for the setting of rooftop HVAC equipment may also be needed. The estimator should consider weight, reach, and needed duration of the crane for installing the rooftop equipment. Again, a rigging subcontractor may be the best option for pricing the work accurately.

Description	Labor-Hours	Unit
Belt-Drive Propeller Fan		
12", 1,000 CFM	.571	EA
14", 1,500 CFM	.667	EA
16", 2,000 CFM	.889	EA
30", 4,800 CFM	1.143	EA
36", 7,000 CFM	1.333	EA
42", 10,000 CFM	1.600	EA
48", 16,000 CFM	2.000	EA
Belt-Drive Airfoil Centrifugal		
12,420 CFM	7.273	EA
18,620 CFM	8.000	EA
27,580 CFM	8.889	EA
40,980 CFM	10.667	EA
60,920 CFM	16.000	EA
74,520 CFM	20.000	EA
90,160 CFM	22.857	EA
110,300 CFM	32.000	EA
134,960 CFM	40.000	EA

Figure 19.9b *(Continued)*

Record drawings, called *as-builts,* that illustrate the actual location of HVAC ductwork piping and appurtenances are often part of the mechanical closeout requirements. As-builts can be simple redline updates on the mechanical drawings or revised electronic files that create new line drawings in CAD. The estimator is directed to research and include all costs associated with as-builts within the estimate.

Most contracts require testing and balancing of air and water systems, especially on commercial projects. This is done by a firm specializing in testing and balancing and independent of the installing contractor. Testing and balancing should not be confused with Start Up and Testing or *commissioning* (introduced in Chapter 18, "Plumbing"). Putting each and every valve, damper, fan, and heating or cooling unit through its paces at turnover is called commissioning. Commissioning is a thorough examination and analysis of the HVAC system to ensure that it is performing in accordance with the design criteria. Each component in the installation is tested and verified to ensure that the work conforms to the design. This requires that the various components are tested, ensuring that valves open and close at the setpoint temperatures, or that ecosystems function as designed. Commissioning is often performed by a specialty engineering firm and can be quite expensive and time consuming. The firm is independent of the project design engineers and installers, thereby ensuring impartiality.

Training of owner's personnel on the new systems and equipment may also be a requirement, especially in commercial projects. Most specifications require that each major system or component be demonstrated as to its operation and maintenance. Frequently, the training is video recorded for future reference by the owner's personnel. In addition to the training, many projects require the HVAC contractor to compile binders or electronic files of all of the technical literature for each major piece of equipment. This may require personnel directly involved in the installation of the equipment. In any event, the process can be time consuming.

Ongoing cleanup of debris generated by this work can be a requirement as well. This would constitute disposing of debris from packaging and duct or insulation scrap to a dumpster or removing it from the site altogether. The estimator is directed to review the mechanical specifications carefully.

SUMMARY

In order to do an accurate takeoff, detailed plans and specifications are required, including an individual mechanical design for heating and air-conditioning systems. Most states require this for commercial projects, but for residential construction it is often left to the individual HVAC contractor to design and install a system that will perform its function in accordance with local codes and standards. In the case of design-build for residences the estimate is often done parametrically by using the square foot area of the home to be conditioned to determine the heat or cooling needed. U-factors, R-factors, and solar heat gain coefficients (SHGC) would also be considered, as well as the location of the home within the various climate zones of the United States.

HVAC systems are becoming more energy efficient every year and more complex to control. To accurately price an HVAC system, it is recommended that the estimator become familiar with the operation of the various systems. Breaking the system down into its various subsystems: heating, cooling, ductwork, hydronic piping, equipment, and controls is often a good start.

It is not essential for the GC estimator to itemize the takeoff and estimate for mechanical systems in the same manner as for other aspects of the project, such as carpentry, painting, or roofing, but it does provide for a more accurate estimate of costs. Listing the basic criteria helps establish a budget estimate for the individual mechanical work. Historical data from projects with similar criteria can then be compared with actual quotes from subcontractors to arrive at reliable costs.

Division 23—HVAC has a tremendous number of miscellaneous items that are necessary for inclusion within a comprehensive estimate. While many of these are subordinate to the heating or cooling system, they are still part of the HVAC scope and can impact the cost significantly.

QUESTIONS/ PROBLEMS

1. Calculate the square foot area of insulation including the appropriate amount for lap and seaming for 80' of 20" × 16" steel duct.
2. A 90' section of 20" × 16", 22 ga ductwork has a weight of 1.406 lbs. per SF. What is the cost for this section if it is $9.70 per lb. fabricated and installed?
3. Using the appropriate figure from the chapter, what is the estimated cost for installation labor for a 1,000 MBH gas-fired cast iron boiler if the hourly billing rate for a boilermaker is $105.00?

4. Explain how an estimator might use specific parameters as a means to generate a budget estimate for the HVAC scope of work.
5. For a design-build residential HVAC system, explain some of the parameters and energy considerations that an HVAC subcontractor would have to consider when sizing an air-conditioning system.

20 | Integrated Automation

Division 25—Integrated Automation is part of the new CSI MasterFormat® structure, added in 2004. In MasterFormat 95, integrated automation was part of Division 13—Special Construction. Integrated Automation is part of the Facility Services subgroup, which includes fire suppression, plumbing, heating/ventilating/air conditioning, electrical, communications, and electronic safety and security. It has since been expanded due to the increasing amount of technology related to automated controls in use in commercial buildings and residences today.

In short, Division 25—Integrated Automation is the application of computers – hardware and software – to control and optimize functions and operations within a structure. The range and depth of integrated automation are advancing constantly and have become mainstream since the second edition of this text.

Computers are being used to control, monitor, and optimize the performance of every function or system within the building. They include functions such as access, interior and exterior lighting, heating and air-conditioning systems, fire identification, valves, switches, energy economization, and the monitoring of an enormous variety of manufacturing and processing functions. Many of the applications are oriented toward improving physical plant efficiency and energy conservation.

This chapter covers only the basic procedures and methods for takeoff and pricing. It is limited to the components of systems normally encountered in residential and light commercial construction.

Estimating Building Costs for the Residential and Light Commercial Construction Professional, Third Edition. Wayne J. Del Pico.
© 2023 John Wiley & Sons, Inc. Published 2023 by John Wiley & Sons, Inc.

INTEGRATED AUTOMATION SYSTEMS

Although systems generally perform the same function in residential and light commercial construction, the grade and type of materials can be significantly different.

Consult the specification sections for the products, methods, and techniques of installation, as well as the related work of other trades that affect pricing. Review Division 1—General Requirements for training the owner's personnel or switching over from existing to new systems that must be done during off-hours.

Integrated automation is often relegated to specific drawings within the set, which can be separate or part of the electrical drawings. It focuses on building operations. The specifications may contain more information for the estimator than the plans.

Occasionally, there are specifications that describe the system only by its expected functions; they leave it to the individual bidder to work out the specific details. This is an outline specification. This is somewhat of a design-build application and can be difficult to estimate unless the estimator has detailed experience in the specific system.

Review all drawings in the bid set, including architectural drawings, for related work that may be shown on other drawings, as well as for detailed information on dimensions and measurements for room sizes, floor-to-floor heights, and location of server racks or head-end equipment. Mechanical and electrical drawings with schedules are helpful in determining types and quantities of materials for takeoff.

It should be noted that a fair amount of the work of Division 25—Integrated Automation can occur off site. This would include custom design and testing of the software.

Taking-off and Pricing

Taking off and estimating these systems requires a detailed working knowledge of the particular manufacturer's system. This is always difficult, as unique or proprietary systems are well guarded from noncertified installers. Customization of the software for a particular application can be difficult to estimate for the estimator not familiar with the system.

However, many of the functions of the integrated automation process are similar, such as wire management within the structure and the installation of hardware (valves, switches, devices, etc.). Estimators should be able to quantify devices and wire, as well as any appurtenances such as cable tray, server racks, and terminal devices.

All takeoff quantities should be listed separately according to type, application, size, function, and any other applicable characteristic. In addition to materials, special items such as shop drawings and demonstrations, as well as training, should be considered as part of the estimate. Additional takeoff guidelines are as follows:

- **Cable tray or surface raceways:** By the linear foot (LF), separated according to type (plastic or aluminum), size (width), and method of suspension (wall mounted or hung from above).
- **Fittings for cable tray:** By the piece (EA) according to type (elbow, tee, reducer, etc.), size (width), method of connection, and material composition (plastic or aluminum).
- **Valves and special appurtenances:** By the individual piece (EA) according to type, manufacturer, model, size, use or application, and other identifying information needed for pricing.
- **Switches, hubs, and modems:** By the piece (EA) according to type, function, location, manufacturer, and model number.
- **Hardware and software:** Quantifying and pricing hardware and software can present a challenge. It is most often referred to as a package with the unit of measure *lump sum* (LS). Costs are driven by the type of system and how many points are being controlled by the software. It is also impacted by the level of software customization required.
- **Racks:** By the piece (EA) according to type, finish, size, model, and application.
- **Terminal control units:** By the piece (EA) according to type, function, location, manufacturer, and model number.
- **Wire:** By the LF, separated according to size (number of conductors), and rating (plenum rated).
- **Connectors:** By the piece (EA) according to type, function, location, manufacturer, and model number.
- **Miscellaneous items:** Items such as J-hooks, anchors for masonry or concrete applications, or wire wraps are taken off and priced by individual piece (EA).

In most cases, the automation contractor's work begins at the interior of the building in rooms where electrical or mechanical hubs are located.

Be sure to include costs for testing and debugging the system, which can be substantial, as well as costs for training personnel in the use and operation of the systems.

Labor

The work of Division 25 is performed by a variety of trades, from electricians to plumbers to door hardware specialists. The list is always growing. Often, the tradesperson is a hybrid of a traditional trade with specific training in that system. They are individuals with specialized training on systems that are highly proprietary to the manufacturer of that system. Figure 20.1 provides a guideline for the installation costs of cable tray.

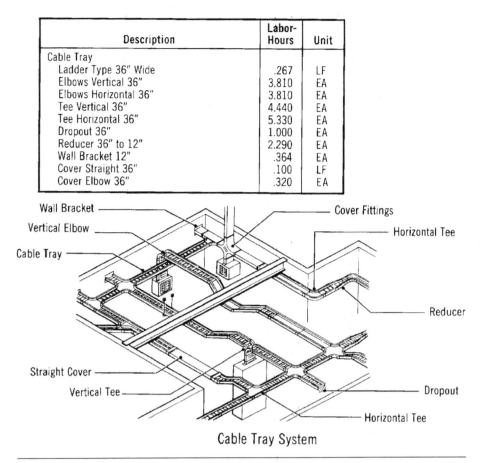

Description	Labor-Hours	Unit
Cable Tray		
Ladder Type 36" Wide	.267	LF
Elbows Vertical 36"	3.810	EA
Elbows Horizontal 36"	3.810	EA
Tee Vertical 36"	4.440	EA
Tee Horizontal 36"	5.330	EA
Dropout 36"	1.000	EA
Reducer 36" to 12"	2.290	EA
Wall Bracket 12"	.364	EA
Cover Straight 36"	.100	LF
Cover Elbow 36"	.320	EA

Cable Tray System

Figure 20.1 Installation Time in Labor-Hours for Cable Tray Systems

Source RSMeans Estimating Handbook, Third Edition. © 2009 John Wiley & Sons, Inc.

SUMMARY

It is not essential to itemize the takeoff and estimate for integrated automation systems in the same manner as for other aspects of the project, such as carpentry, painting, or roofing. Listing the basic criteria helps establish a budget estimate for the work. Historical data from projects with similar criteria can then be compared with actual quotes from subcontractors to arrive at reliable costs. The estimator is urged to solicit pricing from subcontractors specializing in smart systems for both residential and commercial applications based on the project size and complexity.

QUESTIONS/ PROBLEMS

1. Please list the units of measure for: (1) valves, (2) cable tray, and (3) racks.
2. Please opine as to the challenges of estimating integrated automation systems.
3. In the absence of detailed drawings illustrating the integrated automation of various systems, where would the estimator find information about the particular system or systems for the project?
4. Please select and define one of the primary functions of an integrated automation system.
5. How might an estimator use historical data to budget the work of this division?

21 | Electrical

CSI MasterFormat® Division 26 covers electricity distribution, power and lighting, electric heating, panels, raceways, and wiring. Electrical work is performed by individuals or firms with specific training and licensing. Separate permits and insurances are also required. To take off and estimate electrical work, it is extremely beneficial to have a working knowledge of the material components of the particular systems as well as the installation process. Actual electrical expertise is not normally within the realm of the general contractor's estimating experience. Since most general contractors subcontract their electrical work, they do not need to produce a detailed estimate. As seen with fire suppression, plumbing and mechanical sections of this text, a budget or parametric estimate is sufficient to review and analyze proposals from subcontractors. Division 26 can follow a similar approach.

Division 26 work is typically shown on the electrical drawings labeled with the prefix "E." Be sure to review all drawings within the bid set for electrical work that may be shown on other drawings, such as site lighting, utilities plans, or hookup of equipment provided by others. Consult mechanical drawings for related work in other sections, such as power wiring for alarm systems for fire protection and HVAC equipment. While all electrical work should be shown on the electrical plans, it is a good check and balance approach to review fire suppression and mechanical plans to make sure power is provided to necessary equipment. As with mechanical drawings, electrical drawings employ their own trade-specific graphic symbols for conveying information. Review all legends and graphic symbols on the electrical drawing.

Electrical drawings often use schedules that are helpful in determining the materials for the takeoff. Typical schedules include lighting fixtures, panels, equipment, and feeders. Specialized details, such as the electrical

Estimating Building Costs for the Residential and Light Commercial Construction Professional,
Third Edition. Wayne J. Del Pico.
© 2023 John Wiley & Sons, Inc. Published 2023 by John Wiley & Sons, Inc.

riser diagram, illustrate the various components of the system and their configurations. Diagrams of all types are for the graphic representation of information only and are not drawn to scale.

Consult the architectural drawings for dimensions, room sizes, floor-to-floor heights, ceiling heights, location of services entering the building, and coordination with other work within the area. There is no substitute for a thorough review of the plans and specifications.

Review Division 26 technical specifications thoroughly for the products, methods, and techniques of installation, and the related work of other trades. A careful review of Division 1—General Requirements, Temporary Controls and Facilities, is necessary to determine what, if any, special requirements are to be included in the takeoff and estimate. Some classic examples include:

- Securing and paying for electrical permits and inspections
- Temporary lighting and power for the project
- Maintenance and relocation of temporary lighting and power as required by the project schedule
- Temporary connection/disconnection of special construction equipment with electrical needs (hoists, welding machines, floor sanders, etc.)
- Electrical utility company charges for services (utility pole relocation costs, engineering, and design costs, etc.)
- Temporary connection/disconnection of job trailers and temporary light and power

These costs should still be considered part of the electrical portion of the project, even if not specifically referenced. Figure 21.1 illustrates a simplified electrical system for a light commercial project.

This figure shows the basic lighting and power components used for the interior of a typical commercial project.

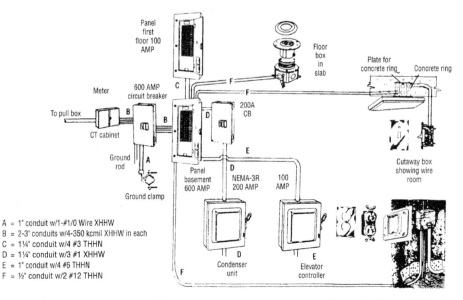

A = 1" conduit w/1-#1/0 Wire XHHW
B = 2-3" conduits w/4-350 kcmil XHHW in each
C = 1¼" conduit w/4 #3 THHN
D = 1¼" conduit w/3 #1 XHHW
E = 1" conduit w/4 #6 THHN
F = ½" conduit w/2 #12 THHN

Figure 21.1 Typical Commercial Electric System

This chapter covers takeoff procedures for the type of electrical systems generally encountered in light commercial and residential construction. Most often the takeoff begins with the service from the utility pole or transformer and proceeds to the final connections and installation of lighting. The starting point for takeoff, however, can be altered for renovation work that starts elsewhere.

Often, architects include the layout of lighting, switches, and receptacles on the architectural drawings, which can be used to establish the cost of the electrical portion of the work. In addition, information on power-consuming equipment, such as furnaces, water heaters, air-conditioning units, and appliances, is needed to determine the electrical estimated cost.

With renovation work, especially renovations of significant scope other tasks may be required. These include safety measures such as disconnection of circuits and equipment scheduled for demolition. This process is referred to as "make safe." As with mechanical and plumbing demolition, the electrical scope may require that items selectively demolished such as light fixtures or abandoned conduits be dropped to the floor for collection and disposal by others, or that the debris generated, be hauled to on site disposal such as dumpsters. This should be clearly identified in any electrical demolition section of the technical specifications.

ELECTRICAL LABOR

Electricians and their helpers, referred to as *apprentices,* are responsible for performing electrical work. Depending on the specific task, productivity can be measured by the output of the individual, as in the case of installing receptacles, switches, and their respective plates. Other tasks, such as pulling feeders and hanging raceways, clearly require multiple personnel for cost-effective installations. Labor to install light fixtures varies depending on the size and type of fixture and the height from the floor.

Occasionally, the use of lifts or ladders might enable the work to be done more efficiently. Using such equipment might also require that an extra person be added to the crew for the support of personnel on the lift. This should be taken into account when establishing crew size. Productivity for simple tasks, such as installing finish devices, plates, or junction or work boxes, can be calculated by totaling the quantity of devices that can be installed by one electrician in a day. The same applies for the wire that carries the electricity, called *conductors.* Base the unit price on the anticipated linear footage of wire that will be pulled by the crew in a day.

Other operations, such as installing panelboards, switchgear, load centers, motor starters, and the like, are based on labor-hours per individual piece. Frequently, bucket trucks or aerial platforms are needed. These costs are time sensitive and calculated by the day, week, or month, depending upon need.

Many estimators separate the *rough* and *finish* phases of the electrical work for greater clarity. Rough electrical work is considered the installation of conduits, wire, device boxes, and panelboards in advance of "closing up" the walls and ceilings with drywall or any other finish. The finish phase occurs after finishes such as final paint, wall tile, and the ceiling grid have been completed. Light fixtures and devices are installed, connections made to equipment, panels are energized, circuits completed, and so forth. This breakdown may help the estimator visualize the scope of work better.

RACEWAYS

Raceways are channels constructed to house and protect electrical conductors. They include conduits, wireways, cable trays, surface metal raceways, and underfloor ducts. As part of the raceway system, fittings are needed to change the direction of, connect, and support the various types of raceway runs. The most common type of raceway is conduit, which can be aluminum, rigid galvanized steel, steel intermediate conduit (IMC), rigid plastic-coated steel, PVC, or electrical metallic tubing (EMT). Conduit can be wall mounted, suspended overhead, encased in concrete, or buried below grade.

Taking-off and Pricing

Raceways are taken off and priced by the linear foot (LF) and classified according to type, size, and application. Individual fittings for wireways, underfloor ducts, surface metal raceways, and larger-diameter conduits are taken off and priced by the piece (EA) and listed according to type, size, and material. In most instances, fittings are not shown on drawings for standard conduit installations. For smaller-diameter conduits, fittings can be accounted for by adding a percentage to the total conduit materials. Percentages will vary with the complexity of the run. The spacing of particular fittings is dictated by local electrical codes or the individual project requirements.

Separate the takeoff and pricing of conduit into three categories: power distribution, branch power, and branch lighting. *Power distribution* includes the main conductors to supply power to the various panels. *Branch power* and *branch lighting* refer to the branches of the panels that provide power and lighting to various locations. Using these categories, all conduit quantities need not be taken off at one time and can be determined system by system. Since drawings are represented graphically in only two dimensions, length, and width. Be sure to include quantities for the vertical portion of the raceway that is not shown.

Raceways installed higher than 15' above the floor should be noted separately because of their reduced productivity. Also note that electrical drawings are "diagrammatic" in nature and not necessarily exactly as the work will be built. For example, the exact location and configuration of raceways may be subject to change as a result of conditions in the field. Figure 21.2 provides labor-hours for installing conduit.

Conduit to 15' high, includes couplings, fittings, and support.

Description	Labor-Hours	Unit
Rigid galvanized steel ½" diameter	.089	LF
1½" diameter	.145	LF
3" diameter	.320	LF
6" diameter	.800	LF
Aluminum ½" diameter	.080	LF
1½" diameter	.123	LF
3" diameter	.178	LF
6" diameter	.400	LF
Imc ½" diameter	.080	LF
1½" diameter	.133	LF
3" diameter	.267	LF
4" diameter	.320	LF
Plastic-coated rigid steel ½" diameter	.100	LF
1½" diameter	.178	LF
3" diameter	.364	LF
6" diameter	.800	LF
EMT ½" diameter	.047	LF
1½" diameter	.089	LF
3" diameter	.160	LF
4" diameter	.200	LF
PVC nonmetallic ½" diameter	.042	LF
1½" diameter	.080	LF
3" diameter	.145	LF
6" diameter	.267	LF

Rigid steel, plastic-coated coupling

PVC conduit

PVC elbow

Aluminum conduit

EMT setscrew connector

Aluminum elbow

EMT connector

Rigid steel, plastic-coated conduit

EMT to conduit adapter

Rigid steel, plastic-coated elbow

EMT to greenfield adapter

Figure 21.2 Installation Time in Labor-Hours for Conduit to 15' High, Including Coupling, Fittings, and Support

Another means for installing wire within protection is the use of underfloor raceways. The estimating procedure is similar. Main trunk lines should be taken off and then the individual fittings and components required for a complete system. Figure 21.3 provides labor-hours for installing underfloor raceway systems.

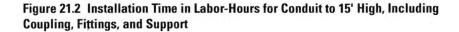

Description	Labor-Hours	Unit
Blank duct	.100	LF
Insert duct	.110	LF
Elbow vertical	.800	EA
Elbow horizontal	.300	EA
Panel connector	.250	EA
Junction box, single duct	2.000	EA
Double duct	2.500	EA
Triple duct	2.950	EA
Saddle support, single duct	.290	EA
Double duct	.500	EA
Triple duct	.720	EA
Insert to conduit adapter	.250	EA
Outlet, low tension (telephone and signal)	1.000	EA
Outlet, high tension (power)	1.000	EA
Offset duct type	.300	EA

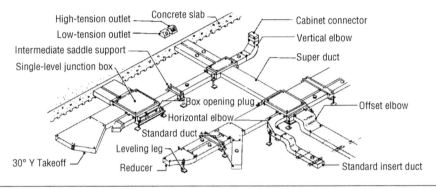

Figure 21.3 Installation Time in Labor-Hours for Underfloor Raceway Systems

Source RSMeans Estimating Handbook, Third Edition. © 2009 John Wiley & Sons, Inc.

CONDUCTORS AND GROUNDING

A *conductor* is a wire or metal bar with a low resistance to the flow of electricity. *Grounding* is accomplished by a conductor connected between electrical equipment, or between a circuit and the earth. Wire is the most common material used to conduct current from the electrical source to electrical use. Copper or aluminum wire conductors with insulating jackets are available in a variety of voltage ratings and insulating materials. Wire is installed within raceways, such as conduit or flexible metallic conduit (sometimes referred to as flex). Flexible metallic conduit is a single strip of aluminum or galvanized steel, spiral-wound and interlocked to provide a circular cross section of high strength and flexibility for the protection of the wire within. Other products similar to flex are covered with liquid-tight plastic and used where protection from liquids or precipitation is required. Other types of conductors include armored cable (BX and MC), a fabricated assembly of cable with a metal enclosure similar in appearance to flex. Nonmetallic sheathed cable (Romex) is manufactured with insulated conductors enclosed in an outer sheath of plastic or fibrous material. It is available with or without a bare ground wire made of copper or aluminum conductors.

Special wires, such as those used in low-voltage control wiring signals and telecommunications, are discussed in Chapter 22. Special connectors or terminations at the end of each wire may be required, and various fasteners, such as staples, clips, and flex fittings, are also necessary. Consult the specifications for the specific conductors required for each application. In the absence of a specification section to define the various conductors, defer to the electrical code having jurisdiction.

Taking-off and Pricing

Wire, flex, and cables are taken off and priced by the LF and divided by 100 to arrive at CLF, which stands for hundred linear feet, since *C* is the Roman symbol for hundred. The total quantity of wire installed within conduits or flex is determined by multiplying the number of conductors by the LF of conduit or flex and converting to CLF. Additional lengths of wire should be added for making connections withing the device or box. All wire and cables should be listed in the takeoff according to type, size (rating), conductor material, and application (feeders, branch power, and branch lighting). The exterior plastic coating on the wire (THWN, THWN-2, THHN) may also be noted in the takeoff for more accurate pricing.

Special fittings for connecting wire or cables, sometimes referred to as *terminations,* are taken off and priced by the individual piece (EA) and listed according to type, size, application, and method of connection. For smaller conductors, the fittings are estimated by adding a percentage of the cost of the conductors. For connectors on larger conductors, such as those used on feeders, the individual termination devices are counted by EA.

In addition to grounding conductors, accessory items such as ground rods, clamps, and exothermic weld metal are taken off and priced by the individual piece (EA) and listed according to type, size, and application. An allowance of 10% for waste and connections on conductor quantities is usually acceptable but may be increased for lengths of wire or cables with numerous interruptions, such as intermediate connections or splices.

Wire should be taken off and separated according to application (feeders and service entrance, branch power, and branch lighting). Use care in the takeoff calculations to allow for sufficient lengths for connections, especially in larger feeders, where insufficient footage is costly. Review the governing codes concerning terminations to ensure that all items have been included. Figure 21.4, parts a–c, provides guidelines for labor-hours of various sizes of conductors and wire types.

WIRING DEVICES AND BOXES

Boxes are used in electrical wiring at each junction point, outlet, or switch to provide access to electrical connections and serve as a mounting for fixtures or switches. They may also be used as pull or splice points for wire in long runs or conduits. A wiring device, such as a switch or receptacle, controls but does not consume electricity. Boxes are most often installed in the rough phase of wiring and the devices follow in the finish phase of the work.

Description	Labor-Hours	Unit
600V copper #14 AWG	.610	CLF
#12 AWG	.720	CLF
#10 AWG	.800	CLF
#8 AWG	1.000	CLF
#6 AWG	1.230	CLF
#4 AWG	1.510	CLF
#3 AWG	1.600	CLF
#2 AWG	1.780	CLF
#1 AWG	2.000	CLF
#1/0	2.420	CLF
#2/0	2.760	CLF
#3/0	3.200	CLF
#4/0	3.640	CLF
250 kcmil	4.000	CLF
500 kcmil	5.000	CLF
1000 kcmil	9.000	CLF

PVC jacket connector

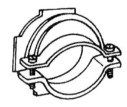

SER, insulated, aluminum

600 Volt, armored

5 KV armored

Cable Connectors

Crimp, 1-hole lug

Terminal lug, solderless

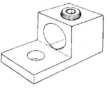

Crimp, 2-way connector

Cable Terminations

Figure 21.4a Installation Time in Labor-Hours for Electrical Conductors: Wire and Cable

Source RSMeans Estimating Handbook, Third Edition. © 2009 John Wiley & Sons, Inc.

Description	Labor-Hours	Unit
CABLE TERMINATIONS		
Wire connectors, screw type, #22 to #14	.031	EA
#18 to #12	.033	
#18 to #10	.033	
Crimp 1-hole lugs, copper or aluminum, 600 volt		
#14	.133	
#12	.160	
#10	.178	
#8	.222	
#6	.267	
#4	.296	
#2	.333	
#1	.400	
1/0	.457	
2/0	.533	
3/0	.667	
4/0	.727	
250 kcmil	.889	
300 kcmil	1.000	
350 kcmil	1.143	
400 kcmil	1.231	
500 kcmil	1.333	
600 kcmil	1.379	
700 kcmil	1.455	
750 kcmil	1.538	

Figure 21.4a *(Continued)*

Boxes often require plaster rings, covers, and various fasteners for support. In addition to receptacles and switches, wiring devices include a variety of specialized controls and finish wall plates that can impact both material and installation costs.

Taking-off and Pricing

Outlet boxes, pull or junction boxes, receptacles, switches, wall plates, and wiring devices in general are taken off and priced by the individual piece (EA). In sufficient numbers, they can be extended to 100-piece counts. Be sure to include the necessary accessories for a complete application. For example, for outlet boxes, include plaster rings and extensions (if required); for pull boxes, include covers; and for receptacles and switches, include plates. The various items should be listed according to type, size, composition, capacity or application, and color (if applicable). In general, boxes and devices have a low unit cost and warrant the inclusion of a waste factor for loss or mishandling. Depending on the size of the project, 5% to 10% is usually adequate for waste.

Review the specs for the exact scope of work for control devices, such as relays and low-voltage transformers, which may affect other trades. Special

Description	Labor-Hours	Unit
Armored cable		
600-volt (BX), #14, 2-conductor, solid	3.333	CLF
3-conductor, solid	3.636	
4-conductor, solid	4	
#12, 2-conductor, solid	3.478	
3-conductor, solid	4	
4-conductor, solid	4.444	
#12, 19-conductor, stranded	7.273	
#10, 2-conductor, solid	4	
3-conductor, solid	5	
4-conductor, solid	5.714	
#8, 3-conductor, solid	6.154	
4-conductor, stranded	7.273	
#6, 2-conductor, stranded	6.154	

Figure 21.4b Installation Time in Labor-Hours for BX Cable

Source RSMeans Estimating Handbook, Third Edition. © 2009 John Wiley & Sons, Inc.

Description	Labor-Hours	Unit
Nonmetallic sheathed cable, 600 volt		
Copper with ground wire (Romex)		
#14, 2-conductor	2.963	CLF
3-conductor	3.333	
4-conductor	3.636	
#12, 2-conductor	3.200	
3-conductor	3.636	
4-conductor	4	
#10, 2-conductor	3.636	
3-conductor	4.444	
4-conductor	5	
#8, 3-avconductor	5.333	
4-conductor	5.714	
#6, 3-conductor	5.714	
#4, 3-conductor	6.667	
#2, 3-conductor	7.273	

Figure 21.4c Installation Time in Labor-Hours for Romex (Nonmetallic Sheathed) Cable

Source RSMeans Estimating Handbook, Third Edition. © 2009 John Wiley & Sons, Inc.

devices may be provided by other trades and installed and wired under the electrical specification section. Examples include relays for heating or cooling units, flow and tamper switches for automatic sprinkler systems, smoke/heat detectors for installation within the heating system, and temperature-sensing controls for heating applications. Figure 21.5 provides installation time for various types of wiring devices.

STARTERS, BOARDS, SWITCHES AND TRANSFORMERS

Be sure to calculate quantities of panelboards, starters for motors, control stations, circuit breakers, safety switches and disconnects, fuses, and meter centers and sockets. These are all part of the completed assembly and necessary for a comprehensive estimate.

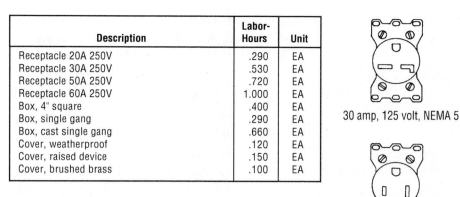

Description	Labor-Hours	Unit
Receptacle 20A 250V	.290	EA
Receptacle 30A 250V	.530	EA
Receptacle 50A 250V	.720	EA
Receptacle 60A 250V	1.000	EA
Box, 4" square	.400	EA
Box, single gang	.290	EA
Box, cast single gang	.660	EA
Cover, weatherproof	.120	EA
Cover, raised device	.150	EA
Cover, brushed brass	.100	EA

30 amp, 125 volt, NEMA 5

50 amp, 125 volt, NEMA 5

20 amp, 250 volt, NEMA 6

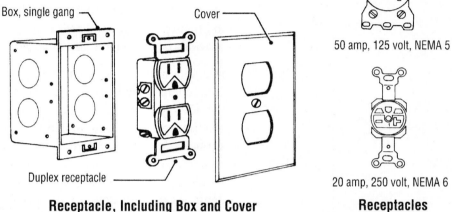

Receptacle, Including Box and Cover

Receptacles

Figure 21.5 Installation Time in Labor-Hours for Wiring Devices

Taking-off and Pricing

The following takeoff guidelines apply.

- **Control stations:** By the individual unit (EA), listed according to type, manufacturer, classification, and application.
- **Circuit breakers:** By the individual piece (EA), listed according to manufacturer, type (number of poles), capacity (rating), voltage, method of installation (plug-in or bolt-on), and classification (NEMA).
- **Panelboards:** By the individual unit (EA), listed according to size (capacity in amperes), type, voltage, and manufacturer. Some standard board and breaker assemblies are available as preassembled units, such as load centers used in residential construction.
- **Starters:** By the individual piece (EA), listed according to size, voltage, NEMA enclosure, and type.
- **Safety switches and disconnects:** By the individual unit (EA), listed according to size, type (duty), number of poles, voltage, NEMA classification, and ampere rating.
- **Transformers:** By the individual piece (EA), listed according to, voltage, phase, and type or class.

- **Fuses:** By the individual piece (EA), listed according to amperes, voltage, and type or class.
- **Meter centers and sockets:** By the individual unit (EA), listed according to size and type for meter sockets, and by bus capacity, number of meter sockets, and type of enclosure for meter centers.

Most of the electrical components in this section can vary in price dramatically with a change to the model number or NEMA classification. Any specific information in the specs that could be used to define the price more accurately should be noted in the takeoff. Most electrical estimators assemble a list of components for pricing by supply houses, similar to carpentry estimators listing materials for pricing by the lumberyard or supplier. (Refer to Chapter 11, "Wood, Plastics, and Composites.") Figure 21.6, parts a and b, provides labor-hours to install motor starters and controls.

In addition to the aforementioned components, special fasteners or auxiliary components may be required. These include perforated support bars (strut), threaded rods, or steel angles. Plywood sheets or concrete pads for mounting electrical equipment, or anchors to attach items to concrete or masonry surfaces, must be included as part of the takeoff and estimate. Coring of concrete or masonry surfaces for passing conduits through may also be required. These should be included in the takeoff according to diameter and depth (in inches). Refer to the specifications to clarify the exact scope of work. Since starters for motors are frequently furnished as part of the mechanical package, verify that they are not supplied by others, in order to avoid costly duplication or omission. One fuse should be counted for each line (or phase) to be protected.

Panelboards are a component of an electrical supply system used to distribute power. It divides an electrical power feed into multiple subsidiary circuits while incorporating a fuse or circuit breaker for the protection of that circuit. Each of these circuits and breakers is housed in a common

Description	Labor-Hours	Unit
Starter 3-pole 2-HP size 00	2.290	EA
5-HP size 0	3.480	EA
10-HP size 1	5.000	EA
25-HP size 2	7.270	EA
50-HP size 3	8.890	EA
100-HP size 4	13.330	EA
200-HP size 5	17.780	EA
400-HP size 6	20.000	EA
Control station stop/start	1.000	EA
Stop/start, pilot light	1.290	EA
Hand/off/automatic	1.290	EA
Stop/start/reverse	1.510	EA

Figure 21.6a Installation Time in Labor-Hours for Starters

Description	Labor-Hours	Unit
Heavy-duty fused disconnect 30 amps	2.500	EA
60 amps	3.480	EA
100 amps	4.210	EA
200 amps	6.150	EA
600 amps	13.330	EA
1200 amps	20.000	EA
Starter 3-pole 2-HP size 00	2.290	EA
5-HP size 0	3.480	EA
10-HP size 1	5.000	EA
25-HP size 2	7.270	EA
50-HP size 3	8.890	EA
100-HP size 4	13.330	EA
200-HP size 5	17.780	EA
400-HP size 6	20.000	EA
Control station stop/start	1.000	EA
Stop/start, pilot light	1.290	EA
Hand/off/automatic	1.290	EA
Stop/start/reverse	1.510	EA

Figure 21.6b Installation Time in Labor-Hours for Motor Control Systems

metal enclosure, with or without a main breaker. Breakers are arranged in two columns and range from 20 to 42 breakers per panelboard. Figure 21.7 provides guidance for estimating the labor associated with the installation of panelboards.

Disconnect devices called *safety switches* are another component of an electrical supply system. Safety switches are typically mounted on or near the entry of the power wiring to a particular piece of equipment. They allow power to be shut off or disconnected at the piece of equipment. They have the added benefit of providing a fused link between the wire providing power to the equipment and the equipment itself. The fuse(s) trips and shunts the circuit when the power being drawn exceeds the rating on the fuse, thereby preventing damage to the equipment. Figure 21.8 provides guidance for estimating the labor associated with the installation of safety switches.

LOAD CENTERS

A *load center* is a specialized type of panelboard used principally for residential applications. These panels are designed for lighter sustained loads than those used in industrial and commercial applications.

Load centers are available from 100 A to a maximum 200 A rating in several configurations: i.e., 120/240V, 1 phase, 3 wire; 120/208V, 3 phase, 4 wire;

Description	Labor-Hours	Unit
Panelboard 3-wire 225 amps main lugs 38 circuit	22.220	EA
4-Wire 225 amps main lugs 42 circuit	23.530	EA
3-Wire 400 amps main circuit breaker 42 circuit	32.000	EA
4-Wire 400 amps main circuit breaker 42 circuit	33.330	EA
3-Wire 100 amps main lugs 20 circuit	12.310	EA
4-Wire 100 amps main circuit breaker 24 circuit	17.020	EA

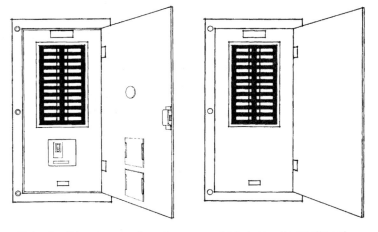

Main circuit breaker panelboard   Main lugs only panelboard

Figure 21.7 Installation Time in Labor-Hours for Panelboards

Source RSMeans Estimating Handbook, Third Edition. © 2009 John Wiley & Sons, Inc.

240V, 1 phase, 2 wire; 120/240V, 3 phase, 4 wire; and 240V, 3 phase, 3 wire. Load centers generally use plug-in circuit breakers. The single-pole breakers range from 15 A to 50 A; double-pole breakers from 15 A to 100 A; and three-pole breakers from 15 A to 100 A. Two- and three-wire switched neutral and ground fault interrupting styles are also available.

Lighting and appliance load centers are limited to a maximum of 42 over-current devices in each cabinet.

Taking-off and Pricing

Load centers are taken off and quantified as individual units, EA, by type and size. While standard breakers may be included, care must be taken to ensure that all special service breakers are noted and priced as additional items. The box or tub with bar, breakers, cover, and trim are generally included as a complete unit, EA, of a load center. The labor portion generally includes receiving, unloading, and uncrating the load center, layout and marking location of the tub, fastening to the wall including leveling, preparation and termination of feeder cable to lugs or main breaker, testing and marking the panel directory. Labor costs for installation can be very similar to panelboards of the same size.

Description	Labor-Hours	Unit
Safety switch NEMA 1 600V 3P 200 amps	6.150	EA
NEMA 3R	6.670	EA
NEMA 7	10.000	EA
NEMA 12	6.670	EA

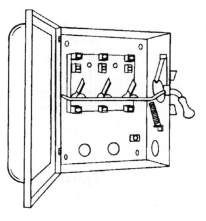

NEMA 1, Nonfusible, 600 Volt

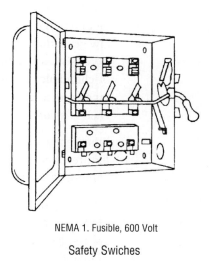

NEMA 1. Fusible, 600 Volt

Safety Swiches

Figure 21.8 Installation Time in Labor-Hours for Safety Switches

Source RSMeans Estimating Handbook, Third Edition. © 2009 John Wiley & Sons, Inc.

Additional items that may be required are taken off and estimated separately. Some of the more common items are structural supports, plywood backerboards, nameplates, and receptacles at the panel.

LIGHTING

Lighting can be a fundamental part of the electrical estimate and can represent a considerable portion of the cost. Types of lighting include interior and exterior, surface-mounted and recessed, emergency and exit fixtures, task lighting, general lighting, and the lamps for the various fixtures. An enormous variety of light fixtures are manufactured to suit every application. Lamps for fixtures include fluorescent, mercury vapor, halogen, metal halide, high-pressure sodium and to lesser degree incandescent. Light fixtures can include whips to connect the fixtures together, or the "lamps" as in the case of LED. Others may require separate pricing for the lamps and specific trims for the light fixtures themselves. Light fixtures are often referred to as luminaires.

On residential projects, it is common for the electrical contractor or design professional to provide the general contractor (GC) or homeowner with an allowance (refer to Chapter 2, "Understanding the Specification for additional information on Allowances") for purchase and delivery of the light fixtures that have yet to be decided on. The estimator should be clear as to what is included in the light fixture allowance such as assembly or lamps.

Taking-off and Pricing

Count, list, and price each fixture for the estimate by the individual unit (EA) according to the manufacturer, model, type, color or finish, location (wall, ceiling, room), and interior or exterior application. The number of lamps can be determined per fixture and listed by individual lamp (EA). Lamps are not always included as part of the fixture's cost, especially with the more expensive, specialized lamps. The level of assembly required for ceiling fans or chandeliers should also be noted, if applicable. Creating a list of fixtures or fans with an assembly time for each that can be multiplied by the quantity of each is a common way in which to track assembly labor-hours.

Emergency and exit lighting should be kept separate from general light fixtures, and it should be quantified and priced by the individual component (EA).

Interior and exterior light fixtures should be taken off and priced separately. Most exterior light fixtures are either wall mounted, pole mounted, or ground based. Some very large fixtures may require cranes or boom trucks to install. Large interior fixtures, such as chandeliers, may also require some type of rigging or hoisting and structural support for attachment. As previously noted, most electrical estimators assemble light fixture types and quantities for pricing by supply houses or, if the quantity is sufficient, directly from manufacturers' representatives. This ensures current pricing, volume discounts, and accuracy. Figure 21.9, parts a and b, provides labor-hours for installation of various types of lighting.

Figure 21.10 provides guidance for the labor costs associated with the installation of parking area lighting.

EMERGENCY LIGHTING AND DEVICES

A section of Division 26 is devoted to emergency lighting, battery backup, and exit devices. When normal power goes out, there needs to be a backup lighting system with exit devices that direct occupants of a building to the exterior or to a safe zone. This emergency system can be completely

Description	Hours	Unit
Ceiling, recess-mounted Alzak reflector		
150W	1.000	EA
300W	1.190	EA
Surface-mounted metal cylinder		
150W	.800	EA
300W	1.000	EA
Opal glass drum 10" 2-60W	1.000	EA
Pendant-mounted globe 150W	1.000	EA
Vaportight 200W	1.290	EA
Chandelier 24" diameter x 42" High		
6-candle	1.330	EA
Track light spotlight 75W PAR halogen	.500	EA
Wall washer quartz 250W	.500	EA
Exterior wall-mounted quartz 500W	1.510	EA
1500W	1.900	EA
Ceiling, surface-mounted vaportight		
100W	2.650	EA
150W	2.950	EA
175W	2.950	EA
250W	2.950	EA
400W	3.350	EA
1000W	4.450	EA

Track lighting spotlight

Exterior fixture, wall mounted, quartz

Round ceiling fixture with concentric louver

Round ceiling fixture with reflector, no lens

Round ceiling fixture, recessed, with Alzak reflector

Square ceiling fixture, recessed, with glass lens, metal trim

Fixtures

Figure 21.9a Installation Time in Labor-Hours for LED/Incandescent Lighting

separate from the main electrical system, as in the case of emergency lights and signs with battery backup. It can also be a component of the main electrical system that is supplemented by a generator once the main source of power is unavailable.

Taking-off and Pricing

The components of the self-contained emergency lighting and exit system are taken off in accordance with the following:

- **Battery packs:** By the individual unit (EA), listed according to type, manufacturer, classification, and application (exposed or remote).

Description	Labor-Hours	Unit
Trotter with acrylic lens 4-32W RS 2' × 4'	1.700	EA
2-32W URS 2' × 2'	1.400	EA
Surface-mounted acrylic wraparound lens		
4-40W RS 16" × 48"	1.500	EA
Industrial pendant-mounted		
4' Long, 2-32W RS	1.400	EA
8' Long, 2-75W SL	1.820	EA
2-110W HO	2.000	EA
2-215W VHO	2.110	EA
Surface-mounted strip, 4' long, 1-40W RS	0.940	EA
8' long, 1-75W SL	1.190	EA

Surface-or pendant-mounted fixture with wraparound acrylic lens, 4-tube

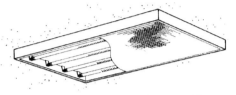

Surface-mounted fixture with acrylic lens, 4-tube

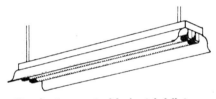

Pendant-mounted industrial fixture

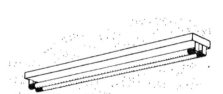

Surface-mounted strip fixture, 2-tube

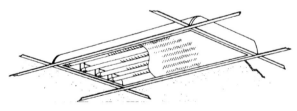

Troffer-mounted fixture with acrylic lens, 4-tube

Figure 21.9b Installation Time in Labor-Hours for LED/Fluorescent Lighting

- **Emergency light head:** By the individual piece (EA), listed according to manufacturer, type (single or dual), capacity (voltage), and method of installation (wall or ceiling mounted).
- **Conduit:** By the LF separated according to diameter and type (steel conduit, EMT, etc.) (if required).
- **Wire:** By the LF separated according to type (stranded or solid), size (gauge), and number of conductors (within a conduit).
- **Exit signs:** By the individual piece (EA) separated according to power source (wired or self-contained), color, size, location, and application (wall, ceiling, or floor mounted).

Description	Labor-Hours	Unit
Luminaire HID 100 Watt	2.960	EA
150W	2.960	EA
175W	2.960	EA
250W	3.330	EA
400W	3.640	EA
1,000W	4.000	EA
Bracket arm 1 arm	1.000	EA
2 arm	1.000	EA
3 arm	1.510	EA
4 arm	1.510	EA
Aluminum pole 20' high	6.900	EA
30' high	7.690	EA
40' high	10.00	EA
Steel pole 20' high	7.690	EA
30' high	8.700	EA
40' high	11.77	EA
Fiberglass pole 20' high	5.000	EA
30' high	5.560	EA
40' high	7.100	EA
Transformer base		EA

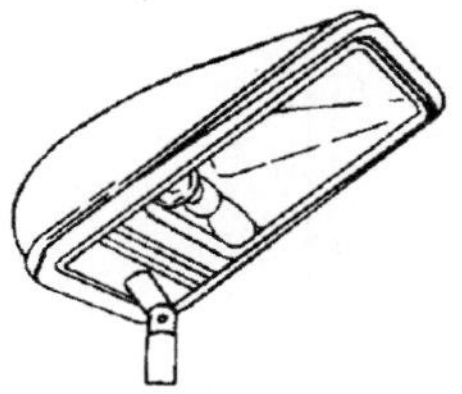

Large luminaire light fixture, 1,000 Watt

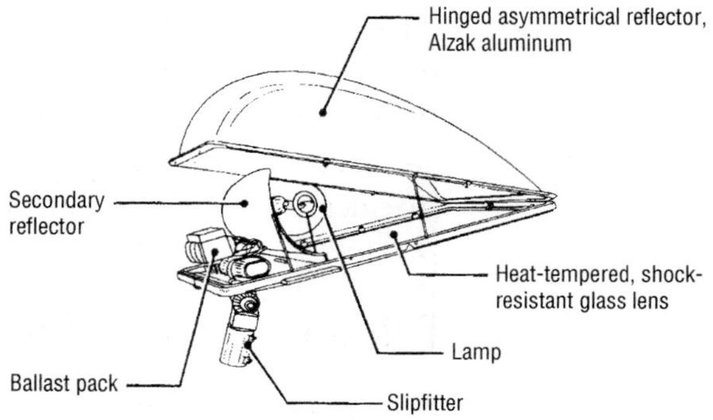

Luminaire Construction Features

Figure 21.10 Installation Time in Labor-Hours for Parking Area Lighting

Source RSMeans Estimating Handbook, Third Edition. © 2009 John Wiley & Sons, Inc.

For emergency lighting systems that are part of the main system, the process is similar but requires that specific lights in each room or floor be tied into an emergency panel. The estimator should use caution not to duplicate the fixture count in this type of emergency lighting. Figure 21.11 provides guidance for the labor-hours required to install an emergency lighting system.

EQUIPMENT HOOKUPS

Appliances and equipment that cannot always be simply plugged into the power supply must be *hard-wired* or attached with fixed connections. This work is typically done by the electrical contractor. Examples include electric ranges and ovens, heating and cooling units, dishwashers, water heaters, boilers, furnaces, garbage disposals, and commercial equipment for specialty operations, such as restaurants, hospitals, or manufacturing. The equipment is typically supplied and set in place by a dealer or another trade. A project can utilize a tremendous amount of equipment that requires power, both interior and exterior. Even something as common as a building elevator has a significant amount of electrical wiring and coordination.

Description	Labor-Hours	Unit
Battery light unit		
6-volt lead battery and 2 lights	2.000	EA
12-volt nickel cadmium and 2 lights	2.000	EA
Remote mount sealed beam light, 25W, 6 volts	.300	EA
Self-contained fluorescent lamp pack	.800	EA
Engine generator 10 kW gas/gasoline		
277/480 volts complete	34.000	System
175 kW complete	96.000	System
Engine generator 500kW diesel		
277/480 volts complete	133.000	System
1,000 kW complete	180.000	System

Figure 21.11 Installation Time in Labor-Hours for Emergency/Standby Power Systems.

Source RSMeans Estimating Handbook, Third Edition. © 2009 John Wiley & Sons, Inc.

Consult the specialty drawings showing the equipment and the related specifications for the exact scope of work. Electrical power drawings should show the electrical interface with various types of equipment. Review mechanical specifications and plans for equipment that requires power for specifics as well as site plans for items such as pumps and signage.

Taking-off and Pricing

Appliances and equipment that require hookup are taken off and priced by the individual piece (EA) and listed according to type and method of installation (such as flexible-connected, direct conduit, overhead, or in-floor), and any other special requirements such as liquid-tight.

Control devices or disconnect switches (see "Starters, Boards, Switches and Transformers" previous in this chapter) may be needed to meet governing codes and may not always be noted in the specifications. It is a good idea to become familiar with the applicable codes governing the electrical installation of the particular equipment.

The estimator should consider the amount of time in labor-hours for an individual electrician or crew of electricians when wiring equipment. The complexity of the installation can often be impacted by the way the equipment was installed. Access can often be a determining factor in time required.

ELECTRICAL SITE WORK

Electrical site work covers the distribution methods used to route power, control, and communications cables onto a facility's property and between its buildings and structures. There are three basic options: direct burial cables, underground in duct banks, and overhead on poles.

Direct burial cables are the least versatile; they are generally used for residential applications where aesthetics is a more important factor than flexibility. Occasionally, direct burial is used in commercial and industrial facilities to route a branch feeder to unique equipment, such as a well pump or roadway lighting. Two techniques are used to place direct buried cables. A trenching machine may dig a narrow slot 2–4' deep for a single cable or a trench 1–2' wide for multiple cables. The trench is usually backfilled with a few inches of sand – below and again above the cables – for protection. Approximately 12" above the cable, concrete planks or plastic marker tape may be placed as a warning to future excavation and digging. Finally, on the finished grade, concrete, monuments, or markers maybe placed every 100' to indicate the path of the buried cable and to denote changes in direction.

Duct banks are a group of two or more underground conduits usually encased in concrete. Although more costly than direct burial systems, duct banks offer three advantages. First, the cables are far better protected from hazards and the elements. Second, groups of several cables can be pulled through each conduit. Finally, new cables may be pulled (or failed cables replaced) quickly and economically to meet future needs.

For duct banks that cover long distances, access must be provided to the run for pulling cables. When only a few conduits are installed, a hand-hole meets this requirement. For multiple conduit banks and for large conduits, a structure called a *manhole* is installed. Hand-holes and manholes may also serve to change the direction of a run or to split up a run. Both are usually made of precast concrete or high-density plastics, and manholes are sized large enough for both personnel and pulling equipment. Pulling eyes (steel inserts) are normally cast into the walls of the precast structure opposite the cable entry points. This is done to facilitate the rigging process for cable pulling. Hangers may also be cast into the boxes to carry or support the installed cables.

Duct banks are generally buried, allowing 2–4' from grade to the top row of conduits; the conduits are separated by plastic spacers and held fast with tie-wires. When the conduit is encased in concrete, it is important to prevent the conduit from "floating" during the concrete placement. A minimum of 2" of concrete is typically required on all sides.

The most common duct bank conduit materials are PVC and fiber duct. Galvanized steel may also be used, especially when power cables and instrumentation will be pulled into separate conduits in close proximity to each other. PVC conduit comes in 20' lengths and is bent in the field with a heater. Couplings and fittings are affixed using PVC cement. Unless the run is very short (under 100'), the installation of duct conduits of less than 2" in diameter is very rare.

Poles and overhead routing represent the most conventional method of distributing power and communication cables. Many cables are built and rated for aerial service. Some cables include "strength members" to carry the tensions of the stretched cables. Still other types of cable, such as service drops and telephone lines, will be supported by a steel messenger wire. The hole for a line pole is usually made with an auger machine. After the pole is set, crushed limestone, concrete, or similar fill is placed and compacted around the pole. In some areas, poles will need lightning rods and grounding. Any special requirements such as grounding should be noted on the plans or in the specifications.

Taking-off and Pricing

Multiple quantity designations for units of measure and pricing may apply to electrical site work:

- For direct burial trenching, the work is measured per linear foot and can be converted to CLF.
- For duct bank conduits, the work is taken off and quantified per LF.
- For sand backfill or gravel bedding, measurements are made in cubic yards (C.Y.). Refer to Chapter 23, "Earthwork."
- Precast structures such as manholes and hand-holes are counted as individual units of each, EA with specific attention to size, access holes, risers, castings, etc.
- Fittings, spacers, and elbows are also counted individually, as each, EA.
- Precast, high density or fiberglass manhole and hand-hole units with cover and inserts (if required) is estimated by each, EA.
- Poles and crossbars are taken off by counting and quantified as individual units, each EA.
- Cable is taken off and quantified by the linear foot and converted to units of 100 linear feet, CLF.
- Concrete for duct banks and concrete pads is taken off by the cubic foot based on length × width × height, all in feet, then converted to CYs.
- Caution tape is taken off by the linear foot and converted to rolls, RLS.

Duct banks may be excavated to the width of the concrete placement so that the sides of the excavation can be used to form the sides of the duct bank. This practice called *earthen forming* eliminates the need for traditional formwork at the side of the duct bank. It should be noted that not all specifications allow this practice. For excavation in sandy or cohesion-less soils where the vertical walls of the excavation are not stable, traditional forming of the sides of the duct bank with wooden forms may be required. Again, formwork may or may not be part of the electrical contractor's scope of work. The estimator is directed to review the *Related Work* section of the technical specifications carefully.

Additional material items that may need to be taken off and quantified are the sand bedding, cast-in-place concrete for the duct bank, conduits, cable and terminations, caution tape, and possibly flowable fill or lean concrete for backfill in the roadways. For further information on estimating excavation, backfill, and compaction the reader is directed to Chapter 23, "Earthwork." For concrete material and estimating placement for duct banks the estimator is directed to Chapter 8, "Concrete."

Electrical sitework may also require the opening of streets or paved areas, requiring both cutting and patching of pavements. Local codes or ordinances regarding street openings may require flagmen or police details as well. Also remember that any excavation requires the notification of Dig Safe® or similar local utility location service.

Note: Code requirements may dictate certain minimum heights for suspended cable. Be sure that these codes are understood before pricing aerial installations.

GENERATOR SET

Generator sets are used as an emergency or stand-by power source in the event of the loss of normal power. They are rated in terms of their capacity in *kilowatts* or *kW*. Generators are available in single-phase or three-phase and may be specified at 120V through 6600V. Typical units are used at 120V single-phase or 480V three-phase.

The motors may be fueled by gasoline, propane gas, natural gas, and diesel fuel.

In general, gas generator sets are available from 10 to 170 kW, and diesel-powered sets range from 15 to 1000 kW. Larger diesel units are available as custom-ordered units. Although some small models may be started with a recoil rope starter, most units are fitted with electric starters and batteries. Control options include local or remote start/stop functions, engine heaters, exhaust mufflers, and annunciators.

Taking-off and Pricing

Generators are taken off and quantified by counting as individual units, EA. The engine generator, control panel, battery and charger, muffler, and day tank are generally included per unit of measure for a complete generator set. Additional items that may be required are taken off and estimated separately. Some of the more common items are transfer switches, weather tight enclosures, mounting pad, conduits, and startup and test by a factory trained technician.

The labor portion of the work typically includes receiving and handling of the equipment, setting, leveling, and fastening in place, wiring connections, and testing.

Most generators with an external fuel source will require piping to the generator from the fuel source. It may also require that the piping be below grade which would require excavation and backfill, another task typically outside the electrical contractor's responsibility. Related work and the responsible party should be identified in the specifications for the generator.

List each generator on a quantity sheet and summarize for estimating. Note all pertinent options and accessories clearly. Also note any special cranes or equipment needed for handling. Determine if any accessories will need to be field assembled.

Many specifications and some manufacturers will require that a trained service representative be present for initial startup to validate the warranty. This is usually an added cost and may also be part of the project closeout package of documents for the electrical contractor.

AUTOMATIC AND MANUAL TRANSFER SWITCHES

Transfer switches are used to change from the normal source of power to an alternate source, such as a generator set or a battery-powered inverter. There are two basic types of transfer switches: manual or automatic. A manual transfer switch is a lever- or handle-operated double-throw device with a pole for each wire. This type of switch is available from 30A to 600A.

Automatic transfer switches are designed to monitor the normal line source and to switch the load electrically to the backup source should the normal source fail. Automatic transfer switches are rated from 30A to 2000A.

A number of extra-cost options and accessories are available for automatic transfer switches. A few of the options are: auto-start relays to signal a generator when normal power fails; sensing relays to prevent transfer until the backup source is at full voltage; status-indicating lights; timed relay transfer to restore the load to normal power after it has been energized for several minutes; time-delay engine-stop relay to keep the generator running unloaded for a few minutes to cool the windings; and test and exercise control switches to permit the generator set to be started periodically and run.

Taking-off and Pricing

Transfer switches are taken off and quantified by counting individual units as each, EA. A complete unit EA should include a transfer switch, complete in enclosure and fasteners for mounting.

Additional items that may be required are taken off and estimated separately. Some of the more common additional items are a load shedding panel and accessories, which is taken off and priced as EA.

The labor for installation generally includes receiving and handling, layout and location of the device, and fastening to the surface. Review the plans and specifications carefully for the type (auto or manual), voltage, amperage, and number of poles for each transfer switch. List each on a quantity sheet, being careful to define any required options or accessories. Summarize the options for estimating.

It is best for the estimator to secure a vendor's quote for accurate pricing, particularly if control options are specified. Also, check the generator package, as the switches can often be purchased from the same manufacturer to ensure compatibility. Frequently, the transfer switch can be included as part of the generator "kit," especially on smaller residential units.

MISCELLANEOUS ELECTRICAL WORK

Some items may be harder to classify and are best categorized in a miscellaneous electrical group. These include permits, utility company tie-in fees, cutting and drilling for electrical access, and temporary power and lighting. These must be included for a thorough estimate if required by the individual project.

As previously noted, make safe operations and demolition can be a large part of the electrical work for renovation projects. Due to the limited amount of materials required for demolition, the lion's share of the work is labor. This can include labor to make safe an area scheduled for demolition by other trades. The scope of the demolition and responsibility for debris cleanup must be carefully defined for an accurate labor-hour determination.

Small plastic nameplates adhered to the front of panels are often required to identify the panel by name or what it controls. In addition to the tags, a printed card is provided within the panel to identify each circuit. Identification systems should be included as a cost in the estimate.

Load testing is also a common cost associated with new electrical distribution systems. This is included within the scope of Division 26 but may be required to be done by an outside testing lab that can produce a report and certify the results.

Cutting, coring, installing sleeves, and fire-stopping for electrical conduits that penetrate rated walls or floors, can also be included within the scope of the electrical work. Ongoing cleanup of debris generated by the new portion of the work can be a requirement as well. This would constitute disposing of debris from packaging and wire scrap to a dumpster or removing it from site altogether. The estimator is directed to review the electrical specifications carefully.

Supports and anchors are often required for the installation of electrical equipment. This can include drilling and setting anchor bolts into concrete or masonry to set the piece of equipment. Other supports such as strut (perforated steel channel) to attach conduit can also be required. For commercial projects, hangers and supports can be specified, but often only shown in typical details on the plans. They are always steel or a metal alloy. This includes everything from C-clamps to hangers. Read the specifications carefully and then allow a quantity based on the spacing required, plus some extra for on-site conditions. The count should be accurate as these supports are often more expensive than hangers for PVC.

Cranes or rigging equipment for setting electrical equipment may also be needed. The estimator should consider weight, reach, and needed duration of the crane for installing the rooftop equipment. Again, using a rigging subcontractor may be the best option for pricing the work accurately.

Training of owner's personnel on the new systems and equipment may also be a requirement, especially in commercial projects. Most specifications require that each major system or component be demonstrated as to its operation and maintenance. Frequently, the training is video recorded for future reference by the owner's personnel. In addition to the training, many projects require the electrical contractor to compile binders or electronic files of all of the technical literature for each major piece of equipment. Compiling closeout documents may require personnel directly involved in the installation of the equipment. In any event, the process can be time consuming.

Record drawings, called *as-builts*, that illustrate the actual location of electrical conduits and devices are often part of the electrical closeout requirements. As-builts can be simple redline updates on the electrical drawings or revised electronic files that create new line drawings using computer-assisted design (CAD). The estimator is directed to research and include all costs associated with as-builts within the estimate.

Taking-off and Pricing

Each miscellaneous item should be taken off and listed separately. Permits and utility company fees are taken off and priced per occurrence and can be listed in the estimate as a lump sum (LS). Fees for electrical permits are typically based on the individual job and paid when a permit is applied for or issued. (Utility companies usually charge a set fee to tie in service or a transformer, per occurrence.)

Cutting and drilling through wood, drywall, or other lightweight material are often considered part of the normal scope of work and, therefore, does not constitute a separate takeoff item. Cutting or coring through masonry or concrete, however, may involve special equipment or additional time, especially for large quantities. Quantify and price the cuts or cored holes by the piece (EA) and list them according to size (diameter or length × width), type of material, thickness of the material, and equipment needed.

Temporary lighting and power, often necessary during the construction process, are typically defined in Division 1—General Requirements. They may also be noted in Division 26—Electrical, under Related Work. Since temporary power panels and lights are reusable, the materials portion of the cost may include only lamps and wire and can be taken off and priced as a lump sum (LS). Installation and removal of temporary power and lighting facilities can be quantified by labor-hours. Maintenance, such as changing lamps and adding or relocating temporary lights and power, is calculated as labor-hours per week or month, depending on the specific needs of the project.

Excavation and backfilling for the installation conduits or direct-burial cables may also be required. Excavation and backfill are typically part of the general or site contractor's work. This is not absolute. Review the specifications for the exact scope of work concerning excavation and backfill for electrical work. If required, follow the applicable procedures identified in Chapter 23, "Earthwork." Minor hand excavation and backfill for underslab conduits may be best estimated using a labor-hour approach.

SUMMARY

While electrical work is clearly performed by individuals licensed and experienced in the trade, it is not uncommon for the general contractor to generate a budget estimate for the electrical scope of the work. The overview provided in this chapter will help you accurately assess the work involved and its costs.

In order to do an accurate takeoff, detailed plans and specifications are required for the electrical scope of work. For commercial projects, this is required by law in most states, but for residential construction it is often left to the individual electrical contractor to design and install a system that will perform its functions in accordance with local codes and standards. In the case of design-build for residences, the electrical estimate is often done parametrically by using the square foot area of the home. Code requirements often substitute for detailed design as a minimum. Considerations for the equipment included within the home, the size of the service, quality of light fixtures, etc. would all affect the cost.

Electrical systems are becoming more energy efficient every year and more complex to control. To accurately price an electrical system, it is recommended that the estimator become familiar with the operation of the various systems such as the generator, light fixtures and controls, and panelboards. Breaking the system down into its various subsystems is often a good start.

It is not essential for the GC estimator to itemize the takeoff and estimate for electrical systems in the same manner as for other aspects of the project, such as carpentry, painting, or roofing, but it does provide for a more accurate estimate of costs. Historical data from projects with similar criteria, especially residential, can then be compared with actual quotes from subcontractors to arrive at reliable costs.

Division 26—Electrical has a tremendous number of miscellaneous items that are necessary for inclusion within a comprehensive estimate. While much of these are subordinate to the main system, they are still part of the electrical scope and can impact the cost significantly.

QUESTIONS/ PROBLEMS

1. Would one expect that four conductors pulled together through a conduit be more or less labor consuming than pulling all four conductors individually? Explain your answer.
2. Using the appropriate figure from the chapter, what is the estimated labor-hours for an electrician to install 950' of #10 AWG wire?
3. Using the appropriate figure from the chapter, determine the estimated cost for installation labor for 270 LF of ½" EMT if the hourly billing rate for an electrician is $95.00.

4. Explain how an estimator might use specific parameters as a means to generate a budget estimate for the electrical work.
5. For a design-build residential electrical system explain some of the parameters and energy considerations that an electrical subcontractor would have to take into account when preparing a price.

22 | Communications, Electronic Safety, and Security

Currently, the last two divisions of the Facilities Services Subgroup of CSI MasterFormat® are Divisions 27—Communications and Division 28—Electronic Safety and Security. While they are two different scopes of work, they share some similarities when estimating. Much of the work of these two divisions was formerly included in Division 16—Electrical in the MasterFormat® 95 version but as a result of technology advances and the endless expansion of the related industries, CSI has provided each topic with its own division.

The work of both Division 27 and Division 28 is performed by individuals or firms with specific training and, frequently, special licensing. While the work of both divisions shares some of the same skills as electrical work, the majority of this work is low voltage. Separate permits and insurances may also be required. To take off and estimate the work of these divisions, it is extremely beneficial to have a working knowledge of the material components of the particular systems. Actual expertise in communication and safety/security systems is not normally within the realm of the general contractor's estimating experience.

Since most general contractors subcontract the work of Divisions 27 and 28, they may not need to produce a detailed estimate. Like the previous divisions in the Facilities Services Subgroup a parametric estimate is often sufficient to qualify bids from subcontractors.

COMMUNICATIONS AND ELECTRONIC SAFETY AND SECURITY

Division 27 and 28 work is typically shown on individual drawings for each scope, though they may be combined on smaller projects. Those drawings are frequently included within the electrical package of drawings. There are fewer standard conventions for labeling these drawings, and they may include labels

Estimating Building Costs for the Residential and Light Commercial Construction Professional, Third Edition. Wayne J. Del Pico.
© 2023 John Wiley & Sons, Inc. Published 2023 by John Wiley & Sons, Inc.

such as COM, TEL, and CA, for Division 27—Communications and FA, SEC, and LV for the work of Division 28—Safety and Security.

Be sure to review *all* drawings within the bid set for the work of Divisions 27 and 28. The work encompassed by these two divisions may include components of mechanical or fire suppression systems that are shown on other drawings. Consider the tamper and flow switches of the fire sprinkler systems that are tied into the fire alarm. Understanding which trade is responsible for the device is critical. Consult the HVAC drawings for duct smoke detectors that are tied into the fire alarm system. Frequently there are jurisdictional questions as to who furnishes and who installs such devices. The estimator should ensure responsibilities have been clarified.

As with mechanical/electrical drawings, communications, safety, and security drawings employ their own trade-specific graphic symbols for conveying information. Review all legends and graphic symbols on these drawings for definitions of the symbols. Security components may be referenced on hardware schedules in the architectural drawing set.

Telephone, data, and fire alarm drawings often use schedules that are helpful in determining the materials for the takeoff. Typical schedules include telephones, panels, terminal devices, equipment, and wire for low voltage or communications only. Line voltage connections are often part of Division 26 scope.

Specialized details, such as a fire alarm riser diagram, illustrate the various components of the system, quantities, and their configurations. Riser diagrams are for the graphic representation of information only and are not drawn to scale. They show the interaction of components within a particular system.

Consult the architectural drawings for dimensions, room sizes, floor-to-floor heights, ceiling heights, location of services entering the building, and coordination with other work within the area. The architectural drawings can be helpful in locating the placement of speakers, audio-video (AV) devices, fire alarm pull stations, horns, and strobes.

For many of the systems of Divisions 27 and 28, the design professionals may only provide an outline specification defining the performance of the system with little or no accompanying plans. This can often present a challenge to the nonspecialized estimator in developing a comprehensive budget for comparison. Any historical data from previous projects may help overcome this type of challenge.

Review Division 27 and 28 specifications thoroughly for the products, methods, and techniques of installation and the related work of other trades. Additional considerations may include:

- Testing of the final system (fire alarm and data/communications wiring)
- Training of owner personnel on the use and maintenance of the system (telephone systems)
- Tie-in charges (fire department monitoring lines for fire alarms)
- Annual maintenance agreements after the warranty period

The following is an outline of the work included within each of the divisions of this chapter.

Division 27—Communications

A. 27 08 00 Commissioning of Communications Systems
B. 27 10 00 Structured Cabling
  • Communication Equipment Room Fittings
  • Communication Backbone Cabling
  • Communication Horizontal Cabling
  • Communication Connecting Cords, Devices, and Adapters
C. 27 20 00 Data Communications
  • Data Communications Network Equipment
  • Data Communications Hardware
  • Data Communications Peripheral Data Equipment
  • Data Communications Software
  • Data Communications Programming
D. 27 30 00 Voice Communications
  • Voice Communications Switching and Routing Equipment
  • Voice Communications Terminal Equipment
  • Voice Communications Messaging
  • Call Accounting and Management
E. 27 40 00 Audio Visual Communications
  • Audio Visual Systems
  • Electronic Digital Systems

Division 28—Electronic Safety and Security

A. 28 08 00 Commissioning of Electronic Safety and Security Systems
B. 28 10 00 Access Control
  • Access Control
  • Access Control Hardware and Software
  • Access Control Interface
C. 28 20 00 Video Surveillance
  • Surveillance Equipment
  • Video Management Systems
D. 28 30 00 Security Detection, Alarm, and Monitoring
  • Intrusion Detection
  • Security Monitoring and Control
  • Tracking System
  • Audio Monitoring
E. 28 40 00 Life Safety
  • Radiation Detection and Alarm
  • Gas and Fuel Oil Detection and Alarm
  • Water and Detection and Alarm
  • Fire Detection and Alarm
  • Mass Notification and Emergency Response Systems
F. 28 50 00 Specialized Systems
  • Information Management and Presentation
  • Detention Security Systems

Figure 22.1 illustrates the typical components of a television system.

Figure 22.2 illustrates the typical configuration and components of a fire alarm system in a riser diagram format.

Figure 22.3 illustrates the typical configuration and components of a burglar alarm system in a riser diagram format.

This chapter covers takeoff procedures for the type of communications, safety, and security systems generally encountered in light commercial and residential construction.

Television Equipment	Crew	Daily Output	Labor-Hours	Unit
TV Systems not including rough-in wires, cables & conduits				
Master TV antenna system				
VHF & UHF reception & distribution, 12 outlets	1 Elec.	6	1.333	Outlet
30 outlets		10	.800	↓
100 outlets		13	.615	
Amplifier		4	2	EA
Antenna		2	4	"
Closed circuit, surveillance, one station (camera & monitor)	2 Elec.	2.60	6.154	Total
For additional camera stations, add	1 Elec.	2.70	2.963	EA
Industrial quality, one station (camera & monitor)	2 Elec	2.60	6.154	Total
For additional camera stations, add	1 Elec.	2.70	2.963	EA
For low light, add		2.70	2.963	
For very low light, add		2.70	2.963	
For weatherproof camera station add		1.30	6.154	
For pan and tilt, add		1.30	6.154	
For zoom lens - remote control, add, minimum		2	4	
Maximum		2	4	
For automatic iris for low light, add		2	4	

Figure 22.1 Television System

Source RSMeans Estimating Handbook, Third Edition. © 2009 John Wiley & Sons, Inc.

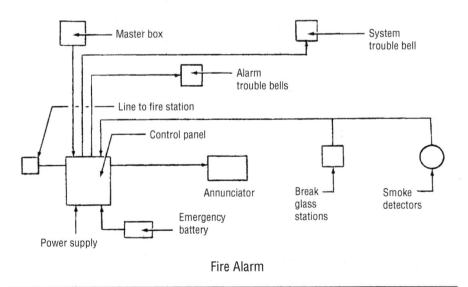

Fire Alarm

Figure 22.2 Fire Alarm System

Source RSMeans Estimating Handbook, Third Edition. © 2009 John Wiley & Sons, Inc.

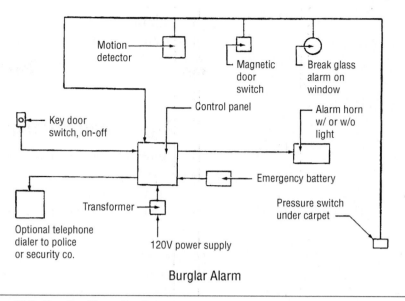

Burglar Alarm

Figure 22.3 Burglar Alarm System

Source RSMeans Estimating Handbook, Third Edition. © 2009 John Wiley & Sons, Inc.

Divisions 27 and 28 should be taken off and estimated separately. Use the outline structure noted here for guidance in organizing the takeoff and estimate. All takeoff quantities should be listed separately according to type, application, size, function, and any other applicable characteristic. In addition to materials, special items such as shop drawings, testing, and training should be considered as part of the estimate.

Additional takeoff guidelines are as follows:

- **Cable tray or surface raceways:** By the linear foot (LF), separated according to type (plastic or aluminum), size (width), and method of suspension (wall mounted or hung from above).
- **Fittings for cable tray:** By the piece (EA) according to type (elbow, tee, reducer, etc.), size (width), method of connection, and material composition (plastic or aluminum).
- **Conduits:** By the LF according to type (steel, IMC, PVC, EMT, etc.), size (diameter), use or application, and other identifying information needed for pricing (such as height off the floor).
- **Switches, terminal devices, etc.:** By the piece (EA) according to type, function, location, manufacturer, and model number.
- **Racks:** By EA according to type, finish, size, model, and application.
- **Transformers:** By EA according to type, rating, function, location, manufacturer, and model number.
- **Wire:** By the LF, separated according to size (number of conductors) and rating (plenum rated). Converted to CLF (100 LF).
- **Connectors:** By EA according to type, function, size, etc.
- **Devices, monitors, annunciators, detectors:** By EA according to type, model, manufacturer, and any defining characteristics.
- **Panels:** By EA according to type, function, size, etc.

In most cases, the work of Divisions 27 and 28 begins at the interior of the building in rooms where electrical communications hubs are located. Special rooms for communications equipment, called *head end rooms*, may also be provided.

Taking-off and Pricing

The takeoff procedure for the work of installing communications and electronic safety and security systems should follow that of other electrical systems. The estimator is directed to Chapter 21, "Electrical Systems," in this text for more information on specific topics illustrated here. Each division should be taken off and priced separately.

Raceways and wiring (conductors) should be taken off and priced by the LF and listed according to identifying characteristics and application. Devices should be taken off and priced by the individual piece (EA) and listed according to type, application, manufacturer, model, and function. Be sure to include all necessary boxes, covers, plates, connectors, termination devices, and equipment for a complete system.

As many specialty systems are designed to perform a specific series of functions, specifications often require testing of the completed system to prove compliance with design criteria. Testing may also be required by an independent firm at the contractor's expense. Owner training on the system may also be required. Reviewing the specifications helps determine the scope and responsibility of testing and training.

For buildings already in use, tie-ins to existing systems may have to be done during off-hours (nights and weekends) to avoid disruption of normal operations. This should be noted in the takeoff for accurate pricing of the labor portion of the estimate.

Much of the work of Divisions 27 and 28 shares common components with Division 25—Integrated Automation and Division 26—Electrical. Refer to Chapter 20, "Integrated Automation," Figure 20.1, for installation time in labor-hours for cable tray raceway and Chapter 21, "Electrical Systems," Figure 21.2, for installation time in labor-hours for various types of conduits, and Figure 21.3, for installation time in labor-hours for underfloor raceway systems. Figures 22.4 through 22.7 provide installation times in labor-hours for components of communications and electronic systems.

Description	Labor-Hours	Unit
Cable assembly 25 pair with connectors 50'	.670	EA
3 pair with connectors 50'	.340	EA
4 pair with connectors 50'	.350	EA
Cable (bulk) 3 pair	.006	LF
4 pair	.007	LF
Bottom shield for 25 pair cable	.005	LF
3–4 pair cable	.005	LF
Top shield for all cable	.005	LF
Transition box, flush mount	.330	EA
In floor service box	2.000	EA
Floor fitting with duplex jack and cover	.380	EA
Floor fitting miniature with duplex jack	.150	EA
Floor fitting with 25 pair kit	.380	EA
Floor fitting call director kit	.420	EA

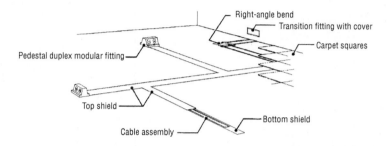

Figure 22.4 Installation Time in Labor-Hours for Undercarpet Telephone Systems

Source RSMeans Estimating Handbook, Third Edition. © 2009 John Wiley & Sons, Inc.

Description	Labor-Hours	Unit
Cable assembly with connectors 40'		
Single lead	.360	EA
Dual lead	.360	EA
Cable (bulk) single lead	.010	LF
Dual lead	.010	LF
Cable notching 90 degree	.080	EA
180 degree	.130	EA
Connectors BNC coax	.200	EA
Connectors TNC coax	.200	EA
Transition box, flush mount	.330	EA
In floor service box	2.000	EA
Floor fitting with slotted cover	.380	EA
with blank cover	.380	EA

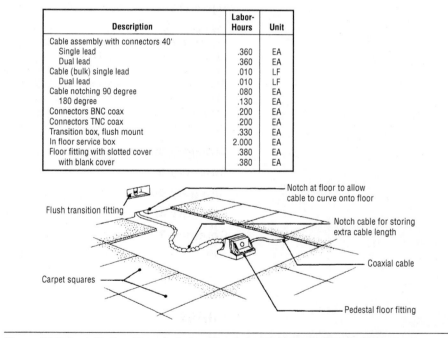

Figure 22.5 Installation Time in Labor-Hours for Undercarpet Data Systems

Source RSMeans Estimating Handbook, Third Edition. © 2009 John Wiley & Sons, Inc.

Description	Labor-Hours	Unit
High-performance unshielded twisted pair (UTP)		
Category 3, #24, 2 pair solid, PVC jacket	.800	CLF
4 pair solid	1.143	
25 pair solid	2.667	
2 pair solid, plenum	.800	
4 pair solid	1.143	
25 pair solid	2.667	
4 pair stranded, PVC jacket	1.143	
Category 5, #24, 4 pair solid, PVC jacket	1.143	
4 pair solid, plenum	1.143	
4 pair stranded, PVC jacket	1.143	
Category 5e, #24, 4 pair solid, PVC jacket	1.143	
4 pair solid, plenum	1.143	
4 pair stranded, PVC jacket	1.143	
Category 6, #24, 4 pair solid, PVC jacket	1.143	
4 pair solid, plenum	1.143	
4 pair stranded, PVC jacket	1.143	
Category 5, connector, UTP RJ-45	.100	EA
shielded RJ-45	.111	
Category 3, jack, UTP RJ-45	.111	
Category 5	.123	
Category 5e	.123	
Category 6	.123	
Category 5, jack, shielded RJ-45	.133	
Category 5e	.133	
Category 6	.133	

Figure 22.6 Installation Time in Labor-Hours for Unshielded Twisted Cable

There are several categories used to describe high-performance cable. The following information includes a description of categories CAT 3, 5, 5e, 6, and 7, and details classifications of frequency and specific standards. The category standards have evolved under the sponsorship of organizations such as the Telecommunications Industry Association (TIA), the Electronic Industries Alliance (EIA), the American National Standards Institute (ANSI), the International Organization for Standardization (ISO), and the International Electrotechnical Commission (IEC), all of which have catered to the increasing complexities of modern network technology. For network cabling, users must comply with national or international standards. A breakdown of these categories is as follows:

Category 3: *Designed to handle frequencies up to 16 MHz.*

Category 5: *(ITA/EIA 568A) Designed to handle frequencies up to 100 MHz.*

Category 5e: *Additional transmission performance to exceed Category 5.*

Category 6: *Development by TIA and other international groups to handle frequencies of 250 MHz.*

Category 7: *Development to handle a frequency range from 1 to 600 MHz.*

Source RSMeans Estimating Handbook, Third Edition. © 2009 John Wiley & Sons, Inc.

Labor

Electricians and low-voltage technicians are responsible for performing much of the work of these divisions. Bear in mind that a licensed electrician is required in most jurisdictions when line voltage power connections are made.

Description	Labor-Hours	Unit
Clock Systems		
Time system components, master controller	24.240	EA
Program bell	1	EA
Combination clock & speaker	2.500	EA
Frequency generator	4	EA
Job time automatic stamp recorder, minimum	2	EA
Maximum	2	EA
Master time clock system, clocks & bells		
20 room	160	EA
50 room	400	EA
Time clock, 100 cards in & out, 1 color	2.500	EA
2 colors	2.500	EA
With 3-circuit program device, minimum	4	EA
Maximum	4	EA
Metal rack for 25 cards	1.140	EA
Watchman's tour station	1	EA
Annunciator with zone indication	8	EA

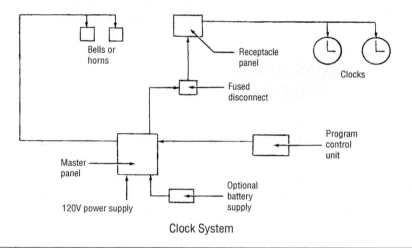

Clock System

Figure 22.7 Installation Time in Labor-Hours for Clock Systems

Source RSMeans Estimating Handbook, Third Edition. © 2009 John Wiley & Sons, Inc.

Depending on the specific task, productivity can be measured by the output of the individual, as in the case of installing devices, or panels. Other tasks, such as pulling cable and hanging raceways, clearly require multiple personnel for cost-effective installations. Labor to install individual detection devices, televisions, monitors, and so on, varies depending on the size and type of device and the height from the floor.

Production may be reduced by the use of lifts or ladders to perform the work. It may also require that an additional individual be added to the crew for the support of personnel on the staging. This should be considered when establishing crew size. Simple tasks, such as installing finish devices, plates, and junction or workboxes, can be estimated by the piece (EA) by calculating the quantity of devices that can be installed by one electrician in a day. The same applies for the conductors. Base the unit price on the anticipated linear footage of wire that will be pulled by the crew in a day. Always remember that pulling wire through a conduit or raceway involves different productivities than loose wire on hangers or cable tray.

Other operations, such as installing panels, annunciator, amplifiers, antennas, cameras, or terminal devices are based on labor-hours per individual piece. Frequently, bucket trucks or aerial platforms are needed for security systems that are mounted on poles at building exteriors. The costs of any lift are time sensitive and calculated by the day, week, or month.

SUMMARY

While communications, safety, and security work are clearly performed by individuals experienced in the trade, it is not uncommon for the general contractor to generate a budget estimate for this scope of the work. The overview provided in this chapter will help the estimator accurately assess the work involved. Note that the work included in Division 27 and Division 28 may also be part of the work of the electrical contractor and inclusive in the electrical subcontract.

QUESTIONS/ PROBLEMS

1. Using the appropriate figure from the chapter, what is the estimated labor-hours for a technician to install 550 ft. of Category 6 #24 four-pair stranded wire?
2. Using the appropriate figure from the chapter what is the estimated labor-hours for a technician to install 100 Category 6 shielded RJ-45 jacks?
3. Using the answers from questions #1 and #2 above, what is the labor cost if the billing rate is $115 per hour?
4. Explain how an estimator might use specific parameters as a means to generate a budget estimate for the security system work.

23 | Earthwork

Division 31 of CSI MasterFormat® is Earthwork, which deals primarily with excavation and backfill, both bulk and general, and all of the ancillary processes that are within the Earthwork scope. It is the first division of the Site and Infrastructure subgroup.

Earthwork includes tasks dealing with soil materials, excavation and backfill, compaction, soil stabilization, hauling, erosion control, grading, clearing and grubbing, excavation stabilization with sheet piling and shoring, and earthwork specialties such as piles and caissons.

Calculating costs for earthwork is often the most difficult part of generating an estimate due to the number of unpredictable factors and contributing information that can be deduced from the site visit and the documents. It also has the unique aspect of being equipment intensive. Very little of the work in this section is performed without some type of equipment.

Even with a detailed and comprehensive site investigation and geotechnical report (see Chapter 7, "Existing Conditions"), it is not always possible to predict with accuracy what lies beneath the surface. The earthwork or *site contractor* must carefully study all available contract documents, as well as any supplemental information provided. It is *strongly* recommended that a site inspection be conducted and that this be done only after becoming familiar with the documents.

CIVIL DRAWINGS

Most earthwork is shown on a special set of drawings, called *civil drawings* or *site drawings,* that pertain exclusively to the site. They are designed under the direction and approval of a civil engineer and focus on the land's changes to

Estimating Building Costs for the Residential and Light Commercial Construction Professional, Third Edition. Wayne J. Del Pico.
© 2023 John Wiley & Sons, Inc. Published 2023 by John Wiley & Sons, Inc.

accommodate a new structure or any engineering improvements. The most common drawings in the civil set of drawings are:

- Existing conditions drawings (existing site survey discussed in Chapter 7, "Existing Conditions")
- Site demolition and/or site preparation plan
- Grading and drainage plan(s)
- Utilities plan(s) (water, sewer, electric, tele/com, and gas)
- Paving and curbing plan
- Septic system or sanitary sewer plan
- Site lighting and site electrical power plan
- Site improvements plan
- Landscape and irrigation plan
- Site details and sections

The quantity of drawings and level of detail vary from project to project and depend, to a large extent, on the owner's budget or the sitework scope. Residential site plans can range from a single drawing to a full set. Commercial projects most often require a comprehensive stamped set of civil drawings as a condition of permit approval. Each drawing in a civil set has special conventions and nomenclature unique to its discipline and provides a specific contribution to the overall civil set. Some drawings are self-explanatory by their title. However, no drawing is used in a vacuum. Each is meant to be complementary and used with the others to help determine accurate quantities for pricing. It is important for the estimator to remember that civil plans, like other disciplines, are intended to provide details and sections to confirm or clarify what is shown on the plan views.

Using Civil Drawings in the Estimate

There are specific types of information provided on each sheet in the civil set of drawings. Not all of the drawings in the civil set are used exclusively in preparing the earthwork estimate; in fact, the civil drawings can illustrate many different divisions and sections of work. For example:

- **Existing conditions drawings:** Used in conjunction with the site grading drawing to determine the quantity of bulk cut or fill required to achieve the proposed final grade.
- **Grading drawings:** Should be consulted along with the utilities or site lighting plan to calculate the trench excavation and backfill required for site features, such as domestic water piping, gas service, electrical power and lighting conduits, and similar utilities.
- **Drainage drawings:** Provide information for taking off precast structures, such as tanks, manholes, catch basins, and reinforced concrete and PVC pipe. Additional information on the details sheets helps verify accurate pricing.
- **Site demolition or site preparation plans:** Provide information on the removal or relocation of existing site features, such as trees or benches, or subsurface structures like abandoned foundations or rock. Site preparation can include erosion and sedimentation control plans.

- **Architectural and structural drawings:** Should be reviewed for details that would be shown in section (such as foundation drains, bottom of foundation footings, and excavations for basement areas).
- **Foundation plans and details:** Show the limits of the excavation (width by depth) for the foundation. They also show interior details, such as footings or depressed areas in the slab that require excavation, backfill, and compaction.
- **Mechanical and electrical drawings:** Must be reviewed for details that show trenching for piping or conduits in the basement or under the slab-on-grade. This work is often overlooked because it is not shown in the civil drawings.
- **Site improvement drawings:** Illustrate walkways, curbing and paving (in the absence of a dedicated plan), bollards, exterior steps, fencing, patios, and virtually any type of exterior improvement.
- **Site detail drawings:** Provide sections and details of features such as trench excavation, subgrade preparation, pipe bedding, and roadway components.
- **Landscape drawings:** Show plantings, mulch beds, lawns, and ground cover features of the site. They frequently include a detailed list of the plantings and staking of plants.
- **Irrigation drawings:** Illustrate how water will be supplied to the landscape and plantings.

Each portion of the documents (plans and specs) provides information the estimator can use to calculate the earthwork required. As an example, consider the following drawings and their respective contributions for defining the excavation and backfill:

- **Utilities drawing:** Shows the location of the water service from the street to the building. The estimator can measure its length. This allows the estimator to calculate the length of the excavation.
- **Site detail drawing:** Shows the cross section of the trench for the water service. It defines the guidelines for the width of the trench and the depth to the top of the pipe. It shows the bedding material depth for the pipe. This allows the estimator to calculate the width and depth of the excavation, which, combined with the length, results in the volume to be excavated from the trench.
- **Technical Specification Section 31—Earthwork:** Defines the materials to be used to backfill the trench, the depth of the excavation to the top of the water service, and the type and classification of the bedding material, which allows the estimator to calculate the cost of bedding materials for the pipe. Division 31 also defines the backfill and compaction requirements, thereby allowing the estimator to calculate the costs for both.
- **Geotechnical report (if available):** Provides borings that tell the estimator the type of soil that will be excavated: clay, sand, organic, and so on. It may also note any rock or existing subsurface materials (foundation, pipe) that may need to be removed or avoided.

Finally, the specifications must be reviewed *thoroughly* for the type and quality of the soil materials and installation. The specifications may offer additional information, such as subsurface compaction requirements, moisture content, and general or supplemental conditions that affect earthwork. Geotechnical reports (if provided) must be studied in detail to alert the estimator to the excavation conditions that can be anticipated.

Division 1—General Requirements must be reviewed for any temporary access roads, or dust control that may be part of the earthwork scope and unit prices (see Chapter 2, "Understanding the Specifications") that affect the earthwork scope of work.

This chapter reviews the more common elements of earthwork takeoff and estimating for the light commercial/residential market. A partial outline of work topics included in Division 31—Earthwork is as follows:

A. 31 10 00 Site Clearing/Site Preparation
- Clearing and Grubbing
- Selective Clearing
- Earth Stripping and Stockpiling

B. 31 20 00 Earth Moving
- Grading
- Excavation and Fill
- Embankments
- Erosion and Sedimentation Controls

C. 31 30 00 Earthwork Methods
- Soil Treatment
- Soil Stabilization
- Rock Stabilization
- Gabions and Riprap

D. 31 40 00 Shoring and Underpinning
- Shoring
- Needle Beams
- Underpinning

E. 31 50 00 Excavation Support and Protection
- Anchor Tiebacks
- Cofferdams
- Cribbing and Whalers

F. 31 60 00 Special Foundations and Load Bearing Elements
- Driven and Bored Piles
- Caissons
- Special Foundations

SITE CLEARING/ SITE PREPARATION

Before the earthwork process can commence, most sites require some level of preparation. That preparation can be in the form of clearing trees and natural growth, as in the case of undisturbed land, or it can be the removal of existing structures, underground piping, foundations, or other existing engineering improvements on previously developed sites. It also includes the stripping and stockpiling of topsoil for reuse. The overall process is called *site clearing* and/or *site preparation*. Part of the site preparation is the containment and control of runoff from precipitation. This will be discussed later in this chapter.

Clearing and Grubbing

Clearing of the site refers to the removal of brush, trees, and other natural growth on the surface. It refers to work that starts inches above the ground. *Grubbing* (sometimes called *stumping*) refers to the removal of tree stumps from the clearing limits to below grade. Most site clearing and grubbing is done with power equipment, although some smaller sites are still cleared with handheld chainsaws and walk-behind equipment. Occasionally, clearing and grubbing require the use of specialized cutting equipment attached to excavating equipment. Carefully inspect the site to identify work that can be done by equipment versus work that must be done by hand.

Taking-off and Pricing

The task of clearing is taken off and priced by the square foot (SF) or area to be cleared or, in the case of large parcels of land, by the acre (ACRE). Either unit of measure is acceptable. Grubbing is also taken off and priced according to the same units but should be kept separate and distinct from clearing. One byproduct of clearing, *brush,* can be chipped into trucks for hauling off site or can be stockpiled for later use as groundcover. Larger trees can be chipped for groundcover or hauled from the site is log lengths for other uses.

Production of both tasks is measured by the crew's performance, including equipment. Units of measure for each task is SF or ACREs per day. Site visits are *strongly* recommended to determine the density of the growth as well as access and terrain. Crews and productivities vary dramatically with species, weather, and size of tree, as well as whether the cleared debris will be removed or left for others. The estimator may want to consider any residual value of the cut timber for resale to sawmills or as firewood.

Grubbing, in addition to the cost of stump removal, has the disposal component. Stumps can be trucked to a special disposal area, and the cost is typically based on the unit of measure of cubic yards (CY) contained within the trailer, or per trailer itself. Disposal fees for the trailers must also be included in the estimate.

Stripping and Stockpiling

Part of the preparation process includes stripping and stockpiling the topsoil. *Topsoil* is classified as the top layer of soil consisting of nonstructural soils high in organic content. A general rule is that if the soil is good to grow things in, it's bad to build on. The depth of the topsoil can be determined from the test boring report or a visual inspection. Topsoil is typically screened for reuse in planting and lawn areas. The screening process can be done on site, or the stripped soil can be trucked off site, screened, and returned to the site when needed. This is sometimes dependent on the site size and available area for stockpiling. The estimator is urged to consider local ordinances on this matter, as some communities prevent the removal of stripped topsoil from the original site.

The goal of the stripping and stockpiling process is to provide a substrate of nonorganic material from which to start the bulk excavation or backfill process.

Taking-off and Pricing

The task of stripping the topsoil is taken off and priced by the SF. The area is then multiplied by the depth and extended to cubic yards (CY) for bulk stripping. When calculating the area of topsoil to be removed, the result should be enlarged to accommodate any clearance needed. If the topsoil is to be screened prior to reuse, that process should be quantified and priced separately by the CY. The estimator must consider a natural reduction in volume after screening due to the removal of organic matter and other debris, called *tailings*. Additional soil may be required to supplant any shortfall caused by the tailings.

Takeoff quantities should account for trucking and fees for disposal of the tailings off-site. These should be listed in the takeoff by the truckload or by the CY. Placement of the screened product should be included as a separate cost in the estimate. This can be quantified by the CY or by the SF (for even depths). It may be part of the landscape package or as earthwork in this division.

For reasonably flat surfaces requiring no significant excavation or backfill, the stripping and stockpiling may require the estimator to include the cost of minor grading to provide a building pad at a specific elevation.

The estimator is directed to review the site plan to determine the stockpile location and may need to include the cost of loading, hauling, and unloading on-site if the distance to the stockpile location is significant. Other considerations include any treatment or stabilization of the screened topsoil, called *loam*, while it is stored prior to reuse.

EXCAVATION AND BACKFILL

In simple terms, *excavation* refers to digging a hole to accommodate a specific engineering improvement. *Backfilling* is placing soil by means of a controlled method to fill an excavation, likewise for the purpose of an engineering improvement, such as foundations, piping for utilities, and precast structures. To a lesser degree, excavation and backfill are required in order to place pavement, walks, and curbs. Excavation and backfill can be classified into two main categories for estimating purposes:

• Bulk excavation
• Trench or general excavation

Excavation, backfill, and earthwork in general are volume calculations that are extended for pricing to CY. (1 CY equals 27 CF.) The reader is directed to Chapter 3, "Calculating Linear Measure, Area, and Volume," for a review of area and volume calculations.

Bulk Excavation

Moving large masses of earth to establish new grades for parking lots, roads, or building pads is referred to as *bulk excavation*. The most common method for determining the quantities of soil to be moved is the *cross-section method,* which involves tabulating cuts and fills for small increments of the total parcel. *Cutting* refers to removing earth in order to achieve the desired grade. *Filling* is the addition of earth to raise the existing conditions to meet the desired grade. Examine the site closely to determine the amount of earth that will have to be handled in order to transform the existing grades to the proposed grades. (See Figure 23.1.)

The cross-section method divides the lot into a series of smaller, equal areas. This is done with a grid drawn on the grading plan. It is helpful if both existing and new proposed grades are shown in the form of *contours,* or lines that indicate the same horizontal elevation. Any point along a contour is the same elevation. Contours are one of the unique conventions of the civil drawings set; they allow the estimator to see the site in three

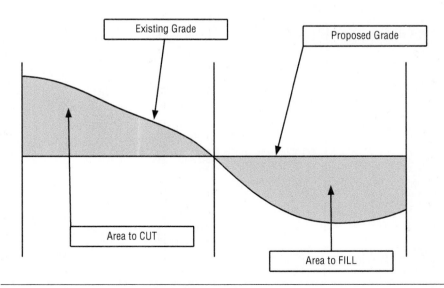

Figure 23.1 Cutting and Filling to Meet Grade

dimensions on a two-dimensional medium – paper. Existing contours are shown on the grading or site plan as dashed lines, with the new or proposed grades shown as solid lines. Figure 23.2 illustrates the cross-section grid in place over the existing grade plan.

The dimensions of the grid should be uniform throughout the plan, and the grids should be square for ease of calculation. The grid spacing depends on the area to be cross-sectioned. If the lot slopes gradually, the spacing can be spread out (larger grid dimensions), but if the lot has dramatic changes in elevation illustrated by a concentration of contours, closer spacing (smaller grid dimensions) may result in a more accurate takeoff.

Each of the smaller areas is identified by a number assigned by the estimator. The horizontal lines are numbered, and the vertical lines lettered so that each grid intersection can be referenced. Be sure to calculate the elevation of the intersection of each of these lines by interpolating between the contours. (*Interpolation* approximates the elevation of a particular point between two known points.) The existing grade is placed in the upper right-hand quadrant and the proposed grade in the upper left-hand quadrant. The difference in feet between the two is noted as either a cut (signified by a negative sign) or a fill (a positive sign) in the lower right or left quadrant. Each grid area is examined to see if it changes from fill to cut or vice versa within the numbered grid. These grids must be calculated separately. For grids that are all cut or all fill, calculate the average cut or fill for the grid, and multiply it by the area of the grid (using a consistent unit of measure, such as square feet). The average is obtained by summarizing all four

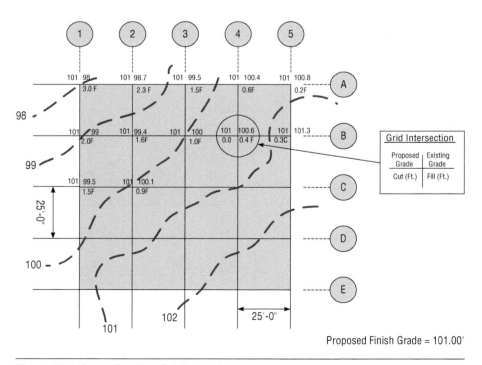

Figure 23.2 Cross-Section Grid Layout

corners of the grid and dividing by 4. The product will be a volume quantity in cubic feet (CF). Then divide by 27 to convert CF to CY. The total of each numbered grid is recorded in a separate column on a cut/fill sheet, as shown in Figure 23.3. (*Note:* The cut and fill sheet calculations shown in Figure 23.3 are independent of Figure 23.2.)

For numbered grids that change from cut to fill, the estimator must determine which portion of the area is cut and which is fill. The same principles are applied to arrive at the respective cut and fill portions. Once all the numbered grids have been calculated, the total cut and fill for each column is tabulated. From this information, the estimator can determine quantities of soil to be *exported* or *imported* to the site, along with the total CY of material to be handled. The terms *export* and *import* have been adopted by the industry for hauling fill from the site and hauling fill to the site, respectively. Figure 23.4 is an example of Grid #5 and the resulting calculations.

Cuts and fills are intentionally kept separate in the tally sheet. For numbered grids that are calculated as half cut and half fill, adding them in the same column would result in a net of zero, despite each quantity having to be "handled."

Readers should bear in mind that large cut and fill operations in bulk excavations are typically performed with the aid of a computer and software. . .expensive software that can only be justified by large site contractors with a significant volume of work. Performing the cut and fill tabulations by hand, as noted here, would be done only for small sites, to enhance the learning experience, or in the absence of the required software. Regardless of the reason, the author is a firm believer that the reader should at least understand the logic behind and method of doing the calculations by hand.

CUT					FILL			
Grid No.	Area (SF)	Ave. Cut Depth (Ft.)	Cut (CF)		Grid No.	Area (SF)	Ave. Fill Depth (Ft.)	Fill (CF)
1	625	1.54	962.50					
2	625	2.11	1318.75					
3	625	3.25	2031.25					
4	625	2.90	1812.50					
					5	625	2.07	1293.75
					6	625	1.85	1156.25
					7	625	1.66	1037.50
8	625	1.22	762.50					
9	625	1.23	768.75					
					10	625	0.92	575.00
11	625	1.81	1131.25					
12	625	2.03	1268.75					
					13	625	0.88	550.00
					14	625	0.78	487.50
					15	625	1.07	668.75
16	625	0.09	56.25					
		Total Cut (CF)	**10,112.50**				**Total Fill (CF)**	**5,768.75**
		Total Cut (CY)	**374.54**				**Total Fill (CY)**	**213.66**

Figure 23.3 Sample Cut and Fill Tally Sheet

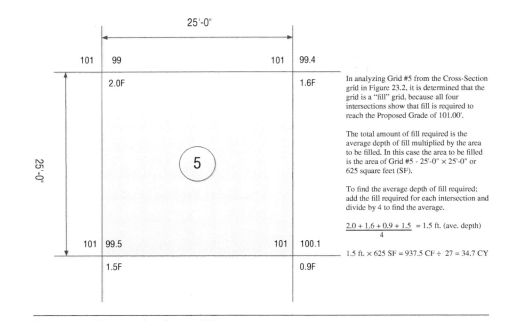

In analyzing Grid #5 from the Cross-Section grid in Figure 23.2, it is determined that the grid is a "fill" grid, because all four intersections show that fill is required to reach the Proposed Grade of 101.00'.

The total amount of fill required is the average depth of fill multiplied by the area to be filled. In this case the area to be filled is the area of Grid #5 - 25'-0" × 25'-0" or 625 square feet (SF).

To find the average depth of fill required; add the fill required for each intersection and divide by 4 to find the average.

$$\frac{2.0 + 1.6 + 0.9 + 1.5}{4} = 1.5 \text{ ft. (ave. depth)}$$

1.5 ft. × 625 SF = 937.5 CF ÷ 27 = 34.7 CY

Figure 23.4 Calculation for Grid #5

Trench or General Excavation

A major portion of earthwork takeoff and estimating includes calculating trench or general excavation. Most of this work is done with excavating equipment, such as backhoes and excavators, although some of the dressing of the excavated area is performed by hand. It is a good idea to keep the quantities for machine and hand excavation separate, as there is a considerable difference in pricing.

When determining the limits of excavation, the estimator must consider the actual size of the basement, foundation, pipe trench, or buried structure, then add a sufficient buffer, or *overdig* to provide access for workers and materials. Also include overdigging to stabilize the slope, which allows the soils to stabilize naturally and prevents earth from sliding into the excavated area. Different soil compositions tend to stabilize at different angles, as measured from the horizontal plane at the bottom of the excavation. This is called the *angle of repose*. The more cohesive the soil, the steeper the angle of repose. The less cohesive the soil, the shallower the angle of repose. The term *cohesive* refers to soils with a large clay content. *Noncohesive soils* consist of sand or gravel materials, an extreme example would be uniformly graded beach sand. In contrast, excavated trenches in extremely cohesive soils can stabilize vertically (90° angle) at the limits of the bucket on the excavator. While the exact angle of repose is not critical for estimating, a general classification of the soil helps determine the slope of the excavation. Figure 23.5 illustrates the angle of repose for various soil types.

Once the angle of repose and the required size of the excavation have been determined, it is fairly easy to calculate the cross-sectional area of the trench

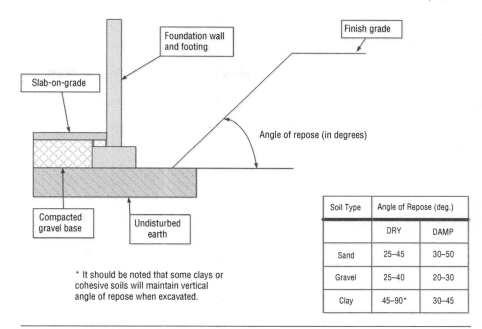

Figure 23.5 Angle of Repose

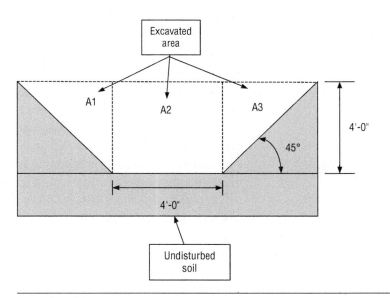

Figure 23.6 Sample Trench

and, subsequently, its volume. Figure 23.6 shows the cross-sectional area of a sample trench. Its corresponding calculations follow.

> *Calculate the volume of earth to be excavated for a trench 4' deep 100' long. The bottom of the trench is 4' wide at its widest point. The angle of repose is 45°, a ratio of 1:1. For every foot of depth, the excavation is 1' wide, as measured from the vertical plane (illustrated by the dashed vertical lines in Figure 23.6).*

The area of A1 = (4' × 4') ÷ 2 = 8 SF. Since A3 is symmetrical to A1, A3 = 8 SF.

The area of A2 = 4' × 4' = 16 SF.

Adding the three areas: A1 + A2 + A3 = 32 SF.

If the total cross-sectional area of the trench is 32 SF and the length of the trench is 100 linear feet (LF), apply the formula for volume:

$$V = A \times L$$

$$V = 32 \text{ SF} \times 100 \text{ LF} = 3,200 \text{ CF} \div 27 \text{ CF per CY} = 118.51 \text{ CY, or}$$
approximately 119 CY

When stabilizing a slope, be sure to address OSHA regulations for worker safety and the actual requirements of the specifications. Some earthwork specs require that a minimum slope of 1:1 (45°) be maintained. However, the required minimum is not always the best answer for the particular situation. In some instances, the excavated trench must be stabilized by artificial means, as discussed later in this chapter.

General Note on Estimating Excavation

For both classifications of excavation, bulk and trench, the estimator must remember that when soil is in its natural state, undisturbed, it is called *in-situ*. Soils in-situ are densely compacted, either because of the long periods during which they have been allowed to settle or because they have been mechanically compacted. Soils in situ have an *inground* volume defined as *bank cubic yards* (BCY). Once that inground volume has been disturbed during the excavation process, the voids between soil particles expand and the soil volume increases. This condition is called *swell*. Different soils expand at different rates or have different *swell factors*. Swell factors are expressed as a percentage increase to the inground or original volume. To clarify: A soil that has an in-situ volume of 100 BCY and a swell factor of 22% now has an excavated volume of 122 CY. Volumes of excavated soils that have been increased by a swell factor are quantified as *loose cubic yards* (LCY). The estimator must recognize that this has an impact on the number of trucks required to haul or export any excess LCY from the site.

The inverse of this process is explained in the next section, "Backfill and Compaction."

Backfill and Compaction

The process of filling an excavated area using controlled methods is called *backfilling*. Remember that where there is excavation, there is usually a backfill component to be considered.

As discussed in the preceding section, soil naturally expands once it has been excavated, which increases the volume the fill will occupy. The inverse is also true: When the volume of a material is compacted, it tends to shrink or lose volume, a process called *compaction*. The actual percentage of swell or compaction depends on the conditions and type of material being excavated. Compaction is also a characteristic of the soil itself. Finely graded soils of a uniform size, such as fine sand, have less volume change. Soils with varying gradations of particle size have more voids, or spaces, between individual particles, and thus more swell and compaction can be expected. Figure 23.7 lists soil characteristics and factors for calculating soil volume in various conditions.

It is essential to become familiar with the requirements for backfilling and compaction as mandated by the specifications. This allows the estimator to

Approximate Material Characteristics*				
Material	Loose (Lb./CY)	Bank (Lb./CY)	Swell (%)	Load Factor
Clay, dry	2,100	2,650	26	0.79
Clay, wet	2,700	3,575	32	0.76
Clay and gravel, dry	2,400	2,800	17	0.85
Clay and gravel, wet	2,600	3,100	17	0.85
Earth, dry	2,215	2,850	29	0.78
Earth, moist	2,410	3,080	28	0.78
Earth, wet	2,750	3,380	23	0.81
Gravel, dry	2,780	3,140	13	0.88
Gravel, wet	3,090	3,620	17	0.85
Sand, dry	2,900	3,250	12	0.89
Sand, wet	3,200	3,600	13	0.89
Sand and gravel, dry	2,900	3,250	12	0.89
Sand and gravel, wet	3,400	3,750	10	0.91

*Exact values will vary with grain size, moisture content, compaction, etc. Test to determine exact values for specific soils.

Typical Soil Volume Conversion Factors			Converted to:	
Soil Type	Initial Soil Condition	Bank	Loose	Compacted
Clay	Bank	1.00	1.27	0.90
	Loose	0.79	1.00	0.71
	Compacted	1.11	1.41	1.00
Common earth	Bank	1.00	1.25	0.90
	Loose	0.80	1.00	0.72
	Compacted	1.11	1.39	1.00
Rock (blasted)	Bank	1.00	1.50	1.30
	Loose	0.67	1.00	0.87
	Compacted	0.77	1.15	1.00
Sand	Bank	1.00	1.12	0.95
	Loose	0.89	1.00	0.85
	Compacted	1.05	1.18	1.00

$$\text{Swell (\%)} = \left(\frac{\text{Wt/bank CY}}{\text{Wt./loose CY}} - 1 \right) \times 100$$

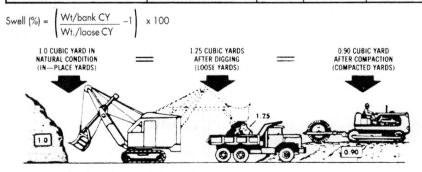

1.0 CUBIC YARD IN NATURAL CONDITION (IN—PLACE YARDS) = 1.25 CUBIC YARDS AFTER DIGGING (LOOSE YARDS) = 0.90 CUBIC YARD AFTER COMPACTION (COMPACTED YARDS)

Figure 23.7 Soil Characteristics

calculate any added fill that must be imported to the site to compensate for the compacted material. Compaction is often specified as a percentage of in-situ (maximum) density. For example, under and around structural elements such as footings, the compaction might be specified as 95% of the maximum density. For nonstructural elements such as beneath planting beds or lawns, 90% may be sufficient.

The estimator should pay special attention to the allowable thickness of the layers of backfill, the materials required, and their density after compaction. In the absence of written specifications on backfill placement and compaction, refer to sections on the architectural drawings that show the thickness of compacted materials under the slab. Trench excavation for utilities or drainage, as well as general excavation for foundations, requires backfill and compaction in some capacity.

When backfilling over or around any engineering improvement, remember to deduct the volume displaced by the feature. For example, once a precast tank is placed in an open excavation, deduct the tank's volume when calculating the amount of backfill. Also note that this calculation occurs only when the amount of material displaced by the engineering improvement is significant. (For example, it would not be necessary to deduct the volume of soil displaced by a 4"-diameter pipe placed within a trench.) A generally accepted rule is to consider deducting the volume of displaced soil when it exceeds 10% of the volume of the excavation.

Taking-off and Pricing

Backfill and compaction can be estimated by the CY, although some estimators prefer to separate backfill from compaction, because backfill is placed by machine and compaction is sometimes done by hand with the aid of a plate compactor or roller. Backfill should still be quantified by CY, and compaction by SF per layer. Most earthwork specifications are fairly definitive about how backfill should be placed. It is typically specified in layers of 6–12" called a *lift*. Each lift is compacted before another lift is placed. This process continues until the backfill is complete.

It may be necessary to add the cost of delays for testing the compacted soils at each lift. Although not usually a major factor, some on-site testing can be time consuming, especially for tests required to comply with governmental regulations. It should also be noted that the swell and compaction factors are not always exactly the same percentage. Figure 23.8 provides guidance on the compaction process.

Fill Materials

Fill, as the name implies, are the materials used in the backfill process. There are numerous types of fills that are employed on the average project during the backfill process. They are defined by their expected purpose. For example,

```
Compaction of fill in embankments, around structures, in trenches, and under slabs is important to control
settlement. Factors affecting compaction are:

    1. Soil gradation
    2. Moisture content
    3. Equipment used
    4. Depth of fill per lift
    5. Density required

            Example: Compact granular fill around a building foundation
                     using a 21" wide × 24" vibratory plate in 8" lifts. Operator
                     moves at 50 FPM working a 50-minute hour to develop
                     95% Modified Proctor Density with 4 passes.
    Production Rate:
    1.75' plate width × 50 FPM × 50 min./hr. × .67' lift
    ───────────────────────────────────────────────────── = 108.5 CY/hr.
                    27 CF per CY
    Production Rate for 4 Passes:
    108.5 CY
    ──────── = 27 CY/hr. × 8 hrs. = 216 CY/day
    4 Passes
```

Figure 23.8 Compacting Backfill

Source RSMeans Estimating Handbook, Third Edition. © 2009 John Wiley & Sons, Inc.

soils used under structural elements of the building, such as the foundation footings and the slab, are classified as *structural fill*. Other fill material used in areas to bring up the grade in less structurally sensitive locations, such as under landscape beds or lawn areas, is called *ordinary fill*. Bedding material for piping and conduits is called *sand base* or simply *sand*. *Crushed stone* is still categorized as fill material but is an aggregate, composed of broken, irregularly shaped pieces of granite or other rock in specific ranges of size such as ½", ¾", and 1½". Crushed stone is used as a bedding material for gravity lines such as drainage or sewer or as a base for precast structures. Pavement base, called *dense grade,* is a composite blend of material used under bituminous paving that allows drainage but has a mix of aggregates, gravel, and cohesive soils that provides a very dense, stable base for pavements.

There are many other fill materials used for specific applications as well as numerous regional materials not noted here. Most fill materials are classified using ASTM (American Society of Testing Materials) criteria and a sieve analysis. They are then specified in the earthwork section by their compliance with an ASTM standard.

The purchase price of fill material changes frequently based on the availability of that particular species of soil. The estimator should price fill materials reasonably close to the time the estimate is prepared for the most accurate cost.

Taking-off and Pricing

All fill materials are taken off or calculated by the cubic foot (CF) and converted to cubic yard (CY). Most are priced by the CY, but some fills, such as crushed stone or sand, can be priced by the ton (TN). Conversions from cubic yards to tons for specific materials can be found online, in *RSMeans Estimating Handbook, Third Edition,* or by the supplier. Each fill material has a different weight per CF that can be used to calculate the weight per CY. For example:

Fill with a weight of 140 lb. per cubic foot can be multiplied by 27 CF per CY to arrive at 3,780 lb. per CY.

Dividing 3,780 lb. by 2,000 lb. per ton = 1.89 TN per CY.

The estimator is reminded to add a percentage to the loose volume (LCY) delivered to the site to allow for compaction. (Refer to Figure 23.7.) Since in-ground (bank) materials are denser than loose materials, their weight per CY is higher.

Grading

The task of altering the ground surface to a designed elevation or contour is called *grading*. Grading can be accomplished by equipment, by hand, or by a combination of the two. Grading done by equipment should be priced separately from grading done by hand. Movement of large quantities of earth should not be mistakenly priced as grading, but as excavation.

Grading should be restricted to dressing previously filled or cut areas. The estimator should refer to the grading and drainage plans to quantify the area to be graded. The estimator is directed to the floor plan in the structural set to determine the area to be graded in preparation for the slab.

Grading can be classified into two main categories: rough grading and finish grading. *Rough grading* is defined as shaping the grade and could involve moving up 12" per pass with the equipment. *Fine grading* is minor fine-tuning of the grade, limited to ±2". Fine grading, especially as a subbase for a concrete slab or sidewalk, is performed with hand tools.

Taking-off and Pricing

Hand grading is taken off by the SF. Machine grading is performed by bulldozer or similar equipment and supplemented by hand grading in areas close to structures. Typically, machine grading is taken off and priced by the SF and converted to a larger unit, such as SY or acres (if the site is large). The cost of grading varies according to the degree of accuracy required and the materials being graded. Use caution in selecting the appropriate equipment for the task and be sure to consider maneuverability and the proximity of buildings or other site improvements. The estimator may want to include costs for additional equipment to spot-fill in low areas or remove excess materials in high areas. Grading accuracy is relative. Grading the substrate for a concrete slab is more critical than an area to receive a lawn. Paved areas for parking and driving must also be more accurate due to grades for drainage. The accuracy can be reflected in additional labor and machine time.

Hauling

As part of the excavation and backfill process, materials frequently need to be moved greater distances than would be economically feasible using the excavating equipment. Moving materials to an alternative location must be done by truck. The quantity of soil to be transported is calculated by the CY. Factors for increased volume from swell must be added in the form of additional truckloads. Soil materials are loaded onto a truck, hauled to a specific location, dumped, and returned to the site for the next *cycle*. Be sure to calculate the distance to be traveled; the approximate time needed to load and dump the material must be calculated to determine the number of truckloads per day, or *cycles per day*. The estimator can use this information to calculate the unit-price cost per CY. Additional costs for dumping excess material and handling at the dump site, as well as dumping (tipping) fees and waiting times, may have to be included.

Rock Removal

Rock can be removed by several methods, although drilling and splitting or blasting are the most common.

Once the rock has been excavated, it is usually removed from the site. Unearthed rock (LCY) can exceed its inground volume by as much as 60%, which must be considered when determining the number of trucks required. Rock excavation is affected by a variety of factors. Price rock excavation only after careful review of the contract documents and consideration of the following factors:

- Classification of rock – hard, medium, or soft
- Proximity of adjacent structures (affected by blasting)
- Depth of drilling
- Quantity of rock to be drilled/blasted
- Type of explosive required
- Special permits, insurance, and safety requirements

Minor rock drilling may be done by a jackhammer with carbide bits. Because each situation is different, the price of drilling and blasting should be secured from a contractor specializing in this work. The unit price of rock removal should be based on the anticipated quantity to be removed and typically decreases as the quantity increases.

Taking-off and Pricing

Takeoff of rock is calculated by its inground size, expressed in CY. When rock is encountered, it is exposed by machine and measured: length × width × depth. This quantification process is called the *pay limit*. Most contracts and specifications have specific language and obligations in the event that rock is encountered and must be removed. Unless it can be specifically quantified on the plans, rock removal is typically handled as a unit price line item. Once the rock is exposed and then quantified, it is multiplied by the unit price to determine the amount paid to the contractor. Carefully review Section 01 22 00—Unit Prices in Division 1—General Requirements of the specifications for additional guidance in estimating rock removal by unit prices.

Dewatering

High levels of groundwater during excavation prevent work from being performed and require pumping or *dewatering* of the excavation in order to perform work in the dry. Simple dewatering can usually be accomplished with the use of a pump and *sump pit,* or an enlarged hole that collects water from the excavation. This method requires localized dewatering in the immediate area of the work and maintaining the water level well below the work. In this case, the cost would include fuel, pump rental, and labor to operate the pump. The cost of dewatering could be calculated based on the time required to dewater the excavation, estimated as an allowance or lump sum, and listed in the takeoff by the day, week, or month.

Some projects require a more elaborate means of dewatering, such as a *wellpoint system,* used to dewater when large areas of earth are scheduled to be excavated. A wellpoint system is a series of small-diameter wells driven or drilled into the water table around the perimeter of the excavation. They are then connected by a pipe called a *header* and tied into a pump. The pump uses vacuum pressure to effectively lower the water table in the immediate area of the excavation. After the work has been completed and the excavated hole backfilled, the wellpoint system is removed. Factors that can affect the cost of dewatering are location, season, anticipated depth, volume of the water table, and the amount of precipitation. Dewatering large areas requires a contractor that specializes in this type of work. Secure a price from an independent contractor for this work.

Dewatering in a wet area can also be a permanent scope of work. This is typically done by wellpoint, and while similar to temporary dewatering, it has some more unique aspects to it due to its permanency. The estimator should not confuse the two scopes of work.

Taking-off and Pricing

Localized dewatering can be calculated in hours, days, weeks, or months based on the time in days over which the work will take place in the wet area. For example, a foundation wall that will take 10 days to form, place, strip, and backfill would be estimated to have 10 days of dewatering with a pump. The same basic concept is applied to the wellpoint system once the initial costs of installation and materials have been estimated. Maintenance and operation of the system may also be a cost.

Permanent systems for dewatering have a similar approach to the take off and pricing of piping systems for wells and electrical to run the system. Due to the liability involved with the performance of the system, the estimator will almost always seek a qualified subcontractor to perform the work.

EROSION AND SEDIMENTATION CONTROL

Erosion control is the practice of preventing, limiting, and controlling the erosion of soil caused by wind and water while land is undergoing construction. Erecting and maintaining effective erosion controls are essential in avoiding the contamination of water and the loss of soil. Most erosion controls on a construction site are temporary, until such time as more permanent measures are completed or until new vegetation is established.

Construction often involves the stripping of natural vegetation in preparation for new improvements. This action leaves underlying soils vulnerable to erosion. Control methods typically include erecting a physical barrier to absorb some of the energy that is causing erosion. Fences that control the runoff of soil are called *silt fences* and are used in conjunction with hay bales as a means of trapping eroded soil. Rapid-growing grasses are used to stabilize sloped areas during construction. Mechanical methods such as the use of *wattles,* or tubes of straw, can be used at catch basins to prevent eroded soils from entering the drainage systems.

This practice of keeping eroded soils on the construction site is called *sedimentation control. Sedimentation basins* are excavated and stabilized retention areas that contain the sediment before it can enter waterways or wetlands. Other mechanical methods, such as *silt sacks* attached to catch basin frames and grates, allow silt to be collected at the catch basin frame.

The estimator is directed to review the site plans and details for the location and extent of the erosion and sedimentation control. The specifications should also be consulted for specific implementation and maintenance procedures.

Taking-off and Pricing

Silt fence and hay bales are taken off and priced by the LF for installation, removal, and disposal. Wattles can be priced by the piece (EA) or by the LF, depending on preference. Silt sacks at catch basins are priced by EA. Plastic, polypropylene, and jute mesh for slope stabilization are taken off and priced by the square yard (SY).

Maintenance is typically calculated by the labor-hour (LH) per period (day, week, or month) required to clean, maintain, and repair the erosion control devices. Consider that certain seasons (periods of high precipitation or wind) require more maintenance than dry, calm seasons.

Remember to consider, as part of the estimate, the cost to remove and dispose of temporary controls when the work is complete. Some products are biodegradable and do not require removal, or they remain in place as part of the permanent system.

There are subcontractors specializing in the erection, maintenance, removal, and disposal of erosion and sedimentation control devices. Subcontractor pricing should be considered when available as they may be more familiar with any unique requirements in an area.

SHORING, UNDERPINNING, AND TRENCH BOXES

Shoring

Shoring is a mechanical means of stabilizing a slope or trench face and may be required either because of safety considerations or if there is insufficient space to stabilize the slope naturally with an incline. Shoring can be in the form of wood, steel, or concrete sheet piling. Wood sheet piling is used for depths up to 20' where there is no groundwater. Tongue-and-groove steel sheet piling is used where groundwater is encountered. Other similar artificial means, such as trench boxes, are frequently employed. Be sure to calculate the area to be retained or held back. The shored area will require overdigging to provide sufficient access.

The estimator is directed to study the specifications carefully for the requirements of the shoring, such as location and whether it has to be removed at completion. Most specifications require that a shoring plan be developed, stamped (sealed) by a professional engineer (PE), and submitted for approval to the architect.

Taking-off and Pricing

Shoring is taken off and priced by the SF of surface area of the slope or trench being retained. The area is calculated by multiplying the perimeter of the area to be retained by the height of the shoring from the bottom of the trench to the top of the excavation. The vertical dimension should include additional length to embed the shoring below the bottom of the trench level. In addition to the actual unit price for installation of the shoring, a separate line item in the estimate should include the materials, labor, and equipment to remove the shoring. Many projects require that the shoring be removed after the work has been performed and it is no longer necessary. The estimated cost for shoring and bracing should be separate from that of the work it is meant to protect.

Costs for preparation, review, and stamping (sealing) of the shoring plan by a PE should be included in the estimate. The specifications may also require that the engineer supervise the installation as well as periodic visits to inspect the integrity of the installation over time. These costs must be captured in the estimate if required.

Underpinning

When the excavation for new work comes too close to, or goes below, the foundation of an existing structure, that structure must be *underpinned*. Underpinning is the process of stabilizing or strengthening the existing foundation of a building. Underpinning can be accomplished by the use of grout, mini-piles, needle beams, or the more traditional method of *mass concrete underpinning*. Mass concrete underpinning is the excavation below the foundation and the form and placement of concrete, thereby effectively increasing the foundation depth and width. It is frequently accomplished by hand.

While underpinning is not always identified in the plans or specifications, it is a matter of safety and responsibility. Most contracts hold the contractor liable for damage to adjacent structures. Underpinning requires the existing foundation support footings to be extended or brought to the depth of the proposed structure. This is often an expensive task and may require the services of a specialized contractor if there is a considerable quantity of underpinning. Small quantities or localized underpinning applications limited to specific areas can often be handled by the site contractor. Frequently, this work is done by hand and for limited access work areas. In this application, productivity can be slow.

Taking-off and Pricing

Mass concrete underpinning can be quantified and priced by a variety of units, depending on the application. Common units include cubic yards (CY), cubic feet (CF), linear feet (LF), and lump sum (LS). Although much of the work can be done by machine, there is always handwork required for the final dressing of the area. Sometimes the hand portion of the work is best quantified and priced per labor-hour since the work is slow and detailed. Crews start with a pair and increase from there if more are required.

Trench Boxes

Trench boxes are different from shoring in that they protect the worker in lieu of supporting the trench face. The soil is not necessarily up against the trench box on the exterior as with shoring. Trench boxes can be used in open areas where a slope can be *benched* or stepped to reduce the possibility of cave-ins. Trench boxes can be moved fairly easily as the work progresses. This type of protection is called *shielding*.

Taking-off and Pricing

Trench boxes are a piece of equipment and as such are either owned by the contractor or rented. In either case, the market rate for the trench box rental is what should be estimated. Any assembly or transportation time should also be calculated along with the cost for each move. Rental periods are by the 5-day week or 21-day month. Equipment time for setting is based on the hour and piece of equipment used along with any hand labor required.

SPECIAL FOUNDATIONS AND LOAD-BEARING ELEMENTS

Caissons

Caissons, as covered in this text, are drilled cylindrical foundation shafts that function similar to short column-like compression members. They transfer superstructure loads through inadequate soils to bedrock or a load-bearing stratum. Caissons can be reinforced or unreinforced, and they either have straight shafts or are tapered to a bell at the bearing level. Shaft diameters range from 20" to 84", with the most common at 36". The maximum practical diameter at the tapered bell is three times the shaft diameter. Plain concrete is the most common material used, with rebar added for heavier loads. Belled caissons provide more bearing surface area than straight-shaft caissons, but belled caissons should be avoided for use at shallow depths and in poor bearing strata conditions. Straight-shaft caissons yield their best value when used for light loads resting on a high-value stratum. *Keyed* caissons are drilled into the bedrock stratum to create a socket, and they are used when extremely heavy loads are being transferred. Handwork at larger-diameter shafts, as well as load testing and visual inspections of the bearing strata, may be required. The concrete filling is typically intended to be placed directly against the earthen shaft of the caisson, so shaft liner forms are typically not used. (See Figure 23.9.)

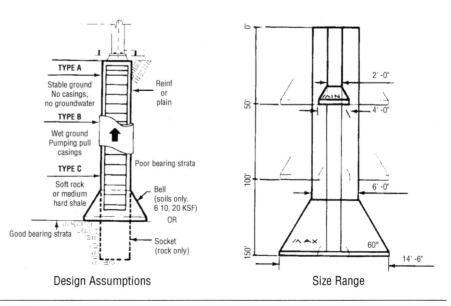

Design Assumptions   Size Range

Figure 23.9 Caissons

Source RSMeans Estimating Handbook, Third Edition. © 2009 John Wiley & Sons, Inc.

Taking off and Pricing

Caissons are taken off and priced by the vertical linear foot (VLF) for the depth of the shaft. Caissons with different shaft diameters should be segregated, as this has a large impact on price. Tapered bell caissons are more expensive due to the added cost for drilling and concreting the bell-end volume. The takeoff and pricing of the bell end are typically listed as EA, based on bell diameter, angle of the bell sides, volume, and depth.

Concrete materials, placement, and rebar for the caisson can be estimated separately by the CY, separated by strength (psi) of the concrete and the quantity of rebar in tons (TNS). This is then added to the cost of the drilling, for a complete cost per VLF. The estimator must remember to include the cost of soil (excavated from the caisson shaft) hauling and disposal if it cannot be left on-site. Cost of inspections and load testing should also be included within the estimate if required.

Piles

Piles are another means of transferring load from the superstructure to the adequate bearing strata required. Piles help to eliminate the effects of the settlement from the consolidation of overlying soils. There are three structural materials used for piles: wood, steel, and concrete. Like caissons, pile size, composition, spacing, and loading are part of the structural design provided by the engineering team. Piles can be driven by means of a pneumatic hammer, called a *pile driver*, or they can be drilled, similarly to caissons. Driving piles causes friction with the soil against the side of the pile, thereby increasing the pile's load-bearing capacity. The installation of drilled piles follows procedures similar to those for the installation of caissons, discussed in the previous section.

Other unique types of piles are *micro piles* used in underpinning and *augercast piles,* for which the drilling is done within a hollow-tube auger. When the appropriate depth is reached, a cement-grout is pumped down the shaft as the auger is removed.

Taking-off and Pricing

Piles are taken off and priced by the VLF, for the depth of the shaft. Piles of different compositions (wood, steel, concrete) are not typically used on the same project, but they should be separated if they are. Piles with different shaft diameters should be segregated, as this has a large impact on price.

Concrete materials, placement, and rebar, if required for the pile, can be estimated separately by the CY, separated by strength (psi) of the concrete and the quantity of rebar in tons (TNS). This is then added to the cost of the pile for a complete cost per VLF. Piles can be *spliced* when the pile length needed exceeds the standard length; this can be difficult, and time consuming for wood piles and should be avoided. Steel piles can be spliced with a *splicer* that eliminates welding. A splicer resembles a tapered end of a pipe that is friction-fit. Steel piles can also be welded.

The estimator must remember to include the cost of soil (excavated from the pile, if applicable) hauling and disposal if it cannot be left on-site. Cost of inspections and load testing should also be included within the estimate if required.

The work of special foundations and load-bearing elements is often highly specialized. The estimator should endeavor to secure a fixed or unit price bid from a specialist in this area whenever possible.

MISCELLANEOUS CONSIDERATIONS

There are several miscellaneous items to consider that may be required in the earthwork takeoff and estimate. These can be specified, implied, or part of typical practice. These include but are not limited to:

- Special fees (bonds) or permits for street openings or water taps
- Charges for transporting equipment (mobilization and demobilization)
- Temporary protection for open excavations
- Surveying and engineering (layout and grades)
- Police/flagman details for street work (traffic control)
- Storage or trailer charges for large items
- Registered engineering layout by a surveyor
- As-built or record drawings
- Assessments or betterment fees for new domestic services
- Barricades or access roads
- Snow plowing and removal (if applicable)
- Securing the services of a licensed arborist to oversee all tree planting, pruning, and care
- On-site geotechnical engineer for inspection of excavations
- Soils analysis: compaction testing and proctor analysis
- Stormwater protection plans
- Frost protection of open excavations in colder climates
- Maintenance for erosion control devices
- Season productivities influenced by temperature

SUMMARY

Costs for earthwork can vary from site to site, as well as from season to season. Weather extremes both hot and cold can impact productivities significantly. Estimating earthwork costs accurately requires a complete understanding of both the existing and the proposed conditions. The earthwork estimator should evaluate all information available before starting the estimate. It is strongly recommended that the estimator visit the site prior to preparing the estimate but after becoming familiar with the documents.

Earthwork encompasses a complete transformation of a parcel or property into a usable site that allows a structure to be integrated into the community it is built within. Earthwork is the precursor to the installation of utilities and drainage structures.

A large part of earthwork is excavation and its counterpart backfill. The cost of excavation is contingent on a variety of factors including depth of the excavation, soil, equipment being used, location, and application. Most excavations have a backfill component that includes compaction. Mechanical compaction to a percentage of the original density of the soil is important to prevent settlement. The estimator must consider swell and compaction of the soil when excavating and backfilling or risk incorrect quantities in the estimate.

There are some special categories of earthwork including rock removal, shoring, underpinning, and soil stabilization. Special foundations can also be part of earthwork. Piles and caissons are fairly common and should have a specialty subcontractor quote whenever possible.

Erosion and sedimentation control is a consideration each time the topsoil is disturbed or removed. The lack of vegetation and topsoil enhances the erosion during rainstorms. This must be contained on the site and prevented from entering public drainage systems.

The estimator may want to consider a checklist of miscellaneous costs to avoid omissions in the estimate.

QUESTIONS/ PROBLEMS

1. Using Figure 23.4 in this chapter with a change to the existing grades as follows (starting in the upper left proceeding clockwise): 99.5', 99.2', 100.5', and 99.4'. The new proposed elevation is 101.7' and the grid dimensions are 25' × 25'. Calculate the amount of cut or fill of the new grid number 5.

2. A trench must be dug 60' long × 5' wide from heel to heel, and 3' deep. Maintaining a 1:1 or 45° slope on each side, what is the quantity of excavated soil in BCY?

3. If the soil in problem #2 is dry gravel, what is the loose volume of soil to be hauled? Use the swell factor from Figure 23.7, "Approximate Material Characteristics."

4. From Figure 23.3, "Sample Cut and Fill Tally," what is the total quantity of soil in CY being handled? If the fill could be used in the cut areas, what would be the net import or export excluding the swell/ compaction factor?

5. Calculate the haul cost (per CY) for soil being exported from a site if each truckload will carry 20 CY, the haul distance is 7 miles at 40 mph, each load has a $40 tipping fee, and the truck and driver is $900 per day. The day is 8 hrs, with 30 mins for lunch and two 15-min breaks. Also, there is a 15-min dump time and 15-min load time for each load. Assume an 85% efficiency rate for the day.

24 | Exterior Improvements

Division 32 of CSI MasterFormat® is Exterior Improvements. The work of this division deals exclusively with the improvements to the surrounding property on which the structure is located.

Topics include paving, curbing, sidewalks, and synthetic surfacing. This division also includes site improvements such as fences and gates, retaining walls, irrigation, and plantings and landscaping.

Division 32 is part of the Site and Infrastructure subgroup and includes a wide variety of items. The contractor must carefully study all available contract documents that are related to the site, as well as any supplemental information provided.

CIVIL DRAWINGS

Most exterior improvements are shown within a special set of drawings that pertain exclusively to the site, called *civil drawings* or *site drawings*. They are designed under the direction and approval of a civil engineer and focus on the land's changes to accommodate a new structure or any engineering improvements. The most common drawings in the civil set of drawings are enumerated in Chapter 23, "Earthwork."

The quantity of drawings and level of detail vary from project to project. Each drawing in a civil set has special conventions and nomenclature unique to its discipline and provides a specific contribution to the overall civil set. However, no drawing is used in a vacuum. Each is meant to be complementary and used with the others to help determine accurate quantities for pricing.

Specific types of information are provided on each sheet in the civil set, although few drawings are used to determine the scope of the exterior improvements. The principal drawing(s) in the civil set that is used to estimate exterior improvements are as follows:

Estimating Building Costs for the Residential and Light Commercial Construction Professional,
Third Edition. Wayne J. Del Pico.
© 2023 John Wiley & Sons, Inc. Published 2023 by John Wiley & Sons, Inc.

- **Site improvement drawings:** Illustrate walkways, curbing and paving (in the absence of a dedicated plan), and the locations of bollards, benches, exterior steps, traffic signs, bicycle racks, and virtually any type of exterior improvement.
- **Paving and curbing drawings:** Illustrate the parking and driveway features of the project, including striping.
- **Site detail drawings:** Provide sections and details of features such as fencing, bollards, flagpoles, unit paving, and paving and curbing sections.
- **Landscape drawings:** Show plantings, mulch beds, lawns, and ground cover features of the site. The frequently include details of the planting and staking of plants.
- **Irrigation drawings:** Illustrate how water will be supplied to the landscape and plantings.

Each portion of the documents (plans and specs) provides information that the estimator can use to calculate the value of the exterior improvements. Consider as an example the following drawings and their respective contributions for defining the paving and curbing improvements:

- **Site detail drawing:** Shows the cross section of the various types of paving (drives, parking, and walks). It defines the thickness of each component in the section. It shows the base course of material and section through the curbing. This allows the estimator to calculate the thickness of each course of paving.
- **Paving and curbing plan:** Allows the estimator to calculate the area of the site to be paved in the varying thicknesses of walks, drives, and parking areas. It also defines the curbing location and quantity.
- **Specification Section 32-00-00—Exterior Improvements:** Defines the type and grade of paving materials and curbing to be used. It also defines the installation requirements. Additionally, this section of specification will provide detailed information as to the other items in this section.

Division 1—General Requirements must be reviewed for Unit Prices (see Chapter 2, "Understanding the Specifications") that affect the exterior improvements scope. This chapter reviews the more common elements of exterior improvements takeoff and estimating for the light commercial/ residential market.

A partial outline of work topics included in Division 32—Exterior Improvements follows.

A. 32 10 00 Bases, Ballasts, and Paving
- Base Courses
- Flexible Paving
- Rigid Paving
- Unit Paving
- Aggregate Surfacing
- Curbs, Gutters, Sidewalks, and Driveways
- Paving Specialties
- Athletic & Recreational Surfacing

B. 32 30 00 Site Improvements
- Fences and Gates
- Retaining Walls
- Site Furnishings
- Screening Device
- Manufactured Site Specialties

C. 32 70 00 Wetlands
- Constructed Wetlands
- Wetlands Restoration

D. 32 80 00 Irrigation
- Irrigation Components
- Irrigation Pumps
- Planting Irrigation

E. 32 90 00 Plantings
- Plantings
- Turfs and Grasses
- Plants
- Planting Accessories
- Exterior Planting Support Structures
- Transplanting

BASES FOR PAVING

All types of paving – flexible or rigid, cast-in-place, or unitary – require a solid support material to absorb the deflection and loads they carry. This solid material is called a *base*. Bases can be composed of a variety of materials often based on local practice.

Bases

Base refers to gravel or a similar material under a *binder* or initial course of pavement. Typically, bases for paving consist of dense grade (see Chapter 23), gravel, or even reclaimed asphalt to be used under the binder course of bituminous or concrete paving. The base can be estimated under the excavation and backfill portion of the work because the work is performed by the site contractor, but its formal location in CSI MasterFormat® is in Division 32. The rough grading or preparation of the base is categorized under the grading scope of work. Fine grading and rolling of the base just prior to base course of paving are typically done by the paving contractor. The estimator should review the paving and curbing plan or the drawing that shows the paving to calculate area and the site details to determine thickness.

Taking-off and Pricing

The base course under the binder course of paving is taken off by the square foot (SF) and extended to the square yard (SY) for pricing of placement. The material portion can also be converted to cubic yards (CY; see Chapter 23, "Earthwork") to determine the quantity of imported materials required, since base material is typically a processed product. The estimator should remember to increase the quantity of imported materials to account for a reduction of in-place materials due to compaction. Readers are again directed to Chapter 23, "Earthwork," for the discussion of compaction and bringing the base up in lifts.

Fine grading of bases for pavement is done by means of a large piece of equipment called a *grader*. The area to be graded is taken off and priced by the SY. The cost of fine grading is additional to placement, rough grading, and any compaction of the initial placement of the base.

The area of the base material is often enlarged slightly to provide stability to the paving. The estimator should consult all plans showing walks, drive and parking areas, and patios with paving. The takeoff and pricing of the walkways, drives, and parking areas are done separately as they often have different base requirements, thicknesses, and some may require handwork versus equipment. Coordinate with other areas of the estimate such as Division 1—General Requirements, to ensure that the cost of establishing grades by a surveyor is included elsewhere within the estimate.

Prior to the placement and rough grading of the base material it is not uncommon for the substrate to be compacted with a roller. This is called *proof rolling* and is done to ensure that substrate is adequate to hold the base and to identify any soft or excessively wet areas.

FLEXIBLE PAVING

Paving refers to surfacing a subbase, typically compacted gravel or stone, with a course of materials, such as bituminous concrete, brick, or concrete. Paving types can be classified as flexible and rigid. *Flexible* paving is defined as a roadway or surface that yields elastically to the traffic load on it. Most flexible pavements have a common component: *asphalt*. Asphalt, the binding component, is mixed with a mineral aggregate, then placed in layers, resulting in a bituminous concrete surface that is flexible. Bituminous concrete or asphaltic paving is usually applied in two courses. The first course is called the *binder,* or base course. The top course is called the *wearing* course. Each course should be taken off and priced separately. Each mix has a different composition and price. It is not uncommon for the binder to be placed early in the project, with the wearing surface applied just prior to project turnover.

Paving large areas is typically done by a full crew using multiple pieces of equipment. This work often has an associated mobilization charge. A more modest crew size can be used for small areas, which also require less and smaller equipment. More time-consuming work, such as leveling existing low spots prior to the top course, applying a tack course, and blending new and existing work, should all be taken off and priced separately after careful consideration of the specific application and site conditions. Trenches that require patching should be taken off and priced by the SY, but this information should be kept separate from the larger production quantities. Paving contractors often have a minimum charge for work such as patches. Include the cost of sweeping between courses, if the binder has been down a long time, before applying the tack coat and wearing surface.

In the absence of a paving and curbing plan, the estimator should consult site drawings that show the parking areas, drive, and bituminous concrete paved sidewalks to determine quantities. Course thicknesses should be shown on the site details plan.

Preparatory Coats and Procedures

There are two types of preparatory coats: the *prime* coat is applied over the gravel base before the binder is laid, and the *tack* coat is a very light application of asphaltic emulsion sprayed on an existing paved surface. A tack coat is used to ensure a good bond between the existing pavement surface and the new layer being installed. Tack coats are used between successive lifts, or *courses*, of pavement. The emulsion consists of some basic ingredients: paving asphalt, water emulsifying agent, and polymers.

The surface to receive the tack coat must be swept thoroughly before the application of the emulsion. This can be done with large street sweepers, walk-behind sweepers, or brooms.

The materials and application procedures for prime and tack coats are typically specified in the paving section of the specifications and may be referenced only by note, if at all, on the plans.

Taking-off and Pricing

The tack coat is taken off by the square foot (SF) of surface area that is to be coated. The total is extended to square yards (SY). This can then be converted to gallons (GL) for pricing the materials. The coverage in SY per gallon, called the *application rate,* is defined in the specifications. Application rates vary with the condition of the existing surface. Figure 24.1 provides takeoff information for prime and tack coats applied in the paving process.

The surface area to receive a tack coat is the same as the area to be swept and should be listed in the estimate by the SY. Quantities can be added for each course. Again, a checklist may be helpful so that the estimator includes all items.

Asphalt Paving Binder Course

The first course of pavement placed on the surface of the gravel base is called the *binder course*. The binder course makes up the bulk of the paved system and carries most of the traffic load. The aggregate in the binder course is

Cost per GAL	Application Rate, GAL Per SY					
	0.05	0.10	0.15	0.20	0.25	0.30
4.00	0.20	0.40	0.60	0.80	1.00	1.20
4.20	0.21	0.42	0.63	0.84	1.05	1.26
4.40	0.22	0.44	0.66	0.88	1.10	1.32
4.60	0.23	0.46	0.69	0.92	1.15	1.38
4.80	0.24	0.48	0.72	0.96	1.20	1.44
5.00	0.25	0.50	0.75	1.00	1.25	1.50
5.20	0.26	0.52	0.78	1.04	1.30	1.56
5.40	0.27	0.54	0.81	1.08	1.35	1.62
5.60	0.28	0.56	0.84	1.12	1.40	1.68
5.80	0.29	0.58	0.87	1.16	1.45	1.74
6.00	0.30	0.60	0.90	1.20	1.50	1.80
6.20	0.31	0.62	0.93	1.24	1.55	1.86
6.40	0.32	0.64	0.96	1.28	1.60	1.92
6.60	0.33	0.66	0.99	1.32	1.65	1.98
6.80	0.34	0.68	1.02	1.36	1.70	2.04
7.00	0.35	0.70	1.05	1.40	1.75	2.10
7.20	0.36	0.72	1.08	1.44	1.80	2.16
7.40	0.37	0.74	1.11	1.48	1.85	2.22
7.60	0.38	0.76	1.14	1.52	1.90	2.28
7.80	0.39	0.78	1.17	1.56	1.95	2.34
8.00	0.40	0.80	1.20	1.60	2.00	2.40
8.20	0.41	0.82	1.23	1.64	2.05	2.46
8.40	0.42	0.84	1.26	1.68	2.10	2.52
8.60	0.43	0.86	1.29	1.72	2.15	2.58
8.80	0.44	0.88	1.32	1.76	2.20	2.64
9.00	0.45	0.90	1.35	1.80	2.25	2.70
9.20	0.46	0.92	1.38	1.84	2.30	2.76
9.40	0.47	0.94	1.41	1.88	2.35	2.82
9.60	0.48	0.96	1.44	1.92	2.40	2.88
9.80	0.49	0.98	1.47	1.96	2.45	2.94

Figure 24.1 Cost per SY of Prime and Tack Coats

Source RSMeans Estimating Handbook, Third Edition. © 2009 John Wiley & Sons, Inc.

typically larger, with a lower asphalt content. The lift is also thicker in the binder course. The binder course is designed to provide a stable base for the wearing surface.

Asphalt Paving Wearing Course

The top course of paving is called the *wearing surface*. This layer is composed of finer aggregates with a higher asphalt content. The thickness of the layer depends on the expected load it will carry. As the name implies, this course bears the brunt of the wear from weather and traffic and can be replaced or resurfaced if the wear becomes excessive.

Taking-off and Pricing

The binder and the wearing course of paving are taken off by the SF and extended to the SY for pricing. The cost of paving can vary dramatically, depending on the type of mix specified, the thickness required, and the number of placement layers. Takeoff and pricing for the installation of binder courses at walks and smaller areas should be performed separately from that done for larger areas such as driveways and parking areas. The unit of measure used in pricing is SY in either case. While both are considered paving, they are frequently performed by separate crews and equipment with different pricing structures.

Additional work for tying in new surfaces to existing surfaces may be required and should be listed separately, taken off and priced by the linear foot (LF) of abutting surfaces. Tie-ins may require sawcutting or jackhammering of the existing surface. Some details require that a lap of the new wearing surface extend over the existing binder at a seam. This constitutes additional work that can be labor intensive.

The estimator should consider additional paving for filling low spots in the base course if settlement occurs. This is typically carried as tons (TNS) of paving. The conversion of tons of asphalt to square yards is dependent on the thickness in inches and the mix type. A general rule is that asphalt weighs between 140 lbs. per CF and 148 lbs. per CF, so it ranges from 3,780 lbs. per CY to 3,996 lbs. per CY. From there the estimator can calculate the volume and weight depending on the thickness of the course. This is a general rule and should only be used for guidance. Exact weights will vary with the recipe of the mix and the type of aggregate used regionally.

Note that frequently the estimator encounters a federal or state specification Department of Transportation (DOT) number in the documents that defines the bituminous concrete mix required. It is a common practice to specify the binder and wearing surface in accordance with local public roadway specifications on a state-by-state basis. The estimator can find the composition of the particular state DOT specification online.

Seal Coats

Despite its flexible, waterproof, and adhesive properties, asphalt is subject to deterioration from salts, chemicals, and weather that attacks and destroys the binding qualities of the pavement. To extend the life of the pavement, a *seal coat* is applied to the surface. A seal coat is made of refined coal tar or asphalt and is applied as an emulsion on the finished surface of the paving. It acts as a barrier to salt, weather, and chemicals.

Seal coat materials and application procedures are typically specified in the paving section of the specifications and may be referenced only by note on the plans.

Taking-off and Pricing

Seal coat is taken off by the SF of surface area that is to be coated. The total is then extended to SY. The coverage in SY per gallon, called the *application rate,*

Description	Labor-Hours	Unit
Subgrade, grade and roll		
Small area	0.24	SY
Large area	.011	SY
Base course		
Bank run gravel, spread compact		
6" deep	.004	SY
18" deep	.013	SY
Crushed stone, spread compact		
6" deep	.016	SY
18" deep	.029	SY
Asphalt concrete base		
4" thick	.053	SY
8" thick	.089	SY
Stabilization fabric, polypropylene, 6 oz./SY	.002	SY
Asphalt pavement, wearing course		
1½" thick	.026	SY
3" thick	.052	SY

Figure 24.2 Installation Time in Labor-Hours for Asphalt Pavement

Source RSMeans Estimating Handbook, Third Edition. © 2009 John Wiley & Sons, Inc.

is defined in the specifications. This can then be converted to gallons (GAL) based on the area to be sealed for pricing the materials. Application rates vary with the condition of the existing surface. Figure 24.2 provides guidance on the installation of asphalt pavement.

The estimator should keep in mind that asphalt mix, and by-products such as prime, tack and, seal coats, are all petroleum based and are tied to the price of oil. Prices for these products change frequently and the individual product should be priced as close to bid day or time as possible.

RIGID PAVING

The second type of paving is defined as *rigid* paving or paving that is not intended to be flexible and allows very little deflection. Examples include concrete patios, sidewalks, precast concrete, and brick pavers. Rigid paving/pavers require a gravel or similar base just like flexible paving. This creates the stratum that absorbs and transfers a large portion of the load into the ground below.

Rigid paving may be shown on site improvement plans or paving and curbing plans. Site details should be reviewed as well for specific type, composition, and thickness. Note that some rigid pavers are installed by hand and as a result can be labor intensive.

Concrete Paving

Concrete paving includes forming and placing concrete for walkways, patios, sidewalks, aprons, and similar features. It involves calculating the CY of concrete necessary, plus a small percentage for waste resulting from spillage of concrete while handling. The composition of the concrete is specified in Division 3—Concrete of the specifications, along with the reinforcing requirements. Concrete mixes for pedestrian paving are often of higher compressive strengths and have additives to resist deterioration. Also, remember that cast-in-place concrete requires some type of formwork. Formwork at the perimeter of the pour is called the *edge form* and is quantified by the LF. Concrete paving also requires finishing at the surface. It can be as simple as a *broom finish,* whereby a broom is used to create a nonslip surface, or *stamped* concrete, where rubber mats are used to stamp an impression in the surface of the concrete.

Taking-off and Pricing

Concrete should be quantified and priced by the CY. Placement and finishing of the surface should be taken off and priced by the SF of surface area finished. The cost varies depending on the type of finish required. The quality and strength of the concrete, including additives or hot water, should be noted, as these factors affect the unit cost. Concrete with color additives should be listed separately in the takeoff as this has an impact on cost.

Edge form at the perimeter should be taken off and priced by the LF. Pricing should include the cost of stripping the form. Materials for edge form can often be used on multiple jobs, and costs may be divided by the number of expected uses, usually limited to 4.

Reinforcing for concrete walkways/patios in the form of welded wire fabric is calculated by the SF, with allowances for overlap determined by the details on the drawings. Rebar should be calculated by weight and extended to tons (TNS) for pricing. Walkways attached to structures may also require dowels to pin the walk to the structure. Holes are drilled into the foundation, and dowels are placed, or epoxied, between the walkway and foundation.

Special curing compounds or protection from rapid evaporation of water should be listed separately and priced by the SF of surface area. In the case of liquid compounds, the SF can be converted to coverage of GAL/SF and to the manufacturer's units, such as 5-gallon containers. Similarly, polyethylene sheeting material to retain moisture can be calculated by the SF and converted to the convenient size roll. Additional labor may be required to manually ensure proper hydration of the concrete.

(Note: Readers are directed to Chapter 8, "Concrete," for a more thorough discussion of estimating concrete.)

Concrete, Brick, and Stone Pavers

Another class of rigid paving is called *unitary paving*. More commonly known as *landscape pavers,* these are individual units in a variety of sizes and thicknesses, which are set in a pattern on a gravel or sand base. They are typically a manmade product, although natural versions are available as well.

Unitary paving should be illustrated on the site improvement plans, along with setting details on the site details drawing if available. Alternatively, the specifications should provide characteristics of the pavers: size, shape, manufacturer, pattern, and so on, to aid in pricing. Pricing for pavers is typically higher than brick due to the manufacturing process. Pavers are denser and less absorbent of moisture to protect against the freeze-thaw cycle of cold weather.

Taking-off and Pricing

Unitary paved walks and patios are quantified by the SF of paved area, converted to the total quantity of bricks needed based on the number of bricks per SF. This varies from manufacturer to manufacturer, based on the size of the paver and the pattern. The associated labor is often calculated based on the number of pavers an installer can lay in a day. A reasonable allowance for waste should be included for breakage and cuts. Rounding the quantity to the manufacturer's standard pallet size may result in a scale of economy discount. Unusually shaped patios or patterns may require more cutting of the brick. Brick is cut by brick saws that are electric or gasoline powered. They have a diamond or carborundum blade that does the cutting. Large installations should include additional blades as part of the cost of the equipment as well as a separate individual just to run the saw and make the cuts for those laying the brick. This results in a more efficient use of labor.

Figure 24.3 provides guidance on the installation of various paving materials.

CURBS AND WALKWAYS

The perimeters of parking areas and walkways are often surrounded by *curbing*. Curbing is used to provide a vertical transition between parking areas or drives and pedestrian or landscaped areas. Curbing is available in many types and compositions; each type is taken off and priced separately.

Curbing can consist of individual units such as precast concrete and granite, or it can be extruded on-site composed of asphalt paving or cast-in-place concrete.

Walkways are used to separate pedestrian traffic from vehicular traffic. They are constructed of asphalt paving, cast-in-place concrete, or unitary pavers such as brick or stone.

The estimator should review the site improvement drawings or, if available, the paving and curbing plan. Also, the site details plan should be consulted for the size and cross section of curbing and the method of stabilizing it once in place. These same plans should show pedestrian walkways in both plan and section.

Description	Labor-Hours	Unit
Brick paving without joints (4.5 brick/SF)	.145	SF
Grouted, 3/8" joints (3.9 brick/SF)	.178	SF
Sidewalks		
Brick on 4" sand bed		
Laid on edge (7.2/SF)	.229	SF
Flagging		
Bluestone, irregular, 1" Thick	.198	SF
Snapped random rectangular 1" thick	.174	SF
1½" thick	.188	SF
2" thick	.193	SF
Slate		
Natural cleft, irregular, ¾" thick	.174	SF
Random rectangular, gauged, ½" thick	.152	SF
Random rectangular, butt joint, gauged, ¼" thick	.107	SF
Granite blocks, 3½" × 3½" × 3½"	.174	SF
4" to 12" long, 3" to 5" wide, 3" to 5" thick	.163	SF
6" to 15" long, 3" to 6" wide, 3" to 5" thick	.152	SF

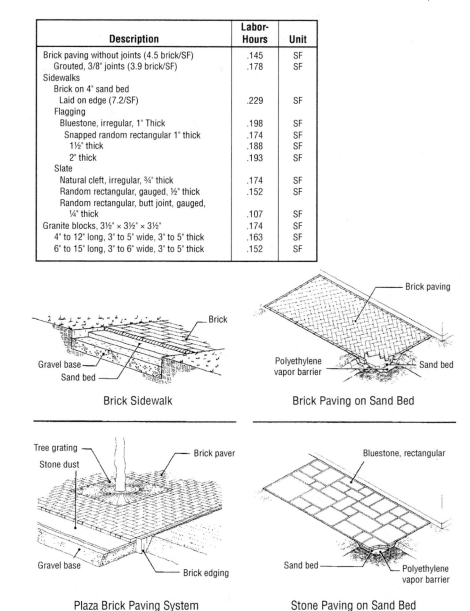

Brick Sidewalk

Brick Paving on Sand Bed

Plaza Brick Paving System

Stone Paving on Sand Bed

Figure 24.3a Installation Time in Labor-Hours for Brick, Stone, and Concrete Paving

Source RSMeans Estimating Handbook, Third Edition. © 2009 John Wiley & Sons, Inc.

Curbs

Curbing is taken off and priced by the LF. Curbing manufactured off-site, such as granite or precast concrete, often requires mortaring of the joints between each section once the curbing has been set. This step should be included as part of the installation cost. Radii for curved precast concrete or granite curbs are listed separately. These are more expensive than straight pieces. The cost of materials should include delivery charges, if needed. In an attempt to keep curbing in place, many plans and specifications require a ready-mix concrete

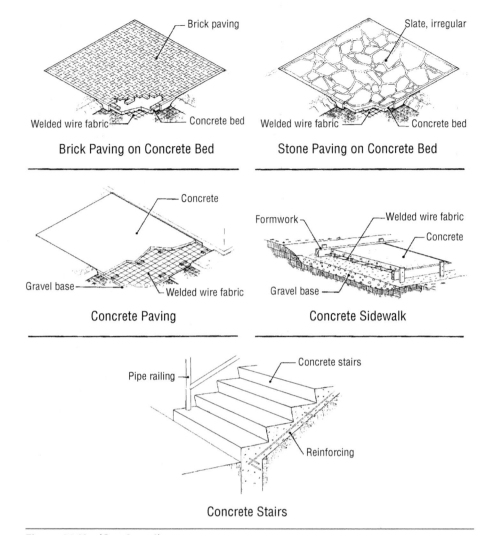

Brick Paving on Concrete Bed

Stone Paving on Concrete Bed

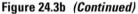

Concrete Paving

Concrete Sidewalk

Concrete Stairs

Figure 24.3b *(Continued)*

setting bed or concrete backup. Review the documents carefully as this can add significant costs to the installation.

Because of its weight, manufactured sections of curbing require equipment for setting. As noted, some drawings (or specifications) provide for concrete to be placed at the base to hold the curb in place. This should be included in the estimated cost. Ready-mix concrete for holding the precast or granite curb in place should be taken off and priced by the CY.

Other types of curbing include cast-in-place concrete, extruded bituminous concrete, and extruded concrete (for curbing and gutters). These are also taken off and priced by the LF. They are typically larger crews with more equipment such as extruding machines.

Taking-off and Pricing

Cast-in-place curbing is taken off and priced by the LF. The cost for bituminous concrete and concrete curbing extruded by machine is based on the production (quantity) rate and type of material used and is taken off by the LF. All extruded curbing, concrete or bituminous, is based on the production (quantity) rate, setup and moves, small quantities, and the accessibility of the work, all of which affect unit costs. The concrete for these types of curbing should be calculated in CY based on the linear footage. The amount of concrete or bituminous material required per linear foot of curb can be calculated using the cross-section on the plan. The estimator may include waiting time for ready-mix concrete trucks, as cast-in-place and extruded curbing tends to move more slowly. Cast-in-place curbing requires the use of forms. The unit cost should include forming, placing, stripping, and rubbing of the exposed surfaces. Figure 24.4 provides guidance on installation for various types of curbing.

Walkways

Rigid paving at sidewalks is most often constructed of cast-in-place concrete. As such, it generally requires some type of formwork and reinforcing. Note that edge form may be required on only one side if the sidewalk is cast up against curbing. Due to its high length-to-width ratio, most sidewalks require expansion joints and control joints at regular intervals. In addition, the surface is typically broom-finished to prevent slipping.

Sidewalks can be asphalt paving or brick pavers as well. Both of these systems along with the cast-in-place require a base on which to install the walking surface. For more information on calculating the base for walkways, refer to "Bases for Paving" previously discussed in this chapter.

Sidewalks are typically shown on paving and curbing or site improvement plans. Site details should be consulted for sections and control and expansion joint locations and type.

Taking-off and Pricing

Edge form is taken off and priced by the LF. For cast-in-place concrete walkways, the welded wire fabric (mesh) and finishing of the surface are taken off and priced by the SF of surface. The estimator must add a lap for the welded wire fabric. For walkways less the 5' in width, the lap is along the length of the walk only. Concrete material is taken off and priced by the CY and is typically higher strength (psi) than concrete use at the interior. The mix may also include accelerators or retarders depending on the ambient temperature conditions. This will typically translate to a higher cost per CY. Curing of the surface is also taken off and priced by the SF and can be accomplished by means of covering to maintain moisture or by a sprayed membrane. Both require that an additional labor cost be added to the placement and finishing. (See Chapter 8, "Concrete," for a more thorough discussion of estimating concrete.)

Walkways surfaced with asphalt paving are taken off by calculating the area, and then converting to a pricing unit of SY. This follows a similar process to

Description	Labor-Hours	Unit
Curbs, bituminous, plain, 8" wide, 6" high,		
50 LF/ ton	.032	LF
8" wide, 8" high, 44 LF/ton	.036	LF
Bituminous berm, 12" wide, 3" to 6" high,		
35 LF/ton, before pavement	.046	LF
12" wide, 1½" to 4" high, 60 LF/ton,		
Laid with pavement	.030	LF
Concrete, 6" × 18", cast-in-place, straight	.096	LF
6" × 18" radius	.106	LF
Precast, 6" × 18", straight	.160	LF
6" × 18" radius	.172	LF
Granite, split face, straight, 5" × 16"	.112	LF
6" × 18"	.124	LF
Radius curbing, 6" × 18", over 10' radius	.215	LF
corners, 2' radius	.700	EA
Edging, 4½" × 12", straight	.187	LF
Curb inlets, (guttermouth) straight	1.366	EA
Monolithic concrete curb and gutter		
Cast-in-place with 6'-High curb and		
6"-thick gutter		
24" wide, .055 CY per LF	.128	LF
30" wide, .066 CY per LF	.141	LF

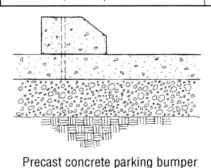

Precast concrete parking bumper

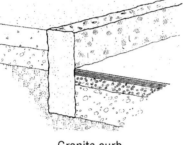

Granite curb

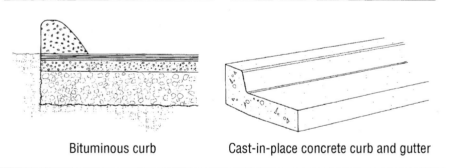

Bituminous curb

Cast-in-place concrete curb and gutter

Figure 24.4 Installation Time in Labor-Hours for Curbs

RSMeans Estimating Handbook, Third Edition. © 2009 John Wiley & Sons, Inc.

estimating asphalt (flexible) paving already discussed. Walkways, however, should be kept separate from driveway and parking area quantities since the mix is often different. Thicknesses are typically less than paving in driveways or parking areas.

For brick or stone pavers, the estimator is directed back to the section on "Concrete, Brick or Stone Pavers." The process is the same as for other applications.

PAVING/PARKING SPECIALTIES

Many site improvement and paving and curbing drawings require parking or paving specialties. These include parking bumpers (wheelstops), bollards, and speed bumps. Each has the purpose of containing vehicles or preventing them from causing damage to nearby structures. The estimator should consult Division 32 of the specifications for specific models, manufacturers, and characteristics of these items, as they can vary dramatically in price.

Parking Bumpers

Parking bumpers, sometimes called wheelstops, are approximately 6" × 10" × 6'-0" long. They are made of reinforced precast concrete and have holes cast from the top to the bottom for attaching to the pavement with rebar, similar to how one would nail one landscape timber to another. Thermoplastic models are available as well.

Bollards

Bollards are used to prevent vehicle damage to site items such as transformers. They can be nondecorative or ornamental. Nondecorative models are steel pipe filled with concrete. They are 4–12" in diameter with lengths of 7–10'. They are set in concrete, with approximately 4' exposed above grade. Decorative or ornamental bollards can be ornate and expensive and are often linked by chains.

Speed Bumps

Speed bumps are made of thermoplastic and vary in length. They are attached to the pavement with spikes and provide the benefit of slowing traffic in congested areas of parking lots or drives.

Pavement Markings

Pavement markings and striping are used to control traffic and designate parking spaces. They are composed of a variety of different coatings ranging from waterborne acrylic to thermoplastic materials with glass beads. Striping can vary in color, with white and yellow being the most popular; they are typically 4–12" wide. Symbols such as directional arrows and wheelchair emblems are also part of the average project. Fire lanes and restricted parking areas can use words and cross-hatching for identification. Note that pavement markings are frequently used in conjunction with parking signage. See Chapter 15, "Specialties," for more information on estimating signage.

Taking-off and Pricing

Parking bumpers are taken off and priced by the individual piece (EA). Size and composition have an impact on price. Bollards are also taken off and priced by EA. They should be segregated by size (diameter and length) and according to whether they are nondecorative or ornamental, as prices may be substantially different for both material and installation. Because bollards and concrete parking bumpers are heavy, some type of equipment may be required to handle, distribute, and install them. Concrete for setting bollards should be calculated and priced by the CY, with consideration for small quantities or short loads (see Chapter 8, "Concrete"). Coordinate excavation for bollards with Division 31—Earthwork, as this task is typically specified in that division. Bollards that require paint should be coordinated with Division 9—Finishes to ensure that the painting scope has been included.

Speed bumps are taken off and priced by EA, based on length, composition, and style.

Pavement striping is taken off and priced by the LF based on composition, color, and width. The estimator should include the cost of layout for parking space lines and cross-hatching. Individual symbols are taken off and priced by the EA based on type, color, and size.

SITE IMPROVEMENTS

Site improvements include a wide range of miscellaneous items and tasks. These can be shown on the site plan or on a separate site improvement plan. The site improvement drawing can be a catchall drawing. It is provided in plan view and shows items such as fencing, playground equipment, trash receptacles, benches, decorative water fountains, planters, fixed outdoor furniture, bicycle racks, flagpoles, and so forth. Division 32—Exterior Improvements technical section of the specifications should provide the manufacturer, model, size, or other characteristics that will allow the estimator to accurately price the product. Whenever possible, secure quotes for specialty equipment typically found on this drawing. Some site improvement items may require assembly or incur shipping costs. This should be noted in the takeoff for accurate pricing.

Fences and Gates

There are numerous types and sizes of fencing used to enclose property. The estimator should consult the site improvement drawing to determine the quantity of each type. Additional details should be reviewed for depth of posts and any anchoring materials such as concrete. Specifications should be reviewed for special considerations such as spacing of posts, requirement for top rails, security tops (barbed wire or razor wire), decorative slats, and similar additional features.

Fences can be PVC or wood as well. Often, fences are more than decoration and may be required to resist certain lateral forces caused by pedestrians. The estimator is urged to review the specifications carefully so that the appropriate fence is priced.

For motorized gates, the estimator should coordinate the pricing with the electrical scope of work for power, control wiring, and tie in to the panel. Excavation and backfill from the structure to the gate to house the conduit may also be required. Consult Division 31—Earthwork for guidance on estimating trench excavation and backfill.

Many projects have temporary fencing to secure the site. This type of fencing is specified in Division 1—General Requirements as a temporary facility. The two types are different and should not be confused. Most temporary fencing has an installation and removal cost along with a monthly rental cost.

Figure 24.5a provides installation times for estimating costs for various types of fencing. Figure 24.5b shows the elements of a chain-link fence.

Retaining Walls

Retaining walls can be constructed of timber, cast-in-place concrete, segmental masonry blocks, or stone. The estimator should consult the site improvement drawings and site details for length, height, composition, and any special filter

Description	Labor-Hours	Unit
Fence, chain link, industrial plus 3 strands Barbed Wire, 2" line post @ 10' O.C.		
1-5/8" top rail, 6' high	.096	LF
Corners, add	.600	EA
Braces, add	.300	EA
Gate, add	.686	EA
Residential, 11-gauge wire, 1-5/8" line Post @ 10' o.c. 1-3/8"		
Top Rail, 3' high	.048	LF
4' high	.060	LF
Gate, add	.400	EA
Tennis courts, 11-gauge wire, 1¾" Mesh, 2½" line posts, 1-5/8" top rail,		
10' high	.155	LF
12' high	.185	LF
Corner posts, 3" diameter, add	.800	EA
Fence, security, 12' high	.960	LF
16' high	1.200	LF
Fence, wood, cedar picket, 2-rail, 3' high	.150	LF
Gate, 3'-6", add	.533	EA
Cedar picket, 3-rail, 4' high	.160	LF
Gate, 3'-6", add	.585	EA
Open rail rustic, 2-rail, 3' high	.150	LF
Stockade, 6' high	.150	LF
Board, shadow box, 1" × 6" treated pine, 6' high	.150	LF

a

Figure 24.5a Installation Time in Labor-Hours for Fencing

Source RSMeans Estimating Handbook, Third Edition. © 2009 John Wiley & Sons, Inc.

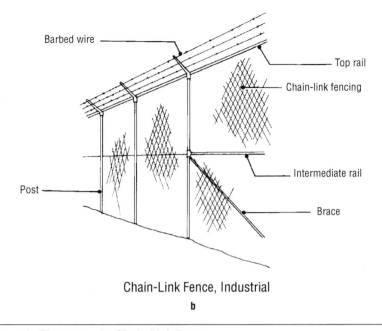

Barbed wire — Top rail

Chain-link fencing

Intermediate rail

Post — Brace

Chain-Link Fence, Industrial

b

Figure 24.5b Elements of a Chain Link Fence
Source RSMeans Estimating Handbook, Third Edition. © 2009 John Wiley & Sons, Inc.

fabric or stone filter media behind the wall. Excavation and backfill for retaining walls are typically estimated under Division 31—Earthwork. Specifications should be consulted for specific characteristics of the wall such as color of segmental blocks, type of stone, manufacturer's model name/ number, and treatment of the wood timber, as this information is not always available from the drawings.

Structural drawings and details should be checked for sections through cast-in-place concrete retaining walls that would indicate reinforcing patterns and sizes of rebar.

Note that any sheet or timber shoring required during the installation of the retaining wall is typically specified and estimated under Division 31— Earthwork, which should be checked to ensure coordination of the work. Stone or gravel fill behind the retaining wall is also estimated in Division 31.

Occasionally, the technical specifications may require a structural engineer to design and stamp the shoring design and retaining wall as part of the work. In addition to Division 32, the estimator is urged to check other parts of the contract documents for design responsibilities.

Taking-off and Pricing

Fencing is taken off and priced by the LF. Fencing should be separated in the estimate according to type (chain link, wood, etc.), size (height), and style (residential, industrial, security, snow fence). Gates should be taken off for each type and quantified as individual units (EA), including any special hardware such as motorized gate operators. Concrete for anchorage of the fence posts should be calculated by the CY, again keeping in mind costs for short loads. For small numbers of posts, hand-mixed concrete from dry bags may be more cost effective.

Cast-in-place concrete retaining walls follow the same procedure for foundation walls detailed in Chapter 8, "Concrete," with the notable exception that the exposed surfaces of the retaining walls may need to be rubbed. *Rubbing* is the process of light hand grinding with a stone float while the concrete is still green. A thick cement slurry is applied with a burlap rag to fill minor imperfections. Rubbing is calculated by the SF of surface to be rubbed and can be labor intensive.

Wood timber walls are taken off by the SF of surface area used to retain the soil. The area is then converted to the size of the timber being used. Remember to include multiple thicknesses of wall and any perpendicular members used as lagging.

Segmental masonry block walls are taken off by the SF of surface area used to retain the soil. Individual manufacturer's literature should be consulted for converting the SF area to the quantity of block required. Given the variety of products available, a general guideline for conversion would not be accurate. The estimator should source information about the specified product.

Natural stone walls should follow the takeoff procedure identified in Chapter 9, "Masonry." The area of the surface should be multiplied by the average depth (thickness) of the wall, a waste factor should be added, and the resultant cubic footage should be converted to tons (TNS) for pricing of both the materials and the labor. Do not forget to include costs for shipping and mortar if appropriate.

Filter fabric behind the wall is calculated by the SF of area and can be converted to the manufacturer's standard sales unit, such as rolls.

Miscellaneous items such as benches, flagpoles, and trash receptacles should be taken off by the unit of measure that most appropriately identifies and quantifies the item. For most of these items, that unit is each (EA). Any level of assembly required or anchoring of the item to surfaces should also be detailed, as this affects the labor portion of the price.

Irrigation Systems

The landscape plan may be accompanied by an irrigation plan for maintaining lawns and plantings. An irrigation plan indicates the placement of sprinkler heads for watering. It may show pipe runs, manifolds with zone valves, and a backflow preventer connected to the source. The main components of an irrigation system are the polyethylene pipe, irrigation heads, control valves, and low-voltage wire. The estimator should consult the plumbing and electrical takeoffs to ensure that connection to the water source and line voltage power to the controller have been included in the estimate. The estimator should consult the specification section on irrigation for items that may not be shown on the plans or for coordination with other trades. Items such as PVC sleeves to pass under lawn or walkways to allow the passage of irrigation pipe and wire may also be needed, yet not shown.

Taking-off and Pricing

The quantity of polyethylene pipe is calculated by the LF and priced according to size (diameter). Control (zone) valves, irrigation heads, and fittings are taken off and priced by the piece (EA). Miscellaneous components, such as the control panel, backflow preventers, flow meters, rain sensors, valve boxes, and so on, are also taken off and priced by EA. Many of the controllers are Wi-Fi enabled and may require access to the internet and programming. Sleeves are taken off and priced by the LF based on the diameter of the sleeve.

Special equipment for laying the pipe below grade is often needed. Trenching machines that excavate, place the pipe, and backfill require an operator and are estimated at a daily rate based on anticipated production. Labor to follow behind the trenching machine to remove tailings or to grade the topsoil should also be included. Special connections to potable water systems include devices such as vacuum breakers and backflow preventers. These items are taken off and priced by EA. For irrigation systems supplied by a well, the specifications or local health department may require water testing as a condition of connection. This should be identified in the specifications.

See Figure 24.6 for installation times for estimating site irrigation costs.

Wetlands

Many residential and even commercial projects can abut wetland areas. As a result, community conservation commissions can impose regulations requiring the reclaiming or restoring of disturbed wetlands. This process is beyond the expertise of all but a few site contractors, and almost all general contractors. This makes estimating the work even more of a challenge.

Frequently, the technical specifications will defer to, or include within the section, an *order of conditions* for how the work will be handled. An order of conditions is a type of permit issued by the conservation commission having jurisdiction over the work. It can require a botanist or similar environmental scientist to oversee the restoration or replication identified in the order. The work can be seasonal and may not coincide with the execution of the general construction contract. It may require lengthy tracking of growth and establishment of the restored area.

Description	Labor-Hours	Unit
Sprinkler system, golf course, fully automatic	.600	9 holes
12' radius heads, 15' spacing		
Minimum	.343	Head
Maximum	.600	Head
30' radius heads, automatic		
Minimum	.857	Head
Maximum	1.040	Head
Sprinkler heads		
Minimum	.267	Head
Maximum	.320	Head
Trenching, chain trencher, 12 HP		
4" wide, 12" deep	.010	LF
6" wide, 24" deep	.015	LF
Backfill and compact		
4" wide, 12" deep	.010	LF
6" wide, 12" deep	.030	LF
Trenching and backfilling chain trencher, 40 HP		
6" wide, 12" deep	.007	LF
8" wide, 36" deep	.010	LF
Compaction		
6" wide, 12" deep	.003	LF
8" wide, 36" deep	.005	LF
Vibrating plow		
8" deep	.004	LF
12" deep	.006	LF
Automatic valves, solenoid		
¾" diameter	.363	EA
2" diameter	.500	EA
Automatic controllers		
4-station	1.500	EA
12-station	2.000	EA

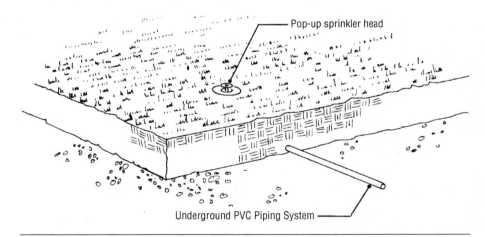

Figure 24.6 Installation Time in Labor-Hours for Site Irrigation

Source RSMeans Estimating Handbook, Third Edition. © 2009 John Wiley & Sons, Inc.

For those with previous experience with projects of this nature, historical data can be used to create a parametric estimate for the scope, assuming the work is very similar. It is, however, prudent for the estimator to secure a price from an expert in this field willing to perform the work. It is the estimator's responsibility to solicit and qualify all proposals for this work, including any inspections by scientists or reports to the authority having jurisdiction over the project. Omissions or errors can be costly and can even include fines.

LANDSCAPING

Landscaping includes a variety of components to improve the appearance of the site. Landscape design is the domain of the landscape architect. The landscape design is illustrated on the landscape plan and included as part of the site drawings. A landscaping drawing is created in plan view and often includes sections or details for plant material installation. As part of the landscaping drawings, a planting schedule lists the plantings by species (Latin classification), size (caliper), and variety (common name). The estimator should consult the landscaping section of the specifications to define the exact scope of work in the landscape section. Tasks such as maintenance cycles, watering, fertilizing lawns, and guying trees are typically specified rather than shown on drawings. Other items such as furnishing, spreading, and rough-grading topsoil (loam) are typically part of Division 31—Earthwork, but may be included in the landscape section.

The testing and treating of soils for the appropriate nutrients may also be specified.

Plantings

The installation of trees, shrubs, and ground cover are a large part of the landscaping scope of work. The scope includes soil preparation, which is the mixing of fertilizers and nutrients to be used in installation of the root ball. Newly planted trees are often required to be staked with wire guys to help ensure vertical growth. Smaller trees, shrubs, and ground cover are dug and planted by hand. Larger trees require excavation and backfill by machine as well as distribution and general handling.

For projects without irrigation systems, the estimator should review the schedule for watering and miscellaneous care that may be part of the estimated costs. Most specifications call for the landscape contractor to be responsible for the plantings until they have established themselves, typically one growing season.

Turfs and Grasses

Areas without hardscape, plantings, and structures are often planted with turfs or grasses. They help stabilize the soil against erosion from wind and water and provide an aesthetic view. Turf, for the basis of this text, is *sod*, or the surface layer of ground consisting of natural grass and grass roots that is installed during the landscape process.

This section of the specifications often includes the preparation of the soil to receive the turf or grass. This includes the fine grading and raking of the topsoil surface to receive the seed or sod, fertilizing, adding of nutrients, and the removal of any unacceptable debris.

Sod and seed, except for large areas, are typically installed by hand. Large areas can be *hydro-seeded*. Hydro-seeding is the process of spraying seed, water, and nutrients mixed at a particular formula over the finely graded loam. Consult the specifications for seeding rates, or quantity in pounds per 100 SF.

For projects without irrigation systems, the estimator should review the specifications for watering and miscellaneous care, including seasonal fertilizing and mowing, that may be part of the estimated costs. Most specifications call for the landscape contractor to be responsible for the lawns until they are established, typically 90 days or more. The added labor for maintenance until the landscaping is established should be included within the estimate.

Planting beds may require *mulch* placed around the plantings to a certain depth. Mulch is a by-product of chipped wood that is often colored. In addition to creating aesthetic appeal, it helps retain moisture in the planting beds.

Taking-off and Pricing

Plantings should be taken off by counting each and listing it according to species, size, and variety (EA). Check the quantities with the planting schedule as a means of verification. These quantities should then be submitted to a nursery (if possible) for up-to-date pricing of stock. The cost of plantings is affected by availability, season, species, and size.

Sod and seeded areas are taken off by the SF and can be extended to SY or even thousands of SF, MSF. Convert the SF area of seed to the rate required in the specifications. Rate refers to pounds per 100 SF. Fine grading and raking of the topsoil are taken off by the SF. A small quantity of topsoil should be included for spotting, or filling areas that have settled. Placement and rough grading of topsoil are often included in the excavation and backfill section of the specifications.

Fine grading of topsoil can be accomplished with a combination of small equipment, such as a skidsteer loader with a rake attachment and hand work. Areas that are inaccessible by machine are graded by hand and should be listed separately. Grading, raking, and rolling of topsoil are quantified by the square foot (SF). Allowances should be made for compaction resulting from rolling of the topsoil.

Additional costs for landscaping work include fertilizers converted from lb./SF to lb. Spreading of bark mulch, wood chips, or similar materials is taken off by the SF or SY. Mulch materials are calculated in CY, based on the depth of the material, traditionally 4–6".

Maintenance Requirements

Always read the specifications carefully for landscape maintenance clauses. Frequently, the specifications call for a maintenance period that includes watering, pruning, and mowing until growth has been established. Plantings of lawns or new seeding may also be deferred to specific weeks or months of the year. Newly established lawns often require mowing and fertilizing on specific schedules that may be after the general construction has been completed. The cost of this work depends on the frequency and season but should be included in the estimate. Many landscape estimators include a small allowance for planting loss during the warranty period. This is up to the individual estimator and may be dictated by the heartiness of the species planted.

LABOR FOR EXTERIOR IMPROVEMENTS

Labor for the work of Division 32—Exterior Improvements is provided by a wide variety of trades. Equipment operators, laborers, landscapers, pavers, and masons are the predominant trades involved in this work. Most of the work of this division is performed by crews, and very little is relegated to the work of a single individual. Hourly rates for each trade can be calculated and added to make up a specific crew.

Most crews include some type of equipment – excavating equipment, paving spreaders, or simple plate compactors, to name a few. The daily output is calculated by the productivity of the crew, including equipment.

MISCELLANEOUS CONSIDERATIONS

There are several miscellaneous items to consider that may be required in the exterior improvement estimate. These include but are not limited to:

- Water trucks for planting/lawn watering
- Charges for transporting equipment (mobilization and demobilization)
- Temporary protection for plantings
- Engineering (layout and grades) for pavements and striping
- Police/flagman details for street work (traffic control)
- Storage or trailer charges for large items
- As-built or record drawings of sitework
- Securing the services of a licensed arborist to oversee all tree planting, pruning, and care

SUMMARY

Costs for exterior improvements can vary from project to project, as well as from season to season (landscape work). Estimating exterior improvement costs accurately requires a complete understanding of the full scope of work. The estimator should evaluate all information available before applying pricing.

Specific items within the Division 32 scope of work can be tied to commodity pricing. The most common is paving, which is impacted by the price of a barrel of oil and will fluctuate almost daily.

It is important for the estimator to consider all of the miscellaneous items that may be required for this work. A checklist can be extremely helpful in capturing all of the tasks and costs associated with Division 32.

QUESTIONS/ PROBLEMS

1. A parking lot is 400' × 100' and requires an 8" (after compaction) gravel base for the paving. The base material has an 18% compaction factor. How many loose (LCYs) cubic yards of material should be estimated to purchase?

2. For the parking lot noted in problem #1 there will be a binder course of 3½" and a wearing course of 2". If the asphalt mix for each course will lose ¼" for each 1" of specified thickness, how many loose tons (2000 lbs. per ton) are required for each course? (Use 140 lbs. per CF.)

3. A pedestrian walkway at the perimeter of the building is 5'-0" wide, 200' long and 5" thick. Calculate (a) the surface area to be finished, (b) cubic yards of concrete, and (c) welded wire fabric required with a longitudinal lap of 6" per sheet.

4. For the walkway in problem #3 if the concrete placement cost is $150 per CY, the surface finishing cost is $1.90 per SF including edging, and the curing cost is $1.40 per SF, what is the estimated labor cost for the walkway?

5. Two (2) laborers at a billing rate of $58.50 per hour each will auger 13 fence posts 12" in diameter to 3'-6" below grade in a typical day. The rental rate for the hand-held auger is $98.00 per day with a $1.20 per hr. operating cost (based on an 8-hr day). What is the estimated cost to auger 400' of stockade fence with posts at 8'-0" on center? (Assume fence is in a straight line.)

25 | Utilities

D ivision 33 of CSI MasterFormat® is called Utilities. This division deals primarily with utilities to and from the structure. It is part of the Site and Infrastructure subgroup in the 2004 reorganization of the divisions. The work of this division was part of Division 2—Site Construction under the 1995 format.

Major categories of utilities include water piping and appurtenances, sanitary sewer and accessories, septic systems, storm drainage structures, piping, and castings. This division includes site lighting and underground electrical and telecommunications services. Testing, inspection, chlorination, and flushing of utilities are also part of this section.

It is important to remember that the work of Division 33 does not include the excavation, bedding, or backfill component of utility excavation work. That is covered in Division 31—Earthwork.

Most of the work of Division 33 requires some type of equipment for handling, distribution, and installation of the utility. Small-diameter PVC conduits used for electric, telecommunications, and gas are the exception.

CIVIL DRAWINGS

Utilities are shown on the civil drawings introduced in Chapter 23. All civil drawings are designed under the direction and approval of a civil engineer and focus on the land's changes to accommodate a new structure or any engineering improvements. The most common drawings in the civil set are enumerated in Division 31—Earthwork. The applicable drawings in the civil set for estimating utilities are:

- Existing conditions drawings (the existing site survey is discussed in Chapter 7, "Existing Conditions")

Estimating Building Costs for the Residential and Light Commercial Construction Professional,
Third Edition. Wayne J. Del Pico.
© 2023 John Wiley & Sons, Inc. Published 2023 by John Wiley & Sons, Inc.

- Site grading and drainage plan
- Utilities plan(s) (water, electric, telecom, and gas)
- Septic system or sanitary sewer plan
- Site lighting and site electrical plan
- Site details (for any of the above)

The level of detail in these drawings should be reasonably comprehensive. Since much of the sanitary sewer and drainage piping is handled by gravity flow, proper design of elevations and inverts (see Chapter 1, "The Working Drawings") is crucial to the system functioning correctly.

Residential site plans can show single-line services from the foundation to the street, while some commercial projects have separate drawings dedicated to each utility. Many civil plans showing utilities have their own collection of symbols and nomenclature, often very different from the building plans. Very subtle changes to a symbol can indicate a totally different thing.

How Civil Drawings Are Used to Estimate Utilities

There are specific types of information provided on each sheet in the civil set of drawings. Not all of the drawings in the civil set are used exclusively in preparing the utilities estimate. The plans listed here are the most common:

- **Existing conditions drawing:** Illustrates the location of existing utilities. It is helpful for any utility that is scheduled to be cut and capped or for new piping connected to existing piping.
- **Site grading and drainage plan:** Provides information for taking off precast structures, such as tanks, manholes, catch basins, and reinforced concrete pipe. Additional information on the details sheets can help clarify for accurate pricing.
- **Site demolition or site preparation plan:** Provides information on the removal of subsurface structures such as abandoned precast structures or piping, in the absence of an existing conditions drawing.
- **Architectural drawings:** Should be reviewed for details that show where the particular utility enters or exits the building.
- **Site electrical drawings:** Should be reviewed for location of the primary and secondary electrical service as well as location of site lighting poles and underground distribution.
- **Site improvement drawing:** May illustrate lighting bollards adjacent to walkways in the absence of a dedicated site lighting drawing.
- **Site detail drawing:** Provides sections and details of ductbanks, light pole bases, thrust blocks, and similar items.

Each portion of the documents (plans and specs) provides critical information the estimator can use to calculate the utility work on the project. As an example, consider the following drawings and their respective contributions for defining the water service work:

- **Utilities drawing:** Shows the location of the water service from the street to the building. The estimator can measure its length. This allows him/her to calculate the length of the pipe and directional fitting needed. It also shows tie-ins, called *taps,* to the existing water service. The locations of miscellaneous items such as hydrants and post indicator valves are also shown and can be counted.
- **Site detail drawing:** Shows the details of the pipe connections, hydrants, and thrust blocks. It can also show a cross section of the trench for size and depth as well as the bedding materials.
- **Specification Section 33—Utilities:** Defines the specific type of pipe, valves, and appurtenances to be used for a variety of systems. It also defines the testing, chlorination, and flushing procedures required for acceptance. It can also spell out the type of bedding and its thickness.

Division 1—General Requirements must be reviewed for record drawings that are required as part of project closeout (see Chapter 2, "Understanding the Specifications").

This chapter surveys the more common elements of utilities takeoff and estimating for the light commercial/residential market. The partial outline of work topics included in Division 33—Utilities is noted as follows.

A. 33 10 00 Water Utilities
  - Water Utilities Transmission and Distribution
  - Water Utility Storage Tanks
  - Water Supply Wells
  - Water Utility Metering Equipment
B. 33 30 00 Sanitary Sewerage
  - Sanitary Sewerage Piping
  - Sanitary Sewerage Equipment
  - Onsite Wastewater Disposal (Septic Systems)
C. 33 40 00 Stormwater Utilities
  - Subdrainage
  - Stormwater Conveyance
  - Stormwater Utility Equipment
  - Stormwater Management
D. 33 70 00 Electrical Utilities
  - Electrical Utility Transmission and Distribution
  - Utility Substation
  - Utility Transformers
  - Site Grounding
E. 33 80 00 Communication Utilities
  - Communication Structures
  - Communication Transmission and Distribution
  - Wireless Communications Transmission and Distribution
  - Utility Transformers
  - Site Grounding

WATER UTILITIES

Water utilities are defined as the piping and appurtenances, including fittings, valves, hydrants, and the like, that are used to bring potable water to the building for domestic and fire protection use. As noted, this category excludes all those items of work covered under Division 31—Earthwork. Water distribution systems are normally tied into a public (or private) water supply. The service is brought onto the site and into the building. Because water distribution piping is under pressure, it does not require that specific inverts be maintained. Water pressure is also the reason that a cast-in-place concrete mass, called a *thrust block,* is placed around fittings that change direction in water piping. In temperate zones where water piping can freeze, it is important that a specific dimension from the surface grade to the top of the pipe be maintained to prevent freezing and rupture. This will vary with climate zone.

Cement-line ductile iron (CLDI) piping for water distribution is rated for working pressure such as Class 150, 200, 250. The number designates the water pressure in pounds per square inch (psi). Polyvinyl chloride (PVC) water pipe is rated by SDR (see later discussion in this chapter under "Sanitary Sewerage Utilities").

In addition to the drawings, the estimator is directed to review Division 33—Utilities thoroughly for the specified materials and installation procedures.

Water Piping and Appurtenances

Water piping used for site distribution is available in a range of diameters, with 4–12" the most common in the residential/light commercial market. Water pipes are also manufactured in a variety of materials, from CLDI to PVC. Residential applications from the street to the house can be soft copper or high-density polyethylene (HDPE) in sizes from ¾" to 1½". Each has specific characteristics and may be mandated by the local authority having jurisdiction over the water utility.

All water piping comes in specific lengths, depending on the product. Individual sections can be attached by inserting the plain end of the pipe into the bell end of the successive pipe. This is referred to as a *push-on* joint. Another type of connection is mechanical, which requires a bolted connection as each successive pipe is added. This is referred to as the *mechanical* joint.

Appurtenances are defined as all of the bends and fittings (tees, wyes, 90°s, 45°s) needed to change direction of the service, gate valves to control flow, valve boxes for access at the street/sidewalk, hydrants for firefighting, and post indicator valves to control flow at fire lines. Also included are the sleeves and saddles used to tap live water lines.

Disinfecting/Testing of Water Services

Once the water system has been completed and is live, it is typically chlorinated to kill any bacteria that may be in the line. After chlorination, a laboratory tests the samples to ensure that the water is within acceptable tolerances for consumption. After the results have been accepted, the last step is to flush out the chlorine and to remove particles of dirt and debris that may have entered the pipe during installation. Flushing involves allowing an open end of the pipe to flush contaminates to daylight.

Taking-off and Pricing

All water pipes are taken off and priced by the linear foot (LF). It should be rounded up to the nearest full section of pipe required. Water pipe should be segregated according to the type (CLDI, PVC, HDPE, or copper), size (diameter), wall thickness, and use (domestic or fire protection). Depending on the material and size of the pipe, costs for cutting it to specific lengths in the field must be included. The cost for cutting a thin-wall PVC pipe with a handsaw is negligible. Cutting larger-diameter pipe that requires special cutting tools can take time and thus is more costly. Cutting ductile iron water pipe can be estimated by the individual cut, EA.

All appurtenances are taken off by counting and priced by the individual piece (EA). They should also be listed by the aforementioned criteria. The estimator should consult the specifications for a particular manufacturer or model numbers of individual items. Chlorination, including testing and flushing, is taken off and priced by the LF of pipe or may use the lump sum (LS) unit of measure for small quantities or where minimum costs apply. Figure 25.1 provides installation times for water distribution systems.

WATER SUPPLY WELLS

Another way to provide water to a structure is by a water well system. The method extracts water from the water table or aquifer below the Earth's surface. It can be tens of feet to hundreds of feet below the surface. A 2–8" hole is drilled to the appropriate depth, the soil is removed, and a pipe with a filter end is lowered into the hole into the water. A pump is attached, creating suction that draws the water up the well to a pressurized storage tank. The water is then made available to the building as valves are opened, allowing the water under pressure to escape. As the pressure decreases, the pump starts to replenish the water supply.

Water Supply Well Piping

Water well pipes are usually steel or PVC. Diameters range from 2" for residential applications to 8" for commercial projects with the need for large volume. Storage tanks vary according to need. Well water straight from the ground may require treatment with chemicals or filters to remove undesirable particles within the water. Most well systems have some type of tank for storage. Wells that supply fire protection systems may have very large capacity storage tanks.

Description	Labor-Hours	Unit
Water Distribution Piping, Not Including		
Excavation and Backfill		
Mains, Ductile Iron, 4" Diameter	0.2	LF
6" Diameter	0.25	LF
8" Diameter	0.3	LF
12" Diameter	0.38	LF
16" Diameter	0.55	LF
Polyvinyl Chloride, 4" Diameter	0.064	LF
6" Diameter	0.076	LF
8" Diameter	0.092	LF
12" Diameter	0.10	LF
Concrete,		LF
12" Diameter	0.292	LF
24" Diameter	0.438	LF
Fittings for Mains, Ductile Iron, Bend, MJ		
4" Diameter	2	EA
8" Diameter	3	EA
16" Diameter	5.5	EA
Wye, 4" Diameter	3	EA
8" Diameter	4.5	EA
16" Diameter	8.25	EA
Increaser, 4" x 6"	2.25	EA
6" x 16"	4	EA
Flange, 4" Diameter	1.600	EA
8" Diameter	3.080	EA
12" Diameter	4.000	EA
Polyvinyl Chloride, Bend, 4" Diameter	.240	EA
8" Diameter	.300	EA
12" Diameter	.800	EA
Tee, 4" Diameter	.267	EA
8" Diameter	.343	EA
12" Diameter	1.200	EA
Concrete, Bend, 12" Diameter	1.75	EA
16" Diameter	5.00	EA
Tee, 12" Diameter	2.65	EA
16" Diameter	7.5	EA
Service, Copper, Type K, 3/4" Diameter	0.04	LF
1" Diameter	0.05	LF
2" Diameter	0.07	LF
4" Diameter	0.168	LF
Polyvinyl Chloride, 1-1/2" Diameter	0.013	LF
2-1/2" Diameter	0.02	LF
Fittings for Service, Copper, Corp. Stop		
3/4" Diameter	.421	EA
2" Diameter	.727	EA
Wye, 3/4" Diameter	.667	EA
2" Diameter	1.140	EA
Curb Box, 3/4" Diameter Service	.667	EA
2" Diameter Service	1.000	LF
Valves for Mains, 4" Diameter	4	EA
8" Diameter	4	EA
12" Diameter	4	EA
Valves for Service, Curb Stop, 3/4" Diameter	.421	EA
1" Diameter	.500	EA
2" Diameter	.727	EA

Figure 25.1a Installation Time in Labor-Hours for Water Distribution Systems

Source RSMeans Estimating Handbook, Third Edition. © 2009 John Wiley & Sons, Inc.

Taking-off and Pricing

Drilled holes for wells are taken off and priced by the vertical linear foot (VLF) below the surface. The pipe is also taken off and priced by the VLF, corresponding to the depth of the hole drilled. Prices will vary with the strata the wells are being drilled through as well as with the diameter and composition of the pipe. Well points or the screen at the deep end of the pipe is taken off and priced by the individual piece (EA), based on type and diameter.

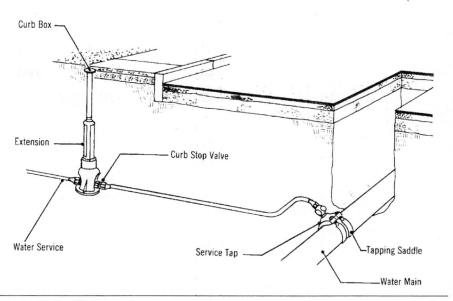

Curb Box

Extension

Curb Stop Valve

Water Service

Service Tap

Tapping Saddle

Water Main

Figure 25.1b *(Continued)*

Water storage tanks are taken off and priced by the individual piece (EA), based on the capacity in gallons. Large capacity tanks for fire protection services may be concrete and require a crane to set. They can also be below grade, which requires excavation, backfill, and even a concrete pad as a means of anchoring the tank. A separate pump may be required for pumping the water in the storage tank through the fire suppression system. These types of pumps would be part of Division 21 or 22. Piping between the well and the pump/tank assembly is taken off and priced by the linear foot (LF).

SANITARY SEWERAGE UTILITIES

A separate underground piping system that carries sewage waste, or *effluent*, from the building to disposal or treatment centers is called a *sanitary sewer*. It is predominantly a gravity-based system but can have portions that are pumped; these are called *force mains*. *Lift stations* are chambers that receive sewage by gravity flow and then pump it to higher elevations.

Except for the oldest of city systems, sanitary sewers are operated independently from storm drainage systems. Most sanitary lines are run near flat, at about 2% pitch. When lines change direction or need access to remove blockages, a precast structure sometimes called a *manhole*, is placed at the change in direction of the pipe so that pipes between manholes are straight. Manholes are accessible from the surface through a cast-iron frame and cover, generically called *castings*.

Other types of precast structures hold the waste or separate harmful components from the waste, such as a *grease trap*. Large precast tanks receive kitchen waste via a sanitary line. The majority of grease is contained within the tank through a series of baffles, and the effluent is allowed to pass on into the main sanitary sewer system. The grease is then pumped from the tank as part of a maintenance program.

As noted, this category excludes all those items of work covered under Division 31—Earthwork.

Sanitary Sewerage Piping

Sanitary sewer piping is most often PVC; 4–12" diameters are the most common. The standard pipe lengths are 10', 14', and 20' with bell and spigot (plain) ends. PVC sewer pipe can have fittings such as couplings, bends, tee-wyes, reducers, and cleanouts. Sewer pipe can also have varying wall thicknesses based on size and expected duty.

Pressure sewer pipe is rated according to *standard dimension ratio* (SDR), which is the ratio of wall thickness to the diameter of the pipe. SDR 11 means that the outside diameter of the pipe is eleven times the thickness of the wall. Another method of defining pipe materials is a measure of wall thickness, called *schedule*. A pipe's schedule bears a relationship to its pressure rating in pressurized systems. For gravity systems, it is used as a way to measure wall thickness.

The estimator should review the Division 33—Utilities specification to determine the materials specified.

Sanitary Sewerage Structures and Castings

A sanitary sewer system employs precast concrete structures for a variety of purposes. As stated, precast manholes are used as access points for directional changes in piping. Other structures such as tanks are for holding effluent before it enters the sanitary line. All precast concrete structures, by their very composition and weight, must be set in place by some type of equipment, such as a backhoe or a crane.

Manholes are a composite of multiple sections to achieve the required vertical depth. The lowest component is called the *base* section. It is flat bottomed and has sides. The next component is called the *riser,* which resembles a concrete ring with varying wall heights. The top section can be one of two major types. The first is called a *flat top,* which is a concrete disk with the approximate outside diameter of the manhole. It has a hole cast in the center of diameters ranging from 24" to 36" or more. The hole allows access to the inside from grade. The second type is called a *conical section* and is tapered. The lower portion has the same diameter as the riser and then tapers to a smaller diameter at the top. The top has a hole for access.

Precast concrete tanks are defined by their size in gallons. Most tanks have a base and top section of approximately equal size. For very deep tanks, risers can be used to increase vertical height. Extremely large tanks (gallons) can be sectional, composed of a series of concrete rings laid horizontally, with closed pieces at either end. Top sections of tanks have hole(s) cast in them for access, frequently at opposite ends. Seams between tank and manhole sections are sealed with a thick asphaltic gasketing material that provides a seal under the weight of the sections above.

Castings are defined as the cast-iron frames, grates, and covers for the access to the precast structure. The frame is set at the top of the structure and adjusted for height with bricks and mortar. The top of the frame, called the *rim,* is intended to be flush with the adjacent surface (paving or ground). The cover fits within the ring on the frame to provide closure of the structure. Frames and covers are heavy and must be set in place by some type of equipment. The brick-and-mortar adjustments are done manually.

Within the base section of the manhole are channels, called *swales,* that direct the flow of effluent. They are constructed of bricks and cast-in-place concrete.

The estimator should review the Division 33—Utilities specification section for information such as the design *load rating* of the structure, or the weight that can be driven over it, the strength of and reinforcing in the concrete, and any necessary accessories such as steps (rungs).

Taking-off and Pricing

Piping for sanitary sewer systems is taken off by the LF. Total length can be converted to standard lengths of pipe for material pricing. The cost for cutting a thin-wall PVC pipe with a gas-powered saw is negligible, especially for smaller diameters. This is considered part of the installation cost. Quantities should be segregated according to diameter and class (SDR or schedule) as this has an impact on price.

Manholes are quantified and priced by the VLF, although individual sections (base, risers, and top) can be quantified and priced by the piece (EA). Either method is acceptable, but the latter is often more accurate. Precast tanks are taken off and priced by the individual piece and should be defined by gallons and, if possible, weight. This helps the estimator in pricing the appropriate piece of equipment for setting the tank. Both should include the required load rating: H20, H15, and so on.

Castings are taken off and priced by the set (frame and cover) and can be quantified as each (EA). Part of the cost of setting the frame and cover is the materials and labor to adjust the vertical height and parge the exterior of the brick with cement. Frames and covers require equipment to distribute and set.

Accessories such as steps are taken off and priced by EA. This cost can be built into the VLF cost of the manhole. Swales at the bottom of the manhole can be labor intensive and are calculated by the labor-hour. The cost can be built into the VLF cost as well. Figure 25.2 provides installation times for sewerage and drainage collection systems.

Description	Labor-Hours	Unit
Catch Basins or Manholes, Not Including Excavation and Backfill, Frame and Cover		
Brick, 4' I.D., 6' Deep	22.857	EA
8' Deep	32.000	EA
10' Deep	40	EA
Concrete Block, 4' I.D., 6' Deep	16.000	EA
8' Deep	22.857	EA
10' Deep	28.675	EA
Precast Concrete, 4' I.D., 6' Deep	10	EA
8' Deep	15	EA
10' Deep	18.75	EA
Cast-in-place Concrete, 4' I.D., 6' Deep	32	EA
8' Deep	48	EA
10' Deep	60	EA
Frames and Covers, 18" Square, 160 lb.	2.400	EA
270 lb.	2.791	EA
24" Square, 220 lb.	2.667	EA
400 lb.	3.077	EA
26" D Shape, 600 lb.	3.429	EA
Curb Inlet Frame & Grate, 24" x 36"	12	EA
Light Traffic, 18" Diameter, 100 lb.	2.4	EA
24" Diameter, 300 lb.	2.759	EA
36" Diameter, 900 lb.	4.138	EA
Heavy Traffic, 24" Diameter, 400 lb.	3.077	EA
36" Diameter, 1150 lb.	8.000	EA
Raise Frame and Cover 2", for Resurfacing, 20" to 26" Frame	2.183	EA
30" to 36" Frame	2.667	EA
Drainage and Sewage Piping, Not Including Excavation and Backfill		
Concrete, up to 8" Diameter	.2	LF
12" Diameter	.320	LF
15" Diameter	.320	LF
18" Diameter	.364	LF
21" Diameter	.400	LF
24" Diameter	.480	LF
27" Diameter	.609	LF
30" Diameter	.636	LF
36" Diameter	.778	LF
Currugated Metal, 8" Diameter	.135	LF
12" Diameter	.218	LF
18" Diameter	.234	LF
24" Diameter	.274	LF
30" Diameter	.431	LF
36" Diameter	.431	LF
Cast Iron, 6" Diameter	.286	LF
8" Diameter	.457	LF
12" Diameter	.561	LF
15" Diameter	.653	LF
Polyvinyl Chloride, 4" Diameter	.064	LF
8" Diameter	.072	LF
12" Diameter	.088	LF
15" Diameter	.117	LF

Figure 25.2 Installation Time in Labor-Hours for Sewerage and Drainage Collection Systems

Source RSMeans Estimating Handbook, Third Edition. © 2009 John Wiley & Sons, Inc.

Septic Systems, Fields, and Structures

A septic system is an on-site sewage disposal system. It takes the *effluent* or wastewater from the building and separates the solids from the liquids and disposes the liquids on-site. The solids are pumped from the tank as part of a maintenance program. Septic systems have some similar components: the storage area (tank) and the *drainfield* (disposal field). From the tank to the drainfield, the effluent can be gravity-drained or pumped. The drainfield is a series of perforated PVC pipes that are typically 4" in diameter laid on top of gravel-filled trenches 2–3' in width in natural gravelly or sandy soil. The soil between the bottom of the trench and the bedrock or water table acts as the filter medium for treatment of the liquid. The soil cleans the liquid before it reaches the water table.

A *gravity drainfield,* as the name implies, uses gravity to distribute the liquid portion of the effluent evenly to the disposal area. The tank is located higher than the disposal area.

Pressure distribution drainfields use pumps to distribute the liquid when the soil is less than optimal or when gravity distribution is not an option. Pumps dose the drainfield evenly and only when the effluent reaches a certain level in the tank.

The size of the drainfield is calculated by the design engineer and depends on the estimated daily wastewater flow and soil conditions. The number of bedrooms (residential) or occupancy (for commercial systems) and the soil type determine the total number of square feet of drainfield area that is needed.

For soils with very high percolation rates, alternative types of disposal fields can be used. The most commonly used types are perforated precast structures approximately 6' in diameter and 6' high, called *seepage pits*. The liquid is gravity-fed to the pit and distributed through the open bottom and perforated sides of the pit. The pits have a layer of stone at the bottom and circumference.

All conventional septic systems have a *septic tank,* which is usually a large precast concrete or fiberglass tank buried outside the building. Effluent from the building is collected in the tank. Heavy solids settle to the bottom, where a bacterial action produces sludge, and lighter solids rise to the top and settle on the liquid. This top layer is called the *scum layer.*

The estimator should review the Division 33—Utilities specification for the materials specified, tanks and pipe sizes. All excavation, backfill, and stone drainfield media are part of Division 32—Earthwork. Pumps for force main system may be specified in Division 22—Plumbing or in Division 33—Utilities. Remember that there are many other tasks that are related to this work: electrical to power and control the pumps, excavation and backfill for the precast structures and piping, testing, and so on.

Taking-off and Pricing

Takeoff and pricing of the piping for the septic system is by the LF. Total length can be converted to standard lengths of pipe for material pricing. Fittings are taken off and priced by the individual piece, EA. The cost for cutting a thin-wall PVC pipe with a gas-powered saw is negligible, especially for smaller diameters. This is considered part of the installation cost. Quantities should be segregated according to diameter and class (SDR or schedule). Precast structures are taken off and priced by the individual piece (EA) or by vertical linear foot (VLF) for the complete structure.

Precast tanks are taken off and priced by the individual piece and should be defined by gallons and, if possible, weight. This helps the estimator in pricing the appropriate piece of equipment for setting the tank. Both should include the required load rating: H20, H15, and so on.

Tanks can require testing to identify any leaks. The estimator should secure a proposal from a qualified firm to do the testing. This can be quantified and priced by the lump sum (LS).

Castings are taken off and priced by the set (frame and cover) and can be quantified as EA. Tanks most often have one frame and cover at each end of the tank; one is used to inspect and service the intake, and the other is to service and inspect the exit to the disposal field. The depth of the tank may also necessitate that there be risers, or concrete rings, from the top of the tank to the surface to allow access without excavation. These are taken off and priced by the VLF or by the individual piece. A final note on buried tanks: Frequently, large tanks require concrete pads on which to anchor the tank to prevent "floating." The estimator is directed to Chapter 8, "Concrete," for estimating concrete pads. Figure 25.3 provides installation times for septic systems.

STORM DRAINAGE UTILITIES

Storm drainage systems are designed by engineers to handle the water from precipitation that lands on impervious surfaces such as paved roadways and parking lots. The system channels runoff water through collection devices called *catch basins*, consisting of an open frame and grate castings, that allow water into a precast structure. The water is then channeled to disposal on-site or to a municipal collection system. There are similarities to the applications in sanitary sewerage and septic systems.

Drainage employs piping and related fittings to carry surface water from the immediate areas adjacent to the building to disposal areas. Frequently, this work also includes precast concrete structures, such as catch basins, manholes, and castings. Check the site plans to determine the limits of the drainage work. Grading and drainage plans show the location of the various types of piping and any changes in direction or elevation. Sections of the piping and trench are typically shown on site detail plans.

Refer to the Division 33—Utilities of the technical specifications for particulars on the type of materials used and the method of installation. This information is essential for accurately taking off and pricing the work. Additional details on precast structures are also shown on the site detail plans, with elevations for inverts of piping and rims of the cover. Sections or details for these items are referred to in order to clarify the construction and materials to be included in the takeoff. All excavation, backfill, and stone bedding materials are part of Division 31—Earthwork.

Description	Labor-Hours	Unit
Septic Tanks Not Including Excavation or		
Piping, Precast, 1000 Gallon	3.500	EA
1500 Gallon	4.0	EA
2000 Gallon	5.600	EA
5000 Gallon	16.000	EA
HDPE 1000 Gallon	4.667	EA
1500 Gallon	7.000	EA
Excavation, 3/4 CY Backhoe	.110	CY
Distribution Boxes, Concrete, 7 Outlets	1.000	EA
9 Outlets	2.000	EA
Leaching Field Chambers, 13' x 3'-7" x 1'-4"		
Standard	3.500	EA
Heavy-Duty, 8' x 4' x 1'-6"	4.000	EA
13' x 3'-9" x 1'-6"	4.667	EA
20' x 4' x 1'-6"	11.200	EA
Leaching Pit, Precast Concrete, 3' Pit	3.500	EA
6' Pit	5.957	EA
Disposal Field		
Excavation, 4' Trench, 3/4 CY Backhoe	.048	LF
Crushed Stone	.160	CY
PVC, Perforated, 4" Diameter	0.153	LF
6" Diameter	0.160	LF

Figure 25.3 Installation Time in Labor-Hours for Septic Systems

Source *RSMeans Estimating Handbook, Third Edition.* © 2009 John Wiley & Sons, Inc.

Storm Drainage Piping

Storm drainage systems use a variety of pipe types: reinforced concrete, called *RCP*; corrugated HDPE and metal; and PVC. Each is available in standard lengths for each type based on size. On-site disposal designs have a variety of different components that allow the runoff to be dosed over a field similar to the way a septic system works. Couplings for piping and inspection fittings, called *cleanouts,* are also needed for a complete system. Reinforced concrete pipe drainage lines may require that the connection seams between pieces are parged with mortar – for larger-diameter pipes from the inside, on smaller-diameter pipes from the outside. This can be labor intensive for long runs of larger diameter pipes. The estimator is urged to read the specifications carefully for mortared seams.

Many projects require that the drainage lines and structures be flushed of sand and debris after completion and as a condition of acceptance. The estimator is recommended to secure a proposal from a qualified firm to do the flushing and provide a report.

Storm Drainage Structures and Components

Drainage work requires precast concrete structures much like sanitary systems. These include items such as manholes, catch basins, and tanks. In addition, there are precast fittings such as *flared ends* and *headwalls* that are used to disperse runoff from the RCP pipe to the surface disposal area.

Catch basins are precast structures that are similar in appearance to manholes. They consist of a base section called a *sump,* a vertical riser, and a top section. A cast-iron frame with a grate is installed on the top section to a height that matches the final surrounding grade. Catch basins are installed at low points in the paved elevations to collect the runoff. Solids such as sand, stones, and debris wash into the catch basin and settle to the bottom. They are cleaned as part of a maintenance program. The storm water in the catch basin rises to the height of the invert and flows out to the disposal site. Catch basins do not require steps as with manhole structures.

Drain manholes are very similar in appearance and construction to sanitary manholes. Readers are directed to the discussion in the previous "Sanitary Sewerage Utilities" section for more information.

Special castings such as *catch basin hoods,* or *traps,* prevent floating debris from entering and clogging the piping system. They also provide a watertight seal against escaping gases that may be in the catch basin. Figure 25.4 illustrates catch basin configurations.

In addition to precast structures, on-site drainage systems employ a variety of high-density, heavy-duty plastic *chambers* used in the construction of stormwater drainfields. They are rounded, as opposed to the square-edge design of precast, as well as lightweight and very durable. Chambers are open on the bottom and perforated on the sides to allow stormwater dosing. They can be installed in a trench or bed field type of design, and, although equipment is required to set them in place, the necessary equipment is considerably smaller because of the reduced weight.

Taking-off and Pricing

Piping for drainage systems is taken off by the LF. The total length can be converted to standard lengths of pipe for material pricing. The cost for cutting a thin-wall PVC pipe with a gas-powered saw is negligible, especially for

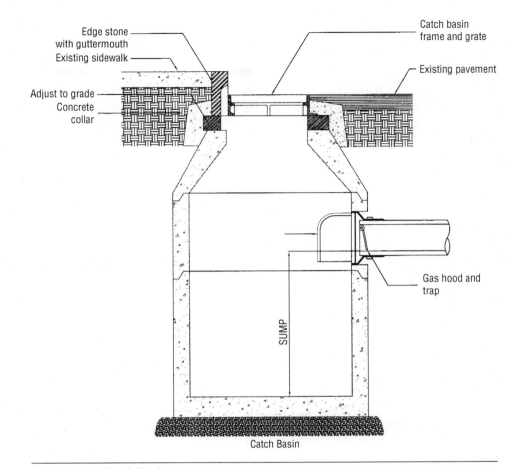

Figure 25.4a Catch Basin

Source RSMeans Estimating Handbook, Third Edition. © 2009 John Wiley & Sons, Inc.

smaller diameters. This is considered part of the installation cost. For larger-diameter pipe cutting can be labor intensive and should be addressed separately, based on type of pipe (RCP, metal) and diameter. Quantities should be segregated according to diameter and type for all piping takeoff.

Manholes and catch basins are quantified and priced by the VLF, although individual sections (base, risers, and top) can be quantified and priced by the piece. Either method is acceptable. Similar items whose sizes are constant should be quantified as each (EA, e.g., 4' catch basins). Chambers are taken off and priced by EA and should be defined by physical size, model number, and manufacturer, and, if possible, weight. This helps the estimator in pricing the appropriate piece of equipment for setting the tank. Both should include the required load rating; H20, H15, and so on.

Miscellaneous items such as flared ends or headwalls are taken off by the individual piece (EA) and separated according to pipe diameter, material composition, and size.

Castings are taken off and priced by the set (frame and cover or frame and grate) and can be quantified as EA. Part of the cost of setting the frame and

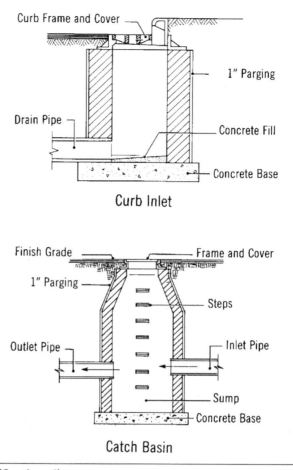

Curb Inlet

Catch Basin

Figure 25.4b *(Continued)*

cover is the materials and labor to adjust the vertical height and parge the exterior of the brick with cement. Equipment is required to distribute and set frames and covers.

Items that vary in size, such as the vertical heights of manholes, must be separated and quantified by EA or VLF. Additional pieces to complete the structure, such as manhole frames and covers, catch basin frames and grates, and special swales or inverts, should be quantified and priced separately or combined as part of the completed unit.

As with sanitary system precast components, structures are large, heavy, and cumbersome to handle. It may require a crane to reach and additional crew personnel may be required to set them in place. This is an added cost that should be accounted for in the estimate.

Refer to Figure 25.2 for installation times in labor-hours for drainage collection system components. Figure 25.5 provides installation times in labor-hours for open-site drainage systems.

Description	Labor-Hours	Unit
Paving, Asphalt, Ditches	.185	SY
Concrete, Ditches	.360	SY
Filter Stone Rubble	.258	CY
Paving, Asphalt, Aprons	.320	SY
Concrete, Aprons	.620	SY
Drop Structure	8.000	EA
Flared Ends, 12" Diameter	.879	EA
24" Diameter	1.96	EA
Reinforced Plastic, 12" Diameter	.280	LF
Precast Box Culvert, 6' x 3'	.343	LF
8' x 8'	.480	LF
12' x 8'	.716	LF
Aluminum Arch Culvert, 17" x 13"	.240	LF
35" x 24"	.48	LF
57" x 38"	.747	LF
Multiplate Arch Steel	.014	lb.

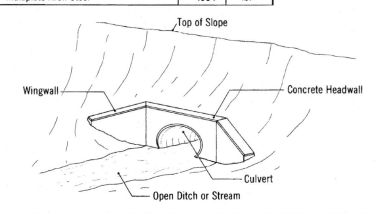

Figure 25.5 Installation Time in Labor-Hours for Open-Site Drainage Systems

Source RSMeans Estimating Handbook, Third Edition. © 2009 John Wiley & Sons, Inc.

Foundation and Subslab Drainage

In areas where the water table is high, or site drainage is inadequate, alternative means of redirecting water before it permeates the building foundation are necessary. Two of these methods are *foundation* and *subslab* drainage systems.

The foundation drainage system is typically installed at the exterior of the basement or foundation walls near the lowest point, at least below the floor level. It follows the outline of the foundation wall. The system consists of a 4" or 6" perforated PVC pipe surrounded by stone and filter fabric. The pipe is pitched to a collection structure or outflows to a daylight disposal area. As the water table rises to the level of the perforations, it flows though the pipe to the disposal area.

A subslab drainage system works the same way. Smaller-sized perforated pipes, 2–4" in diameter are laid parallel to each other with spacing of 12–36" in a bed of stone wrapped in filter fabric. Nonperforated header pipes are run perpendicularly at the end of the perforated pipes to collect and channel the water to the disposal area.

The excavation, backfill, and stone bases are included in the work of Division 31—Earthwork. The piping and the filter fabric are part of Division 33—Utilities. Part of the cost of the work may include line and grade to ensure that the drainage is pitched correctly. This can be done by a separate crew or as part of the installation crew.

Taking-off and Pricing

Piping for foundation and subslab drainage systems is taken off and priced by the LF. Total linear footage of pipe can be converted to full-size lengths of pipe and listed according to size (diameter). Many types of piping can be installed by hand, as in the case of smaller-diameter PVC pipe for foundation and subslab drainage systems. Cutting these smaller-diameter pipes is part of the installation cost and is not a separate cost.

The PVC foundation drain and the subslab drainpipes require fittings for connecting various components to complete the system. Quantities for different types and sizes of fittings (couplings, tees, wyes, etc.) should be taken off and priced by the individual piece (EA) and kept separate according to type and size (diameter).

Filter fabric is taken off and priced by the SF and can be converted to rolls, based on the individual product and the manufacturer's roll size. The estimator should add sufficient waste for lapping the fabric.

Outside engineering line and grade for the piping is typically based on full or half days for the crew. When line and grade work is performed by the same crew doing the work, it is considered part of the installation cost.

ELECTRICAL/ COMMUNICATIONS UTILITIES

Buildings require electrical, data, and communications technology wiring. This can be provided underground via conduits or overhead from utility pole to structure. This discussion deals exclusively with electrical and communications utilities provided via underground services. The conduits are primarily 4" in diameter for electrical and 3" or smaller for telecommunications, cable TV, and data wiring. In most cases, a spare or empty conduit of the same size for each service is provided for growth or in the event new wiring is required if the primary run is damaged.

Electrical Primary and Secondary Services

Most underground electrical/communications services are provided by wires pulled through conduits. For electrical services, there are two general classifications: the *primary* service is the power as it leaves the utilities substation, and the *secondary* service is the wiring as it enters the building. In between these two, the power may be modified through the use of a transformer. Primary services run from the utility source, most often a utility pole, underground to a transformer, where the voltage is changed as required for the structure, and then from the transformer to the building as the secondary service. Even though the work is specified in Division 33—Utilities, it often is done by licensed electricians due to the requirement that the installer be licensed. The scope of this work is typically included with in the electrical subcontractor's scope of work.

Concrete-Encased Duct Banks and Pads

Most underground conduits that will carry power wires for primary and secondary services are encased in concrete for protection. This encasement is called a *duct bank* and can house the communications conduits as well. Excavation, trench preparation, and backfill are considered part of Division 31—Earthwork, discussed in Chapter 23. Concrete for the duct bank and cast-in-place concrete pads for transformers is usually specified in Division 3—Concrete, but the placement can be within the scope of Division 33—Utilities. Check the specifications carefully for correct assignments of costs.

Traditionally, there is no formwork required for the sides of the duct bank. The trench is excavated to the appropriate depth and width required, and the sides of the excavation act as the formwork to retain the concrete. This is called an *earthen form*. It is recommended that the estimator check the Division 33—Utility specification section to confirm that an earthen form method is acceptable. Duct banks may also contain rebar for reinforcing. Site detail drawings typically provide cross sections of the duct bank for clarity, and the site electrical drawing provides the distance for calculating the required amount of concrete. The estimator may consider a concrete pump as part of the placement cost of the concrete for hard-to-access locations that prevent direct-chute placement.

Pads for transformers consist of cast-in-place concrete, and they often contain both rebar for reinforcing and welded wire fabric for thermal expansion/contraction. Transformer pads are formed using edge form. Pad thicknesses vary by design, from 8" to as much as 18". Site electrical and site detail drawings should be consulted for sizes (length × width × thickness) and reinforcing requirements. See Chapter 8, "Concrete," for specifics relative to estimating formwork.

Communications Services

Communications services can be classified as telephone, data, cable TV, and fiber optics wiring to the building. These wiring systems originate at the source of the service on or near the property line, then travel underground to the building. As with electrical services, the wire is pulled through conduits, most often PVC, that are within a duct bank. A spare conduit for each service is also frequently provided. The scope of this work is typically included with in the electrical subcontractor's scope of work despite it being specified in Division 33—Utilities.

Figure 25.6 provides guidance for concrete encasement for conduits.

Number of Conduits	1	2	3	4	6
Encasement Dimensions (W x H)	12" x 12"	20" x 12"	30" x 12"	20" x 20"	30" x 20"
2" Diameter Conduit	0.034	0.055	0.084	0.092	0.137
3" Diameter Conduit	0.030	0.047	0.071	0.074	0.111
4" Diameter Conduit	0.025	0.036	0.055	0.053	0.076

Multiply LF of ductbank by number in table to determine concrete in CYs required for encasement. No waste included.

Figure 25.6 Concrete For Conduit Encasement

Taking-off and Pricing

All conduits are taken off and priced by the LF. Material costs can be determined by converting the linear footage to the nearest full length of pipe. Quantities should be separated according to type (PVC and schedule), diameter, and use (telephone, secondary electric, etc.).

Many types of conduits can be installed by hand, as in the case of smaller-diameter PVC pipe for electrical and communications systems. Cutting these smaller-diameter pipes is part of the installation costs and is not a separate cost.

Electrical and communications conduits require fittings for connecting various components to complete the system. Quantities for different types and sizes of fittings (couplings, sweeps, etc.) should be taken off and priced by the individual piece (EA) and kept separate according to type (steel, PVC) and size (diameter).

See Chapter 8, "Concrete," for specific information regarding the takeoff and pricing of cast-in-place concrete, reinforcing, and formwork used in duct banks and pads. The estimator may consider a concrete pump as part of the placement cost of the concrete for hard-to-access locations that prevent direct-chute placement.

Site Lighting

Many projects require site lighting for safety and security. Site lighting can be in the form of high poles with light fixtures. It can also include low-level fixtures, called *lighting bollards,* which illuminate walkways and paths of egress. Other site lighting can include focal lighting for landscaping and signage. Site lighting originates at the building and then connects each light fixture by looping from one to the next underground. Very rarely is direct burial of the power wire preferable. Like all wire in the ground, this wiring is in conduits; however, it is rarely, if ever, in a duct bank. Conduits are small-diameter PVC (1", 1½", 2") and are of the bell and spigot type, although fittings such as sweeps are used to make a gradual transition from the underground conduits to the light fixtures/poles.

Large-site light poles often require concrete bases to anchor the fixture in place. These bases can be made of precast or cast-in-place concrete. The estimator is directed to multiple locations within the specifications for the correct assignment of the bases. Specification sections in Division 3—Concrete, Division 26—Electrical, and Division 33—Utilities should be consulted.

The excavation, backfill, and sand bedding are included in the work of Division 31—Earthwork.

Taking-off and Pricing

All conduits are taken off and priced by the LF. Material costs can be determined by converting the linear footage to the nearest full length of pipe. Quantities should be separated according to type (PVC and schedule) and diameter.

Many types of conduits can be installed by hand, as in the case of smaller-diameter PVC pipe for site lighting. Cutting these smaller-diameter pipes is part of the installation costs and is not a separate cost.

Conduits for site lighting require fittings for connecting various components to complete the system. Quantities for different types and sizes of fittings (sweeps, etc.) should be taken off and priced by the individual piece (EA) and kept separate according to type (steel, PVC) and size (diameter).

Site light poles can be aluminum, wood, or steel. The fixtures at the top of the poles, called *luminaires,* can have single or multiple heads. Walkway bollards and landscape light are available in an enormous variety. The fixtures and poles are typically the work of the electrician and are included within Division 26—Electrical.

Precast concrete light pole bases are taken off and priced by EA. Also note that precast bases are heavy and must be set in place by a piece of equipment such as a backhoe. For estimating the cost of cast-in-place bases, see Chapter 8, "Concrete."

General Information

Many project specifications include comprehensive testing of piping and structures after the work has been completed, as a means of confirming compliance. These tests require the services of a certified laboratory or testing agency. Be sure to secure quotes for testing installed piping and structures. The cost should be carried as an independent contractor price. Typically, the cost of testing the pipe should be included in Division 33—Utilities, because it is considered part of the actual installation work. Electrical and communication conduits are rarely if ever tested, but may require they be installed by a licensed electrician and inspected by the utility or local official.

Use caution when estimating piping on plans of large engineering scale (1" = 50' or greater), as the chance for error increases dramatically. Piping is sold in specific lengths, and the takeoff should be rounded to the nearest full-length required.

In most jurisdictions, pipe within the confines of the foundation walls itself, such as underground waste lines for sanitary sewer services or conduit piping for underground electrical work, is the responsibility of the respective trade and not the work of the utility contractor. Similarly, piping for underground gas services may be the responsibility of the gas utility. These items may not always be apparent on the drawings, but they should be researched and, if necessary, qualified in the bid.

Last, the wire within the conduits, whether electrical or telecommunications, is almost exclusively the work of an electrician. The estimator is directed to review the documents carefully and consult the applicable code for the installation of the conduits.

MISCELLANEOUS CONSIDERATIONS

There are several miscellaneous items to consider that may be required in the utilities' takeoff and estimate. These include but are not limited to:

- PVC cement for connecting PVC pipe and fittings
- Spacers for ductbank conduits
- Mortar for RCP joints and connections and parging brick
- Bricks for swales and collars at castings
- Rubber boots/sleeves for connecting PVC to precast structures
- Accessories for the interior of manholes – steps, pull rings, etc.
- Precast concrete risers
- Special fees or permits for street openings
- Temporary protection for open excavations while work is in progress
- Disposal of cutoff sections of pipe and debris
- Engineering (layout and grades for inverts)
- Police details for street work (traffic control)
- Registered engineering layout by a surveyor
- As-built or record drawings of utilities
- Flushing and cleaning costs of drainage and sewer system piping and precast structures
- Pressure tests of piping and tanks

SUMMARY

Costs for utility work can vary from project to project, as well as from season to season. Estimating utility work costs accurately requires a complete understanding of both the existing and the proposed conditions. The utilities estimator should evaluate all the information that is available before applying pricing. The estimator should note that Division 33—Utilities has a direct relationship with Division 31—Earthwork in as much as there is an excavation, backfill, and compaction cost related to each underground utility.

Much of the work of gravity systems require line and grade control for the placement of the pipe and the structures. This can be a separate cost provided by an engineering or surveying firm or performed by the crew installing the work. Except for the smallest diameter PVC pipe or conduit, some piece of equipment is required to set and handle. Cutting of smaller-diameter (12" or less) PVC pipe or conduits do not require an additional cost in the estimate; however, the cost of cutting reinforced concrete pipe (RCP) can be labor intensive and time consuming.

RCP joints and connections often require mortar at the circumference. This is typically handwork and can be labor intensive for larger-diameter piping.

Testing and flushing of gravity systems has a cost that should be included within the estimate for each system.

QUESTIONS/ PROBLEMS

1. Using the appropriate figure from the chapter, how many labor hours are required to set 355 LF of 12" diameter PVC sanitary sewer lines?
2. Using the solution to problem #1, if one (1) machine hour is required for each 3 hrs. of labor and the machine time cost $198.55 per hour, what is the cost of the machine time?
3. A 48" diameter RCP drainpipe is 200' long. The pipe comes in 8'-0" bell and spigot sections. If each mortar joint requires 0.08 labor hours to mortar one (1) foot of circumference, how long will it take to mortar all 200'?
4. For the RCP in problem #3, calculate the cost of installation if the excavator will set 7 sections per day at a machine cost of $1,195.00 per day and an hourly operating expense of $46.88 per hour. The crew of two (2) laborers has an hourly cost of $66.50 each and the operator has an hourly cost of $88.50. What is the cost for the installation (exclusive of mortaring the joints)?
5. How many days will the installation take for problem #4, exclusive of mortaring the joints?

26 | Profit and Contingencies

Two crucial, but difficult to calculate, costs must be evaluated in the recapitulation, or *recap*, phase of estimating, just prior to submitting the bid or the budget. These are profit and contingency. The former is a must, and the latter must be considered on a project-by-project basis. These topics are rarely addressed in even the best estimating texts on the market today. The most reasonable explanation for this lack of information is that determining the appropriate profit and/or contingency is a process that relies more on experience or judgment than on calculating a quantity and pricing it, such as the cost of concrete. If 10 different estimators were queried, they would most likely say that profit is assigned as a percentage of the cost of the work. However, the actual decision-making process that led to that particular percentage would be different from one estimator to the next. The same applies to contingencies.

This chapter introduces specific considerations to review before determining an appropriate profit and contingency (if applicable) on a project-by-project basis. Although profit is the *last* number to be added to an estimate (with the exception of a performance and payment bond premium) before the bid is submitted, we discuss profit before contingency, since not all projects warrant a contingency.

PROFIT

One of the main yardsticks for measuring the success of a construction project is profit. Without it, the project would be considered a failure. *Profit* can be loosely defined as the amount of money left after all of the bills have been paid. It is a necessary component for making a business viable, fiscally healthy, and capable of growth. Profit is the basis of our business economy, and no project should be bid without a profit. Doing work at cost is a waste of precious time and less than a sound business decision. Predicting the correct

Estimating Building Costs for the Residential and Light Commercial Construction Professional, Third Edition. Wayne J. Del Pico.
© 2023 John Wiley & Sons, Inc. Published 2023 by John Wiley & Sons, Inc.

amount of profit that a project can support is one of the most difficult tasks for the construction professional. Too small a profit, and the return does not warrant the risk taken. Too large a profit, and the bid can be lost to greed. Ideally, the amount of profit to be added should be the maximum the project can support, but just slightly less than the next bidder.

It is generally acknowledged among construction estimating professionals that the cost of materials, labor, and equipment calculated by the professional contractor's estimator will be roughly the same for most contractors bidding the same project. Some items will be higher, and some will be lower. However, in the end, all costs should be about equal. This also applies to subcontractors. On bid day, subcontractors submit quotes to the majority of general contractors bidding a project. Some subcontractors may have been solicited by a particular general contractor, while others just "cover all the bases" by submitting bids to all of the bidders. If the statement that "cost is cost" is true, then it could be inferred that adding overhead and profit to the estimate can be the deciding factor in winning or losing a bid.

In some companies, determining profit is the responsibility of the estimator, while other contractors consider it to be the domain of senior management or company principals only. Irrespective of the party assigned this duty, it is clear that by the end of the estimate preparation the person with the best understanding of the risk involved and the uniqueness of the project is the estimator. As a result, the estimator is the most likely candidate to contribute to the decision-making process in determining the profit.

Many texts recommend 10% as the appropriate profit percentage to be added, regardless of the project. Others vary between 8% and 15% of the cost of the work. Still others exceed 20%. While acknowledging that all three percentages may be acceptable for *some* projects and *some* contractors, determining profit is clearly not a one-size-fits-all process. There are local economic factors that contribute to the decision-making process. The estimator and senior management best be aware of what the economy is like in the general area in which they work.

Factors in Determining Profit

The actual mechanical process of assigning profit to a project can be done in two different ways:

- As a percentage of the cost of the work
- As a fixed fee based on time (schedule)

Regardless of the selected method, many factors affect the determination of profit. Some are tangible, and some less so. As with all portions of the estimate, careful consideration should be given to the reasons behind each decision. While acknowledging that there are no clear answers or step-by-step procedures for arriving at the correct percentage of profit for a project, there are a series of considerations that should be reviewed when determining the appropriate amount to apply. The following sections show the thought process that must take place before the bid is finalized and should provide guidance for properly assigning the right amount of profit. These guidelines are not presented in any specific order. Their order of importance varies depending on the individual project.

Risk vs. Reward

All construction projects entail a certain degree of risk, which can manifest itself in many forms. As a means of offsetting the risk, specific management techniques are used to "share" it. For example, a general contractor might secure a performance and payment bond from a subcontractor who has a large share of the work in order to assign some of the risk to another party. Risk comes in many forms. Risk can be tight schedules, environment, technically complex builds, or difficult clients, for example. However, from a business standpoint, a project with a high degree of risk requires more management time, and resources, and generally creates more of a strain on a company's infrastructure. As a result, the company should be compensated for the risk endured. In other words, risk must be rewarded, which, in the construction business, is defined as profit. Profit is the reward for risk assumed, managed, and triumphed over. The reward should match the risk, supported by the general theory that the more risk involved, the higher the profit should be. There is no magic formula to calculate profit as a function of risk, yet almost all reasoning for applying a specific profit to a project can be traced back to the risk involved. One must carefully evaluate the risk that the project will impose on the company and define or quantify it in terms that can be used to determine a profit.

Another way that contractors measure risk is by dollars. Each dollar of cost that passes through a contractor's books represents a dollar's worth of risk, and as a result each dollar of cost is marked up for profit.

Reputation in the Marketplace

All completed projects have an impact on your company's standing in the marketplace. Many larger construction firms with marketing departments actively pursue projects that will enhance the chances of future work or that have high visibility in their sphere of influence. While it does not take a marketing genius to figure out that a high-profile project will receive more attention, this may not always be a good thing. Projects that are "built in the newspaper" or under the watchful eye of the public can be management nightmares. Along with the normal management team, damage control specialists may be needed to help keep public opinion and rumors in check, as these types of projects have a momentum and dynamic all their own. They can have a tremendous impact on a firm's reputation and on future business opportunities. Be sure to consider what the successful completion of the project will mean for the company and its reputation in the marketplace. Conversely, it is always wise to also speculate on how a failure would affect the firm's reputation.

Impacts from the Schedule

A project's schedule greatly affects the amount of profit that should be added to the estimate. Projects with durations in excess of a year will affect the company's balance sheet for multiple years and will need to carry enough profit for the firm beyond the current year. Losses will affect more than the current year's balance sheet as well. It is a recognized fact that projects extending beyond one year in duration are more difficult to manage because of potential changes in the marketplace that cannot be accurately predicted at bid time. Wage increases, inflated materials costs, availability of resources, and the economy in general are some of the variable factors. In addition, long projects develop "project fatigue" that reduces the acuity needed to get the work done on time.

Carefully review project durations that appear to be too short. Those with unrealistic schedules often require infusions of capital and extra management to be completed on time. While these costs can be accounted for in the estimate, the contractor also needs to consider the fact that they may not be able to perform other work at the same time, which means lost business opportunities. Projects with tight schedules require subcontractors that can accommodate those schedules and perform at the level required.

By the time the estimate has been completed and you are ready to add profit, a construction schedule should have been developed – and refined. This is necessary to determine project duration for time-sensitive costs. (See Chapter 6, "General Requirements.") The schedule should enable the estimator to support or reject the owner's time allotted for performance under the contract provisions.

Contract Documents and Team Relationships

The level of design development in the bid documents also has an impact on profit. The more complete the design, the less the risk to the contractor from a contractual viewpoint. Poorly developed plans and specs on the verge of being defective documents can be a nightmare. This is especially true if the design team refuses to acknowledge the deficiency. While the level of design development affects the amount of justifiable profit, it also may necessitate adding a contingency. (This is addressed from another perspective later in this chapter under "Contingencies.") A contractor's prior history and working relationship with the architects and engineers for the project are also critical. Successful relationships with architects, engineers, and even owners play a significant role in assigning profit. If the contractor is viewed as part of the "team," rather than as an adversary, it will have a direct impact on the profit line. The contractor's expertise is seen as critical to a successful project and must be rewarded by allowing reasonable profit. Business savvy owners and architects recognize a contractor's need to make a profit. Contractors involved with architects, engineers, or owners who have a reputation for taking a "hard-line" approach often add greater profit margins to their estimates to compensate for the extra work caused by these adversarial relationships.

Contract Clauses

Many contractors and estimators interpret the tone of the contract (included in the project manual) as a precursor of the way the project will be administered. Are the general and supplemental conditions peppered with unfavorable contract clauses or punitive language toward the contractor? Does the contract have liquidated damages or penalties of any kind? If so, are they reasonable? Is there exculpatory language that absolves the owner and architect from responsibility for delay to the project? While owners often downplay the use of penalty clauses, they are there for a reason. Should the relationship deteriorate, the owner has the right to exercise his or her contractual rights, and they will. Review the contract clauses carefully, and, if necessary, seek legal advice on specific language that may be a concern. If the final decision results in bidding the project, make sure that adequate profit is included to compensate for the risk.

Impact on the Company's Resources

When determining profit, be sure to address the following questions as they pertain to your company's needs and resources:

- Does the company have access to capable subcontractors and suppliers to perform the work?
- Are the majority of subcontractors carried in the bid unknown and untested?
- Does the company have enough of its own labor resources to self-perform work or augment underachieving subcontractors?
- Does the management staff have the skill sets necessary to manage the project? Is the project manager experienced on this type of project, or will the staff be learning as they go?
- Will the firm have to hire new individuals to supervise or manage the project? If so, would this be considered an additional risk as a result of the unknown factors involving new employees?
- Does the company have the working capital to finance the work between owner payments? Not having adequate finances to capitalize a project puts tremendous tension on relationships with subs and suppliers who are key to a successful project.
- If the project can be managed by the current personnel and infrastructure, what effect will it have on company morale? Projects that are always in crisis destroy morale.
- Will the project tax the company to the extent that no other projects will get their fair share of attention or management, or worse – that the company will not be able to manage other projects at all?

All of these are necessary considerations to determine the impact on the company's resources, a key factor in assigning profit.

Repeat Business

Many estimators consider the potential for repeat business when applying profit. This is a very real and important consideration. Estimators and management teams often reduce their profit in the hopes that the owner will reward this behavior with repeat business. This is a common and sound business practice for many sectors of the construction industry. Bear in mind, however, that too small a profit may not make future projects with a particular owner attractive for your company. It may also set a precedent for future contracts. For the contractor who does frequent business with a client, remember that "a contractor is only as good as his/her last job." Reducing profit in hopes of repeat business can often have a negative effect because projects with insufficient or marginal profit lines are frequently relegated to the "back burner" in favor of more lucrative ones. This can end the repeat business cycle that one was hoping to develop.

Many projects are awarded on a lowest bid basis as in many public arenas. As a result, there is no "goodwill" and repeat business is a result of being lowest bid. That is a particularly challenging task since many public projects have a surplus of bidders as compared with the select list of private projects. Being the low bidder is often interpreted as the estimator having missed something or underbid the project. This is not always the case.

Project Location

Many desirable projects may be outside your company's normal sphere of influence. While this does not necessarily mean you should not bid on the project, the estimator must acknowledge that there are inherent risks that come with working outside the typical business area. These include travel time and related costs, subcontractors', and suppliers' abilities to service the location, and the general unknowns of doing business with new building departments and inspectional services, as well as public utilities.

For any contractor looking to grow, it may require that they expand beyond their normal areas. This may require a smaller profit margin to "break into" the area and should only come after the contractor has researched the area and the competition thoroughly.

Bidding Strategies

Some estimators employ a bidding strategy for winning work, which encompasses a wide range of techniques meant to provide an advantage, such as tracking the workload of the competition to determine potential competitive threats. Most contractors strive to create a market niche for themselves. The theory is that as you do repetitive work, the learning curve disappears, and, as a result, the firm becomes more financially successful in that market. In doing so, the company fits into a niche that is shared by its competitors. Contractors who competitively bid on projects find themselves frequently competing against the same firms. The ability to track who is busy and who is "hungry" is helpful since the hungry contractor is more of a threat than the busy one. The estimator can take this knowledge and apply it when they add profit. Other bidding strategies include unique means and methods, such as prefabrication, or assembly off-site, or the reduced costs of owned equipment, which often helps the bidder be more competitive. A successful bidding strategy provides an edge over the competition and can allow a higher profit margin.

Specialization

One-of-a-kind projects with no comparison model warrant an increased profit. While there are very few projects in the residential/light commercial sector of the construction industry that have never been done before, unique projects often involve highly specialized contractors, thereby limiting competition. The fewer the competitors, or the more specialized the work, the larger the profit that can be expected. This is not as a result of the contractor taking advantage of the client but simply charging for a more acute and well-honed skill set. It is called specialization, and it is done in every industry.

Current Workload

Frequently, the profit margin is determined by how much work a construction company currently has in backlog. Contractors with sufficient work add larger profit margins, using the logic that if more work is going to be added, it will have to be highly profitable, as it taxes the company's infrastructure. Contractors with minimal work under contract are prone to take projects with little or no profit, as any cash flow is preferable to a negative profit-and-loss statement. While this is true, accepting low-profit work can be an extremely dangerous practice and is not advocated in any situation other than the most dire of circumstances. The contractor then tries to add profit by negotiating the subs and vendors prices down. This leaves subs with insufficient margins of their own or diminished interest in the project. This process can often backfire. Subcontractors and vendors often recognize this practice and build a little extra in their estimate to give back during negotiations. This results in the contractor not getting the best price from them up front.

Demographics

By virtue of their locations (and the requirements of the market), certain projects warrant a larger profit margin. For example, assuming that all other (construction) costs are equal, a residential project in an area with higher real estate values is typically assigned a larger profit percentage than a similar residence built in a lower-priced area. Projects far from the beaten path may also warrant higher profit margins just for the added inconvenience and difficulty presented by the location.

Economic Conditions

The economic conditions, both macro and micro, are major indicators of the construction market in an area. When market conditions are poor and the general feeling is one of malaise due to inflation or high interest rates, projects can become scarce. Owners table expensive expansion projects and stick to just what is required to keep them in the black. This can have a devastating impact on contractors and subcontractors. Similar to how animals behave when there is a scarcity of food, contractors and subcontractors can abandon all common sense in the hopes of securing work. This can mean cutting prices or forgoing any profit just to keep the lights on. If one is in business long enough, one is bound to experience a downturn in the economy. Savvy contractors that have saved or invested will have money to weather the storm, however that may not be enough. The contractor that will survive is the contractor that can react the fastest. Indirect overhead must be cut as a start and new markets, or types or work explored in advance. One must be prepared for the inevitable.

General Comments on Profit

The *estimated* profit on a project is already earmarked by the contractor on the day the project is awarded. It is the duty of the project team to maintain that profit margin. For most contractors, the goal is to improve the profit margin in the estimate. This includes negotiating with subcontractors, vendors, and suppliers to improve prices received on bid day. While this is viewed as an unethical practice by some, it is done by the majority of GCs and subcontractors.

A contractor can also improve the profitability of a project by avoiding waste of materials and improving the efficiency of self-performed work. This is the responsibility of project managers, superintendents, and foremen. Ordering materials earlier, or with sufficient time, helps avoid expediting fees and special shipping costs.

Another way to improve the estimated profit margin is by shortening the schedule. A stipulated sum contract that was estimated to take 24 months but finishes in 21 months has three months' worth of project overhead that can converted to profit because it is not used. The same is true about the indirect overhead. This can add a significant amount of profit to the bottom line and free up resources for another project.

Bear in mind that the 10% profit estimated on bid day has a habit of shrinking as all of the bills for the project are paid. It frequently ends as a net profit in the single digits. Starting with an excessively low profit to begin with may result in zero net profit or a loss. This is a ridiculous business practice and is damaging to our industry. Sound judgment would dictate that owners or design professionals that don't recognize a contractor's need to make a reasonable profit should be avoided.

CONTINGENCIES

A *contingency* or adding money to address an unknown condition or as a result of underdeveloped documents, is the most misunderstood line item in an estimate. The estimator should try to anticipate any and all costs that are represented by the plans and specifications. However, for project documents that are at various stages of design development (35%, 50%, 75%, etc.) the estimator often adds a contingency to cover what is yet to be designed. These estimates are for cost control of the design and not as an offer to perform the work.

The contingency makes the estimate more representative of the final number of the incomplete design. In this application, the dollar amount or percentage of the contingency is with the full knowledge and concurrence of the owner or design professional. As the design development progresses, the contingency amount should start to go down. With final bid documents (99% complete documents), the contingency should be removed from the bid and transferred to the owner for paying change orders.

For other types of estimates such as a firm fixed price (hard) bid there is no need for a contingency to be added. It is a cost not warranted and self-defeating. Firm fixed price bid requires that the estimator bid what is on the plans and in the specifications. The standard is "what could have reasonably been inferred from the documents at the time of bid." Period. For changes that happen after the contract has been signed and the work has commenced, there is a contractual mechanism for remedy. It is the contract modification, otherwise known as the change order.

There are two schools of thought from the contractor's perspective on contingencies.

Always Add a Contingency

Contingencies should be added for costs that cannot otherwise be recovered. Justifications for contingencies include callbacks that are not the contractor's responsibility but are sometimes done to further a firm's reputation. Consider the residential developer/contractor who repairs damage caused by an unidentifiable party. Repairing the damage is a good business move and keeps the client happy. It also portrays the contractor as a reputable businessperson who stands behind his/her work even when there is a question of who is responsible.

Other scenarios include adding contingencies for "weak" or incomplete documents. It is not uncommon for architectural services to be kept to a minimum in the design stage of a residence. The homeowner's logic may be that if an architect's services can be kept to designing only the essentials, a "good" contractor can flush out the details and make the design work. This is not a sound practice. Even the most conscientious estimator cannot anticipate every condition, unforeseen or otherwise. Again, adding a contingency to help the homeowner or client over some unexpected costs goes a long way toward future business. However, it is important to know when enough is enough. If this contingency amount and terms for use is shared with the owner or the other party to the contract (in the case of the subcontractor and GC) then it is a real contingency. If the contingency's existence is hidden, is not a real contingency but another profit line in the estimate.

Never Add a Contingency

Some people feel that adding any money to an estimate that is not applied to a tangible, defined cost or scope of work is a sign of a weak estimate. It is the purpose of an estimate to accurately anticipate *all* costs to be incurred in a project *based on the documents*. The contract documents, plans, and specifications act as the basis for the estimate. The drawings represent the quantity, and the specifications represent the quality. If an item or scope of work in question is not shown on the drawings or called out in the specifications, it is extra to the contract.

If the estimator is unsure of a detail on the plans it can be resolved through the *RFI* process. A RFI or request for information is a formal process for asking and answering a question regarding the documents and the intention of the design. The answer is intended to clarify or supplement the information on the bid documents. An extra to the scope is dealt with as an equitable adjustment to the contract. Adding money to the estimate for work that is not defined by the documents is often considered irresponsible estimating and nonresponsive to the documents. It will negatively affect the outcome of competitive (hard) bids and should be discouraged.

How to Decide If Contingencies Are Necessary

Other, more general questions arise as a result of the adding contingencies. For example:

- What if the amount of the contingency is insufficient and does not cover the cost of the work it anticipates?
- What if the amount is too much and reduces the competitiveness of the bid?
- Does performing work at no cost to the owner under the guise of a contingency create a dangerous precedent for future uncovered problems?
- Should the owner be aware of the contingency and its amount?
- If the contingency is not spent, is it returned to the owner?
- Are there rules for using the contingency that must be approved by both parties?
- An allowance is different than a contingency. The two should not be confused.

Owner's Contingencies

From the owner's perspective, the term *contingency* is a different matter. In setting the project budget, prior to the award of the contract, it is essential that the owner recognize the potential for changes that can occur during construction. In acknowledging this fact, it is in the project's best interest that the owner set aside a contingency to pay for changes that inevitably occur. This contingency is called a *management reserve*. The design professional should know the project better than anyone else at this stage and is in the best position to determine the amount of contingency required. Contingencies can run the gamut from single-digit percentages to 100% of the contract amount. Sound guidance is essential to ensure there are sufficient funds to pay for changes that inevitably occur.

SUMMARY

It is clear from the discussion in this chapter that there are no hard-and-fast rules for assigning profit or contingencies to an estimate, but merely considerations that must be carefully reviewed for each and every project. Ample thought beforehand, paired with increased experience, will help estimators arrive at the appropriate profit margin for the individual company and project.

In addition to the estimated profit there are other means of increasing the net profit on a project. These often subsidize the estimated profit.

Contingencies have their place in specific types of estimates; however, they are never applied to subsidize poor estimating practice. A real contingency requires that both parties to the contract are aware of its amount and that there are rules requiring the concurrence of both parties as to how it is used. Contingencies are used in specific types of estimates and at specific times in design development.

QUESTIONS/ PROBLEMS

1. Explain why profit is important in an estimate and why work should not be done at cost in a fixed price contract.
2. Why is a contingency added at various stages of design development?
3. Opine as to how an unnecessary contingency could impact a competitive bid.
4. How can questions at the bid phase be addressed?
5. Why is the design professional in the best position to advise the owner on the contingency amount?

27 | Estimating by Computer

Since the mid-1980s, construction companies have increasingly relied on computers for an ever-growing number of purposes. Contractors have even managed to make computers as much a part of the job site as cranes and concrete ready-mix trucks. The introduction of this time-saving tool to an industry always striving for faster ways to produce was a match made in heaven. Integration into large construction firms was almost immediate, with smaller contractors wading in one step at a time.

Contractors currently use computers to communicate with field offices via email, compile labor-hours and costs through job cost systems, track inventory, plan and monitor critical path method (CPM) schedules, and a multitude of other equally important functions. It has become an indispensable part of their daily routine. However, nowhere in the industry has the efficiency of computers simplified and improved the construction process as much as it has in estimating. The ability to accelerate what once were slow, tedious calculations, combined with the accuracy offered by computers, has allowed estimators to perform these basic functions more efficiently and cost effectively. Leaving number crunching to the computer allows more time for strategizing and exploring new methods to perform the actual construction work.

The estimating and bidding process has always been rife with last-minute changes. Bids submitted late from suppliers and subcontractors to prevent "bid shopping" have always proved problematic. How can last-minute changes be made accurately? Computer estimating builds in the flexibility of making changes – even at the last minute – without having to retrace steps or redo calculations. By changing just one number, the entire estimate can be recalculated automatically. Computer estimating is also essential for performing "what-if" calculations when multiple scenarios

Estimating Building Costs for the Residential and Light Commercial Construction Professional,
Third Edition. Wayne J. Del Pico.
© 2023 John Wiley & Sons, Inc. Published 2023 by John Wiley & Sons, Inc.

need to be priced to determine the best approach and corresponding markup. Prior to computers, this process often took a multiperson staff or, at the very least, extra time to check the accuracy of the calculations.

This can now be done in a matter of minutes. Even the cost of performance and payment bonds can be calculated by using a computer-generated algorithm. In the past, this had always been done by hand by dividing the bid price into tiers, then applying the bond premium to each tier.

The evolution of computers in construction has not been without some minor problems, however. Remember, a computer is a tool to increase accuracy and productivity. In that respect, it is no different from an electric saw or drill, and it performs only as well as the level of expertise of the individual operating it allows. In short, good tools don't make good craftspeople. The same applies to estimating. A computer enhances your estimating ability but cannot replace it.

It is also difficult to think that in today's day and age that anyone is still estimating by pencil and paper. Yet many are reluctant to change a tried-and-true method. This is especially true among small residential builders and remodelers. For many, the expense to purchase and commitment of time to learn how to use it efficiently are just too much. This should not be the case.

ESTIMATING SOFTWARE

Computerized estimating provides the distinct advantages of flexibility, efficiency, and accuracy. These are desirable, if not necessary, attributes to succeed in winning projects. Reduced to its simplest terms, an estimating program is a series of mathematical formulas, mainly involving multiplication, that multiply a quantity by a price. A database may be provided as part of a software package, or you can create your own unit prices from a historical database.

Estimating software packages for the construction industry are as varied as the firms that use them. Selecting and purchasing the right software for your company or application can be a daunting task. Some systems are extremely complex and require compatible hardware. These packages often offer links to other modules that provide complete integration. For example, once you complete an estimate and the bid has been won, many packages offer the ability to roll over the estimated costs into a job cost reporting system for performance comparison. Some systems are linked to takeoff software that can be directly input. Other systems are little more than enhanced spreadsheets with built-in macros.

Selecting the Best Software

It is important that the estimating software fit the intended usage. Selecting a complex system that exceeds the company's needs can prevent it from being used to its full capacity and can be costly by requiring support services, training and frequent updates. However, purchasing software too simple for the application can relegate it to the dustheap next to the thermal paper fax machines. While these examples are at extreme opposite ends of the spectrum, they occur with more frequency than one would expect.

Here are some simple suggestions to apply when shopping for estimating software:

- Evaluate the application needs to see if a simple spreadsheet program such as Excel will suffice or whether a more complex system is required.
- Anticipate that the software should meet approximately 85% of the company's application needs. No "canned" software packages will satisfy them 100%, not even today.
- View demonstrations of as many different packages as practical. Ask if the software comes with a demonstration or trial period to ensure that it meets the company's needs.
- For additional verification of its usefulness, request references from contractors in the same market or type of work who currently use the particular software.
- The selected software should allow for some growth. Although a difficult parameter to judge, growth depends on the business cycle of the company. Companies whose sales volume is expanding at an exponential rate may be unable to satisfy this requirement, as they will outgrow software products rapidly as a result of constantly changing needs.
- Make sure the software fits the budget and is cost effective.
- Train the personnel who will use the software – multiple personnel, if possible.
- First investigate non-trade-specific software that may suit your needs before looking into programs designed specifically for a trade, such as electrical, plumbing, or site contractors.
- Assess whether systems can be integrated with other departments, such as job cost reporting, general ledger accounting, or payroll.

While many of these suggestions appear to be common sense, they sometimes get overlooked in the search for ways to increase productivity immediately. Remember, estimating software is not a panacea. It cannot cure bad estimating practices or lack of estimating ability. Regardless of the advertising claims, it will not instantly increase the volume of work. It may, however, eventually allow the estimator to bid more, thereby increasing opportunities.

Do not forget that becoming fluent with any software takes time, and time is an investment of money in the company. There is no avoiding this. Do not spend the money if you are not willing to commit the time to master the learning curve!

Types of Software

Computerized Spreadsheets

Estimating spreadsheets come in a variety of generic applications. The most popular is Microsoft Excel™. This software requires a basic understanding of both the estimating process and the particular application. It is important to note, however, that the takeoff is still performed manually. The spreadsheet can be modeled after a cost analysis form (See Figure 27.1) in a columnar format. Use the following headings as titles for each column, starting from left to right:

- Identification number of the task
- Description of task
- Quantity of task units being priced
- Unit cost for material component
- Total cost for material component
- Unit cost for labor component
- Total cost for labor component
- Unit cost for equipment or subcontractor component
- Total cost for equipment or subcontractor component
- Summary of totals for material, labor, equipment, and subcontractors

Spreadsheet software has become extremely flexible since the first edition of this text. The average estimator can create a unique estimating tool in a short time. The learning curve to efficiently use the tool is also very short. The spreadsheet can be modified as the user determines what works best. While this may not be the purest definition of estimating software, it can often help the user experiment with the type of software he or she is looking for.

Figure 27.2 illustrates the format for a simple estimating spreadsheet built from Microsoft Excel.

Packaged Databases

Construction cost databases may be helpful in meeting a company's estimating needs. They can offer material, labor, and equipment costs, most often in a unit price format, often according to CSI MasterFormat® divisions. Databases can be used as baseline for determining the costs of various tasks or activities. Packaged databases and combined estimating software programs are based on average costs and may require adjustments for a specific location. Note that not all contractors require a packaged database to efficiently estimate by computer. Many contractors and subcontractors whose projects are of a limited scope, or who perform repetitive types of work, have no need for a full 50-division database. In addition, because the costs in packaged databases are subject to change, they will need regular updates.

Comprehensive as they may be, remember that databases may not represent the costs incurred by *your* firm. Some packages allow the user to adjust unit prices and make refinements based on a company's own historical performance of a task or localized pricing of products. Work can be broken down into the CSI MasterFormat® system, by division and section number. The cost for each section can be summarized and linked to a recapitulation or *recap sheet,* which summarizes each section number within the estimate.

COST ANALYSIS

PROJECT

LOCATION

TAKEOFF BY ___ QUANTITIES BY ___ PRICES BY ___ EXTENSIONS BY ___

CLASSIFICATION

ARCHITECT

SHEET NO.

ESTIMATE NO.

DATE

CHECKED BY

DESCRIPTION	SOURCE/DIMENSIONS	QUANTITY	UNIT	MATERIAL		LABOR		EQUIPMENT		SUBCONTRACT		TOTAL
				UNIT COST	TOTAL	UNIT COST	TOTAL	UNIT COST	TOTAL	UNIT COST	TOTAL	TOTAL

Figure 27.1 Cost Analysis Form

SECTION	DESCRIPTION	QUANTITY	UNIT	------MATERIAL------		------LABOR------		---EQUIPMENT----		----TOTAL---	check
				UNIT COST	TOTAL	UNIT COST	TOTAL	UNIT COST	TOTAL	TOTAL	
09 21 00	Gypsum Drywall Systems										
1.01	Partition Type 1	756	SF	$1.54	1,164.24	$1.90	1,436.40	$0.12	90.72	$2,691.36	
1.02	Partition Type 2	900	SF	$1.90	1,710.00	$2.20	1,980.00	$0.11	99.00	$3,789.00	
1.03	Partition Type 3	860	SF	$2.23	1,917.80	$2.21	1,900.60	$0.90	774.00	$4,592.40	
1.04	Partition Type 3A	182	SF	$3.21	584.22	$3.00	546.00	$1.10	200.20	$1,330.42	
1.05	Partition Type 4	200	SF	$2.10	420.00	$2.00	400.00	$0.12	24.00	$844.00	
1.06	Partition Type 5	451	SF	$2.20	992.20	$2.30	1,037.30	$0.00	0.00	$2,029.50	
1.07	Partition Type 6	80	SF	$1.55	124.00	$1.10	88.00	$0.00	0.00	$212.00	
	Gypsum Drywall Systems Totals				$ 6,912.46		$7,388.30		$ 1,187.92	$ 15,488.68	15488.7

Figure 27.2 Sample Estimating Spreadsheet

Not all estimators are looking to generate a competitive bid from the software. Many estimators prepare budgets for appropriations of funds or as part of a capital improvement budget and may be only being looking to "get close." They can close the price gap with a contingency. This is where published or packaged databases have their greatest applicability.

Figure 27.3 illustrates a recap sheet used for summarizing costs contained within the actual cost analysis sheets.

Historical Database

The most accurate costs can be achieved by documenting a contractor's own expenditures on the projects he or she builds and reconfiguring them to be used as a database. This is referred to as a *historical database*, and it is always preferable to packaged software for the estimator bidding work. If historical costs are collected accurately and unit prices are derived correctly, there is no better source of costs than this true representation. A historical database is a collection of "snapshots" of costs along the schedule of a project that has been completed. The costs for work are actual, not presumed, and relate to performance and techniques of one particular crew or individual. A crew or individual that will be doing the work that is being estimated. The thought of collecting and analyzing cost performance data may seem less than exciting, but it is not always as time consuming as one would imagine. Most general contractors have a specialty that they perform and they subcontract the rest. It is the self-performed work that requires the detailed analysis.

One of the byproducts of tracking costs by computer is the ability to track labor-hours. While the importance of historical costs of materials should not be diminished, the real value is the historical cost of labor and its derivative, productivity. Since the majority of production employees are paid by the hour, there is a direct correlation between dollars and time. Although wages change, productivity in performing the same task usually does not fluctuate. This allows the estimator to factor in wage increases by using the hours recorded in the database to predict future costs. The following scenario illustrates this fact.

> *A carpenter was documented over time to be able to hang an average of 1,200 SF of ½" GWB per 8-hour day, under specific conditions. When estimating a second project with similar conditions but a higher wage rate, it could be predicted that the same carpenter will produce the same average. The increase in wages could be adjusted in the unit price for the labor.*

The following is a simple example of how a wage increase affects unit price.

Bid Date: July 30, 2023
Time: 2:00PM

New and Renovated Office Building
Boston, MA

Base Bid: $6,429,634.20
Addenda: #1, #2

Sect.	Description	Materials	Labor	Equipment	Sub	Totals	Remarks
01 00 00	Project Overhead Summary Sheet				$ 405,726	405,726	See Project Overhead Summary Sheet
02 41 00	Selective Demolition				52,200	52,200	Acme Demolition Company
02 81 00	Asbestos Abatement				6,950	6,950	ACM Abaters, Inc.
03 11 00	Cast-in Place Concrete (Formwork)			3,600	108,000	111,600	Cape Pumping Co./Smith Foundations, Inc.
03 21 00	Reinforcing Steel/WWF	17,557				17,557	Rusty's Steel Company, Inc. deliv. bar/WWF
03 31 00	Cast-in Place Concrete (Redi-mix)	26,554				26,554	Tri-County Concrete
03 35 00	Cast-in Place Concrete (Flatwork)			7,169	26,350	33,519	Jones Concrete Finishing Co.
04 22 00	Concrete Unit Masonry				276,500	276,500	ABC Masonry Contractors
05 12 00	Structural Steel				156,200	156,200	State Iron Works, Inc
05 21 00	Steel Joists					–	State Iron Works, Inc
05 31 00	Metal Decking					–	State Iron Works, Inc
05 50 00	Metal Fabrications				80,735	80,735	Columbus Metal Fabricators
06 10 00	Rough Carpentry	7,973	13,248			21,221	Self-performed work
06 20 00	Finish Carpentry	16,785	53,486			70,271	Self-performed work
07 10 00	Dampproofing				17,600	17,600	All Weather Dampproofing, Inc.
07 21 00	Building Insulation	1,841			24,190	26,031	Self-performed work
07 53 00	Elastomeric Roofing/Flashings				126,780	126,780	Sky Roofers, Inc.
08 11 00	Metal Doors and Frames	31,025	8,694			39,719	Megga-Hardware/Self-performed work
08 11 16	Aluminum Doors and Frames				17,584	17,584	Cape Storefronts, Inc.
08 51 13	Aluminum Windows				21,238	21,238	Modern Aluminum Window, Inc.
08 71 00	Finish Hardware	37,000	19,668			56,668	Megga-Hardware/Self-performed work
08 81 00	Glass and Glazing				5,500	5,500	Able Glass Co.

Figure 27.3 Estimate Summary Sheet (Recapitulation)

Sect.	Description	Materials	Labor	Equipment	Sub	Totals	Remarks
09 21 00	Gypsum Drywall Systems				113,910	113,910	Eastern Drywall Systems Co.
09 30 13	Ceramic Tile				32,000	32,000	Western Ceramic Tile
09 51 00	Acoustical Ceilings				83,670	83,670	Capital Ceiling Systems
09 65 00	Resilient Flooring and Base				79,800	79,800	Top-Notch Floors
09 68 00	Carpet				112,000	112,000	Top-Notch Floors
09 91 00	Painting & Coatings				99,780	99,780	Commercial Painters, Inc.
10 11 00	Visual Display Boards				13,500	13,500	Office Interior Contractors
10 14 00	Identification Devices				3,500	3,500	Office Interior Contractors
10 21 13	Toilet Compartments				12,500	12,500	Office Interior Contractors
10 28 13	Toilet Accessories		3,542		3,500	7,042	Office Interior Contractors
10 44 00	Fire Protection Specialties		540		2,000	2,540	Office Interior Contractors
11 40 00	Food Service Equipment				123,490	123,490	The Kitchen Suppliers Co.
12 34 00	Manufactured Plastic Casework				29,750	29,750	New England Casework
12 21 00	Window Blinds				4,214	4,214	Clear-Vue Window Decor, Inc.
13 12 10	Exterior Fountains				34,000	34,000	Pete's Fountain Co., Inc.
14 24 00	Hydraulic Elevators				155,456	155,456	Uptown Elevators, Inc
14 42 00	Wheelchair Lifts				16,883	16,883	Uptown Elevators, Inc
21 00 00	Fire Suppression				65,000	65,000	Safety Fire Protection Contracting Co.
22 00 00	Plumbing				138,900	138,900	Best Plumbing Co.
23 00 00	HVAC				890,990	890,990	New England Pipe HVAC Co., Inc.
26 00 00	Electrical				292,300	292,300	Sparky's Electrical
27 30 00	Voice Communications				19,455	19,455	Best Telephone and Data, Inc.

Figure 27.3 *(Continued)*

Bid Date: July 30, 2023
Time: 2:00PM

New and Renovated Office Building
Boston, MA

Base Bid: $6,429,634.20
Addenda: #1, #2

Sect.	Description	Materials	Labor	Equipment	Sub	Totals	Remarks
28 31 00	Fire Detection and Alarm				28,345	28,345	Sparky's Electrical
31 11 00	Clearing and Grubbing				23,560	23,560	Super Sitework Contractors, Inc.
31 23 00	Excavation and Backfill				329,899	329,899	Super Sitework Contractors, Inc.
31 23 19	Dewatering				13,400	13,400	Super Sitework Contractors, Inc.
31 64 00	Caissons				234,500	234,500	Super Sitework Contractors, Inc.
32 11 00	Gravel Base Course				56,900	56,900	Super Sitework Contractors, Inc.
32 11 26	Bituminous Concrete Paving				93,450	93,450	Hot Stuff Paving Co., Inc.
32 13 13	Concrete Paving (Sidewalks)			1,277	5,785	7,062	Jones Concrete Finishing Co.
32 16 13	Bituminous Concrete Curb				–	–	Hot Stuff (included in 32 11 26)
32 17 23	Pavement Markings				–	–	Hot Stuff (included in 32 11 26)
32 31 00	Fencing and Gates				5,500	5,500	All State Fence Co, Inc.
32 80 00	Irrigation				29,555	29,555	Green Side Up Landscaping Co.
32 92 00	Turfs and Grasses				26,975	26,975	Green Side Up Landscaping Co.
32 93 00	Plants				13,245	13,245	Green Side Up Landscaping Co.
33 11 00	Water Distribution Piping				78,000	78,000	Super Sitework Contractors, Inc.
33 31 00	Sanitary Sewer System				156,000	156,000	Super Sitework Contractors, Inc.
33 41 00	Site Storm Drainage Systems				198,570	198,570	Super Sitework Contractors, Inc.
33 71 00	Electrical Service						Sparky's Electrical (included in 26 00 00)
	Sub Total	138,735	99,178	12,046	4,945,835	5,195,794	5,195,794
	All Risk Insurance					17,666	
	General Liability					41,566	
	Sub Total					5,255,026	
	Main Office Overhead	11.3%				593,818	
	Sub Total					5,848,844	
	Profit	9.0%				526,396	
	Sub Total					6,375,240	
	Performance and Payment Bonds	0.93%				54,394	
	BID					**6,429,634**	

Figure 27.3 *(Continued)*

Project 1 – Cost for carpenter per day:

8 hours × $32.50 per hour = $260 per day ÷ 1,200 SF per day = $0.22 per SF labor cost.

Project 2 – Cost for carpenter per day with increased wage rate:

8 hours × $35.00 per hour = $280 per day ÷ 1,200 SF per day = $0.24 per SF labor cost.

This same example can be used for increases in other labor-related costs, such as insurance rates, benefits, or tax increases. Using a simple spreadsheet application, such as Microsoft Excel™, a file can be created to adjust unit prices as labor costs change or as productivity rises or falls with the change in difficulty of the task.

Materials are somewhat easier to price. The graphic representation of a task on the drawings is finite. In other words, the estimator can quantify the material by using measurements from the drawings. Requesting a proposal for materials from local suppliers is usually sufficient to capture the costs of the total materials needed for the estimate. The ambiguous portion of the materials cost is waste. A well-documented database can provide sufficient detail for the comparison between the net and gross quantities of purchased materials. Over multiple projects, the estimator can adjust the factor for waste based on the actual conditions.

Creating Estimating Templates

Most contractors and subcontractors create similar estimates over and over. Estimating by computer allows the estimator to build a template for a particular type of estimate. Each time a new project is to be bid, the estimator can copy the template for that type of estimate and make the necessary adjustments for the individual projects. This is a common time-saving feature in construction estimating programs, even those created from generic spreadsheet software. Figure 27.4 illustrates a template estimate.

For example, consider a drywall and metal stud subcontractor who is estimating interior drywall partitions for an office building. The unit price for each type of partition could be built into a template by the SF or LF, so all that is required is performing the takeoff, entering quantities of the various types of partitions, and making adjustments for specific project conditions, including overhead and profit. By perfecting this template, the contractor could bid more work, thereby increasing the chances for winning work.

QUANTITY TAKEOFF SOFTWARE

Since 2004, when the first edition of *Estimating Building Costs* was written, there have been enormous advances in software for performing takeoff. In the past, takeoff by computer required a digitizer (wand) attached to a computer, some expensive hardware and software, and a lot of training. Takeoff was done by attaching the paper plans to an integrated work surface and using a

Project Name: New Office Building Drywall
Location: Boston, MA
Architect: ABC Architects, Inc.
Scope: Section 09 21 00 - Gypsum Drywall Systems

Estimate No. 1
Bid Date: 02/08/23

Pages: 1 of 1
Date: 30/07/23
Estimator: WJD

SECTION	DESCRIPTION	QUANTITY	UNIT	-----MATERIAL-----		------LABOR------		----EQUIPMENT----		----TOTAL---		check
				UNIT COST	TOTAL	UNIT COST	TOTAL	UNIT COST	TOTAL	TOTAL		
09 21 00	**Gypsum Drywall Systems**											
1.01	Partition Type 1	756	SF	$1.54	1,164.24	$2.90	2,192.40	$0.15	113.40	$3,470.04		
1.02	Partition Type 2	900	SF	$1.90	1,710.00	$2.50	2,250.00	$0.25	225.00	$4,185.00		
1.03	Partition Type 3	860	SF	$2.23	1,917.80	$3.21	2,760.60	$0.90	774.00	$5,452.40		
1.04	Partition Type 3A	182	SF	$3.21	584.22	$3.33	606.06	$1.20	218.40	$1,408.68		
1.05	Partition Type 4	200	SF	$2.10	420.00	$2.55	510.00	$0.35	70.00	$1,000.00		
1.06	Partition Type 5	451	SF	$2.20	992.20	$2.30	1,037.30	$0.90	405.90	$2,435.40		
1.07	Partition Type 6	80	SF	$1.55	124.00	$2.10	168.00	$0.92	73.60	$365.60		
	Gypsum Drywall Systems Sub Totals				$ 6,912.46		$9,524.36		$ 1,880.30	$ 18,317.12		18317.1
	Sales Tax on Materials	6.25%								$ 432.03		
	Sub Totals									$ 18,749.15		
	Main Office Overhead	10.55%								$ 1,978.04		
	Sub Totals									$ 20,727.18		
	Profit	15.00%								$ 3,109.08		
	Sub Totals									$ 23,836.26		
	Performance and Payment Bond	1.50%								$ 357.54		
	Total									**$ 24,193.81**		

Figure 27.4 Estimating Template

Project Overhead Summary Sheet

Project: New & Renovated Office Building
Architect: ABC Architects, Inc.
Duration: 18 months (78 weeks)
Liquidated Damages: N/A

1000	Description	Quantity	Unit	Cost	Total	Remarks
1.01	Office Trailer - G.C.	18	mon	345.00	6,210.00	Share trailer with Owners' Rep.
	a. Furnishings	1	LS	800.00	800.00	
	b. Setup and delivery	1	EA	340.00	340.00	
	c. Breakdown and return	1	LS	340.00	340.00	
	d. Trailer cleaning	18	mon	45.00	810.00	
1.02	Storage Trailers	14	mon	120.00	1,680.00	
	a. Delivery and Pickup	1	LS	240.00	240.00	
1.03	Office Trailer - Clerk or Arch.				–	See spec section 1500- C.2
	a. Furnishings	1	LS	530.00	530.00	See list in Section 01500;C.2
	b. Fax machine	1	EA	450.00	450.00	
	c. Copy machine	1	EA	800.00	800.00	
	d. Answering machine	1	EA	75.00	75.00	
	e. Desk phone	1	EA	20.00	20.00	
	f. Computer/printer					
	g. Supplies	1	LS	500.00	500.00	
	h. Toilet hookup and dismantle		LS	600.00	–	
	I. Software		LS	600.00	–	
1.04	Telephone service				–	
	a. Install and removal	3	LINES	125.00	375.00	See Section 01500-Temp Facilities
	b. Monthly serv.- GC	36	mon	175.00	6,300.00	
	c. Monthly serv.- Clerk	18	mon	150.00	2,700.00	
1.05	Temporary Electric				–	
	a. Office Trailers hookup	3	EA	500.00	1,500.00	
	b. Trailer consumption	18	mon	175.00	3,150.00	
	c. Project consumption	18	mon	150.00	2,700.00	
1.06	Water cooler/consumption	18	mon	30.00	540.00	
1.07	Thermometer	1	EA	20.00	20.00	
1.08	Temporary Toilets	36	mon	85.00	3,060.00	
1.09	Temp. Construction Fence	985	LF	4.00	3,940.00	
1.10	Staging					
	a. Set up/dismantle	1	LS	2,000.00	2,000.00	
	b. Monthly rental	18	mon	200.00	3,600.00	
1.10a	Ramps to trailers		LS	–	–	
1.11	Manlifts				–	
	a. Delivery and pickup				–	
	b. Monthly rental			–	–	none required
1.12	Small tools and equipment	1	LS	2,000.00	2,000.00	

Figure 27.5 Project Overhead Summary Sheet

1000	Description	Quantity	Unit	Cost	Total	Remarks
1.13	Temporary water				–	
	a. Hook up/dismantle	1	LS	200.00	200.00	Hook up and dismantle by Plumber
	b. Consumption	18	mon	50.00	900.00	
	c. Fees	1	EA	250.00	250.00	Meter rental from Municipality
1.14	Temp. Heat				–	
	a. Trailers				–	In section 1.05 above
	b. Project	5	mon	700.00	3,500.00	
1.15	Temporary Protection	6	mon	300.00	1,800.00	
1.16	Winter Protection					
	a. Plowing		mon			
	b. Enclosures	1	LS	8,000.00	8,000.00	Enclose & remove staging for masonry
	c. Heat			–	–	
1.17	Fork lift or Lull				–	
1.18	Crane	4	days	1,100.00	4,400.00	Hoist for roofer
1.19	Project Photos	1	LS	200.00	200.00	
1.20	Tree Protection				–	Carried in sitework proposal
1.21	General cleaning - ongoing	20	WKS	1,439.00	28,780.00	25% of the schedule
1.22	Final Cleaning	15000	SF	0.20	3,000.00	
1.23	Materials handling & distribution	40	Mnhrs	36.00	1,440.00	Receiving and handling material
1.24	Project Sign	1	EA	750.00	750.00	Included in sub bid from Acme Signs
1.25	First Aid kits	2	EA	75.00	150.00	
1.26	Temporary Fire Protection	1	LS	150.00	150.00	
1.27	Dumpsters- 30 CY	45	EA	620.00	27,900.00	1 dumpster per 2000 S.F floor space.
					–	
1.28	Pest Control		LS	500.00	–	
1.29	Cutting and Patching					All cut and patch over 6" dia by GC
a.)	Labor to core holes	120	Mnhrs	36.00	4,320.00	Direct labor by employees
b.)	Purchase coring machine/bits	1	LS	2,000.00	2,000.00	
1.30	Permits				–	
	a. Building permit	1	LS	8,500.00	8,500.00	Based on budget value
	b. Occupancy permit				–	
	c. Miscel. fees				–	
1.31	Police Details	2	Days	248.00	496.00	at Street opening and patch
1.32	Layout				–	
	a. Registered	7	Days	1,080.00	7,560.00	Control provided by Surveyor
	b. Own forces				–	By Superintendent
1.33	Testing				–	
	a. Soil testing			–	–	By Owner
	b. Concrete testing			–	–	By Owner

Figure 27.5 *(Continued)*

1000	Description	Quantity	Unit	Cost	Total	Remarks
1.34	Miscel. Hardware	18	mon	120.00	2,160.00	
1.35	Pickup Trucks		mon	–	–	
	a. Gasoline usage	78	Wks	50.00	3,900.00	
1.36	CPM Schedule-Initial Devel.		LS			By Project Manager
	a. Update CPM	17	mon	100.00	1,700.00	By Project Manager
1.37	Dewatering				–	
	a. Localized dewatering				–	By Site Contractor
1.38	Special safety equipment	1	LS	400.00	400.00	Rebar caps 300
1.39	Attorneys Fees				–	
1.40	Interior Barricades	4	mons	300.00	1,200.00	Restrict access between reno/new
1.41	As-Builts				–	
	a. Microfilm				–	
	b. Printing /Reproduction				–	
	c. Mylars	1	LS	3,000.00	3,000.00	Complete set
1.42	Project Closeout	2	Phases			By Project Manager
1.43	Site Security					
	a. Watchman					
	b. Custodial Overtime					
1.44	Utility Company Charges					
	a. Electric					
	b. Water Taps					
	c. Sewer					
	d. Gas					
	e. Cable/Tel.					
	f. Assements					
1.45	Special Requirements					
1.46	Insurance					
	a. Builders Risk	1	LS	–	–	Included in Estimate Summary Sheet
	b. 3YR extended comp opps	1	LS	–	–	Included in Estimate Summary Sheet
	c. 10M add umbrella	1	LS	–	–	Included in Estimate Summary Sheet
1.47	Punchlist	10	WKS	1,623.00	16,230.00	
1.48	Personel				–	
	a. Superintendent	78	WKS	1,920.00	149,760.00	100% of schedule
	b. Asst. Superintendent		WKS			
	c. Project Manager	28	WKS	2,000.00	56,000.00	35% attention
	d. Adminsinstrative Staff	28	WKS	800.00	22,400.00	35% attention
	TOTAL				405,726.00	

Figure 27.5 *(Continued)*

combination of point-and-click tasks with some data entry. New advances in data file storage capacity, the Web, and software no longer require all the original components. In fact, takeoff can now be done on a laptop while traveling at 30,000 feet in an airliner, with the only requirement being the appropriate takeoff software and internet access.

Many architects, engineers, and owners are providing bid documents as electronic files for general contractors and their subcontractors to use in bidding projects. For quite a few contractors, the plans are never even printed unless they are awarded the project! The architect posts the plans to an *ftp site* (File Transfer Protocol site). The contractor logs on with the architect's permission, downloads the files, and performs the takeoff and resultant calculations to arrive at the quantities necessary. The estimator then transfers the quantities to the estimating program to generate a bid. Some takeoff programs are actually integrated with the estimating program so that the takeoff becomes the estimate once prices have been applied.

Again, a contractor has to be of sufficient size to justify the expense of takeoff software, and the software has to be used if it is going to create benefits. No contractor would buy a new tool and then leave it in the box – unused!

As a general note, the author has tried several quantity takeoff software packages with varying degrees of success. Most programs are available with a free trial period to allow estimators to see if it will meet their needs. Programs without the free trial might not be worth the investment. A general bit of advice: If one cannot invest the time to learn the software, then one should not invest the money to purchase it.

PROJECT OVERHEAD SUMMARY SHEETS

Computerized project overhead summary sheets allow the estimator to view different project overhead costs based on different schedules. All projects have two components for overhead: direct costs and indirect costs. Indirect overhead is usually applied as a percentage of the total overhead of the main office. Direct overhead costs are those nonproduction costs that are unique to each project. They include such items as site supervision, temporary facilities, building permits and fees, and dumpsters. Some of these costs are fixed, such as for a building permit, but many are time sensitive and vary with the amount of exposure on the project. The project overhead summary sheet can be used as a checklist for reference. For a more detailed discussion of indirect and direct overhead costs, refer to Chapters 5, "Understanding the Cost Elements of a Unit Price" and Chapter 6, "General Requirements," respectively. Figure 27.5, parts a and b, illustrates a project overhead summary sheet using a spreadsheet.

SUMMARY

In summary, computers have a tremendous value in construction applications. Estimating by computer provides flexibility, accuracy, and speed in producing bids. The computer's value is limited only by the construction and estimating knowledge – as well as computer skills – of its users. Those looking to purchase computer estimating and quantity takeoff programs should do so only after careful review and understanding of the company's needs.

For many smaller volume contractors, a simple spreadsheet crafted from Microsoft Excel™ maybe all that is needed. As the company grows and the volume of work increases, the company can look into more trade specific software.

QUESTIONS/ PROBLEMS

1. What is the main purpose of estimating software?
2. Why is it important to carefully research software for estimating?
3. What is a packaged database? Who typically uses packaged databases?
4. What is a historical database? Who typically uses historical databases?
5. Why are computerized project overhead sheets important to a bidder?

28 | Conceptual Estimating

Thus far, the discussion of estimating practice has been limited to one type of estimating, the unit price method. As mentioned earlier, the unit price estimate is the takeoff or quantification of the plans in incremental parts. It is a dissection of the project into each component almost to the individual nut and bolt. This is the method most often used by contractors in the development of competitive bids. The unit price method of estimating requires a comprehensive set of plans and specifications that provides details on every aspect of the project.

It should come as no surprise that comprehensive bid documents are not always available. In fact, they are *rarely* the norm. It is not uncommon to receive a set of plans that is incomplete and without specifications, yet the estimator is asked to prepare a price. This is especially common in the residential and light commercial market.

Many architects and engineers are required to work within a budget for construction, so in order to keep a design on target for cost, interim estimates are prepared as a way to gauge the direction of the design. This process avoids putting the project out to bid when documents are complete, only to have the bids come in substantially higher than available funding. Estimates done at specific points in the design development often help the owner and design professional maintain a budget.

Other reasons for early estimates with incomplete documents might include the development of preliminary budgets to start the financing process or to determine if, based on the cost of the project, it makes financial sense to proceed. There are many reasons to account for this kind of estimate.

The preparation of estimates with incomplete or missing drawings and specifications is called *conceptual estimating*. This chapter explores some of

Estimating Building Costs for the Residential and Light Commercial Construction Professional,
Third Edition. Wayne J. Del Pico.
© 2023 John Wiley & Sons, Inc. Published 2023 by John Wiley & Sons, Inc.

the more common conceptual estimating methodologies used today and what can be expected in the way of accuracy.

CONCEPTUAL ESTIMATING

Conceptual estimating is the domain of the experienced estimator. It is not a guess at cost. It should be a logical, defendable calculation. It requires that estimators fill in the blanks created by the missing plans, details, and specifications. They must understand and apply a mix of industry standards, building codes, experience, and owners' requirements to arrive at an anticipated approximate cost of construction for the completed project. It may also include some assumptions about the direction of the project design, including escalation costs. The estimator and/or owner and design team set parameters as a basis for pricing, which explains why conceptual estimating is sometimes referred to as *parametric estimating*.

It should be noted that conceptual estimates are not intended to be firm prices but preliminary pricing at various stages of design development. That having been said, it is all too common for an eager contractor to negotiate a contract based on incomplete documents and a conceptual estimate. This is manageable if the contractor has some flexibility in revising the price as the design matures. In fact, this is frequently done using a Guaranteed Maximum Price (GMP) or a Design-Build delivery method contract. It is also common under the often-misused fast-track methodology, whereby the building design is being completed even though construction has commenced.

In addition to experience, the estimator is required to have a fully developed database of historical costs or access to detailed published construction costs. While accurate assumptions and pricing are important, there is less reliance on contemporaneous costs such as subcontractor proposals unless they are specifically invited to participate in the preparation of the estimate. Frequently, the architect prepares the conceptual estimate, which may be part of the architect's fee or an added service. This is acceptable so long as the architect has the estimating experience and up-to-date cost data.

This discussion considers three types of conceptual estimating methods, their use, their limitations, and the level of accuracy that can be anticipated. Each can provide significant value to the estimating process when prepared and used appropriately:

- Order-of-magnitude estimate
- Square-foot estimate
- Assembly or systems estimate

Order-of-Magnitude Estimate

The order-of-magnitude estimate, sometimes called a rough order of magnitude (ROM), is the 10,000-foot view or big picture estimate. It frequently relies on parameters that are not typical construction parameters. For example, the ROM for a dormitory may be expressed in dollars per bed, or the cost of an apartment building would be reported in dollars per apartment.

ROMs are planning tools that rely heavily on historical or published cost data. They rarely have any bid documents at all. They are used as a basis in early decision-making. The estimator may have little more to go on than criteria provided by the developer or investor. The developer might make the following request of the estimator:

What would it cost to build a 200-bed nursing care facility?

In response, the estimator would prepare a list of questions to help define or qualify the facility. He or she might ask the following:

- Location: urban or suburban
- Class of construction: wood frame, steel, brick, and glass, etc.
- Labor rates: union or open-shop labor
- Quality: luxury, average, or economy finishes
- Amenities: pharmacy, nursing stations, public dining room, etc.
- Beds per room

With some basic questions answered, the estimator would use historical cost data for completed projects of a similar nature to generate the cost of this project. This is really the domain of the seasoned professional with experience in this type of project. In the absence of historical cost experience, the estimator would rely on published cost models such as those provided by the RSMeans data from Gordian®.

Total time invested in the estimate may be no more than an hour or two. The level of accuracy would be based on the data used, but an accuracy of ±25% is the average range.

Square-Foot Estimate

Another method of conceptual estimating uses the gross square foot (SF) area of the structure as the reporting parameter. This is likewise meant as a planning tool. Square-foot estimates also rely heavily on historical costs from projects of a similar nature, type, and size. They are based on the premise that all building types can be distilled to a cost per SF of gross floor space. This is an especially useful tool for budget planning and decision-making and even prequalifying clients.

Consider a homebuilder who builds several models with a range of amenities. One particular model – let's call it the "Early American" – is a two-story colonial with four bedrooms, two and a half baths, a two-car garage, and a full basement. It will be 2,000–2,250 SF (exclusive of the garage and basement). The base model includes regular appliances, plastic laminate countertops, white bathroom fixtures, and forced warm-air heating with AC. The luxury model includes designer appliances, ceramic tile, choice of color for the bath fixtures, and forced hot-water heating. Over a very successful previous year, our builder has been able to accurately track the cost of each Early American home built with the amenities selected for each one. Using the data derived from these costs of previous Early American models, the builder is able to establish with precision a range of costs that can be reported as a cost per SF.

The builder can, with reasonable confidence, translate the SF area and specific amenity requests of a potential client into a range of costs based on the cost per SF obtained from previous builds. For example:

If costs for the Early American models ranged from $182.00/SF to $200.00/SF, then a 2,100-SF model would cost $382,200 to $420,000.

If the potential clients were satisfied with this range, the builder could proceed to the next step of developing a more definitive estimate. If they were not satisfied, a minimum of estimating time was invested.

This method can be used for numerous types of structures. Again, in the absence of historical data, the builder would rely on published cost data such as *square foot cost data* published annually by the RSMeans data from Gordian®. Accuracy is again dependent on the quality and detail of the historical or published cost data, but generally the accuracy is limited to ±20%.

Contractors can often use multiple subcontractor bids from projects completed or even projects bid but not won to develop square-foot cost data for plugs (place holders) in unit price estimates. These costs can be used to generate square foot ranges – low, average, and high costs for, say, a wet-pipe sprinkler system for a particular type of building. Next time the estimator bids this type of project, a plug can be added to the estimate until the subcontractor's bids are received. They can also be a means to review the new prices.

Assemblies or Systems Estimate

A more detailed type of conceptual estimate is called the *assemblies* or *systems* estimate. In this methodology, individual systems or groups of tasks assembled together are estimated as a single unit. This methodology can be used at almost any level of design development. It requires that there be decisions made on major systems and building components, even though they may be subject to change as the design develops. It allows the estimator the maximum amount of flexibility for making changes to systems without an entire revision of the estimate. For example, a heating system for a home could be priced as two different systems – forced warm air or forced hot water (baseboard radiant). The entire system can be priced by the square area being heated and the merits of each could be discussed along with the price of the system.

In contrast to the CSI MasterFormat® that is employed as the organizational structure for unit price estimating, the assemblies estimate uses its own formatting, called *Uniformat II.* Uniformat II organizes the project into major groups. Each group and the systems included are outlined as follows:

- **A—Substructure:** Includes the work of the foundation: slab, basement or stem walls, footings, grade beams, piles, caissons, excavation, backfill, and compaction for the foundation systems.
- **B—Shell:** Includes the exterior envelope and closure of the structure, including exterior walls, openings, roof systems, structural steel or wood frame, and floor systems.
- **C—Interior Construction:** Includes partitions, ceilings, doors, finishes, and interior systems.
- **D—Systems:** Subdivided into groups of interior building systems such as:
  - **Fire Suppression:** Includes fire suppression systems throughout the building.
  - **Conveying:** Includes elevators, escalators, material moving systems, and dumbwaiters.
  - **Plumbing:** Includes plumbing fixtures and trims, piping; hot and cold water, waste and vent, and plumbing appliances such as hot-water heaters.
  - **HVAC:** Includes heating, ventilation, and air-conditioning systems and their subsystems such as controls.
  - **Electrical:** Includes power and lighting distribution and panels, fire detection systems, communications, and data wiring.
- **E—Equipment and Furnishings:** Includes security, mercantile, ecclesiastic, theater, library, medical, food service, darkroom, and athletic equipment, as well as window treatments.
- **F—Special Construction:** Includes special systems of buildings such as special-purpose rooms, radiation protection, liquid and gas storage, and ice rinks.
- **G—Sitework:** Includes rigid and flexible pavements, landscape, utilities, grading and drainage, and miscellaneous site items.

Assemblies estimating uses a combination of industry standards and parameters to establish the value of a building system. Consider the following example:

> *The estimator is preparing an assemblies estimate for the HVAC on a two-story office building of 20,000 SF. Based on the occupancy and industry standards for the area in which the office is being constructed, it is determined that 1 ton of cooling will handle 300 SF of floor space. If the total floor area of 20,000 SF is divided by 300 SF per ton, the estimator arrives at 66.66, rounded off to 67 tons of cooling required. The historical cost data reveals that for projects of a similar size and type, HVAC cooling costs $5,000 per ton, including RTUs, ductwork, insulation, and piping. If those costs were extended for this project:*
>
> 67 tons × $5,000 per ton = $335,000

Assemblies estimating saves time while still providing a reasonable degree of accuracy: ±15%.

Another example is to develop an assembly cost for a partition type. Instead of estimating each task separately, the assembly would create one price per linear foot (LF), or SF, for the metal stud framing, gypsum wallboard, tape and finish of both sides, and the insulation within the partition.

An estimate of this type helps design-build contractors decide on the type of system to suggest to the client.

SUMMARY

In summary, not every estimate prepared has detailed plans and specifications as required for the unit price estimate. When documents are at various stages of design development, estimates are prepared for cost control. Conceptual estimating is a valuable tool for early decision-making or to decide if a project is financially viable. It can also be used to stimulate investors or lenders. For many projects it is the basis of negotiation between the contractor and the owner.

Conceptual estimating is a very important part of the estimator's profession, although it requires experience as a unit price estimator to be able to develop the conceptual estimate. The three different methodologies allow different levels of accuracy and detail with varying degrees of estimating time invested. All three are meant to be used as planning tools and not as competitive bid tools.

QUESTIONS/ PROBLEMS

1. What is the purpose of conceptual estimating?
2. Provide a brief explanation of ROM estimating.
3. Provide a brief explanation of square foot estimating.
4. What is the difference between SF estimating and assemblies estimating?
5. Why is unit price estimating the most accurate when done correctly?

Solutions to Questions And Problems

CHAPTER 1: THE BASIS OF THE ESTIMATE

1. What role do the plans play in the estimating process?

    The plans serve as the graphic representation of the project. They provide dimensions to derive quantities and counts. They are the quantitative portion of the estimate.

2. Identify the six (6) main graphic formats that a reader would expect to see in a set of plans.

    The six main graphic formats are: plan view, elevation, section, details, schedules, and diagrams. They are used to convey information on plans. One feature may be shown in multiple views as a means of confirming another view.

3. Name five (5) things a reader would expect to see in the title block of a drawing.

    The title block contains the information about the design professional, the drawing name, the drawing number and prefix, the date of the drawing, and revisions, project number, and the draftsperson.

4. Define extension lines and dimension lines. Ensure that the definition includes how they are used together.

    The extension lines are the limits of the dimension line. They are typically 90° to the main object line. The dimension line extends from extension line to extension line and provides the dimension of the feature.

5. What is the purpose of a revision marker on a set of plans?

    Frequently after the plans have been released, changes are required for clarification, correction, or to supplement an omission. This is done by issuing a modified plan that has the revised detail(s) enclosed in a scalloped line with a triangle that represents the number of the revision. The revision markers are noted in the title block along with the date issued. This allows the reader to know if he/she are viewing the latest version of that drawing.

Estimating Building Costs for the Residential and Light Commercial Construction Professional, Third Edition. Wayne J. Del Pico.
© 2023 John Wiley & Sons, Inc. Published 2023 by John Wiley & Sons, Inc.

6. Explain why design professionals use abbreviations on a set of plans.

   Abbreviations are substituted for words or phrases that are used frequently on plans. It helps avoid clutter on the plans and makes them easier to read.

7. What is a contour? Include in your explanation what information can be derived from a contour.

   A contour is a line or series of lines that connect points of equal elevation. It is a technique used on site plans that allows the design engineer to convey the elevation of a feature or plot of land above a known datum. It shows three dimensions length width and height on a two-dimensional medium.

8. What can a reader expect to see on an *Existing Conditions* plan?

   Existing Conditions plans shows the current conditions and features of the land scheduled for improvement. It helps the estimator determine the work needed to go from the current state to the proposed improvements.

9. Define the term *benchmark* and explain its purpose for the estimator.

   A benchmark is a known elevation used as a reference point on a set of civil or site plans. The benchmark is established with reference to a datum such as sea level. Benchmarks are used to establish grades for calculating changes in vertical elevation.

10. Explain the limitations of a geotechnical report for an estimator.

    Geotechnical reports are the results of subsurface investigations of the existing soils. They come with the disclaimer that the soil conditions are only representative of what is in the boring and not indicative of the surrounding area. It is no guarantee that the conditions are ubiquitous on the site. The estimator is cautioned that he/she uses the information at their own risk.

CHAPTER 2: UNDERSTANDING THE SPECIFICATIONS

1. Identify the key functions of the specifications. Include their relationship to the plans in the bidding process.

   The specifications are part of the bid documents used by the estimator. They define the quality of the materials and labor illustrated on the plans. They are the qualitative portion of the bid documents that are used in conjunction with the quantitative plans to arrive at an accurate estimate of the work. The bid is reflective of both the quantity shown on the plans and the quality defined in the specifications.

2. Name the four categories of CSI MasterFormat® including a brief description of what can be found in each.

   CSI MasterFormat® groups the information in the project manual into 4 main categories: (1) the Bidding Requirements define what is required of the bidder to be considered responsive. It includes bid forms, instructions to the bidders, and any information available to the bidders, (2) contract forms are added to provide the bidder a copy of the contract he or she will be expected to sign if awarded the project, (3) the General Conditions of the Contract for Construction is a legal instrument that supports the contract. It provides more detail as to the rights and responsibilities of the parties, and (4) the technical specifications define the scope, products, and execution of the work. It defines the administrative requirements, the quality or governing standards, the products, and the installation.

3. Explain what information is contained in Division 1 of the CSI MasterFormat®.

    Division 1 is the General Requirements. It contains much of the non-production type or direct overhead costs associated with the project in such a manner as to allow the estimator to price their cost. It also identifies other contractual obligations such as project time.

4. Define the term *nonresponsive* as it applies to the submission of a bid.

    Nonresponsive in relation to the submission of a bid indicates that the bid does not meet the requirements to be eligible for award. It is typically applied to a bid form that is incorrectly filled out or is missing information or contains qualifications not requested.

5. Provide a detailed explanation as to why the contract form is included in the bid documents.

    The contract itself is included so the bidder can analyze the contract they will be expected to sign should they be awarded the project. If a bidder is unable to agree with the terms of the contract, they should consider refraining from bidding the project.

6. Explain why the General Conditions of the Contract for Construction is included in the contract by reference. Explain its purpose.

    The General Conditions of the Contract are a supporting document to the Contract for Construction. They are guidelines that define the rights and responsibilities of the parties and the complex legal relationship between the parties. They provide procedures and mechanisms for resolving disputes and clarifying information. They are broken down into 15 articles that address specific conditions and terms. Its purpose is to provide additional substance to the contract itself.

7. What does Article 7 of the General Conditions of the Contract for Construction govern?

    Article 7 defines the procedures for handling changes in the work during the life of the contract. It defines procedures for work that is in dispute as to scope, price, or time.

8. Identify the three (3) parts of a technical specification section. Provide a brief description of what information is contained in each part.

    The three (3) parts of a technical specification section are: (1) the General which provides a summary of the work and related work, it contains the flow-down provision that ties this section to Division 1, (2) the Products and accessories that are incorporated within the work of this section, and (3) the Execution defines the methods, techniques, and quality of the workmanship of this section.

9. Explain in detail how conflicts between plans and specifications should be handled.

    Conflicts between plans and specifications, or any other discrepancy identified during the bid phase, should be brought to the attention of the design professional for clarification. The resulting clarification is then shared with all bidders, which assures each bidders has the same information.

10. Explain the purpose of Bid Alternates. Include in the explanation why they are used and how they are selected.

> *Bid Alternates are a way of determining the value of a separate item or process. They can be additive or deductive thereby allowing the owner/awarding authority to meet a budget. Alternates are intended to be selected in the order in which they are offered to prevent any type of bid rigging during award. In other words, if an owner or awarding authority wants Add Aternate#3, they must also accept Alternates #1 and #2. This prevents the misuse of the system by selecting the alternates that result in the preferred bidder. By requesting pricing of Alternates in the bid phase, the pricing remains competitive.*

CHAPTER 3: CALCULATING LINEAR MEASUREMENT, AREA, AND VOLUME

1. Convert the following dimensions to their decimal equivalent.
   a. 19'-10" = *19.83'*
   b. 0'-3½" = *0.29'*
   c. 33'-7¾" = *33.65'*

2. Calculate the perimeter of a hexagonal shape if each side is 14'-7". Provide the answer in decimals.

> *A hexagon has six (6) sides; therefore, 14'- 7" × 6 sides = 87.5'.*

3. What is the hypotenuse of a right triangle with a base of 12'-7" and an altitude of 8'-5"? What is the area of the same triangle?

> *The hypotenuse of a right triangle uses the formula $C = \sqrt{A^2 + B^2}$ therefore, converting the values to decimal $C = \sqrt{8.42^2 + 12.58^2} = \sqrt{228.14}$ or 15.10'. The area of the same triangle is calculated by using the formula $A_T = ½ (b \times h)$ where b = base and h = height, or $A_T = ½ (12.58' \times 8.42') = 52.96$ SF.*

4. What is the circumference of a round patio that has a 12'-8" radius? Provide the answer to two places after the decimal.

> *The formula for the circumference of a circle is $2 \times \pi \times R$ where $\pi = 3.14$ or $2 \times 3.14 \times 12.67' = 79.56'$*

5. What is the area of a sector that has a radius of 9'-5" and an interior angle (between two radii) of 67°?

> *The formula for the area of a sector is $n/360° \times A_c$ where n = the interior angle between the radii in degrees and $A_c = \pi \times R^2$ or $A_s = n/360 \times A_c$, where n = 67/360 = 0.186 × 3.14 × (9.42)² = 51.83 SF*

6. A concrete form is 36' long, 2'-4" deep, and 3'-0" wide. Calculate the amount of concrete needed to fill the form. Round the answer up to the nearest cubic yard (CY).

> *Concrete is a volume calculation therefore 36' × 2.33' × 3' = 251.64 CF. Converted to CY = 251.64 CF ÷ 27 CF per CY = 9.32 CY rounded to 10 CY.*

7. There are seven (7) concrete columns that require their surface be painted. The dimensions of each are 20'-6" high and 8'-0" in diameter. What is the total surface area requiring paint? Round the answer up to the nearest square foot (SF).

> *To calculate the surface area of a column, start by determining the circumference using the formula: $C = \pi \times D$, where; D = the diameter*

of the circle. Therefore C = 3.14 × 8' = 25.12' now multiply be the height of the column; 25.12' × 20.5' = 514.96 SF × 7 columns = 3,604.72 SF rounded to 3,605 SF.

8. Calculate the net area of brick for an exterior wall that is 70'-0" long by 20'-8" high and has four (4) windows that are each 8'-0" × 6'-0". If there are 6.55 brick required per SF of wall, how many brick are required?

   Start by calculating the gross area of the wall: 70' × 20.67' = 1,446.9 SF then deduct the windows: 8' × 6' × 4 windows = 192 SF therefore the net area is 1,446.9 – 192 = 1,254.9 SF. Now multiply by 6.55 bricks per SF to find the bricks needed. 1,254.9 SF × 6.55 bricks/SF = 8,120 bricks (exclusive of waste).

9. A parking lot is being repaved, the area of the parking lot is 7,680 SF and the thickness of the new surface is 2½". How many tons of paving are required if 1 CY of asphalt paving weighs 2.025 tons?

   Start by calculating how many CYs are in the parking lot. 7,680 SF × 0.21' (thick) = 1,612.8 CF ÷ 27 CF per CY = 59.74 CYs. Now extend the CYs to tons: 59.74 CYs × 2.025 tons/CY = 120.97 tons or rounded to 121 tons.

10. What is the volume of soil to be excavated from the trench that is 4'-0" deep, 100 ft long, and a width of 4'-0" at the bottom of the trench? A 45° (1:1) slope must be maintained at the sides of the excavation for safety. Round up to the nearest whole cubic yard (CY).

    Using the formula and procedure identified in the chapter:
    A_2 = 4' × 4' = 16 SF, plus the sloped cross sections; A_1 = (4' × 4') ÷ 2 = 8 SF + 8 SF for A_3, which is the mirror image on the opposite side of the trench. This results in a cross-section area of 32 SF. Now multiply by the length of the trench; 100' × 32 SF per ft. = 3,200 CF ÷ 27 CF per CY = 118.52 CYs or rounded to 119 CYs.

CHAPTER 4: THE QUANTITY TAKEOFF AND PRICING PHASES

1. Why is it essential to use industry-accepted units of measure when doing the takeoff?

   Using industry-accepted units of measure helps to reduce chance for error when the takeoff units are priced. Pricing and takeoff units should be the same. They should be industry standard units of measure.

2. What is an element of a task? Explain your answer with an example.

   A task can have three elements: material, labor, and equipment. These are the individual components that make up a task. They can have one, two or all three elements in a single unit price for a task. For example: a unit price can have a material and labor element in a line.

3. What are the three reasons for adding waste in the estimate?

   Waste is the difference between what is needed and what is purchased. The three primary reasons for waste are: (1) to adjust for the standard sales unit of the product, (2) to account for waste due to handling of the material, and (3) to achieve a specific lap. A single product can have waste factors for one, two or all three reasons.

4. Why is it necessary for an estimator to adjust the quantities to the standard sales unit?

   Products are sold in standard units. Waste rounds ups to the nearest whole standard sales unit. For example: If a task requires only 30 SF of plywood, one will still need to estimate and purchase a single sheet that has 32 SF because that is the standard sales unit.

5. Explain the premise behind compaction and swell of soils and why additional materials may be required.

   When soils are in their natural inground state and undisturbed, they are at maximum density. Over time all of the smaller particles of soil have filled in between the larger particles. When the soil is disturbed by excavation the voids between the particles reappear and cause the volume of the soil to increase. This is swell. When loose soils are used to fill an excavation, the reverse occurs. The loose volume becomes compacted by mechanical means and the inground compacted volume becomes less, so more loose material is needed to fill the volume of the excavation. This is called compaction.

6. What is the difference between production and nonproduction costs?

   Production costs are the materials, labor, equipment, and subcontractor costs that move the project forward toward completion. Nonproduction costs are overhead, or the cost of supporting the production costs. Nonproduction costs (overhead) can be classified as direct or indirect costs.

7. Explain in detail the difference between direct and indirect overhead costs.

   Direct overhead are the non-production costs that are related to one and only one project. For example, a building permit has a cost that is applicable to only one project. It is considered a direct overhead cost. Indirect overhead costs are the costs of being in business. Each project is expected to absorb a percentage of those costs. Larger projects (based on cost) absorb more than smaller projects. The office rent is an example of an indirect overhead cost.

8. Explain the various sources of cost data and when each is appropriate to use.

   Contractors almost always use a mix of historical and contemporaneous cost data when bidding a project as it is the most up to date and reflects what expected costs will be. Historical data is the result of tracking costs elements from a previous project of a similar nature. These costs are then used to determine what the expected cost will be on the project being bid. Contemporaneous costs are prices quoted in real time from suppliers, vendors, and subcontractors willing to supply materials or perform the work at that price. Published cost data is less timely and is most often used for generating budget estimates for funding or to determine what the work is worth. Published cost data is typically used by owners or awarding authorities who are looking to approximate the cost of the work. Many contractors use published cost data to fill in missing numbers or as plugs until the quotes are submitted.

9. Provide an explanation of how indirect overhead percentages used in an estimate are derived.

    Indirect overhead percentages are a function of the gross volume of work performed. A contractor tracks their indirect overhead costs over time and at the end of the year they calculate the costs as a percentage of the work they have completed during that same period. If they perform $10M worth of work in the calendar year and the overhead cost $1.5M then their overhead costs are approximately15% of work completed.

10. What is the purpose of using a bid form when requesting an offer from a contractor?

    Bid forms are used to solicit a true apples-to-apples comparison of costs. Bid forms have lines to be filled in and do not allow the bidders to add qualifications. It assumes that all bidders have priced the same scope, so it comes down to the price bid. This makes comparing bids easier for the owner or awarding authority.

CHAPTER 5: UNDERSTANDING THE COST ELEMENTS OF A UNIT PRICE

1. The cost of an element in an estimate has modifiers. Identify and explain each modifier for labor.

    Modifiers to labor can be a formal benefit package if the trade is union, fixed taxes, and insurances such as Workers Compensation, (protects the worker against injury on the job), FICA, FUTA, and SUTA, (which are social security and Medicare, federal and state unemployment programs), General liability insurance, (protects the employer from the actions of the employee), retirement programs, Family and Medical Leave Act, indirect overhead, and profit to name a few. Each of these costs are required by law or as a company policy and are paid by the employer, which requires them to be captured in the labor cost billed to the client.

2. Explain the term *productivity*. Why is it important in estimating?

    Productivity is the amount of worked produced per unit of time. Time is typically measured as an 8-hour workday. The estimator must evaluate the anticipated productivity for a task in order to estimate the labor cost correctly. Productivity of the same task will change with the context under which the task is performed.

3. Opine as to how subcontractors are accounted for in an estimate. What is the estimator's responsibility with regards to a subcontractor's scope of work?

    Subcontractors often perform a large part of the work in a general contract. As a result, it is the GC estimator's responsibility to ensure that a subcontractor's bid is properly reviewed for scope, price, and time before being included within their GC bid. Many GC estimators prefer to estimate the anticipated price of a subcontractor's work and add a place marker called a plug to the estimate until the quotes have been submitted. Estimating the cost of the plug has the benefit of educating the estimator as to the scope of work that should be in the subcontractor's bid.

4. What is the purpose of the Davis Bacon Act with regards to labor on a project?

   The Davis Bacon Act is intended to help close the gap between the higher union wage and benefit and lower open shop wages when competing for federally funded projects. Awarding Authorities are issued a predetermined minimum wage scale by the Department of Labor and Industries to include as part of the project manual. Contractors and subcontractors can pay higher wages but must meet these minimum requirements. Failure to comply can result in debarment, fines, and/or imprisonment.

5. Why is direct or project overhead not included as part of a billable hourly rate.

   Project overhead is defined in Division 1—General Requirements and standard estimating practice dictates that the work of Division 1 is estimated much like anything else, quantities × the cost per unit. In addition, many of the direct overhead costs are time sensitive or vary with context. This cannot be captured in a billable hourly rate because there is no direct relationship between wage and project overhead.

6. Calculate the billing rate for a bricklayer with the following: wage of $37.50 per hr., $15.90/hr. benefits, FICA at 7.65%, FUTA at 0.87%, SUTA at 7.2%, workers' compensation at 31.77%, General Liability at 0.91%, main office overhead of 16.5%, and a profit of 15%.

Bricklayer Wage	$37.50
Benefit Package	$15.90
FICA at 7.65%	$2.87
FUTA at 0.87%	$0.33
SUTA at 7.2%	$2.70
Workers Comp at 31.77%	$11.91
General Liability at 0.91%	$0.34
Subtotal	$71.55
Main Office OH at 16.5%	$11.81
Subtotal	$83.36
Profit at 15%	$12.50
	$95.86

7. Calculate the billing rate for a bricklayer tender (helper) with the following: wage of $30.50 per hr., $12.45/hr. benefits, FICA at 7.65%, FUTA at 0.87%, SUTA at 7.2%, workers' compensation at 38.5%, General Liability at 0.91%, main office overhead of 16.5% and a profit of 15%.

Bricklayer Tender Wage	$30.50
Benefit Package	$12.45
FICA at 7.65%	$2.33
FUTA at 0.87%	$0.27
SUTA at 7.2%	$2.20
Workers Comp at 38.5%	$11.74
General Liability at 0.91%	$0.28
Subtotal	$59.76
Main Office OH at 16.5%	$9.86
Subtotal	$69.62
Profit at 15%	$10.44
	$80.07

8. Using the results from questions #6 and #7, calculate the daily billing rate for a crew of three bricklayers and two bricklayer helpers.

    The daily billing rate for a crew of three bricklayers and two bricklayer tenders would be: 3 × $95.86 = $287.58 plus 2 × $80.07 = $160.14 or $447.72 per hr. × 8-hrs = $3,581.76 per day.

9. If the daily output for the crew in #8 is 1,655 bricks, what is the labor cost to install single brick?

    It is the total cost of the crew per day divided by the daily output (in bricks) or $3,581.76 ÷ 1655 brick = $2.16 per brick.

10. Using a factor of 6.55 brick per SF and a material cost of $0.87 per brick, and 5% sales tax, calculate the material and labor cost for laying up 40,000 SF of brick veneer.

    $2.16 + $0.87 = $3.03 × 1.05 (for tax) = $3.18 ea. × 6.55 brick per SF × 40,000 SF = $833,160.

CHAPTER 6: DIVISION—1 GENERAL REQUIREMENTS

1. Define project overhead. As part of the definition, explain how project overhead is calculated. Include within your answer the different types of overhead.

    Project overhead is direct overhead for one and only one project. It is most often the nonproductions costs associated with a project as defined by the owner in Division 1—General Requirements. There are three (3) types of project overhead. (1) Fixed overhead such as the cost of a building permit. It is paid for once, and that is it. (2) Time-sensitive overhead costs are those costs that vary with time. The longer the cost is on the project, the more it costs. For example, consider a rented trailer. The longer it is on site the more it costs, its cost is a function of time. (3) A variable cost such as snow plowing or winter conditions. The more snow, the more the cost. The estimator can anticipate certain conditions at best.

2. What are some of the advantages of using computer software to generate schedules?

   Frequently, a contractor will consider various means and methods to complete the work when preparing the estimate. As a result, they may have different time requirements. While the concept of scheduling is extremely simple the actual task takes time to create and evaluate the different relationships between tasks. A computer completes the computation at the speed of light where as doing it by hand takes significantly longer. The acceleration of computation time is the principal advantage of scheduling by computer.

3. Explain how indirect overhead is different from direct overhead.

   Indirect overhead is the cost of doing business and is absorbed by all projects under contract. It is a function of cost and is added to the end of the estimate as a percentage of costs. Direct overhead is calculated based on the Division 1- General Requirements. Some are fixed, some time sensitive and some are variable based on difficult to predict conditions. It is rarely if ever applied as a percentage of project costs, except at the preliminary estimating stage.

4. Is project management a direct overhead cost or indirect overhead cost? Explain your answer.

   Project management costs can be accounted for in either direct overhead or indirect overhead categories. It is a dependent on the policy of the company. A project manager that is responsible for one and only one project is best costed as a direct overhead cost. For a project manager that is managing multiple projects, costs could be split between the project as a direct cost for each project, or it could be included as part of the indirect costs and applied as part of the indirect overhead percentage.

5. Please explain why the critical path method of scheduling is uniquely suited to construction projects.

   CPM schedules are more detailed, and the activities are shown linked. This linking of activities creates the interdependence between activities, which ultimately identifies a critical path through the project. It is exactly like a construction project. If one task on the critical path is delayed, it will cause the delay of all succeeding critical path tasks, and the project as a whole, if corrective action is not taken. The CPM schedule also illustrates the fact that on a real construction project, multiple tasks can occur independently at the same time, while other tasks must occur in a certain order and are highly dependent on completion of the preceding task. It also allows the scheduler to look at a variety of scenarios based on different means and methods.

6. Explain what is meant by "means and methods." How does this impact an estimate?

   Plans and specifications define what work needs to be completed to satisfy the contract, however it does not mandate exactly how the work is to be done. The sequence and specific techniques to complete the work are up to the contractor. For example, some work can be built in place and other tasks may be best suited to prefabrication off site. That is the choice of the contractor. There are multiple ways in which to complete the work.

7. Explain the difference between a fixed direct overhead cost and a variable direct overhead cost.

   Fixed direct overhead appears in the estimate as a one-time cost. For example, a building permit cost is a one-time cost. Variable costs are based on less definitive conditions. For example, winter conditions are a requirement for many projects in colder climates but predicting the anticipated snowfall or winter temperatures to assign a cost in the estimate for plowing or heat is less based on fact than educated assumptions. The basis of the cost can vary.

8. Explain why some Division 1 language is intentionally ambiguous.

   Many Division 1 requirements use broader, more general words than in other technical sections. Words such as sufficient and adequate leave room for interpretation. Consider the interpretation as the technical writer has intended whenever possible or seek clarification from the design professional.

9. Explain the relationship between the General Conditions of the Contract for Construction and the General Requirements of the project.

   The General Conditions are for use in the interpretation of the contract language. The General Requirements are the nonproduction costs as assigned to the project by the owner or design professional in such a way that the contractor can apply a price to them. There may be language in the General Conditions that helps support the General Requirements. For example, the GCs require that the contractor control project time and the GRs require that the contractor produce a schedule to demonstrate the use of project time.

10. Explain how a change order can impact project overhead.

    A change order that has an extension of time included within it can prolong the duration of the project and as a result require the contractor and/or subcontractor to spend more days on the project. This extends the need for project overhead and subsequently the cost of that overhead.

CHAPTER 7: EXISTING CONDITIONS

1. Define the process of selective demolition. Explain the difference between selective demolition and removal and salvage from an estimating viewpoint.

   Selective demolition is the targeted removal of a building element, component, or finish with no concern for reuse. It is scheduled for disposal. This task can be accomplished by hand or with tools and equipment. Removal and salvage of a specific item or element requires greater care and often longer labor hours for the removal of the element with the intent of reuse or salvage for resale.

2. Select one of the six processes identified for lead paint removal and explain it in detail.

   Encapsulation is a process that leaves the well-bonded paint in place after the loose and flaking paint has been removed. Before the work can begin, the adjacent area is protected to catch the loose paint scrapings. The scraped surface is then prepared by washing with a detergent and thoroughly rinsing. The prepared surface is then painted with up to 10-mils of paint. A reinforcing fabric can be added for added protection. The scraped paint is disposed of as hazardous waste. Protective clothing and respirators are required. (Answers will vary dependent on type of lead paint removal selected.)

3. How is removal and salvage calculated by the estimator?

   Estimating removal and salvage requires that the estimator have a particular plan or method for the removal in mind when performing the estimate. Part of the scope of this section frequently includes the protection and storage of the work until its final disposition. Storage on urban projects with restricted space may require hauling to off-site storage and back to the site later for installation. Removal and storage tasks are labor intensive and can often take between half and twice the normal installation time of the element.

4. How are Existing Condition drawings helpful to the estimator?

   Existing Conditions drawings often show the estimator the current status of an item scheduled for demolition. It is used as a reference or beginning for the work that has to be performed and even helps decide the means and methods to be used.

5. Explain in detail why some selective tasks are quantified as a lump sum unit of measure.

   The lump sum (LS) unit of measure allows the estimator to combine series of tasks that are included in the process of removal. There are some tasks or segments of work that contain more than one operation but are better left as a whole for the purposes of estimating. An example is removing a metal window and frame from an existing concrete block wall, cutting the jambs to the floor, and removing debris to prep the opening for a new door.

6. Opine as to why it is important for an estimator to secure a subcontractor quote for much of the work or Division 2.

   Much of the work, especially remediation scopes, is not only highly specialized, with special licenses and insurances, but can be challenging to estimate without prior experience of performing the work. It often can be labor intensive and require follow-up paperwork to prove compliance.

7. Explain in detail one of the challenges in calculating the number of dumpsters needed and the weight of the debris within it.

   Dumpsters are estimated by the size and rental period of the container. There are additional charges for weight in excess of the specified limit for the container. A common method to estimate dumpsters is to dedicate specific containers to certain types of debris. For example, a dumpster can be dedicated to masonry debris. The estimator can calculate the amount (volume) of the masonry debris in the demolition. This can be converted to weight. For general debris and ongoing cleanup, many contractors allow a specific square foot area of a project per dumpster. This is often based on historical costs.

8. Explain the reasons and under what conditions negative air would be required.

Negative air is a term used to define air being removed from a space and passed through a filter to outside atmosphere. It is often used to prevent contaminants in the air from migrating from a work space to a nonwork space. Air is drawn from the work space under vacuum, filtered, and disbursed to the outside.

9. Select one of the categories of demolition and explain what it includes.

Whole Building Demolition is the complete removal of a structure above grade without salvage concern for any specific building element, component, or material during the demolition. This work is performed with large pieces of construction equipment that break up the structure, load it onto trailers, and transport it to disposal sites. The limits of work end at grade (ground) level of the building. This type of demolition is equipment intensive, and labor is generally limited to the operators of the equipment, truck drivers, and supervisors. (Answers will vary.)

10. What is the purpose of a test boring report? How can it be used by the estimator? What are the disadvantages of a test boring report?

The purpose of the investigation is to define the conditions of the soil that will support the structure, and to provide bidders with an idea of the subsurface conditions that may be encountered during excavation. Included are conditions such as soil types, strata, rock locations, and levels of the water table, all of which can have a tremendous effect on the pricing of sitework. The detail and specificity of the geotechnical reports can run the gamut from test pits to extensive soil analysis and deep borings. The geotechnical report can aid the estimator in pricing the excavation scope of work. The identification of soils by species and grading helps the estimator to determine the ease of excavation and the daily outputs. Make careful note of the disclaimer that accompanies each report. It explicitly states that the information contained in the boring reports is for the convenience of the contractor and that the geotechnical firm assumes no responsibility for the representation of the soil conditions of the site as a whole.

CHAPTER 8: CONCRETE

1. A building has a perimeter of 300' and a footing size of 32" × 18". Assuming the bottom of the footing is at the same elevation along the perimeter, calculate the quantity of cubic yards required to fill the form. Round your answer up to the nearest whole yard.

Start by converting the dimensions to feet. 300' × 2.67' × 1.5' = 1201.5 CF. Now convert to CYs: 1201.5 CF ÷ 27 CF per CY = 44.5 CYs rounded to 45 CYs.

2. If a crew of two laborers will place 50 CY per day, how many labor-hours should be allotted to place the concrete from the answer in the previous question?

45 CYs ÷ 50 CYs per day = 0.9 days × 8-hrs/day = 7.2-hrs × 2-laborers = 14.4 labor-hrs. rounded to 15-labor hours.

3. On top of the footing in question #1, there is a 4'-0" foundation wall form that will be filled with concrete to 3'-6". Calculate the amount of concrete required if the wall is 10" thick. Round your answer up to the nearest whole yard.

    Start by converting the dimensions to feet. 300' × 0.83' × 3.5' = 871.50 CF. Now convert to CYs; 871.50 CF ÷ 27 CF per CY = 32.27 CYs or rounded to 33 CYs.

4. Calculate the placement cost per CY of a 20,000 SF, 5" thick slab-on-grade if the crew cost $3,650 per day and the crew will place 165 CY per day. Assume 5% waste and round to the whole day.

    Start by calculating the volume required. 20,000 SF × 0.42' = 8,400 CF or 311.11 CYs × 1.05 to account for the waste = 326.66 CYs. Now divide the total by the daily output; 326.66 CYs ÷ 165 CYs per day = 1.98 days rounded to 2 days. If the crew cost $3,650 per day, then 2 days × $3,650. per day = $7,300. Dividing $7,300 by 327 CYs = $22.33 per CY.

5. How many days would a pump be needed to place the concrete in problem #4? Round your answer up to the nearest whole day.

    The task of placing the concrete would take 2 days so the pump would be required for 2 days.

6. A footing for a retaining wall is 95' long × 7' wide and has 6 pieces of continuous #5 rebar in the footing and a lap of 20d. Using *Figure 8.13 Reinforcing Steels Weights and Measure* calculate the weight of the rebar in tons.

    6 pcs × 95' each = 570'. Since its #5 bar the diameter is 5/8" so the lap is 20d or 20 × 5/8" = 12.5" ÷ 12" per foot = 1.041' lap. Since the typical stock length of a #5 is 20'-0" that would mean each lap requires an additional 1.041'. As a percentage this can be translated to 1.041' ÷ 20.0' = 0.052 or 5.2%. If 5.2% is added to the total; 570' × 1.052 = 599.64'. Since rebar is estimated by weight, then the next step is to convert the length to weight. Using Figure 8.13 the weight per foot for a #5 bar is 1.043 lbs. Therefore 599.64' × 1.043 lbs./ft. = 625.42 lbs. or 0.32 tons.

7. If an ironworker will place 157 lbs. per hr. how many labor hours should be estimated to set the rebar in Problem #6?

    625.42 lbs. ÷ 157 lbs. per hr. = 3.98 or rounded to 4-hrs.

8. A 12,000 SF slab requires WWF. The sheets are 5' × 10', and the specifications require a side and end lap of 6". How many sheets will be required and what is the percentage lost to lap?

    Start by calculating the net area of the sheet, which is each dimension less the 6" lap. 4.5' × 9.5' = 42.75 SF per sheet. Dividing the total area of 12,000 SF by 42.75 SF per sheet = 280.7 or 281 sheets. The percentage lost to lap can be calculated by 50 ÷ 42.75 = 1.169 or 17%.

9. A subfloor needs to be leveled with gypsum cement underlayment. The subfloor is 25' wide and 100' long. From side to side (across the 25' dimension), the floor has a ¾" difference in elevation. If the specifications call for a minimum thickness of ½", what is the total quantity in cubic feet (CF) required to level the floor while maintaining the minimum thickness?

The high side would require the minimum thickness of ½" and the low side would be at the maximum of ¾" + ½" = 1-1/4" therefore the average across the 25' width is (1¼" + ½") ÷ 2 = 7/8" (0.875").
Now convert the inches to feet or 0.875" ÷ 12" per ft = 0.073' × 25' × 100' = 182.5 or 186 CF.

10. If the underlayment in Problem #9 is estimated at $85.00 per CF in place what would the price be to level the floor? Assume a 5% waste factor.

Adding 5% waste to the volume in #9: 182.5 CF × 1.05 = 191.63 or rounded to 192 CF × $85/CF = $16,320.

CHAPTER 9: MASONRY

1. A CMU building is 77'-2" × 40'-0" with a height of 22'-0" to the top course of the 8" CMU. There are three (3) 12' × 12' overhead doors, three (3) 3' × 7'-2" passage doors and eight (8) 6' × 4' windows. What is the net area (SF) of the block walls? With a 5% waste factor included how many blocks are required?

The perimeter of the building is 234.32' × 22' high = 5,155.04 SF of gross area. Deducting the windows and doors; 3 × 12' × 12' = 432 SF, plus 3 × 3' × 7.16' = 64.44 SF plus 8 × 6' × 4' = 192 SF for a total of 688.44 SF. The net area is therefore 5,155.04 – 688.44 = 4,466.6 SF. Now to convert the area to block. Since a block is 0.89 SF, then 4,466.6 SF ÷ 0.89 SF per block = 5,018.65 block × 1.05 = 5,270 blocks.

2. Using the net area derived in question #1, how many cubic yards of grout are required to fill the cells solid if the cores are grouted at 32" o.c. and the block is 40% solid. Use the appropriate table from Chapter 9 and round your answer up to the nearest whole cubic yard. Cite the table used.

Using Figure 9.15 Volume of Grout Fill for Concrete Block Walls, under 8" CMU per SF, 40% solid, at 32" o.c. the factor is 0.09 CF per SF of wall. 4,466.6 SF × 0.09 CF per SF = 401.99 CF or 14.88 CYs rounded to 15 CYs.

3. Calculate the amount of lateral reinforcement in LF, needed if the wall is reinforced at every other course. Do not deduct for the openings.

Some interim calculations are required starting with height 22' × 12"/ft = 264" high ÷ 8" per course = 33 courses. Every other course is 16.5 or 16 courses × 234.32' perimeter = 3,749.12 or 3,750.' (The top course is omitted hence 16 courses.)

4. Using the appropriate table from Chapter 9, what is the labor cost in dollars if the cores are grouted at 32" o.c. and it takes 0.038 labor-hours per SF of wall? The cost per labor-hour is $96.50.

The net area of the building is 4,466.6 SF × 0.038 labor-hrs. per SF = 169.73 or 170-labor hrs. Therefore $96.50 × 170 labor-hrs. = $16,405.

5. For the building in question #1, there is a continuous bond beam at 10'-0" and the top course of the building. Calculate the total material cost of the grout to fill just the bond beam courses if there 0.34 CF per LF and the cost per CF is $11.50.

Using the perimeter of 234.32' × 2 courses = 468.64 less the 3 – 12' high overhead doors = 432.64' or 433' × 0.34 CF per LF = 147.22 CF. Now determine the cost; 148 CF × $11.50 = $1,702.

6. Using the appropriate figure in Chapter 9, calculate the amount of brick required to cover (veneer) the front elevation (40') only, full height with a standard face brick (8" × 2¼") with a 3/8" mortar joint, set in a running bond. The front elevation has one (1) 3' × 7'-2" passage doors and three (3) 6' × 4' windows. Do not include bricks for the window sills. Cite the table used.

   Using Figure 9.7 Brick Quantities per SF from Chapter 9, a standard face brick with a 3/8" mortar joint has 6.55 bricks per SF in a running bond. The front elevation is 40' × 22' or 880 SF less the 3 openings each 6' × 4' = 72 SF plus 1 door 3' × 7.16' = 21.48 SF for a net area of 880 SF– 93 SF = 787 SF now to covert the net area to bricks: 787 SF × 6.55 bricks per SF = 5,155 bricks (exclusive of waste).

7. If the crew will lay up 1500 bricks per day complete, how long will it take to lay up this wall? Round your answer up to nearest whole day.

   5,155 bricks ÷ 1,500 bricks per day = 3.43 days rounded to 4 days.

8. Using the appropriate table in Chapter 9, what would be the difference in the number of bricks needed if the wall was changed to a Modular Norman brick with the same joint thickness. Cite the table used.

   Using Figure 9.7 Brick Quantities per SF Modular Norman brick has 4.57 bricks per SF, therefore 787 SF × 4.57 bricks per SF = 3,597 bricks (exclusive of waste).

9. Using the appropriate table in Chapter 9, calculate the labor-hours required to lay up the Modular Norman brick in question #8. Round up to the nearest whole labor-hour.

   Using Figure 9.9a Labor Hours Required for the Installation of Brick Masonry, Modular Norman brick require 0.125 labor-hrs. per SF so 0.125 labor-hrs. per SF × 787 SF = 98.37 or 99 labor-hrs.

10. Using the appropriate table from Chapter 9, how many labor-hours are required to wash the face of the brick? Round your answer up to the nearest whole labor-hour.

    Using Figure 9.9a Labor Hours Required for the Installation of Brick Masonry requires 0.014 labor-hrs. per SF to wash the brick veneer so 787 SF × 0.014 labor-hrs. = 11.02 or 12-labor-hrs.

CHAPTER 10: METALS

1. A wide flange member with the designation W27 × 161 is 54' long. If the cost for fabrication, priming, and delivery is $1,230 per ton, what is the cost of this member?

   The second number in the designation is the weight per foot. Therefore 54' × 161 lbs. per ft. = 8,694 lbs. ÷ 2000 lbs. per ton = 4.35 tons. So, 4.35 tons × $1,230 per ton = $5,350.50.

2. An erection crew has a daily cost of $5,678.00 and will erect 28 tons of steel in a day. What is the cost to erect 467 tons of steel and how many days should it take?

   467-tons ÷ 28-tons per day = 16.67 or 17-days × $5,678. per ton = $96,526.

3. Using the structural steel plan in Figure 10.3 of this chapter, how many squares of decking are required? Round up the answer to the nearest whole number of squares.

    Seven (7) bays × 20' + 2 bays × 8' = 156' long. Three (3) bays × 15' + 2 bays × 8' = 61' wide. Therefore 156' × 61' = 9,516 SF ÷ 100 SF per SQ = 95.16 or 96 SQs (exclusive of waste or lap).

4. Using Figure 10.3 Structural Steel Plan, calculate the weight in tons of all the wide flange members along the numbered column lines only.

    Lines #1 and #8: 3 × 15' × 26 lbs. per ft. × 2 lines = 2,340 lbs.
    Lines #1 and #8: 2 × 8' × 26 lbs. per ft. × 2 lines = 832 lbs.
    Lines #2 thru #7: 61' × 76 lbs. per ft. × 6 lines = 27,816 lbs.
    for a total of 30,988 lbs. or 15.5 tons.

5. If the span in Figure 10.5 Open Web Joist Plan is increased from 30' to 38' and the species of K-series joist is changed from 16K4 to 28K6 what is the weight of the joist in tons?

    10 pcs × 38' = 380' then using figure 10.6 a 28K6 weighs 11.4 lbs. per ft. therefore 380' × 11.4 lbs. per ft. = 4,332 lbs. or 2.17 tons.

6. In question #5, if the erection crew can set 1,700 lbs. of bar joist per hour, how long will this installation take? Round up to the nearest whole hour.

    4,332 lbs. ÷ 1,700 lbs. per hr. = 2.55 hrs. rounded to 3 hrs.

7. Using Figure 10.11, Coating Structural Steel, and Figure 10.3, Structural Steel Plan, calculate the cost of the material (gallons of paint) to prime by spray two coats on all of the beams shown in plan view on Figure 10.3. If a gallon is $86. and the steel is considered light structural steel.

    From problem #4: 30,988 lbs. for the numbered column lines. For the lettered column lines; 4 lines A-D 156' × 4 lines × 22 lbs. per ft. = 13,728 lbs. plus 3 lines × 140' × 22 lbs. per ft. = 9,240 lbs. plus the 30,988 lbs. = 53,956 lbs. ÷ 2000 lbs. per ton = 26.98 tons.
    1ˢᵗ coat – 26.98 tons × 0.9 gallon = 25 gals.
    2ⁿᵈ coat – 26.98 tons × 1.0 gallon = 27 gals.
    52 gals. × $86. per gal. = $4,472.

8. Considering the quantity of steel in tons in question #7 and Figure 10.11, how many labor-hours will it take to spray the two coats?

    1ˢᵗ coat – 26.98 tons × 1.6 labor-hrs. per ton = 43.17 labor-hrs.
    2ⁿᵈ coat – 26.98 tons × 1.3 labor-hrs. per ton = 35.07 labor-hrs.
    Total = 78.24 or 79 labor-hrs.

9. Calculate the vertical length of the 6" × 6" × ½" columns in section A of Figure 10.3 Structural Steel Plan if the TOS for the W24 × 76 beam is elevation 174.00 and the elevation of the leveling plate for the column is 152.40 (depth of W24 can be found on line).

    TOS = 174.00 – 152.40 = 21.6' less the depth of the W24 × 76, which is 23.92" or 1.99' (sourced from internet). Therefore, 21.6' – 1.99' = 19.11'.

10. Using the appropriate figure from the chapter, how many labor-hours should it take to install 80 LF of 1½" diameter two-rail steel railing? If the labor cost per hour is $99.50, what is the installed cost of 80 LF? Cite figure used.

    Using Figure 10.10, 80' × 0.20 labor-hrs. per ft. = 16 labor-hrs. × $99.50 per labor-hr. = $1,592.00.

CHAPTER 11: WOOD, PLASTICS, AND COMPOSITES

1. What are the four (4) major categories of Division 6 for estimating purposes?

    Division 6 can be broken into 4 categories for estimating purposes. They are: 1) Rough carpentry and framing, 2) finish carpentry, 3) architectural millwork, and 4) casework and cabinetry.

2. What is the difference between a nominal and actual dimension?

    A nominal dimension is for naming purposes and an actual dimension is what the piece measures. For example: 2" × 4" is a nominal dimension, whereas the piece actually measures 1-1/2" × 3-1/2".

3. What are the four (4) main assemblies in a frame? Explain each briefly.

    The framing of a structure can be broken down into four main systems, or assemblies. An assembly is a series of components that together form a larger system or portion of work. The four main assemblies in a frame are: 1) Floor framing, which is the floor joists and sheathing, 2) Wall/partition framing, exterior walls, and interior partitions, 3) Roof/ceiling framing, roof rafters, ceiling joists and roof sheathing and 4) Exterior trims and miscellaneous items, such as stairs.

4. What is the subfloor? How is it taken off and priced? Explain in detail.

    The subfloor is typically a single layer of sheathing 3/4" thick nailed perpendicular to the floor joists. It is sold in 4' × 8' sheets. The length of the floor frame is multiplied by its width to arrive at the area. The area is then divided by 32 SF to determine the number of sheets required.

5. A foundation is 64' long × 28' wide with a girder running down the center (parallel to the 64' dimension). Calculate the number of 2" × 10" floor joists if they are spaced at 16" o.c. Calculate the band joist and the solid blocking over the girder. Note the length and quantity of each member. (Assume no openings in the deck.)

    Start by determining the number of joists required 64' ÷ 1.33' o.c. = 48 joist + 1 to start = 49 joists. Next calculate the length of the joists. Since there is a girder midspan of the 28', the joist can be 14'. There are 49 joists × 2 sides = 98 joists each 14' long. The band joist and solid blocking run perpendicular to the floor joists along the 64' length. Therefore, 64' × 3 locations = 192 LF ÷ 16' = 12 pieces each 16' long.

6. How many sheets of ¾" tongue-and-groove (T&G) subfloor sheathing are required to cover the floor frame in question #5? If a sheet of ¾" T&G subfloor is $80.50 what is the total material cost for the subfloor sheathing?

    L × W = Area; therefore, 64' × 28' = 1,792 SF ÷ 32 SF per sheet = 56 sheets × $80.50 per sheet = $4,508.

7. For the building in question #5 calculate the length of the rafter and the quantity of the rafters if the roof pitch is 6:12 (each side of the ridge), with ½" sidewall sheathing and a 10" overhang (measured from the face of the sheathing). It has a 2" × 12" ridge. What is the stock length needed?

    The quantity would be the same as the floor joists: 98 joists. The length would be calculated using the Pythagorean theorem.

First determine the run and rise of the rafter. The run is 14' + 0.83'(overhang) + 0.042' (½" wall sheathing) less 0.063' (¾" for half thickness of ridge) = 14.81' for the run. Since the rise is ½ the run with a 6:12 pitch, then the rise is 7.40'. Now plug these numbers into the formula C = √A² + B² where A = 7.40 and B = 14.81 or √ 54.76 + 219.34 = 16.56' or 16'-6¾" length,

so 18' stock is needed. Therefore, the roof would require 98 pieces of 18' rafters.

8. Calculate the amount of board feet (BFM) in 236' of 2" × 8" (actual dimension) of poplar.

    2" × 8" = 16 square inches ÷ 12 square inches per BF = 1.33 BFM per LF × 236' = 313.88 BFM

9. The average daily output for a crew of three carpenters to picture frame trim (no sill or apron) windows is 422 LF. If the billing rate for a carpenter is $65.50 per hour, what is the cost to trim out a window that is 4' × 6'?

    Crew cost/day = 3 carpenters × 8 hrs. × $65.50 per hr. = $1,572. Now divide by the daily output = $1,572 ÷ 422 LF per day = $3.73 per ft. finally a 4' × 6' window has a perimeter of 20'; therefore, 20' × $3.73 per ft = $74.60 to trim the window.

10. Calculate the header stock to be purchased if the rough opening for a window that is 6'-2" × 4'-6"? The wall is 2" × 4" construction and requires a minimum of 3" of bearing.

    The width of the rough opening is 6'-2" + 3" bearing on each side = 6'-8" length of the header. Since the wall is 2" × 4" construction, two pieces are needed. Ideally, one 14' would be purchased to minimize the waste.

CHAPTER 12: THERMAL AND MOISTURE PROTECTION

1. Identify some graphic views and the work of Division 7 that an estimator may find on those graphic views.

    Details and sections of the roof system for flashing and sheet metal details, insulation at the roof envelope, and roof accessories. Wall sections and details for information about thermal insulation, air barriers, and vapor barriers at the exterior walls. Wall and foundation sections and details for surfaces to be waterproofed, dampproofed, or insulated below grade.

2. Explain how iron oxide waterproofing works.

    Many types of cementitious waterproofing have iron particles (fillers) added to the mix. The iron-oxidizing agent in the mix causes the iron fillers to rust and expand to fill any pores in the coating.

3. A 2" rigid insulation is required to be placed under the entire slab of a building that has an outside of foundation dimension of 100' × 180'. The wall is 1' thick and the slab is 4" thick. The insulation is also required to be continuous from the underside of slab to the top of footing, which is 3'-9" below the top of slab. The slab and top of wall are at the same elevation. Calculate the total area of 2" rigid insulation required.

Start by calculating the area of the slab inside the wall, which is 1' less all around or 98' × 178' = 17,444 SF. Now calculate the vertical insulation, which is the inside perimeter × height. The perimeter is (98' + 178') × 2 = 552'. Next calculate the height of the insulation, which is 3'-9" less the slab thickness of 4" or 3'-5" = 3.42'; therefore, 3.42' × 552' = 1,887.84 SF plus 17,444 SF = 19,332 SF without waste.

4. If a laborer will place 500 SF of vertical insulation per day and 2000 SF of horizontal insulation per day, how many laborers will be required to complete the task in question #3 in a single workday?

    17,444 SF ÷ 2,000 SF per day per laborer = 8.77 or 9 laborers for the slab and 1,888 SF ÷ 500 SF per day per laborer = 3.77 or 4 laborers for the vertical insulation so in total 13-laborers to do the task in a day.

5. A wood-framed roof has an 8:12 pitch on each side of the ridge. It is 60' long and 32' wide with a 12" overhang along the 60' lengths. Calculate the following: (a) SF of asphalt felt required; (b) squares of roof shingles; (c) quantity of 10' length of drip edge for fascia and ridge; and (d) LF of ridge cap. Allow 5% waste for each material.

    Start by calculating the length of the rafter. Determine the run first 32' ÷ 2 = 16' + 1' (overhang) + ½" sheathing less ½ of ridge, which is ¾" = 16'-11¾" or 16.98'; therefore, the rise 8 ÷ 12 = 0.67 × 16.98' = 11.38'. Use the Pythagorean theorem: $C = \sqrt{A^2 + B^2}$, where A = 11.38 and B = 16.98 or $\sqrt{129.50 + 288.32}$ = 20.45'. The area of the roof is 20.45' × 60' × 2 sides = 2,454 SF or 24.54 SQs. So, (a) 2,454 SF × 1.05 = 2,577 SF, (b) 24.54 × 1.05 = 26 SQs, (c) [(20.45' × 4 + 60' × 2) × 1.05] ÷ 10' = 22 pcs., and (d) 60' × 1.05 = 63 ft.

6. If a crew of three roofers will install 16-SQ of shingles, including drip edge, cap, and underlayment per 8-hr day, how many days will the roof in Problem #5 take to complete?

    25.77 SQs ÷ 16 SQs per day = 1.6 days or 2-days.

7. Using the appropriate figure from the chapter, how many gallons of sealant will be required to fill 788 LF of ½" wide × ½" deep joint? Round your answer up to the nearest gallon. Cite the table used.

    Using Figure 12.15: 77 LF of ½" × ½" requires 1 gallon of sealant therefore 788 LF ÷ 77 LF per gallon = 10.23 or 11 gals.

8. The net area of a building sidewall is 4,765 SF. If the plans call for ½" × 6" clapboard with 4" exposure, how many LF will be required with 4% waste? Round your answer to the nearest LF.

    There are 3 LF per 1 SF of clapboard siding at 4" exposure. Therefore, 4,765 SF × 3 LF per SF = 14,295 LF × 1.04 (waste) = 14,867 LF.

9. An EIFS crew consists of 3-plasters at $96.50 per hour and two plasterer tenders at $73.40 per hour. If they will install complete 550 SF of EIFS in a day, what is the cost per SF?

    Three plasters × $96.50 per hour = $289.50 and two tenders × $73.40 per hour = $146.80 or $436.30 per crew hr. × 8 hrs. = $3,490.40 per day. Now divide by the daily output: $3,490.40 ÷ 550 SF per day = $6.35 per SF.

10. If the building in problem #5 has a flat ceiling with an R-38 insulation and the cost of the material is $75.99 per bundle containing 47.9 SF, what is the material cost and how many bundles are required?

    The ceiling area (to outside of frame) is 60' × 32' = 1,920 SF ÷ 47.90 SF per bundle = 40 bundles × $75.99 = $3,039.60.

CHAPTER 13: OPENINGS

1. Schedules are frequently used in this division to identify doors, windows, or hardware. Select one (1) of these categories and explain what a reader would expect to see in that schedule.

   Schedules are intended to be a central location where all of the information that would typically appear on a set of plans for that scope of work can be located. For example, a window schedule would have the letter designation for the window, the model and manufacturer of the window, the operation type, the rough opening in width by height, the glazing, the jamb and head details, and a remarks column.

2. Explain the difference between a knockdown metal door frame and a welded door frame from an estimating perspective.

   A knockdown frame is a three-piece set that is installed after the drywall has been hung and typically taped. It is a less labor-intensive process than a welded frame, which consists of all three pieces welded together. It has to be set in the wall or partition as the wall or partition is built. It has to plumbed, squared, and braced in the opening as it is constructed. It is more labor-intensive to install.

3. Identify the characteristics that an estimator should consider when pricing a wood door.

   The characteristics of a wood door that impact material price are: size, thickness, composition (species of wood), fire rating, operation (swing or slide), unfinished or finished, and prep on the door for hardware or vision panels.

4. How is the labor estimated for a wood, metal, or plastic window?

   Labor is estimated by the time it takes to install the windows in labor-hours per unit. The crew can be a single individual or multiple people. The trade to install a window is typically a carpenter.

5. Describe the estimating process for a specific hardware set.

   The estimator can list the individual pieces included within a set of hardware and then add both material and labor costs for the individual component. The material and labor cost can be totaled for the set and then multiplied by the quantity of that set on the project.

6. Define the unit of measure called the *united inch*. What product is it typically applied to?

   The united inch is the unique unit of measure used in pricing storm windows. It is the width and height in inches added together and used as a linear measurement.

7. How do U-factors and solar heat gain coefficients (SHGCs) impact the cost of a window unit?

Both U-factors and solar heat gain coefficients are energy efficiencies for glazing in windows. The more energy efficient a window is, typically the more expensive it is.

8. Using the information below and the hourly rate of $88.50 for a carpenter, what is the cost of installing Hardware Set #2?

Labor-hours to install Hardware Set #2	
1 ½ pairs 4½" Stanley HD Series Hinges (1 ½ PR)	0.60 hours
Hang and swing door	0.25 hours
Corbin 123L-5500 Series Mortise Lockset (1 EA)	1.10 hours
LCN 4010 CUSH Series Closer (1 EA)	1.80 hours
Ives 436 B Floor Stop (1 EA)	0.45 hours
Ives #20 Silencers (3 EA)	0.15 hours
Adjust hardware and door	0.40 hours
Subtotal for labor-hours for HS #2	4.75 hours

The total labor hours to install Hardware Set #2 is 4.75; therefore, 4.75 labor-hours × $88.50 per labor hour = $420.38

9. Explain why it is preferable for a door to be premachined and finished by the manufacturer.

Premachined doors are doors that are prepped at the distributor or door vendor and are delivered to the site ready to install hardware. They are prepped in a shop where tolerances and quality control are better. Much of the prep is done by machine, which is faster and less expensive than being done by hand on site by a carpenter. The same is true about the finish of the doors.

10. Why is it important to secure subcontractor pricing for materials and labor for specialty doors?

Specialty doors can have proprietary qualities that impact the price of both the cost of materials and the labor to install the doors. Subcontractors representing the product can have the correct tools and prior experience that make that installation faster and more efficient. Subcontractor installations also include warranties.

CHAPTER 14: FINISHES

1. Calculate the amount of ⅝" gypsum wallboard required for a partition that is 276' long by 12' high with two layers of board each side full height. Provide your answer in square feet, SF.

276' long × 12' high × 4 layers = 13,248 SF

2. If the two carpenters on the crew will hang 1,500 SF per day (total), how many days will the hang in Problem #1 take? Round your answer to the full day.

13,248 SF ÷ 1,500 SF per day = 8.9 days or 9 days.

3. The partition in problem #1 requires 3" sound attenuation batts within the metal studs. If the cost for materials and labor is $1.91 per SF, what is the cost of the acoustical treatment? Assume the installers will use the lift on site.

276' long × 12' high = 3,312 SF × $1.91 per SF = $6,325.92

4. The lobby to a mall has 6" × 6" quarry tile in an area 136' × 42'-8". If the tile material cost $4.22 per SF, thin set at $0.83 per SF, and grout at $1.22 per SF, what is the material cost for the work with 5% waste and a 6% sales tax?

    First calculate the area for the tile: 136' × 42.67' = 5,803.12 SF × 1.05 (5% waste) = 6,093.28 or 6,094 SF. Now calculate the cost: tile at $4.22 + thin set at $0.83 + grout at $1.22 = $6.27 per SF × 1.06 (6% sales tax on materials) = $6.65 The final step is to multiply the material cost by the area: 6,094 SF × $6.65 per SF = $40,525.10

5. The tile laying crew (one tile setter and one helper) will set 900 SF per day. How long should the tile in Problem #4 take to set? (Round to the full day.) If the tile setter is billed at $86.50 per hour and the helper is billed at $60.25 per hour, what is the labor cost per SF for the job?

    6,094 SF ÷ 900 SF per day = 6.77 or 7-days so 8-hrs × ($86.50 + $60.25 = $146.75 per hr.) = $1,174.00 per day × 7-days = $8,218 ÷ 6,094 SF = $1.35 per SF.

6. There are seven (7) offices along the outside wall of an office space. They are 11'-6" wide and total 138'-0" long. Calculate the amount of carpet to be installed (net area) and the amount of carpet materials to be carried in the estimate if the carpet is sold 12'-0" wide rolls. Include 5% waste.

    The gross area to be purchased is 138' × 12' = 1,656 SF × 1.05 = 1,739 SF. The net area for installation is 138' × 11.5' = 1,587 SF ÷ 9 SF per SY = 176.33 or 177 SYs.

7. The following trim will need to be back-primed with paint (all 4-sides); 220' of 1" × 6", 280' of 1" × 8", 135' of 1" × 10", and 96' of 1" × 12". Calculate the surface area to be painted. If a single gallon of primer will cover 220 SF of surface area (one coat), how many gallons are required to prime the wood trim with two coats?

    Start by calculating the area of each size trim 220' × 1.04 SF per ft. (sum of all 4 side of 1" × 6" actual dimensions) = 229 SF, plus 280' × 1.34 SF per ft. = 376 SF plus 135' × 1.67 SF per ft. = 226 SF plus 96' × 2 SF per ft = 192 SF for a total of 1023 SF ÷ 220 SF/gal. × 2 coats = 9.3 or 10-gals.

8. The exterior elevation of a house shows eight (8) windows, each with a pair of shutters. Historical data for the painting company reveals that it takes 0.3 hrs. to remove a single shutter from the house, 0.85 hrs. to prep a shutter, 1.2 hrs. to paint each shutter with two coats (both sides), and 0.25 hrs. to rehang each shutter. If the painter's billing rate is $66.50 per hr., what is the total cost to complete the shutters on this elevation?

    Start by summarizing the labor time for a shutter: 0.3 + 0.85 + 1.2 + 0.25 = 2.6 labor-hours per shutter × 16-shutters = 41.6 labor-hrs. × $66.50 per hr. = $2,766.40

9. Using the partition in Problem #1, calculate the material cost to prime and paint two coats of interior paint on the exposed surface of the GWB full height each side. Primer coverage is 180 SF per gallon at a cost of $22.00 per gallon, first coat of finish latex coverage is 220 SF per gallon

and the second coat of finish latex coverage is 270 SF per gallon. The finish latex paint cost $31.50 per gallon.

13,248 SF ÷ 2 = 6,624 SF of surface area therefore primer 6,624 ÷ 180 SF /gal. = 36.8 or 37-gals. First coat 6,624 SF ÷ 220 SF/gal. = 30.1 or 31-gals. Second coat 6,624 SF ÷ 270 SF/ gal. = 24.5 or 25 gals. Therefore:

Primer: 37-gals. × $22.00 per gal. = $814.00

1st coat 31-gals. × $31.50 per gal. = $976.50

2nd coat 25-gals. × $31.50 per gal = $787.50

Total material cost = $2,578.00

10. The interior painting in problem #9 will be rolled. A single painter will roll primer at 900 SF per day, first coat of finish at 750 SF per day and second coat of finish at 1,000 SF per day. How many days rounded to the full day will it take to complete the paint in Problem #9 if there are two painters on the crew? At the billing rate of $66.50 per hr. what is the labor cost for the painting?

Primer: 6,624 SF ÷ 900 SF/day = 7.36 or 8 days

1st coat: 6,624 SF ÷ 750 SF/day = 8.83 or 9 days

2nd coat: 6,624 SF ÷ 1,000 SF/day = 6.7 or 7 days

24-days ÷ 2 painters per day = 12 days × 16 hrs./day × $66.50 per hr. = $12,768.

CHAPTER 15: SPECIALTIES

1. The project has twelve (12) marker/tackboards all 4'-0" high and 5–10' long, 3–12' long and 4–8' long. If it takes 8 hours for a crew of two (2) carpenters to install all 12, what are the labor-hours required per SF?

Start with hours per day: 2-carpenters/day × 8 labor-hrs./day = 16 labor-hrs.

Total length: 5 × 10' = 50', 3 × 12' = 36', 4 × 8' = 32' or 118' × 4' = 472 SF, therefore 16– labor-hrs. ÷ 472 SF = 0.034 labor-hrs. per SF.

2. If the two carpenters on the crew in problem # 1 are billed at $86.50 per hr., what is the cost per SF?

0.034 labor-hrs. per SF × $86.50 per labor-hr. = $2.94 per SF.

3. What sources should an estimator consider searching to determine how a bank of metal lockers are assembled?

The estimator can locate the specified product on the manufacturer's website. Most manufacturers provide details as to the level of assembly required for the product. If that is not successful, contact the vendor for more information on assembly. They may have the information, as this is a fairly common question.

4. Historical data indicates that it takes one (1) carpenter 8 hours to install all of the toilet accessories for a six (6) -fixture toilet room (3-water closets and 3 lavs). If there are four (4) more toilet rooms with four (4) fixtures each, what is the amount of labor-hours required by the carpenter to complete the installation of the accessories in the remaining four toilet rooms? (Assume same accessories per fixture.)

Carpenter labor-hrs.: 8 labor-hrs. ÷ 6 fixtures = 1.34 labor-hrs. per fixture. Therefore 4 toilets × 4 fixtures each × 1.4 labor-hrs. per fixture = 22.4 or 23 labor-hrs.

5. What is the designer's purpose for including the name and contact information for the vendor of a specific product within a technical specification section?

   When the design professional takes the time to include the name and contact information for a particular product, it is a good indicator that it is the preferred product, and the estimator should include that cost in the estimate.

CHAPTER 16: EQUIPMENT, FURNISHINGS, SPECIAL CONSTRUCTION, AND CONVEYING EQUIPMENT

1. What is the typical unit of measure for window treatments? How is it obtained?

   Window treatments are measure by the square area (SF) of the window being covered. It is the width × the length or height of the window converted to SF.

2. What is the typical unit of measure for residential appliances? What should be included in the labor costs?

   The typical unit of measure for residential appliances is each, EA. The labor costs should include distribution time from delivery vehicle to location, unpacking, setting in place and leveling, if required, connection to power or water/waste source, and the hauling away of debris. Additional costs may include fastening or modifying cabinetry the appliance is attached to. Multiple trades may be required.

3. What is the suggested unit of measure for special purpose rooms such as a sauna?

   Saunas are taken of and estimated by the square foot of floor space and listed in the takeoff as SF.

4. How would an estimator develop the labor portion of an estimate for manufactured wood casework?

   The estimator would assign a labor-hrs. cost to each carcass of manufactured casework in a project. Then summarize the total number of labor-hours required to complete the installation.

5. Explain in detail how an estimator might develop a budget estimate or plug from historical cost data for an elevator.

   Assuming there was sufficient data from previous similar projects with elevators, the estimator could use the number of stops (floors) as a unit of measure. This could include the hoistway and electrical costs as well if that information was available. Parametric estimates are budgeting purposes only and should not be used as a substitute for contemporaneous subcontractor bids.

CHAPTER 17: FIRE SUPPRESSION

1. Provide a brief explanation as to difference between a wet pipe and a dry pipe sprinkler system from an estimator's perspective.

   From an estimating perspective, a dry-pipe sprinkler system has more components than a wet pipe system. It has a compressor to supply air, additional control equipment, potentially different types of piping, and different types of sprinkler heads to name a few. It can also require a nitrogen supply if the dry system uses nitrogen in lieu of air.

2. Prepare a checklist of possible interfacing trades for a fire pump on the upper level of a building.

   Trades involved with the installation of a fire pump on the upper level of a building could be plumbers to install the pump, electrician for power and control wiring, fire alarm tech to install in the fire alarm system, rigging and hoisting crew with crane to set the pump, potentially a glazing crew to remove and reinstall the window if the pump is a replacement.

3. Explain in detail as to how an estimator would prepare a parametric estimate for a wet pipe sprinkler system for a project.

   Assuming the estimator had historic data from projects of a similar nature or subcontractor quotes for projects of a similar nature, the estimator could calculate the lowest, average, and highest cost per SF of floor space protected by the system. That range could be used to predict or develop a parametric estimate for the new project. Parametric estimates are budgeting purposes only and should not be used as a substitute for contemporaneous subcontractor bids.

4. What is a typical unit of measure for a water storage tank?

   Water storage tanks use gallons (Gals.) as the standard unit of measure. The size would be the number of gallons a tank would hold.

5. Provide a list of some of the nonproduction costs associated with a fire suppression system.

   A permit, the design fees and hydraulic calculations for the system, a flow test if required, scaffolding or lifts to access the work, cleanup of debris, inspectional services charges if required.

CHAPTER 18: PLUMBING

1. Provide a brief explanation of parametric estimating and how it can be used to generate a plumbing estimate for comparison with subcontractor bidders.

   Parametric estimates for plumbing systems can be broken down to a cost per SF, or per toilet fixture, or appliance. The prime or general contractor can use bids or previous contract amounts as a way to determine a cost per unit. The bids would have to be for project of a similar nature or at least similar fixtures. Parametric estimates are budgeting purposes only and should not be used as a substitute for contemporaneous subcontractor bids.

2. For below-grade systems such as waste and stormwater, explain what other costs may be associated with the work.

   The principal costs other than plumbing would be the excavation, sand bedding if required, and the backfill and compaction of the trench after the work has been installed. For waste systems, tanks maybe required outside the structure and for drainage of stormwater that goes to daylight, concrete headwalls at the end of the line.

3. What part of the technical specification section 22 00 00—Plumbing would an estimator review to determine the submittals required? What section for cleanup of debris generated by the plumbing work?

Part 1—General of the technical section deals with the submittals for a particular scope of work. This has a cost associated with it. It can be part of the project manager's cost, or it can be an administrative duty performed by an assistant project manager. Regardless, there is a cost associated with the production of submittals. Cleanup is typically specified in Part 3—Execution. Both general ongoing cleanup and final cleanup, both of which have a labor cost.

4. Define the process of "make-safe" for plumbing?

    The term "make safe" is a common expression for the cutting and capping of live water and waste, gas, drain, and similar piping that is scheduled to remain or be the connection point for new work. It can also be work that is abandoned in place. This allows the demolition process to proceed without fear of cutting into a live line in an active system.

5. Using the appropriate table from the chapter, calculate the cost of installation (setting fixture and trim) for a wall-hung urinal if the hourly billing rate for the plumber is $110.75 per hour. Cite the figure used.

    Using Figure 18.6 Installation Time in Labor-Hours for Plumbing Fixtures: One (1) wall hung urinal requires 5.33 labor-hours for a plumber to install therefore: 5.33 labor-hours × $110.75 per hr. = $590.30.

CHAPTER 19: HVAC

1. Calculate the square foot area of insulation including the appropriate amount for lap and seaming for 80' of 20" × 16" steel duct.

    The perimeter of a 20" × 16" steel duct is 20" + 16" = 36" × 2 = 72". Now add approximately 6"-8" to lap and staple. Therefore 72" + 8" = 80" or 6.67' × 80' = 533.6 SF. There is typically a percentage added for length so as the duct is installed and connected to the next piece in line there is sufficient overlap in the direction of the run of ductwork.

2. A 90' section of 20" × 16", 22 ga ductwork has a weight of 1.406 lbs. per SF. What is the cost for this section if it is $9.70 per lb. fabricated and installed?

    The perimeter of a 20" × 16" duct is 72" or 6' × 90' = 540 SF × 1.406 lbs. per SF = 759.24 lbs. × $9.70 = $7,364.63.

3. Using the appropriate figure from the chapter, what is the estimated cost for installation labor for a 1,000 MBH gas-fired cast iron boiler if the hourly billing rate for a boilermaker is $105.00?

    Using Figure 19.6a for a gas-fired, 1000 MBH, cast-iron boiler the installation labor is approximately 64 labor-hrs. × $105.00 = $6,720.00.

4. Explain how might an estimator use specific parameters as a means to generate a budget estimate for the HVAC scope of work.

    Similar to other mechanical systems, if the estimator had historical data from previously completed projects of a similar nature or from previous but recent bids, the estimator could break the project down into tons (TNS) of cooling or MBH of heating as a parameter. Another unit of measure that could be used is the square foot area of conditioned space of the building. A range of pricing could be created by dividing the low,

high, and average price by the SF area, then applying that to the SF area
of the new project. Parametric estimates are budgeting purposes only and
should not be used as a substitute for contemporaneous
subcontractor bids.

5. For a design-build residential HVAC system, explain some of the
parameters and energy considerations that an HVAC subcontractor would
have to consider when sizing an air conditioning system.

   *Naturally, a primary parameter is the area of conditioned space, in
   SF. Additional considerations would be the orientation of the house, and
   the amount of fenestration that is facing the sun. Other important factors
   are the R-value and type of insulation in the exterior walls, U-factors and
   SHGCs of the glass, climate zone, and local conditions.*

CHAPTER 20: INTEGRATED AUTOMATION

1. Please list the units of measure for: (1) valves, (2) cable tray, and
(3) racks.

   *The take-off units for (1) valves is each, EA based on size, type, and
   composition, (2) cable tray is by the linear foot, LF, and (3) racks are by
   the individual piece, EA based on size and number of tiers.*

2. Please opine as to the challenges of estimating integrated
automation systems.

   *Estimating integrated automation systems often requires intimate
   knowledge of the actual system and the proprietary components and
   software functions. Due to the competitive nature of the automation
   industry, much of the proprietary components and functions of a system
   are a closely guarded secret. This creates a challenge in calculating the
   cost of the components for a particular system. Other challenges include
   pricing and programing of software to run the systems.*

3. In the absence of detailed drawings illustrating the integrated automation
of various systems, where would the estimator find information about
the particular system or systems for the project?

   *Often in the absence of a set of detailed drawings, the design
   professional will include a narrative as to the scope of the integrated
   automation system. This narrative can be directed at one manufacturer
   of a system or can be more general as to the functions and systems of the
   building it is intending to control. The narrative allows bidders at the
   subcontractor level to design and price a custom system.*

4. Please select and define one of the primary functions of an integrated
automation system.

   *One of the primary functions of an integrated automation system is
   energy management. It reduces human input and controls functions
   based on use and occupancy. It relies on specific conditions in lieu of
   human control.*

5. How might an estimator use historical data to budget the work of this division?

   Using very general parameters, an estimator can use historical data to generate a budget estimate based on the number of systems or points of control the scope of work includes. For integrated automation systems it is recommended that an estimator secure multiple subcontractor quotes for both the design and implementation of the system.

CHAPTER 21: ELECTRICAL

1. Would one expect that four conductors pulled together through a conduit be more or less labor consuming than pulling all four conductors individually? Explain your answer.

   Assuming the conduit is adequately sized for the four conductors, it is a fact that pulling four conductors at once is less labor intensive than pulling four conductors individually. In fact, pulling the conductors individually through the same conduit, may not be an option due to the potential for restrictions caused by the wires already in the conduit.

2. Using the appropriate figure from the chapter, what is the estimated labor-hours for an electrician to install 950' of #10 AWG wire?

   Using Figure 21.4a, Installation Time in Labor-Hours for Electrical Conductors: Wire and Cable, #10 AWG is 0.8 labor-hours per CLF (100 LF). Therefore 950 LF ÷ 100 LF = 9.5 CLF × 0.8 labor-hours per CLF = 7.6 labor-hours.

3. Using the appropriate figure from the chapter what is the estimated cost for installation labor for 270 LF of ½" EMT if the hourly billing rate for an electrician is $95.00?

   Using Figure 21.2 Installation Time in Labor-Hours for Conduit to 15' (assuming the conduit is 15' AFF or less) using 0.047 labor-hours per LF × 270 LF = 12.7 labor-hrs. × $95 per LH = $1,206.50

4. Explain how an estimator might use specific parameters as a means to generate a budget estimate for the electrical work?

   Parametric estimates for electrical systems can be broken down to a cost per SF. The prime or general contractor can use bids or previous contract amounts as a way to determine a cost per SF. The bids would have to be for project of a similar nature. Parametric estimates are budgeting purposes only and should not be used as a substitute for contemporaneous subcontractor bids.

5. For a design-build residential electrical system explain some of the parameters and energy considerations that an electrical subcontractor would have to take into account when preparing a price.

   There are several key parameters in design-build electrical systems: (1) size of the service in amps; (2) overhead or underground service; (3) appliances requiring electricity: water heater, heat pump, range, dryer, etc.; (4) lighting requirements; (5) light fixture allowance; (6) general power and lighting requirements; (7) automation requirements; and (8) site lighting and power requirements.

CHAPTER 22: COMMUNICATION, ELECTRONIC SAFETY, AND SECURITY

1. Using the appropriate figure from the chapter, what is the estimated labor-hours for a technician to install 550 ft. of Category 6 #24 four-pair stranded wire?

   Using Figure 22.6 Installation Time in Labor-Hours for Unshielded Twisted Cable: Category 6 #24 four-pair stranded wire requires 1.143 labor-hours per CLF (100 LF). Therefore 550 LF ÷ 100 LF = 5.5 CLF × 1.143 labor-hours per CLF = 6.3 labor-hours.

2. Using the appropriate figure from the chapter what is the estimated labor-hours for a technician to install 100 Category 6 shielded RJ-45 jacks?

   Using Figure 22.6, Installation Time in Labor-Hours for Unshielded Twisted Cable: Cat 6 shielded RJ-45 jacks require 0.133 labor-hours per jack (EA). Therefore, 100 EA × 0.133 labor-hours per jack = 13.3 labor-hours.

3. Using the answers from questions #1 and #2 above, what is the labor cost if the billing rate is $115 per hour?

   Combining the labor-hours from questions #1 and #2: 6.3 + 13.3 = 19.6 or 20 labor-hours × $115 per hr. = $2,300.

4. Explain how an estimator might use specific parameters as a means to generate a budget estimate for the security system work?

   Parametric estimates for communication systems can be broken down to a cost per SF. The prime or general contractor can use bids or previous contract amounts as a way to determine a cost per SF. The bids would have to be for project of a similar nature. It could also be broken down into a cost per device. Parametric estimates are budgeting purposes only and should not be used as a substitute for contemporaneous subcontractor bids.

CHAPTER 23: EARTHWORK

1. Using Figure 23.4 in this chapter with a change to the existing grades as follows (starting in the upper left proceeding clockwise): 99.5', 99.2', 100.5', and 99.4'. The new proposed elevation is 101.7' and the grid dimensions are 25' × 25'. Calculate the amount of cut or fill of the new grid number 5.

   Applying the same methodology as in Figure 23.4, the sum of the four corners of the grid is: 99.5 + 99.2 + 100.5 + 99.4 = 398.6. Calculating the average elevation for the grid is: 398.6 ÷ 4 = 99.65. Now subtract this average from the new proposed elevation: 101.7 − 99.65 = 2.05 or an average fill of 2.05' over the 25' × 25' grid area. This results in 1,281.25 CF or 47.45 CYs of fill (exclusive of an allowance for compaction).

2. A trench must be dug 60' long × 5' wide from heel to heel, and 3' deep. Maintaining a 1:1 or 45° slope on each side, what is the quantity of excavated soil in BCY?

   Applying the methodology illustrated in Figure 23.6 Sample Trench in this chapter, start by calculating the cross-sectional area of the trench. A2 = 5' × 3' = 15 SF, A1 = A3 therefore maintaining a 45° slope:

$3' \times 3' \times \frac{1}{2} = 4.5\ SF \times 2\ (A1 + A3) = 9\ SF.$ *Adding* $A1 + A2 + A3 =$ *15 + 9 = 24 SF.* $60' \times 24\ SF = 1,440\ CF \div 27\ CF\ per\ CY = 53.33\ CYs$ *(exclusive of an allowance for swell).*

3. If the soil in problem #2 is dry gravel, what is the loose volume of soil to be hauled? Use the swell factor from Figure 23.7 – *Approximate Material Characteristics.*

    Applying the swell factor for dry gravel in Figure 23.7 of 13%: 53.33 CYs × 1.13 = 60.27 CYs.

4. From Figure 23.3, "Sample Cut and Fill Tally," what is the total quantity of soil in CYs, being handled? If the fill could be used in the cut areas, what would be the net import or export excluding the swell/compaction factor?

    From Figure 23.3, "Sample Cut and Fill Tally," the total cubic yards being handled is the sum of the cut and the fill totals: 374.54 CYs + 213.66 CYs = 588.2 CYs. Assuming the cut material could be used to fill the low areas, then 374.54 – 213.66 = 160.88 CYs of surplus material to be exported from the site (exclusive of any allowance for the compaction of the fill materials or swell of the excess to be exported).

5. Calculate the haul cost (per CY) for soil being exported from a site if each truckload will carry 20 CYs, the haul distance is 7-miles at 40-mph, each load has a $40 tipping fee, and the truck and driver is $900 per day. The day is 8 hrs. with 30 mins for lunch, and two 15-min breaks. Also, there is a 15-min dump time and 15-min load time for each load. Assume an 85% efficiency rate for the day.

    Start by calculating the time for one dump cycle. At 40-mph and a distance of 7-miles for the dump site; 60 minutes/hr. ÷ 40-mph = 1.5-mins. per mile × 7-miles × 2 (round trip) = 21-mins. plus 15-mins to load plus 15-mins to dump = 51-mins per complete cycle. Therefore, in 7-hrs (net) day: 420-mins ÷ 51-mins per cycle = 8.23 cycles × 0.85 (efficiency rate) = 6.99 or 7 trips per day. Now calculate the costs per cycle: $900 per day ÷ 7-trips = $128.57 per trip + $40. Tipping fee = $168.57 ÷ 20 CYs per trip = $8.43 per CY.

CHAPTER 24: EXTERIOR IMPROVEMENTS

1. A parking lot is 400' × 100' and requires an 8" (after compaction) gravel base for the paving. The base material has an 18% compaction factor. How many loose (LCYs) cubic yards of material should be estimated to purchase?

    Start by calculating the volume of the gravel base: 400' × 100' × 0.67' = 26,800 CF × 1.18 (increase for compaction) = 31,624 CF ÷ 27 CF/CY = 1,171.25 or 1,172 CY.

2. For the parking lot noted in problem #1 there will be a binder course of 3½" and a wearing course of 2". If the asphalt mix for each course will lose ¼" for each 1" of specified thickness, how many loose tons (2000 lbs. per ton) are required for each course? (use 140 lbs. per CF)

    Start by calculating the loose volume with the compaction factor. For the binder course: 40,000 SF × 0.30' = 12,000 CF × 1.25 (compaction factor) = 15,000 CF × 140 lbs. per CF = 2,100,000 lbs. ÷ 2000 lbs. per ton = 1,050 tons. For the wearing surface: 40,000 SF × 0.17' = 6,800 CF × 1.25 = 8,500 CF × 140 lbs. per CF = 1,190,000 lbs. ÷ 2000 lbs. per ton = 595 tons.

3. A pedestrian walkway at the perimeter of the building is 5'-0" wide, 200' long and 5" thick. Calculate (a) the surface area to be finished, (b) cubic yards of concrete, and (c) welded wire fabric required with a longitudinal lap of 6" per sheet.

   (a) Surface area = 5' × 200' = 1000 SF, (b) Cubic yards 5' × 200' × 0.42' = 420 CF ÷ 27 CF/CY = 15.6 or 16 CYs., (c) 200' ÷ 10' = 20-sheets × 1.05 = 20.3 or 21 sheets. (There is a loss of 6" for each 10' sheet, or 5%.)

4. For the walkway in problem #3 if the concrete placement cost is $150 per CY, the surface finishing cost is $1.90 per SF including edging, and the curing cost is $1.40 per SF, what is the estimated labor cost for the walkway?

   16 CYs × $150 per CY = $2,400 + 1,000 SF × $1.90 per SF = $1,900 + 1,000 SF × $1.40 = $1,400 for a total of $5,700.

5. Two (2) laborers at a billing rate of $58.50 per hour each will auger 13 fence posts 12" in diameter to 3'-6" below grade in a typical day. The rental rate for the hand-held auger is $98.00 per day with a $1.20 per hr. operating cost (based on an 8-hr day). What is the estimated cost to auger 400' of stockade fence with posts at 8'-0" on center? (Assume fence is in a straight line.)

   Start by calculating the number of fence posts in the example: 400' ÷ 8' o.c. = 50 posts + 1 post to start = 51 posts. Now calculate the costs per post: 2 laborers × $58.50 per hr. × 8-hrs. = $936 per crew day, plus $98. per day auger rental + 8 hrs. × $1.20 = $9.60. Therefore, the total cost per day for crew and equipment is $936 + $98 + $9.60 = $1,043.60. If the cost per day is divided by the daily output: $1,043.60 ÷ 13 posts/day = $80.28 per post; therefore, 51 posts × $80.28 = $4,094.28.

CHAPTER 25: UTILITIES

1. Using the appropriate figure from the chapter, how many labor-hours are required to set 355 LF of 12" diameter PVC sanitary sewer lines?

   Using Figure 25.2, Installation Time in Labor-Hours for Sewage and Drainage Collection Systems, 1 LF = 0.088 labor-hours; therefore, 355 LF × 0.088 labor-hours/ per LF = 31.24 labor-hours, or 32 labor-hours.

2. Using the solution to problem #1, if one (1) machine hour is required for each 3 hrs. of labor and the machine time cost $198.55 per hour, what is the cost of the machine time?

   31.24 labor-hours ÷ 3 labor-hours per machine hour = 10.41 machine-hours × $198.55 = $2,066.90 or if the hours were rounded to the next whole hr.: 11 × $198.55 = $2,184.05.

3. A 48" diameter RCP drain pipe is 200' long. The pipe comes in 8'-0" bell and spigot sections. If each mortar joint requires 0.08 labor-hours to

mortar one (1) foot of circumference, how long will it take to mortar all 200'?

>*Start by calculating the circumference. The 48" diameter is the inside measurement of the pipe: C = π × D = 3.14 × 4' = 12.56' per section. Next calculate the number of sections requiring mortar: 200' ÷ 8' per section = 25–sections + 1 at the end × 12.56' = 326.56' × 0.08 labor-hours = 26.13 labor-hours. (This example assumes that the first and last circumferences will be mortared to a headwall.)*

4. For the RCP in problem #3, calculate the cost of installation if the excavator will set 7 sections per day at a machine cost of $1,195 per day and an hourly operating expense of $46.88 per hour. The crew of two (2) laborers has an hourly cost of $66.50 each and the operator has an hourly cost of $88.50. What is the cost for the installation (exclusive of mortaring the joints)?

>*Start by calculating the equipment and crew costs per day: $1,195. + $375.04 (8 hrs. × $46.88 per hour) = $1,570.04. Next is the crew cost per day: 2 laborers × $66.50 = $133.00 per hr. + $88.50 per hr. = $221.50 per hour × 8 hours/day = $1,772.00 per day. Adding the crew to the equipment: $1,570.04 + $1,772.00 = $3,343.04 per day. Using the 25–sections from Problem #3: 25 sections ÷ 7 sections per day = 3.6 –days. Therefore: 3.6 days × $3,343.04 per day = $12,034.95. If the cost was rounded to the full day: 4 days × $3,343.04 = $13,372.16.*

5. How many days will the installation take exclusive of mortaring the joints?

>*Using the solution to #4: 3.6, or 4 days.*

CHAPTER 26: PROFIT AND CONTINGENCIES

1. Explain why profit is important in an estimate and why work should not be done at cost in a fixed price contract.

>*Profit is essential to maintain and to grow a construction company. Without profit a project will not contribute to the company for the resources and time expended. With a fixed price contract, a project has no buffer without profit. A project performed at cost with one mistake will suffer a loss.*

2. Why is a contingency added at various stages of design development?

>*Contingencies in design development are added to represent costs that will increase or be added as the design matures.*

3. Opine as to how an unnecessary contingency could impact a competitive bid.

>*Competitive firm fixed-price bids are intended to only reflect what has been shown or specified in the bid documents. Adding money for work not specified or not shown is considered nonresponsive and will make a bid less competitive. Work not shown or specified or work that could not be reasonably inferred necessary by the documents, that is later required can be added by modifications to the contract.*

4. How can questions at the bid phase be addressed?

>*Questions or required clarifications that arise during the bid phase can be addressed by a query to the design professional. That query is*

called an RFI (request for information). RFIs are written and the response are shared with all bidders through an addendum to keep a "level playing field." This prevents one bidder from obtaining an unfair advantage due to the answer to the RFI.

5. Why is the design professional in the best position to advise the owner on the contingency amount?

The theory is that as the design matures during design development, the design professional(s) become increasingly more knowledgeable about the intricacies revealed during investigation for design. This puts the design professional in the best position to know the potential pitfalls requiring a contingency and the amount of that contingency.

CHAPTER 27: ESTIMATING BY COMPUTER

1. What is the main purpose of estimating software?

Computerized estimating provides the distinct advantages of flexibility, efficiency, and accuracy. It is intended to streamline the process and allow for last minute changes to the estimate. It reduces the hand calculations and errors, providing more time for preparing additional estimates. It is not intended to take the place of estimating knowledge or proficiency.

2. Why is it important to carefully research software for estimating?

It is important to select the best software for the intended purpose. Selecting a complex system that exceeds the company's needs can prevent it from being used to its full capacity and can be costly by requiring support services, training, and frequent updates. However, purchasing software too simple for the application can reduce its use as well. Software is expensive and should be selected based on needs while allowing a little room for growth.

3. What is a packaged database? Who typically uses packaged databases?

Construction cost databases may be helpful in meeting a company's estimating needs. They can offer material, labor, and equipment costs, most often in a unit price format, often according to CSI MasterFormat® divisions. Databases can be used as baseline for determining the costs of various tasks or activities. It is most often used by owners when preparing budgets for funding. It is used by contractor for checking costs and filling in last-minute "holes" in the estimate.

4. What is a historical database? Who typically uses historical databases?

A historical database is a collection of "snapshots" of costs along the schedule of a project that has been completed. The costs for work are actual, not presumed, and relate to performance and techniques of one particular crew or individual(s). A crew or individual that will be doing the work being estimated. The most accurate costs can be achieved by documenting a contractor's own expenditures on the projects he or she builds and reconfiguring them to be used as a database.

5. Why are computerized project overhead sheets important to a bidder?

Computerized project overhead summary sheets allow the estimator to view different project overhead costs based on different schedules.

All projects have two components for overhead: direct costs and indirect costs. Indirect overhead is usually applied as a percentage of the total overhead of the main office. Direct overhead costs are those nonproduction costs that are unique to each project. They include such items as supervision, temporary facilities, building permits and fees, and dumpsters. Some of these costs are fixed, such as for a building permit, but many are time sensitive and vary with the amount of exposure on the project. Project overhead electronic spreadsheets allow for last-minute edits and "what-if" scenarios.

CHAPTER 28: CONCEPTUAL ESTIMATING

1. What is the purpose of conceptual estimating?

   Projects are always required to maintain cost control. Many architects and engineers are required to work within a budget for construction, so in order to keep a design on target for cost, interim estimates are prepared as a way to gauge the direction of the design. This process avoids putting the project out to bid when documents are complete, only to have the bids come in substantially higher than available funding. Estimates done at specific points in the design development often help the owner and design professional maintain a budget. It also allows the owners to determine if the project is financially viable or if the rate of return is as anticipated.

2. Provide a brief explanation of ROM estimating.

   The order-of-magnitude estimate, sometimes called a rough order of magnitude (ROM) is the 10,000-foot view or big picture estimate. It frequently relies on parameters that are not typical construction parameters. For example, the ROM for a dormitory may be expressed in dollars per bed, or the cost of an apartment building would be reported in dollars per apartment. ROMs are planning tools that rely heavily on historical or published cost data. They rarely have any bid documents at all. They are used as a basis in early decision-making. The estimator may have little more to go on than criteria provided by the developer or investor.

3. Provide a brief explanation of square foot estimating.

   Another method of conceptual estimating uses the gross square foot (SF) area of the structure as the reporting parameter. This is also meant as a planning tool. Square-foot estimates also rely heavily on historical costs from projects of a similar nature, type, and size. They are based on the premise that all building types can be distilled to a cost per SF of gross floor space. This is an especially useful tool for budget planning and decision-making and even prequalifying clients.

4. What is the difference between SF estimating and assemblies estimating?

   Assembly estimates are based on individual systems or groups of tasks assembled together, then estimated as a single unit. This methodology can be used at almost any level of design development. It requires that there be decisions made on major systems and building components, even though they may be subject to change as the design develops. It allows the estimator the maximum amount of flexibility for making changes to systems without an entire revision of the estimate. It is a more detailed and flexible type of estimating than SF estimating in that is breaks the project down into systems that can be changed as the project design matures.

5. Why is unit price estimating the most accurate when done correctly?

The unit price estimate is the takeoff or quantification of the plans in incremental parts. It is a dissection of the project into each component almost to the individual nut and bolt. This is the method most often used by contractors in the development of competitive bids. The unit price method of estimating requires a comprehensive set of plans and specifications that provides details on every aspect of the project. It is intended to reflect a cost in the estimate for each task shown on the plans and described in the specifications.

Index

Estimating Building Costs for the Residential and Light Commercial Construction Professional,
Third Edition. Wayne J. Del Pico.
© 2023 John Wiley & Sons, Inc. Published 2023 by John Wiley & Sons, Inc.